BASIC MATHEMATICS FOR COLLEGE STUDENTS

★

Third Edition

Gregory Fiore
Dundalk Community College

Teresa Healy
Brookdale Community College

Virginia Lee
Brookdale Community College

An Imprint of Addison Wesley Longman, Inc.

Reading, Massachusetts • Menlo Park, California • New York • Harlow, England
Don Mills, Ontario • Sydney • Mexico City • Madrid • Amsterdam

Custom books consist of products that are produced from camera-ready copy. Peer review, class testing, and accuracy are primarily the responsibility of the author(s).

Custom Publishing Manager: Andrea Sheehan
Production Administrator: Liz Faerm

BASIC MATHEMATICS FOR COLLEGE STUDENTS
Third Edition
Customized for Brookdale Community College

ISBN: 0-673-67631-5

3 4 5 6 7 8 9 10--99 98

This book is dedicated to our students, past, present, and future.

CONTENTS

PREFACE

Basic Mathematics for College Students was prepared especially for the students of Brookdale Community College. Several years ago, as a result of increased dissatisfaction with standard approaches and books in this field, the mathematics faculty at Brookdale decided to design a curriculum that would fulfill several goals:

- Take advantage of students' prior math experiences. All students come to Brookdale having taken mathematics before and we wanted to take advantage of their knowledge base.
- Provide students with the skills necessary to make informed decisions based on quantitative information.
- Provide students with a firm foundation for and a smooth transition to algebra.

It quickly became clear that we would have to write our own materials for much of this new course. A small group of faculty developed the original materials and they were class tested and revised over a two-year period. Now those materials have been extensively revised and expanded and represent about half of this book. The other half is from a text by Gregory Fiore, a dynamic and committed teacher from Dundalk Community College in Maryland. His examples and exercises showing how mathematics is used both in other disciplines and in everyday activities are a major strength of this text.

The text has many examples to provide students with a model for good problem-solving techniques. Whenever possible, alternate solutions are shown so that students see that they can and should make choices. Many examples and exercises are taken from newspaper articles so that students will become familiar with interpreting quantitative information in a realistic setting. A very important feature of the book is the *Writing to Learn* exercises included in most of the exercises sets. These exercises give students the opportunity and responsibility to write about mathematics. The ability to explain or create examples is critical to complete understanding.

ACKNOWLEDGMENTS

Many people have contributed to this project over the past several years and we would like to take this opportunity to thank them.

The group who worked on the first version of these materials, in addition to the authors: Ellen Rebold and Fran Ventola.

The group who developed support materials in the same spirit: Lois Calamia, Andy Page, Barbara Tozzi, Fran Ventola, and Linda Wang.

Teachers who used the early materials and provided valuable comments, suggestions, and critiques: Shirley Blatchley, Lois Calamia, Cathy Folio, Lois Higbie, Ellen Rebold, Gene Schmid, Barbara Tozzi, Fran Ventola, Linda Wang, and all of our adjunct faculty members.

Colleagues who read our materials carefully and critically: Carole Carney, Elaine Klett, Majid Masso, and Andy Page.

Colleagues who helped with proofreading, editing and checking answers: Kerry Behler, Shirley Blatchley, Carole Carney, Lois Higbie, Elaine Klett, Majid Masso, Charlie Meehan, Andy Page, Ellen Rebold, Gene Schmid, Barbara Tozzi, Fran Ventola, and Linda Wang.

Special thanks to:

- Joan Longo and the staff at the Materials Design and Production Department of Brookdale, for their advice, for working with our newspaper clippings, and for letting us use some of their equipment. Their advice was invaluable.
- Caralee Woods, HarperCollins Custom Books Manager, for helping us through this new experience of custom publishing.
- Gregory Fiore, a kindred spirit, for letting us use his materials.

Finally, a very special thank-you to our families, Adam and Hank Green, and Kate, Brian, Andy and John Healy, for their support during this project.

Using Mathematics to Solve Problems

USA SNAPSHOTS®

A look at statistics that shape our lives

Median income of families with children

Family type	Median income
Married couple	$41,260
Father only	$25,211
Mother only	$13,092

Source: U.S. Bureau of the Census, 1990 data

By Sam Ward, USA TODAY

There are many numbers that can be used to describe you. Some of them are your height, your weight, your age, and your income. When scientists or sociologists want to describe a group of people with a single number, they often use an average. A *median* is a type of average. It is found by taking all of the data (numerical information you are studying) and arranging it from the lowest value to the highest value. The median is the middle value. You'll work with medians in this chapter.

In this chapter, we'll begin our review of mathematics. You'll solve practical problems that come from everyday life and you'll learn rules and develop skills that will be used later in your study of algebra.

1.1 NUMBERS IN THE NEWS

OBJECTIVES

1. Read and interpret bar graphs.
2. Read and interpret line graphs.
3. Read and interpret a table.
4. Read and interpret news articles.

NEW VOCABULARY

bar graph	forecast	downward trend
ticks	double bar graph	table
trend	line graph	interpolate
upward trend	double line graph	

As adults in today's society, we are often asked to make decisions based on numerical information. This information might be presented in the form of tables, graphs, charts, or articles. The ability to read, understand, interpret, and draw conclusions from these sources is critical to making informed decisions.

1 BAR GRAPHS

A **graph** is a picture of data. It conveys a lot of information at a glance. A graph helps you to see trends and relationships. It helps you to compare, interpret, and analyze data.

A **bar graph** is used to compare quantities. A bar graph has a horizontal axis and a vertical axis. In Figure 1.1 the horizontal axis stretches left to right and is labeled in years. The vertical axis stretches up and down and is length of life. The vertical axis is marked at intervals with short lines called **ticks**. In Figure 1.1, the distance between two consecutive ticks represents 5 years.

FIGURE 1.1 Life Expectancy in the United States for Females Born in a Given Year, 1850 - 1990

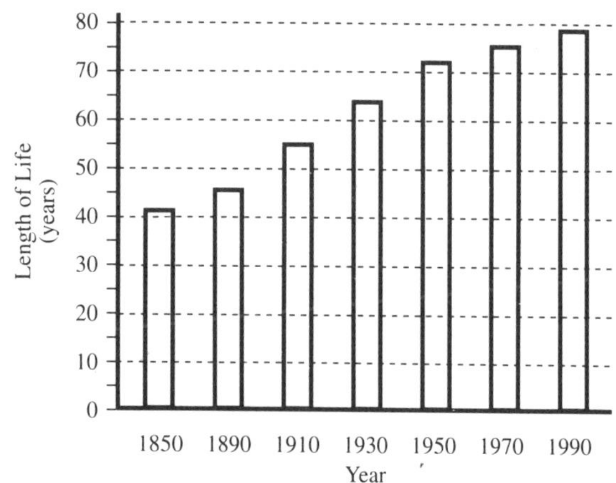

SOURCE: Department of Health and Human Services, National Center for Health Statistics.

EXAMPLE 1 Use Figure 1.1 to answer each question.

a. How long was a woman born in 1850 expected to live?

Align a ruler with the top of the bar at 1850. Move right to left until you reach the vertical axis. A woman born in 1850 was expected to live about 41 years.

b. Estimate the female life expectancy in 1990, and compare it to 1850.

A woman born in 1990 is expected to live about 79 years. In 1850 she was expected to live 41 years. From 1850 to 1990, the life expectancy of a woman almost doubled, rising $79 - 41 = 38$ years.

c. When was the average life expectancy 62 years?

Align your ruler with 62 years on the vertical axis. The bar closest to this height is at 1930. The life expectancy for a woman was 62 years if she was born about 1930.

♦ You Try It Answer the following questions using Figure 1.1.

1. How long was a woman born in 1910 expected to live?
2. Estimate female life expectancy in 1950 and compare it to 1910.
3. When was the average life expectancy 76 years?
4. What is a reasonable guess for life expectancy in 2010?

A bar graph is also used to show **trends** or patterns. In Figure 1.1, the life expectancy for a female rose continuously from 1850 to 1990. This is called an **upward trend**. It rose the fastest from 1890 to 1950, since the height of the bars increased the most in that period. It began to level off from 1950 to 1990.

A bar graph can also be used to **forecast** or make predictions about future trends based on recent trends. The leveling off from 1950 to 1990 may cause one to forecast a slight decrease in life expectancy in the near future.

Figure 1.2 is a **double bar graph.** It is used to compare two quantities to themselves, and to each other. Different shadings are used to distinguish each bar in a pair. A key is used to display what each differently shaded bar stands for.

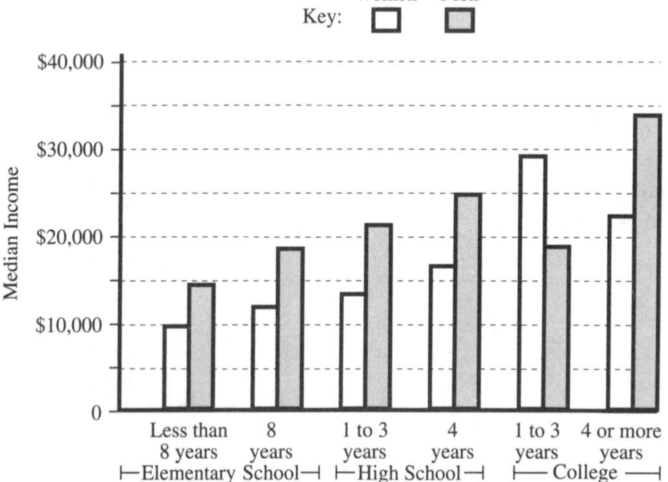

FIGURE 1.2 Median Annual Income for Different Educational Levels, 1987

SOURCE: Department of Commerce, Bureau of the Census.

Median is a type of average. It is found by listing all the values from smallest to largest and finding the *middle* value.

EXAMPLE 2 Use Figure 1.2 to answer each question.

a. What is the median income for a man with 1 to 3 years of high school?

About $21,000 a year. (See third shaded bar from the left.)

b. What is the median income for a woman with 4 or more years of college?

About $23,000 a year. (See unshaded bar farthest right.)

c. How much more does a woman make than a man with 1 to 3 years of college?

A woman makes about $29,000 a year with 1 to 3 years of college.
A man makes about $19,000 a year with 1 to 3 years of college.
A woman makes about $29,000 – $19,000 = $10,000 more per year.

d. How much more does a man make with 4 or more years of college versus 4 years of high school?

A man makes about $34,000 a year with 4 or more years of college.
He makes about $25,000 a year with 4 years of high school.
The difference is $34,000 – $25,000 = $9,000 more per year.

♦ You Try It

Answer the following questions using Figure 1.2.

5. What is the median income for a man with 8 years of elementary school?
6. What is the median income for a woman with 4 years of high school?
7. How much more does a man make than a woman when each has less than 8 years of elementary school?
8. How much more does a woman make with 4 years of high school compared to 8 years of elementary school?

2 LINE GRAPHS

A **line graph** is used to show trends. One common trend is showing how a quantity changes over time. Each dot on the graph represents a certain quantity at a given time. The dots are connected with straight lines, called *line segments*. The line segments show how a quantity increases or decreases over a given time period.

A line graph has a horizontal axis and a vertical axis. In Figure 1.3, birthrate in the United States (vertical axis) is graphed against time measured in 10-year intervals (horizontal axis). There is a "squiggle" at the bottom of the vertical axis. This tells you there are no birthrates from 0 through 15 on this graph.

FIGURE 1.3
Birthrates in the United States, 1910 - 1990

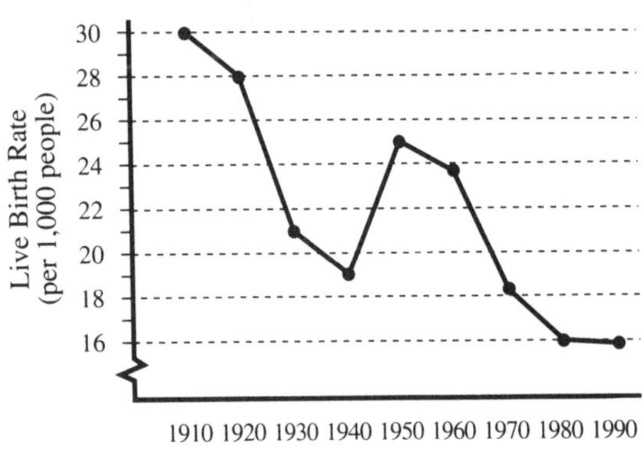

SOURCE: Department of Health and Human Resources.

EXAMPLE 3 Use Figure 1.3 to answer each question.

a. How many live births were there in 1920 per 1,000 people?

Locate 1920 on the horizontal axis. Move upward until you reach the graph. Then move left. You reach 28 on the vertical axis. There were about 28 live births per 1,000 people in 1920.

b. When were there 18 live births per 1,000 people?

Locate 18 on the vertical axis. Move right until you reach the graph. Then move down. In 1970 there were 18 live births per 1,000 population.

c. To **interpolate** is to estimate a value *between* two given values. Interpolate the birthrate for 1945.

1945 is midway between 1940 and 1950. Mark this spot with your pencil. Using a ruler, move upward from 1945 to the graph. Then move left. There were approximately 22 live births per 1,000 people in 1945.

d. What was the only 10-year period to have an increase in birthrate?

The only increase in birthrate occurred from 1940 to 1950. This period is represented by the only upward sloping line segment on the graph. It increased 19 per 1,000 in 1940 to 25 per 1,000 in 1950.

e. What upward and downward trends do you see in this graph?

Where the graph slants upward indicates an upward trend. This means the birthrate is rising. Where the graph slants downward indicates a **downward trend**. This means the birthrate is falling.

From 1910 to 1940 there was a downward trend in birthrate.
From 1940 to 1950 there was an upward trend.
From 1950 to 1990 there was a downward trend.

♦ You Try It

Answer the following questions using Figure 1.3.

9. How many live births were there in 1960 per 1,000 people?
10. When were there 28 live births per 1,000 people?
11. Interpolate the birthrate in 1975.
12. What was the drop in birthrate from 1960 to 1970?
13. Between which two dates did the birthrate decline for 40 years?

A **double line graph** shows the change in two quantities over time. It is often used to compare two trends. Figure 1.4 shows the percent of the labor force in farm occupations and nonfarm occupations from 1830 to 1990.

FIGURE 1.4
Percent of Persons in Farm Versus Nonfarm Occupations in the United States, 1830 - 1990

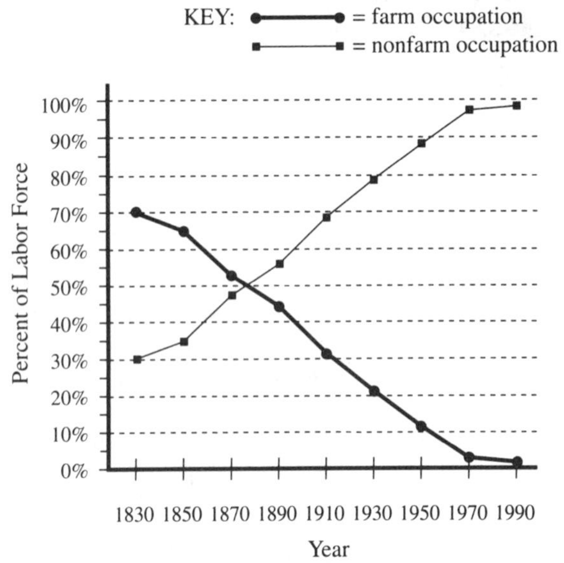

SOURCE: Department of Commerce, Census Bureau.

EXAMPLE 4 Use Figure 1.4 to answer each question.

a. What percent of the labor force was employed in farm occupations in 1830?

About 70%. Start at 1830 on the time axis. Move vertically upward to the thicker line. Move left to 70%.

OBSERVE The two occupations are opposites of one another, so the sum of the two percents above any date is 100%. For example, above 1830, percent farm + percent nonfarm occupations = 70% + 30% = 100% of the labor force.

b. What percent of the labor force was employed in farm occupations in 1990?

About 2%.

c. Estimate when 80% of the labor force was employed in nonfarm occupations.

About 1930. Locate 80% on the percent axis. Move right to the thinner line. move down to the time axis to 1930.

d. Estimate when the percent of farm occupations and nonfarm occupations were equal.

About 1875. This occurs where the two graphs intersect. Using a ruler, move downward from the point of intersection to 1875. Here, 50% of the labor force was employed in farm occupations, and 50% had other employment.

e. What has been the trend in farm occupations from 1830 to 1990?

The graph for farm occupations slants constantly downward. There has been a downward trend in farm occupations from 1830 to 1990.

♦ You Try It Answer the following questions using Figure 1.4.

14. What percent of the labor force was employed in farm occupations in 1910?
15. What percent of the labor force was employed in nonfarm occupations in 1950?
16. Estimate when 10% of the labor force was employed in farm occupations.
17. What was the trend in nonfarm occupations from 1830 to 1990?

---

3 TABLES

A **table** consists of rows and columns of numbers. Tables are used when more accurate data is wanted than that found in a graph.

FIGURE 1.5
Desirable Weights and Weight Ranges Versus Height

Height		*Weight*	
in	*ft in*	*Men (lb)*	*Women (lb)*
58	4′10″	—	102 (92–119)
60	5′ 0″	—	107 (96–125)
62	5′ 2″	123 (112–141)	113 (102–131)
64	5′ 4″	130 (118–148)	120 (108–138)
66	5′ 6″	136 (124–156)	128 (114–146)
68	5′ 8″	145 (132–166)	136 (122–154)
70	5′10″	154 (140–174)	144 (130–163)
72	6′ 0″	162 (148–184)	152 (138–173)
74	6′ 2″	171 (156–194)	—
76	6′ 4″	181 (164–204)	—

SOURCE: *Recommended Dietary Allowances*, 9th ed., 1980.

EXAMPLE 5 Use Figure 1.5 to answer each question.

a. According to the National Center for Health Statistics, the average American male is 5'10" tall and the average female is 5'4" tall. What are their desirable weights?

5'10" male: Desirable weight is 154 pounds; desirable weight range is from 140 pounds to 174 pounds.

5'4" female: Desirable weight is 120 pounds; desirable weight range is from 108 pounds to 138 pounds.

b. What is the highest desirable weight for a 6'2" man?

The weight range for a 6'2" man is 156 pounds to 194 pounds. Therefore, his highest desirable weight is 194 pounds.

c. What is the desirable range of heights for a 135-pound woman?

A woman's weight of 135 pounds is contained in the weight ranges corresponding to heights of 5'4" through 5'10".

♦**You Try It** Answer the following questions using Figure 1.5.

18. Find the desirable weight and weight range for a 5'8" woman.
19. What is the lowest desirable weight for a 5'6" man?
20. What is the desirable range in heights for a 158-pound man?

4 ARTICLES

Every day in newspapers and magazines we are presented with numerical information in charts, graphs, and news stories. It is important to read carefully and understand the kinds of numbers that are being presented. Is the information given in hours or minutes? Is your chart using decimals or percents? Did the table refer to millions or billions? Reading mathematical information is a skill requiring precision and care. You must always note the proper units for your numbers.

An **article** poses particular challenges because the information is embedded in prose. You must identify the numbers of interest in the story by careful reading.

FIGURE 1.6

Women more likely to live after chest pain

By Doug Levy
Special for USA TODAY

Women treated for chest pain live longer than men and are about half as likely to die of a later heart attack, a study in today's *Journal of the American Medical Association* says.

But women whose first sign of heart disease is a heart attack fare no better than men.

Ten years after treatment, 70% of women and 59% of men diagnosed with chest pains were alive, says Dr. Anthony Orencia, Northwestern University.

The study was based on 1,140 women and 1,501 men diagnosed with heart disease in Rochester, Minn., between 1960 and 1979.

The risk of sudden heart death in patients who had not previously been told they had heart disease was nearly the same for men and women.

The study underscores the need for more research on heart disease in women, Orencia says. "We can only speculate why women with (chest pain) survive longer compared to men of the same age."

Among the possibilities: the chest pain often turns out to be heartburn. Orencia also says men may avoid treatment after chest pain episodes.

Heart disease may be riskier for women because it typically develops later, when the body may be weakened by age or other diseases. Researchers also have found doctors more easily diagnose underlying heart disease in men.

EXAMPLE 6 Use Figure 1.6 to answer each question.

a. How many people were in the study referred to in this article?

There were 1,140 women and 1,501 men for a total of 2,641 people in the study. (See paragraph 4.)

b. Of the people diagnosed with chest pains, what percentage of men and women were alive after ten years?

70% of women and 59% of men with chest pain were alive after ten years. (See paragraph 3.)

♦You Try It Answer the following questions using Figure 1.6.

21. What is the difference in life expectancy between women and men whose first symptom is a heart attack?

22. What are some of the differences in heart disease in men and women?

♦**Answers to You Try It** **1.** A woman born in 1910 was expected to live 54 years. **2.** A woman born in 1950 was expected to live about 72 years, 18 years longer than a woman born in 1910. **3.** The life expectancy was 76 years in about 1970. **4.** A woman's life expectancy in 2010 would be about 82 years. (There are other reasonable answers.) **5.** The median income for a man with 8 years of elementary school is $19,000. **6.** The median income for a woman with 4 years of high school is $16,000. **7.** A man with less than 8 years of elementary school makes $5,000 more than a woman with the same education. **8.** A woman with 4 years of high school makes $4,000 more than one with only 8 years of elementary school. **9.** There were 24 live births per 1,000 people in 1960. **10.** In 1920 there were 28 live births per 1,000 people. **11.** The birthrate in 1975 was about 17 per 1,000 people. **12.** The birthrate from 1960 to 1970 dropped about 6 per 1,000 people. **13.** The birthrate declined for 40 years between 1950 and 1990. **14.** In 1910, 32% of the labor force was employed in farm occupations. **15.** In 1950, 89% of the labor force was employed in nonfarm occupations. **16.** 10% of the labor force was employed in farm occupations in about 1950. **17.** There has been a constant upward trend in nonfarm occupations from 1830 to 1990. **18.** The desirable weight for a 5'8" woman is 136 pounds. The desirable range is 122 - 154 pounds. **19.** The lowest desirable weight for a 5'6" man is 124 pounds. **20.** The desirable range in heights for a 158-pound man is 5'8" to 6'2". **21.** There is no difference between women and men if the first symptom is a heart attack. **22.** Heart disease typically develops later for women; underlying heart disease is easier to diagnose in men.

SECTION 1.1 EXERCISES

I *Use Figure 1.7 to answer questions 1–8.*

FIGURE 1.7
U.S. Unemployment Rate for Selected Years

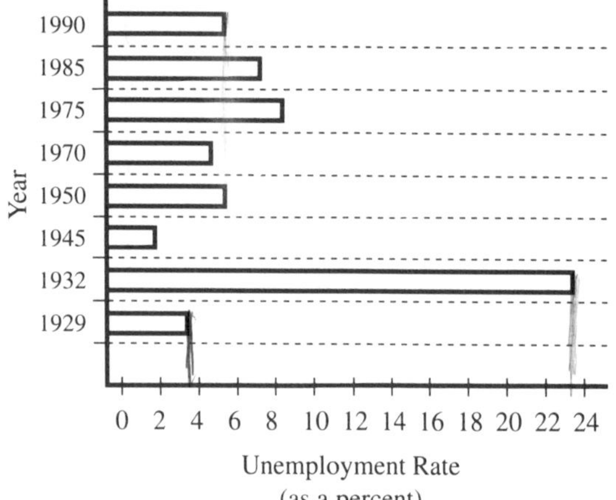

SOURCE: Department of Labor, Bureau of Labor Statistics.

1. What was the unemployment rate in 1970?

2. What was the unemployment rate in 1945?

3. In what year was the unemployment rate closest to the 1950 rate?

4. In what years were the unemployment rates under 4%?

5. In what year was the unemployment rate the lowest, and what was that rate?

6. In what year was the unemployment rate the highest, and what was that rate?

7. **a.** What were the unemployment rates in 1929 and 1932?

   b. About how many times greater was the 1932 rate?

8. About how many times smaller was the 1945 unemployment rate compared to 1932?

Use Figure 1.8 to answer questions 9–16.

FIGURE 1.8

Daily Recommended Calories for Men and Women by Age

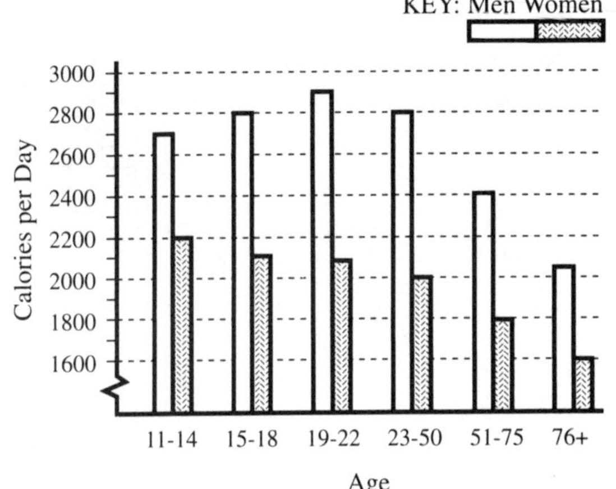

SOURCE: National Academy of Sciences.

9. How many calories are recommended daily for a 21-year-old man?

10. What age range recommends about 1,800 calories daily for a woman?

11. a. In what age range are men expected to have the most calories per day?

    b. How many calories is this?

    c. Describe the trend in calories needed by men from ages 11 through 76+.

12. a. In what two age ranges do women have the same number of calories recommended daily?

    b. How many calories is this?

    c. Describe the trend in calories needed by women from ages 11 through 76+.

13. a. In what age range is the difference between calories recommended for men and women the greatest?

    b. What is this difference?

14. a. In what age range is the difference between calories recommended for men and women the smallest?

    b. What is this difference?

15. What total number of calories should two eighty-year-old grandparents consume daily?

16. A woman nursing a baby needs 500 additional calories per day. If the woman is 26 years old, how many calories should she consume daily?

2 *Use Figure 1.9 to answer questions 17–24.*

FIGURE 1.9
Percent of College Graduates that Are Women in Selected Years

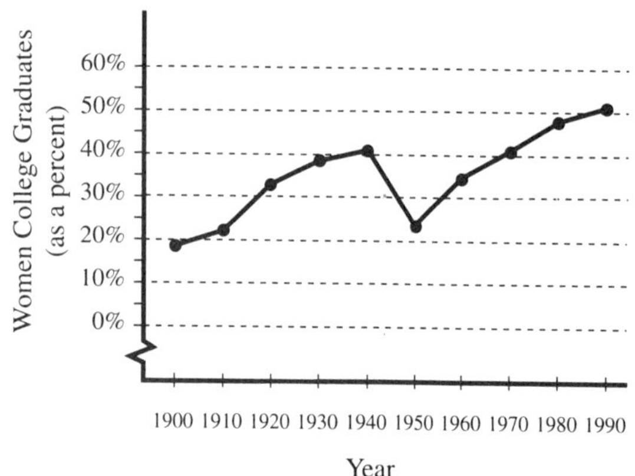

SOURCE: Department of Education, Center for Educational Statistics.

17. About what percent of college graduates in 1990 were women?

18. About what percent of college graduates in 1990 were men?

19. Between which two dates did the percent of women college graduates drop? By how many percentage points did it drop?

20. On which two dates did women make up about 41% of all college graduates?

21. What percent of college graduates were women in 1970?

22. What percent of college graduates were women in 1900?

23. When were the upward trends in Figure 1.9?

24. a. When were the downward trends in Figure 1.9?
    b. Can you offer any explanation for the downward trend?

Use the graph in Figure 1.10 to answer questions 25 - 32

FIGURE 1.10

Comparison of Sales in 1991 and 1990

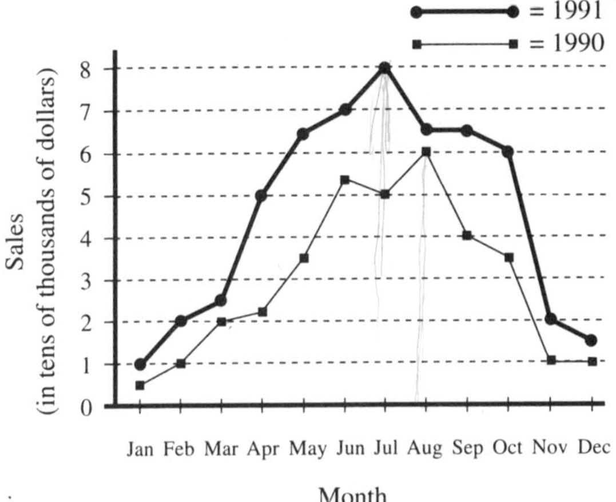

25. What were the total sales in March 1990?

26. What were the total sales in October 1991?

27. a. In which month in 1990 were sales the greatest?

b. What were the sales for that month?

28. a. In which month in 1991 were sales the greatest?

b. What were the sales for that month?

29. How much more in sales were made in July 1991 compared to July 1990?

30. In which months in 1990 were sales about $10,000?

31. In what time periods in 1990 were the trends
a. upward?
b. downward?
c. steady (no change)?

32. In what time periods in 1991 were the trends
a. upward?
b. downward?
c. steady?

3 *Use the table in Figure 1.11 to answer questions 33 - 36.*

FIGURE 1.11

World Population Growth from 1650 to 1989

	Population in Millions				
Continent	*1650*	*1750*	*1850*	*1950*	*1989*
North America	5	5	39	219	420
South America	8	7	20	111	287
Europe	100	140	265	530	685
Asia	335	476	754	1,418	3,131
Africa	100	95	95	199	642
Australia/Pacific Islands	2	2	2	13	26
Antarctica	....	uninhabited	....		
World Total	550	725	1,175	2,490	5,192

SOURCE: Rand McNally & Co.

33. What was the population of Europe in 1850?

34. Name the only two continents to experience a drop in population. Give the dates over which these drops occurred.

35. Which continent had a 1989 population about 7 times its 1650 population?

36. Which continent more than tripled in population from 1950 to 1989?

Use Figure 1.12 to answer questions 37 - 41.

FIGURE 1.12
Income Needed to Get a 30-Year Mortgage

Interest Rate	*Amount Borrowed* $50,000	$75,000	$100,000	$150,000
	Annual Income Needed			
9%	$17,242	$25,863	$34,484	$51,726
9.5%	18,018	27,028	36,037	54,055
10%	18,085	28,208	37,611	56,415
10.5%	19,602	29,403	39,203	58,805
11%	20,407	30,611	40,814	61,221
11.5%	21,221	31,831	42,441	63,662
12%	22,042	33,063	44,084	66,125

SOURCE: National Association of Realtors.

37. What annual income is needed to borrow $75,000 at 10.5% for 30 years?

38. About how much can be borrowed at 9.5% if your annual income is $36,150?

39. What additional annual income is needed to borrow $75,000 versus $50,000 at 10%?

40. What additional annual income is needed to borrow $100,000 at 10% versus $75,000 at 12%?

41. A family has two annual incomes of $28,950 and $27,896. What rate must they get on a loan to borrow $150,000?

4 *Use Figure 1.13 to answer questions 42 - 45.*

FIGURE 1.13

Pushing parents' role in education

By Dennis Kelly
USA TODAY

Parents are so crucial to school success that 95% of them back written plans for parental involvement and 86% of teachers think parents should be fined if their kids are chronic truants, new surveys say.

The first, commissioned by *Newsweek* for the National PTA, finds parents are involved (74% discuss school with kids every day). But more parents feel competent to help their kids in subjects like math (58%) and English (55%) than in science (44%) and computers (31%). The survey of 1,148 adults also finds:

▶ Fewer black (41%) and Hispanic (35%) parents think public school funding is adequate than do white (50%) parents. And black and Hispanic parents have higher educational aspirations for kids: 53% of black parents and 50% of Hispanic parents expect their kids to earn advanced degrees; just 34% of white parents do.

▶ 45% of parents help with homework every day; 95% like the idea of written agreements on what parents should do.

"It shows parents want some assistance and guidance in how they can become involved in their child's education," says Pat Henry, PTA president.

The other survey — of 1,000 public school teachers done for Metropolitan Life — finds support for President Clinton's education agenda: 81% back national standards for what students should know; 61% support a national exam system.

It also shows 86% favor fines for parents whose children are chronically truant; 54% say the highest priority for education policy should be boosting parents' roles in children's learning. Only 26% gave top priority for preschool programs like Head Start.

42. What percent of parents discuss school with their children every day?

43. What percent of parents feel competent to help their children in math?

44. What percent of parents help with homework every day?

45. What percent of black, white, and Hispanic parents expect their kids to earn advanced degrees?

Use Figure 1.14 to answer questions 46 - 48.

FIGURE 1.14

Medals could come with cash attached

Starting in 1994, Olympic medals will bear price tags.

Even fourth-place finishers won't be empty-handed.

The U.S. Olympic Committee is designating $7.6 million from its 1993-96 budget for a larger Operation Gold program. The structure awaits final approval June 4.

Providing "a way to pay for training, to stay in the sport longer" is the USOC's motivation, executive director Harvey Schiller says.

In non-Olympic years, a smaller pay scale is proposed, with $5,000 earmarked for world titles. Under the old plan, a top-eight world ranking was rewarded with $2,500, whether an athlete was No. 1 or No. 8.

— ***Steve Woodward***

▶ **Sensitive issue, 8C**

The payout for future Olympics

The U.S. Olympic Committee plans to raise its bonuses for top athletes. The only current USOC performance-based award is $2,500 for a top-eight finish.

By Julie Stacey, USA TODAY

46. How much money is being designated for the Operation Gold planning?

47. What is the proposed bonus award for world titles in non-Olympic years?

48. Prior to 1994, what was the Olympic Committee award for a top-eight finish in world ranking?

EXTEND YOUR THINKING ♦ ♦ ♦

WRITING TO LEARN ♦ ♦ ♦

Use Figure 1.15 to answer questions 49 - 50.

FIGURE 1.15

49. What is the point that this graph was constructed to illustrate?

50. Is it effective? Why or why not?

1.2 REDISCOVERING MATHEMATICS: NUMBERS AND OPERATIONS

OBJECTIVES

1. Identify types of numbers and use appropriate symbols for comparing numbers.
2. Translate expressions.
3. Evaluate expressions containing variables.
4. Determine if an equation is true or false.

NEW VOCABULARY

whole number	sum
number line	difference
fraction	product
mixed number	quotient
decimal	commutative
variable	expressions
	equation

Education in the 1990's changes as our society changes. All college students must become mathematically literate and they must be able to use and produce mathematical information in order to function as enlightened citizens. The ability to work with and appreciate all kinds of numerical information is essential to being productive members of our society.

1 TYPES OF NUMBERS

Whole numbers are numbers that you have always used for counting, along with zero.

0, 1, 2, 3, 4, 5, . . .

Counting numbers are the nonzero numbers listed.

Numbers between 0 and 1 can be represented as **fractions** or **decimals**.
Numbers between any two whole numbers will have a fraction or decimal part.

EXAMPLE 1 $\frac{1}{3}$ is a fraction between 0 and 1.

This is a number line and it shows the approximate location of $\frac{1}{3}$:

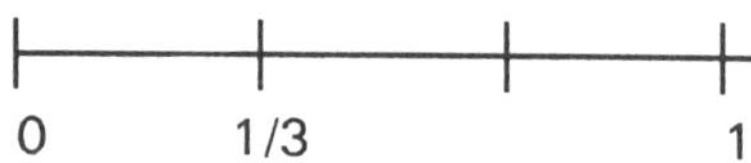

EXAMPLE 2 0.86 is a decimal between 0 and 1 .

♦ **You Try It** **1.** Write three numbers between 0 and 1. Pick at least one fraction and one decimal.

Symbols for Comparing Numbers

Mathematics is a language and often replaces quantitative words or phrases with symbols. When comparing two quantities, we may use the phrases *is equal to, is between, is greater than* or *is less than*. There are mathematical symbols for these phrases.

The symbol for the phrase *is equal to* is =.
The symbol for the phrase *is less than* is <.
The symbol for the phrase *is greater than* is >.

EXAMPLE 3 **a.** 5 is equal to $3 + 2$ is written $5 = 3 + 2$.
b. 2 is less than 8 is written $2 < 8$.
c. 8 is greater than 2 is written $8 > 2$.

Notice that the symbol always "points" to the smaller number.

There is no single symbol for *is between*. Instead, if we want to say that $\frac{1}{3}$ is between 0 and 1, we use the fact that $0 < \frac{1}{3}$ and $\frac{1}{3} < 1$ and combine these two statements to write:

$$0 < \frac{1}{3} < 1$$

EXAMPLE 4 Since $0 < 0.86$ and $0.86 < 1$, we can write $0 < 0.86 < 1$, which translates " 0.86 is between 0 and 1" .

Mixed Numbers Numbers between any two whole numbers other than 0 and 1 are often called **mixed numbers.** They are a combination, or a mix, of a whole number and a fraction or a decimal.

EXAMPLE 5 $11\frac{1}{3}$ is a mixed number between 11 and 12 .

On a number line, we have

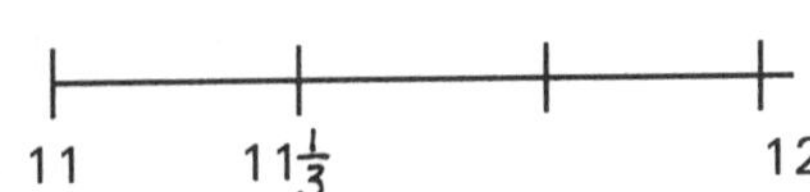

To write the statement "$11\frac{1}{3}$ is between 11 and 12" using mathematical symbols:

$$11 < 11\frac{1}{3} < 12 .$$

EXAMPLE 6 2.12 is between 2 and 3 . On a number line, we have

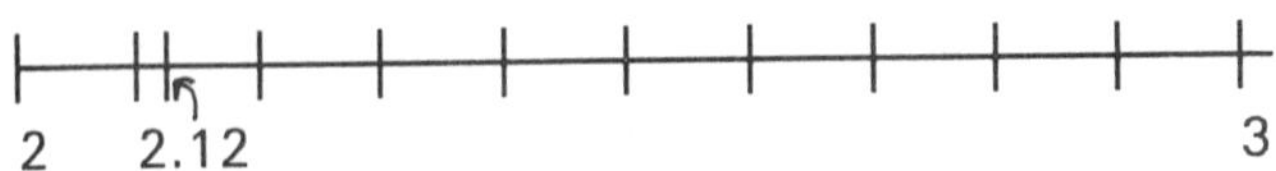

We can write the statement "2.12 is between 2 and 3" as $2 < 2.12 < 3$.

◆ **You Try It**

2. Write three numbers between 8 and 9. Give at least one example using a fraction and one using a decimal.

Write the statement using < or >.

3. 5.25 is less than 6.

4. $9\frac{1}{9}$ is greater than 9.

5. 88.89 is less than 89.

6. 88.89 is greater than 88.

7. 88.89 is between 88 and 89

8. $18\frac{1}{2}$ is between 18 and 19.

2 MATHEMATICAL OPERATIONS

In this section we will focus on solving problems that involve numbers. To solve quantitative problems, you must possess the ability to read the problem carefully and then be able to analyze the information contained in the problem. To help you develop these skills, we will begin by looking at the various arithmetic operations (addition, subtraction, multiplication, and division) and the words that lead you to choose one operation over another. Certain words are indications of a specific mathematical operation. This chart lists some of the words for each operation.

OPERATION	KEY WORDS
Addition	the sum of, total, increased by, more than, added to, plus
Subtraction	the difference of, decreased by, less than, subtracted from
Multiplication	the product of, double, triple, multiplied by
Division	the quotient of, divided by

Symbols for Operations

Rather than writing out lengthy phrases for these operations, we use a symbol to save time. The standard symbol for addition is + and the standard symbol for subtraction is −. There are several ways to indicate multiplication: you can use a small, raised dot between the two numbers; you can use an × between the two numbers; or one or both numbers can be placed inside parentheses, like this:

$2 \cdot 3$ or 2×3 or 2(3) or (2)3 or (2)(3)

Think about these options. Which would you choose to use? Why?

___

___

Can you see disadvantages in any of the choices?

___

___

Division can also be written more than one way. For example, 12 divided by 6 can be

written $12 \div 6$ or $\frac{12}{6}$ or 12/6 or $6\overline{)12}$

Which would you choose to use? Why?

___

___

Before we begin to solve problems, we will work on translating several English phrases into the appropriate mathematical symbols. Here are some examples.

WORDS	SYMBOLS
the sum of 9 and 2	$9 + 2$
decrease 20 by 4	$20 - 4$
the quotient of 81 and 3	$81 \div 3$
the quotient of 3 and 81	$3 \div 81$
8 divided by 12	$\frac{8}{12}$
the product of 6 and 7	$6 \cdot 7$
double 3.12	2(3.12)
five more than seven	$7 + 5$
four less than eight	$8 - 4$

OBSERVE Be careful with subtraction and division. **ORDER COUNTS!**
Four less than eight **must** be $8 - 4$, NOT $4 - 8$.
8 divided by 12 **must** be $8 \div 12$ NOT $12 \div 8$.
If the order does **not** matter in an operation, like in addition and multiplication, we say that the operation is *commutative*. Subtraction and division are not commutative.

♦ **You Try It** *Rewrite the English phrases using symbols. DO NOT do the arithmetic.*

9. triple 11.23

10. 4 is added to $11\frac{15}{23}$

11. the difference of 65 and 11

12. the quotient of 67 and 6

13. the product of 16 and 11

14. seven more than your age

Variables

Look at problem #14 in the You Try It again. Will everyone translate it exactly the same way? Why could the answers be different? Is there a way for everyone to write the same symbols?
When we don't know the value of a particular quantity and we want to use it, we can use a symbol to represent it. Usually we choose a letter (often x or n) and we call the symbol a **variable**. The numerical value of a variable will depend on the situation.

> **A variable** is a letter that represents a numerical quantity. Its value depends on the situation.

EXAMPLE 7 We can write the statement "Deduct 5 pounds from your weight" in symbols as " $x - 5$ ", as long as everyone knows that x stands for weight.

EXAMPLE 8 If the statement is "Divide a number by 12" but you have not been told the number, you can use a variable to represent the number. Suppose you choose n to represent the number. The statement "divide a number by 12" would then be written as $\frac{n}{12}$.

3 EVALUATING EXPRESSIONS

When we put together numbers, variables and operations we create mathematical *expressions*. Examples of expressions are $3 + y$, $6 - 1$ and $a + b$. When the operation is multiplication and we are multiplying a variable with a number or another variable, we do not need a symbol for multiplication. We can write $3(x)$ as $3x$. We can write $a(b)$ as ab. (We certainly cannot write 2(4) as 24 !). Remember too that we often choose to write $a \div b$ as a/b or $\frac{a}{b}$. When a particular value is given to the variable in an expression, we can perform the operation indicated using the numbers that are given to replace the variable. We call this procedure **evaluating an expression.**

EXAMPLE 9 Evaluate $a + 3$ when $a = 12$.

Replace a with 12.
$a + 3 = 12 + 3 = 15$

EXAMPLE 10 Evaluate $2y$ when $y = 6$.

$2y$ means $2(y)$. Replace y with 6.
$2y = 2(6) = 12$.

EXAMPLE 11 Evaluate xy when $x = 6$ and $y = 2$

xy means $x(y)$. Replace x with 6. Replace y with 2.
$xy = x(y) = 6(2) = 12$

♦ You Try It

Evaluate each expression.

15. $2x$ when $x = 11$.

16. $a + 17$ when $a = 23$.

17. $21 - b$ when $b = 7$

18. $\frac{18}{x}$ when $x = 2$.

4 EQUATIONS

An *equation* is a statement that says that two mathematical quantities are equal.

An equation may be *true*, like $2 + 3 = 5$, or *false*, like $12 = 4 + 1$. Sometimes an equation contains a variable and you cannot tell whether the equation is true or false unless you know the value of the variable.

EXAMPLE 12 If $x = 7$, is the equation $x - 4 = 3$ true or false?

Replace x with 7 and evaluate the left side of the equation:

$$x - 4 = 3$$

$$7 - 4 \stackrel{?}{=} 3$$

$$3 = 3$$

Thus, $x - 4 = 3$ is *true* if $x = 7$.

EXAMPLE 13 If $k = 14$ and $y = 7$, is the equation $k - 5 = 3y$ true or false?

Replace k with 14 and y with 7:

$$k - 5 = 3y$$

$$14 - 5 \stackrel{?}{=} 3(7)$$

$$9 \neq 21$$

Thus, $k - 5 = 3y$ is *false* if $k = 14$ and $y = 7$.

♦ **You Try It**

19. If $t = 8$, is the equation $12 = 16 - t$ true or false?

20. If $a = 4$ and $b = 15$, is $5a = b + 5$ true or false?

♦**Answers to You Try It** **1.** $0.87,\ 0.3,\ \frac{3}{4}$ **2.** $8.54,\ 8.9,\ 8\frac{2}{5}$ **3.** $5.25 < 6$ **4.** $9\frac{1}{9} > 9$ **5.** $88.89 < 89$ **6.** $88.89 > 88$ **7.** $88 < 88.89 < 89$ **8.** $18 < 18\frac{1}{2} < 19$ **9.** $3(11.23)$ **10.** $4 + 11\frac{15}{23}$ **11.** $65 - 11$ **12.** $67 \div 6$ **13.** $16(11)$ **14.** $x + 7$ when x is your age **15.** 22 **16.** 40 **17.** 14 **18.** 9 **19.** False **20.** True

SECTION 1.2 EXERCISES

1 *Use the correct symbol, < or >, between the pair of numbers to write a true statement.*

1. 0 8

2. $\frac{2}{7}$ 2

3. 101 100

4. 999 990

5. 6 $6\frac{1}{3}$

6. $11\frac{1}{2}$ 12

7. $99\frac{1}{4}$ 100

8. 1000.5 1001

9. 77.78 78

10. 3.8 4

2 *Rewrite the English phrase or sentence using symbols. If you use a variable, define it.*

11. the sum of 11 and $12\frac{1}{2}$

12. double 3.2

13. 18 is divided by 6

14. the quotient of 28 and 13

15. triple 18

16. 5 pounds more than your weight

17. double your height

18. 10 years older than the age of the next person you meet

19. 19 plus 7

20. the product of 11 and 3

21. 8 more than a number

22. the difference of 99 and 1

23. 5 less than 18

24. the difference of two numbers

25. the quotient of 11 and a number

26. the quotient of a number and 11

27. 8 is more than a number.

28. 8 is less than a number.

29. 3 is less than 3.3.

30. 4 is greater than 3.2.

31. 5 is between 4 and 6.

32. $\frac{1}{2}$ is between 0 and 1.

33. One number is the product of two different numbers

34. 8 is equal to the sum of 6 and 2.

35. 56 is the product of 8 and 7.

36. 2 is the quotient of 18 and 9.

37. 14 is the difference of some number and 23.

38. 99 is the quotient of two unknown numbers.

3 *Evaluate the expression given.*

39. $x + 2$ when $x = 7$

40. $x - 2$ when $x = 7$

41. $x + y$ when $x = 3$ and $y = 5$

42. $x - y$ when $x = 11$ and $y = 3$

43. $9 - x$ when $x = 3$

44. xy when $x = 4$ and $y = 9$

45. WH when W = 12 and H = 5

46. AB when A = 3 and B = 8

47. $3x$ when $x = 16$

48. $8a$ when $a = 11$

49. $\frac{x}{2}$ when x = 6

50. $\frac{12}{x}$ when $x = 6$

51. $\frac{a}{b}$ when $a = 12$ and $b = 3$.

52. $\frac{x}{y}$ when $x = 56$ and $y = 8$

4 *Using the given values for the variables, decide if the equation is true or false.*

53. $5 + x = 12$ if $x = 7$

54. $6 = \frac{t}{2}$ if $t = 12$

55. $9k = 40$ if $k = 4$

56. $60 - y = 40$ if $y = 20$

57. $4x = 15 - y$ if $x = 2$ and $y = 7$

58. $41 - w = 3y$ if $w = 20$ and $y = 8$

59. $3a + 10b = 27$ if $a = 9$ and $b = 0$

60. $\frac{h}{d} = 3$ if $h = 12$ and $d = 4$

EXTEND YOUR THINKING ♦ ♦ ♦

♦ SOMETHING MORE

Use Figure 1.16 to answer question 61.

FIGURE 1.16

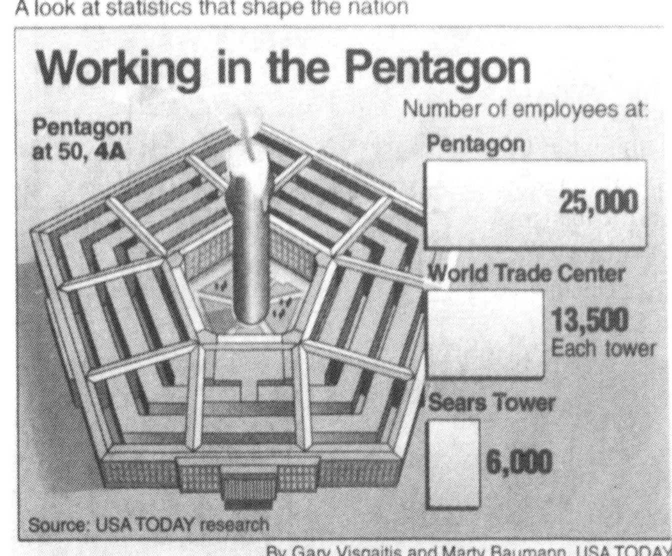

61. Use the symbol < or > and the appropriate numbers to describe the relationships below.

a. Combined total of employees at one tower of the World Trade Center and the Sears Tower — Number of employees at the Pentagon

b. Twice the number of employees at one tower of the World Trade Center — Number of employees at the Pentagon

c. Difference of the number of employees at the Pentagon and one tower of the World Trade Center — Number of employees at the Sears Tower

62. Write 3 numbers between 0 and 1.

63. Write 3 numbers between 100 and 101.

64. What two consecutive whole numbers is 11.11 between? (Note: *consecutive* means one after the other, with no whole number in between.)

65. Between what two consecutive whole numbers is $\frac{7}{8}$?

66. Between what two consecutive whole numbers is 78.11?

67. Write two mathematical expressions with the the variable x. One expression should involve addition and one should involve multiplication.

68. Use the information in Figure 1.17 to pick a combination of the other positions that equals the total number of defensive linemen. Use variables and write an equation to show the relationship.

FIGURE 1.17

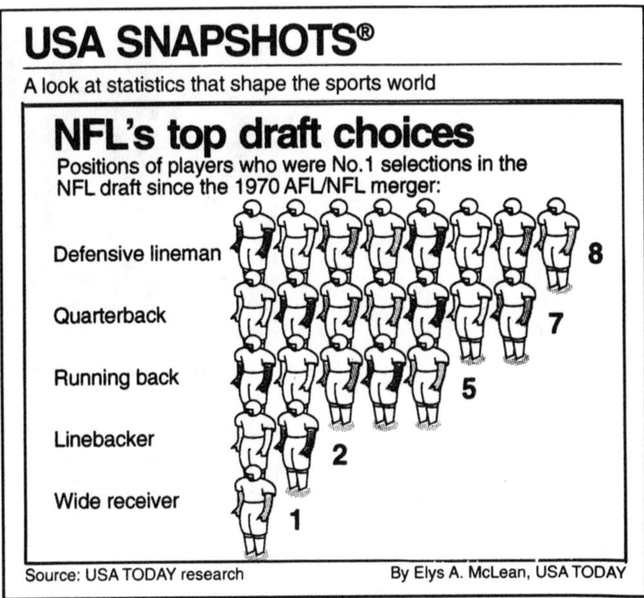

1.3 ROUNDING AND ESTIMATION

OBJECTIVES

1. Use place value to round whole numbers.
2. Use rounding to estimate answers.

1 PLACE VALUE

A whole number written using the digits 0, 1, 2, 3, 4, 5, 6, 7, 8, 9 is written in **standard notation.**

Our number system is a **place value** system. This means the position of a digit in a number determines its place value.

The **place value key** below gives the first fifteen place value names for whole numbers.

Place Value Key

hundred trillions	ten trillions	trillions	hundred billions	ten billions	billions	hundred millions	ten millions	millions	hundred thousands	ten thousands	thousands	hundreds	tens	ones
_	_	3,	1	3	0,	6	9	7,	8	5	3,	7	2	4
trillions period			billions period			millions period			thousands period			ones (units) period		

EXAMPLE 1 Use the place value key above to find the place value name for the digit 2 in each number.

4,172 ⟶ 2 is in the units or ones place. It is read 2 ones.

5,627 ⟶ 2 is in the tens place. It is read 2 tens.

219 ⟶ 2 is in the hundreds place. It is read 2 hundreds.

2,803 ⟶ 2 is in the thousands place. It is read 2 thousands.

♦ You Try It *Use the place value key to find the place value name for the digit 7 in each number.*

1. 375 **2.** 7,240 **3.** 687 **4.** 14,708

ROUNDING WHOLE NUMBERS

You buy a coat for $82. In conversation, you round the price to the nearest ten dollars, and say the coat cost you $80. The Census Bureau says the population of the United States is 253,783,496. Since this number is closer to 254,000,000 than 253,000,000, a newspaper reports the population as 254 million.

It is sometimes more convenient to express numbers as approximate values. **Rounding off** is finding approximate values for exact values. Your round off, or simply round, to a given place value.

EXAMPLE 2 Round 28 to the nearest ten.

On the number line, 28 is closer to 30 (3 tens) than to 20 (2 tens).

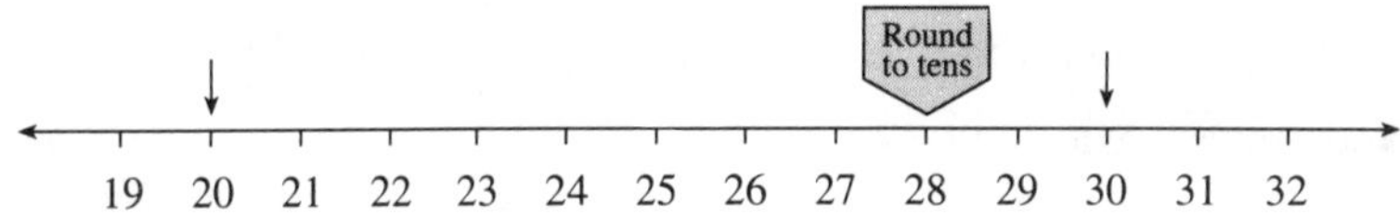

Therefore, 28 rounded to the nearest ten is 30.

EXAMPLE 3 Round 23 to the nearest ten.

On the number line, 23 is closer to 20 (2 tens) than to 30 (3 tens).

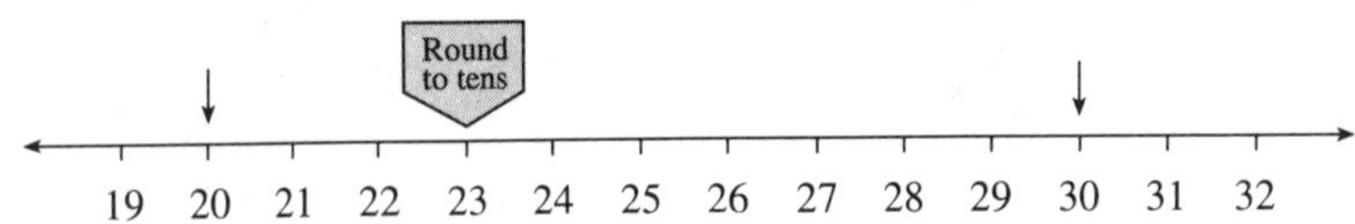

Therefore, 23 rounded to the nearest ten is 20.

♦ You Try It

Use the number line to round to the nearest ten.

5. 77 **6.** 72 **7.** 49 **8.** 44

Round 25 to the nearest ten. Since 25 is midway between 20 and 30, you could round up or down. The accepted convention is to round up. Therefore, 25 rounded to the nearest ten is 30.

> To round whole numbers
>
> **1.** Circle or underline the digit in the place you are rounding to.
> **2.** If the digit on its right is
> **a.** 5 or larger (5, 6, 7, 8, or 9), add 1 to the circled or underlined digit.
> **b.** 4 or smaller (4, 3, 2, 1, or 0), do not change the circled digit.
> **3.** Replace all digits to the right of the circled digit with zeros.

The symbol $\doteq$ means "is approximately equal to." It is used to say the rounded number is approximately equal to the original number.

EXAMPLE 4 Round 573 to the nearest ten.

Circle the tens digit. — 7 does not change because the digit on its right, 3, is less than 5.

$5⑦3 \doteq 570$

Replace the 3 with a 0.

$573 \doteq 570$ to the nearest ten. This is pictured on the number line. You can see that 573 is closer to 570 (57 tens) than 580 (58 tens).

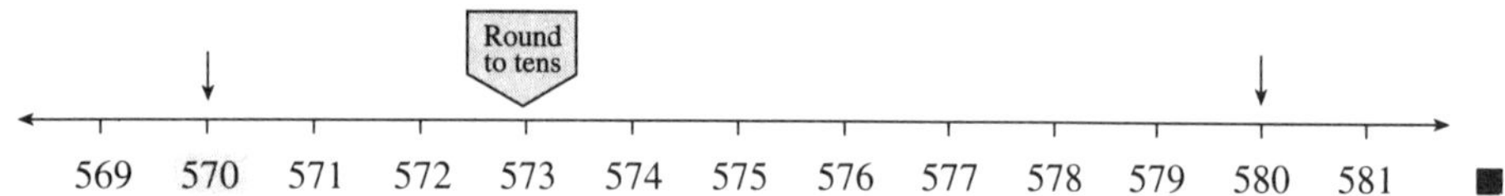

■

EXAMPLE 5 Round 8,352 to the nearest hundred.

Circle the hundreds digit. — Add 1 to the 3 since the digit on its right, 5, is 5 or larger.

$8,③52 \doteq 8,400$

Replace 5 and 2 with 0's.

$8,352 \doteq 8,400$ to the nearest hundred. On the number line, 8,352 is closer to 8,400 (84 hundreds) than 8,300 (83 hundreds).

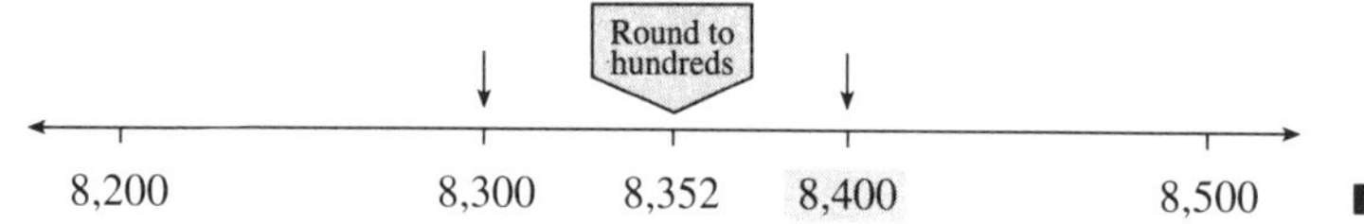

■

EXAMPLE 6 Round 7,499 to the nearest thousand.

Circle the thousands place. — 7 remains the same because the digit on its right, 4, is less than 5.

$⑦,499 \doteq 7,000$

Replace 4, 9, and 9 with 0's.

$7,499 \doteq 7,000$ to the nearest thousand. Observe, 499 is less than half a thousand. So 7,499 is closer to 7,000 than 8,000.

♦You Try It

Round to the indicated place.

9. 362 to tens **10.** 879 to tens **11.** 5,813 to hundreds
12. 3,054 to hundreds **13.** 3,299 to thousands **14.** 46,922 to thousands

EXAMPLE 7 Round 4,972 to the nearest hundred.

Circle the hundreds digit. — Add 1 to the 9 since the digit on its right, 7, is 5 or larger. Since 1 + 9 = 10, write 0 in the hundreds place and carry 1 to the thousands place.

4,⑨72 ≐ 5,000

Replace 7 and 2 with 0's.

4,972 ≐ 5,000 to the nearest hundred. Observe, 4,972 is closer to 5,000 (50 hundreds) than 4,900 (49 hundreds).

♦You Try It

Round to the indicated place.

15. 8,972 to hundreds **16.** 16,499 to hundreds

2 ESTIMATION A young couple decides to build a home. The land will cost them $28,560. The house will cost $93,825. They make a quick mental estimate of the total cost as follows.

land:	$28,560	⟶	$30,000 estimate for land
house:	$93,825	⟶	+$90,000 estimate for house
actual cost:	$122,385		$120,000 estimate of cost

$28,560 is close to $30,000. $93,825 is close to $90,000. The $120,000 estimate is close to the actual cost, $122,385. Note that the couple rounded both costs to the ten-thousands place.

Estimation is not meant to be exact. You estimate because you want a quick, approximate answer. An estimate is a "ballpark figure."

> To estimate an answer to an addition or subtraction problem
>
> **1.** Round all numbers to the same place value.
> **2.** Add or subtract the rounded values.

Choose to round to a place that gives a fast, though practical, answer. There are many different ways to estimate, and many different answers depending on how you estimate.

EXAMPLE 8 Five passengers are on an elevator. Their weights are 247, 115, 189, 36, and 78 pounds, respectively. Estimate their total weight in the elevator.

Since most of the weights are in the hundreds, round all the weights to the hundreds place. Then add.

Actual weights: 247 + 115 + 189 + 36 + 78

↓ ↓ ↓ ↓ ↓

Estimated weights (rounded to the hundreds place): 200 + 100 + 200 + 0 + 100 = 600 lbs is the estimated weight in the elevator

Verify that the actual total weight is 665 pounds.

OBSERVE In Example 8, 36 is less than 50, or half a hundred. Therefore, 36 rounded to the nearest hundred is 0 hundreds, or 0.

♦**You Try It**

17. Six people are in a van. They weigh 180, 214, 48, 136, 75, and 155 pounds, respectively. Estimate the total weight.

EXAMPLE 9 The population of California in 1940 was 6,907,387. In 1980, it was 23,667,826. Estimate the growth in population from 1940 to 1980.

Both numbers are in the millions. Round to the millions place. Then subtract.

1980:	23,667,826	round to	24,000,000
1940:	6,907,387	millions →	− 7,000,000
			17,000,000 estimated growth

Verify that the actual growth in population was 16,760,439.

OBSERVE If you round both numbers in Example 9 to ten-millions, the estimate becomes 20,000,000 − 10,000,000 = 10,000,000, which is much less accurate. You must use your judgment when selecting the place value to round to.

♦**You Try It**

18. The population in Texas in 1970 was 11,198,655. In 1980 it was 14,227,574. Estimate the growth in population from 1970 to 1980.

Estimation can help you catch errors.

EXAMPLE 10 Use estimation to decide if the answer below is reasonable.

$$647 + 3{,}219 + 1{,}159 + 986 + 860 + 2{,}180 \stackrel{?}{=} 6{,}751$$

$$\downarrow \quad \downarrow \quad \downarrow \quad \downarrow \quad \downarrow \quad \downarrow$$

Estimate: (round each number to hundreds) $600 + 3{,}200 + 1{,}200 + 1{,}000 + 900 + 2{,}200 = 9{,}100$ estimate of the sum

The estimate, 9,100, is not reasonably close to 6,751. Rechecking your work shows the actual sum is 9,051.

OBSERVE In Example 10, the fact that the answer and the estimate do not agree does not tell you which one is wrong. It only tells you to recheck your work.

♦**You Try It**

19. Use estimation to decide if the answer is reasonable. $4{,}695 + 815 + 2{,}751 + 367 + 7{,}216 \stackrel{?}{=} 15{,}844$

♦ **Answers to You Try It** **1.** tens **2.** thousands **3.** ones **4.** hundreds **5.** 80 **6.** 70 **7.** 50 **8.** 40 **9.** 360 **10.** 880 **11.** 5,800 **12.** 3,100 **13.** 3,000 **14.** 47,000 **15.** 9,000 **16.** 16,500 **17.** about 800 lb (round to hundreds) **18.** about 3,000,000 (round to millions) **19.** estimate = 16,000 (round to thousands); answer = 15,844, so the answer is reasonable

SECTION 1.3 EXERCISES

I *Round off each number to the indicated place.*

1. 56 tens

2. 72 tens

3. 175 hundreds

4. 416 hundreds

5. 1,782 tens

6. 2,428 tens

7. 549 hundreds

8. 751 hundreds

9. 4,366 hundreds

10. 7,836 hundreds

11. 3,499 tens 3,500

12. 8,994 tens

13. 34 hundreds

14. 63 hundreds

15. 3,723 thousands

16. 9,398 thousands

17. 4,996 tens

18. 7,993 tens

19. 555 thousands

20. 499 thousands

21. 56,834 ten-thousands

22. 7,488 ten-thousands

23. 3,835,921 millions

24. 59,489,999 millions

25. 99,986 tens

26. 9,997 tens

27. 2,499,845 thousands

28. 3,858,705 thousands

Complete the following table by rounding off each number to the three indicated places.

	tens	hundreds	thousands
29. 69			
30. 736			
31. 1,658			
32. 9,989			
33. 7,293			
34. 263			
35. 34,608			
36. 9,999,745			
37. 262,911			
38. 79,567			

39. Round a car's odometer reading of 112,378 miles to the nearest thousand miles.

40. Round the $98,612,450 cost of the new stadium to the nearest million dollars.

41. A sporting event grossed $12,488,122. What is this rounded to the nearest one hundred thousand dollars?

42. Light travels 186,282 miles per second. Round this to the nearest thousand miles per second.

2 *Estimate the answer by first rounding each number to the indicated place.*

43. $36 + 47$ tens

44. $68 + 73$ tens

45. $268 + 431$ hundreds

46. $691 + 380$ hundreds

47. $6{,}192 + 8{,}398$ hundreds

48. $2{,}820 + 3{,}746$ hundreds

49. $72 - 48$ tens

50. $97 - 51$ tens

51. $6{,}924 - 2{,}318$ thousands

52. $3{,}094 - 950$ thousands

53. $1{,}230 - 467$ hundreds

54. $2{,}880 - 1{,}343$ hundreds

55. 459 + 120 + 785 + 98 hundreds

56. 77 + 830 + 117 + 457 + 99 hundreds

57. 75 + 36 + 40 + 98 + 3 + 25 tens

58. 108 + 56 + 84 + 9 + 33 + 2 tens

59. 1,416 + 780 + 128 + 45 + 849 hundreds

60. 6,280 + 12,098 + 2,754 + 860 + 9,812 thousands

61. Today the printing center spent $712 on paper, $340 on ink, $180 on envelopes, $572 on copying, and $301 on minor repairs. Estimate the amount spent by rounding to hundreds.

62. A real estate agent sold homes this month for $96,423, $87,280, and $156,254. Estimate total sales by rounding to ten-thousands.

63. A family started a cross-country trip with an odometer reading of 36,278 miles. The reading at the end of the trip was 47,819 miles. Estimate the total miles traveled by rounding to thousands.

64. The distance from Baltimore to Atlanta is 675 miles. The distance from Baltimore to Boston is 399 miles. Estimate how much closer Boston is to Baltimore by rounding to hundreds.

65. This month the Lopez family spent $918 on the mortgage, $514 on food, $78 on gas, and $267 on utilities. Estimate the total spent by rounding to tens.

66. Estimate the total area of the six New England states if the actual areas in square miles are 5,009 for Connecticut, 8,257 for Massachusetts, 33,215 for Maine, 9,304 for New Hampshire, 1,214 for Rhode Island, and 9,609 for Vermont. Round to thousands.

Estimate each answer two ways, as indicated.

67. 55 + 78 + 31 + 8 + 93

a. tens **b.** hundreds

68. 89 + 72 + 97 + 8 + 43 + 76

a. tens **b.** hundreds

69. 461 + 790 + 834 + 805
   a. hundreds
   b. thousands

70. 943 + 392 + 45 + 958 + 457 + 98
   a. hundreds
   b. thousands

71. 8,285 + 854 − 6,190 + 4,120
   a. thousands
   b. ten-thousands

72. 4,921 + 9,605 − 2,890 + 812 − 499
   a. hundreds
   b. thousands

SKILLSFOCUS (Section 1.2) *Evaluate each expression if $x = 4$ and $y = 2$.*

73. $x + y$ **74.** $x - y$ **75.** xy **76.** $\frac{x}{y}$

EXTEND YOUR THINKING

TROUBLESHOOT IT

Find and correct the error.

77. 654 rounded to the nearest ten is 660.

78. 29,951 rounded to the nearest hundred is 29,900.

WRITING TO LEARN

79. Explain the difference between = and ≐.

80. Explain why 36 rounded to hundreds is 0 (see Example 8).

81. The attendance for game #1 was 67,306 people. Game #2 was attended by 74,691 people. Estimate how many more people attended game #2 by
   a. Rounding to ten-thousands.
   b. Rounding to thousands.
   c. Explain which answer is more practical and why.

YOU BE THE JUDGE

82. Jack claimed 4,449 rounded to the nearest thousand is 5,000. He reasoned that 4,449 rounded to the nearest ten is 4,450. 4,450 rounded to the nearest hundred is 4,500. And 4,500 rounded to the nearest thousand is 5,000. Is he correct? Explain your decision.

1.4 PROBLEM SOLVING USING MATHEMATICS

OBJECTIVES

1. Estimate and find the exact answer to application problems.
2. Find an average of a group of numbers.
3. Find the area of a rectangle by using a formula.

NEW VOCABULARY

average
mean
formula
area

1 PROBLEM SOLVING

College students and all informed citizens are constantly receiving and reacting to information. Much of the information is given in numerical form. The numbers may be presented in a statement, in a table, or in a graph. In this section, the problems will contain numerical information that you will need to analyze. You MUST develop a strategy for working on a problem. We will begin the process by following these four steps.

Steps for Problem Solving

1. **READ** carefully. Identify what each of the numbers in the problem measures. For example, do you have 60 dollars, 60 miles or 60 chickens?

2. **WRITE**, in your own words, the question that is being asked. Often it is helpful to include a sketch.
   It is also important to note the units for the answer. Again, do we want dollars, miles, or chickens?

3. **ANALYZE** how you can use the information in the problem to answer the question.
   - Should you use all or only some of the values stated in the problem?
   - Must you do one or more arithmetic operations (addition, subtraction, multiplication, or division) to answer the question?
   - Make an outline of the steps you need to follow to solve the problem.

4. **CHECK** your answer to see if it makes sense. For example, if you are computing a clerk's annual salary and you get $500,000, it is likely you made an error!

The four arithmetic operations are used to solve countless important (and some unimportant) problems that exist in the world around us. You must always READ the problem carefully, WRITE down all the relevant information, ANALYZE what mathematical operations you must use to answer the questions raised in the problem, and CHECK that your answer is reasonable. Carefully read through the examples below and then try the problems on your own.

Estimation and Reasonable Answers

Certainly precision and accuracy are valued in the mathematics classroom. There are times in "real life", though, where being in the "ballpark" gives you an idea if your answer could be right. "Ballpark" figures are estimates. If someone wanted to give you an idea of how many people attended the last Giants game, they might not give you the exact answer but an estimate, a number that is reasonably close.

Why estimate? We estimate to make things easier to calculate mentally. It's easier to subtract 10 thousand than 10,437 if we are without a calculator. A general idea for an estimate is to find a number close to the real value which makes the calculation easier.

REASONABLE ANSWERS mean answers that make sense. Estimating is one tool for finding reasonable answers. The context of the problem also helps to test for reasonable answers. If you make $500 a week, you certainly must make less than that in a day and even less than that in an hour. If it takes you 15 minutes to walk a mile, it will take you much longer to walk 10 miles. Don't just do a mathematical operation and assume the problem is done.

Use estimates before you calculate AND use reason to judge your answers.

EXAMPLE 1 A truck is delivering 40 crates of Twinkies to local stores. Each crate contains 24 boxes and each box contains 16 Twinkies. How many Twinkies are in the truck?

READ, WRITE, ANALYZE

1 truck ..... 40 crates on the truck ..... 24 boxes in one crate. . . . .16 Twinkies in one box

There is an old saying that "you can't add apples and oranges." In math we say you can only add (and subtract) like items. You cannot add or subtract crates and boxes and Twinkies. What should we do?

The question asks for the total number of Twinkies in the truck.

One box holds 16 Twinkies and there are 24 boxes in each crate. That means each crate holds 16 + 16 + 16 + ... : we need to add 24 16's, but **repeated addition is multiplication** so we really want to multiply

16 Twinkies per box × 24 boxes per crate

This will give the number of Twinkies in one crate.

So far, then: 16 Twinkies per box × 24 boxes per crate = 384 Twinkies in a crate

Now, since there are 40 crates on the truck and each one has 384 Twinkies, we need to multiply again:

384 Twinkies per crate × 40 crates = 15,360 Twinkies on the truck

The complete solution to the problem looks like this:

16 Twinkies per box × 24 boxes per crate × 40 crates = 15,360 Twinkies on the truck

Check: Since you start out with 16 in a box and then you take 24 of these 16 and then 40 of these, it is reasonable that the answer be quite a large number.

Answer: There are 15,360 Twinkies on the truck.

EXAMPLE 2 In order to repair your car you need $445. You only have $72. How much more do you need to get your car fixed?

READ, WRITE, ANALYZE

\$445 .... \$72 .... Both values are dollars. We want to know how much **MORE**, which means we want to know the difference between the these two. "Difference" is a word that indicates subtraction.

Solution: \$445 – \$72 = \$373

Check: Since we know the answer is the difference of these two numbers, the answer must be more than \$345, since \$345 is \$100 less than \$445.

Answer: You need \$373 more to get your car fixed.

EXAMPLE 3 An economy car gets 27 miles to the gallon for highway driving. How many gallons would it take to go 1,188 miles?

READ, WRITE, ANALYZE

27 miles per gallon .... 1,188 miles driven .... A picture might be helpful.

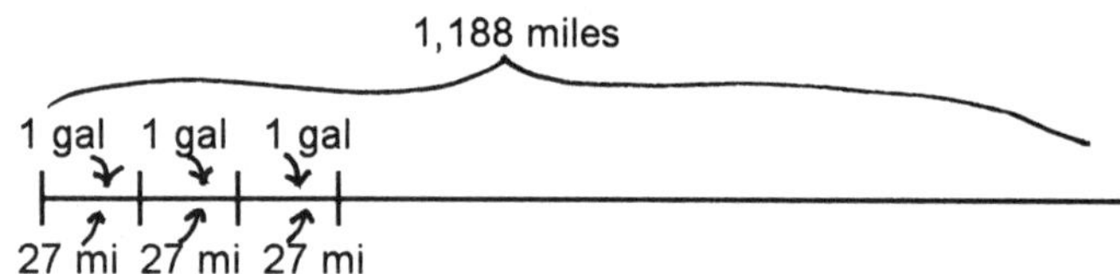

Every time you travel 27 miles you use 1 gallon of gas. You want to know how many 27's there are in 1,188. To do this, you take the total distance driven (1188 miles) and **divide** it by 27 miles. We need to do the division

$$1{,}188 \text{ miles} \div 27 \text{ miles per gallon}$$

Solution: 1,188 miles ÷ 27 miles per gallon = 44 gallons

Work:

```
    44
27)1188
   108
    108
    108
```

Check: One way to check the value of your answer is to recognize that 1,188 miles is *about* 1,200 miles and 27 miles per gallon is *about* 30 miles per gallon, so 1,188 miles ÷ 27 miles per gallon is *about* 1,200 miles ÷ 30 miles per gallon. 1,200 ÷ 30 = 40, so the answer is *about* 40 gallons.

Besides checking the value of the answer, it is also important to decide whether the answer is reasonable. To do this, it is useful to bring in some experience. Think about a trip you are familiar with, say your ride to school. Suppose it's 10 miles. 1,188 miles, then, would be more than 100 trips to school. That would take a few months. How many times would you fill your gas tank if you only drove to school? Suppose your gas tank holds 15 gallons. It would be reasonable to think you would fill your tank about 3 times to drive 100 trips to school at 10 miles each. That would be about 45 gallons of gas, so the answer we got, 44 gallons, is reasonable.

The point is that you need to check more than the arithmetic of the problem; you also need to determine whether you chose the right arithmetic to do. In this problem, the most likely error is to choose multiplication instead of division. If you multiply 1,188 by 27, the answer is 32,076. Do you think this would be a reasonable number of gallons for the trip?

Answer: It would take 44 gallons for this car to go 1188 miles.

♦ **You Try It**

1. Jerry buys 3 shirts at $19 each and 6 pairs of socks at $3 per pair. He pays for his purchases with a $100 bill. What is his change?
2. How many minutes are there in one day?
3. Tom owns three rental units. They each rent for $585 per month and all three are occupied for an entire year. What is Tom's income for the year from the three units?

2 AVERAGE

An *average* is a single number that should represent a group of numbers. The most commonly used average is called a **MEAN** and you find it by adding up all of the numbers and then dividing by the number of values.

What does it mean for a single number to REPRESENT a group of numbers? Think of the expression, "the average American". The phrase implies a person who is representative of all Americans. It does not refer to just one person but all Americans at once. How old would the average American be? This is certainly open to discussion but it has to be a number between 0 and 125, the ages of the youngest and oldest Americans. Most of us would pick an age that would be in between, something like 44. What is the average number of children in the American family? Cartoons have shown the average 2.2 children to be two whole children and a piece of a third. The idea of such an average is to convey information about the whole group at once. Many families have 1, 2 or 3 children. Some have more; some have none. When we add up the number of children in families and divide by the number of families, our result is 2.2.

EXAMPLE 4 In a bowling tournament, one contestant made scores of 215, 222, and 235. What was her **average** for the tournament?

Solution: $\frac{215+222+235}{3}=\frac{672}{3}=224$

Check: Since the average should be representative of all the numbers and should be between the smallest and the largest, the average in this problem should certainly be between 215 and 235. The answer of 224 is reasonable.

Answer: The bowler's average score is 224.

EXAMPLE 5 A group of 6 friends was discussing parking tickets. The number of tickets each received was 0, 1, 2, 0, 2, 13. What was the average number of parking tickets?

Before doing the calculations, make a reasonable guess for the average. ______

Now do the calculations.

Solution: $\dfrac{0+1+2+0+2+13}{6}=\dfrac{18}{6}=3$

Check: The average should be between 0 and 13 and the answer we got is between 0 and 13. It just may not be what you expected.

Answer: The average number of parking tickets is 3.

How close was your guess?

♦ **You Try It**

4. You survey another group of 6 students to find out how many parking tickets they received. This time the answers are 2, 2, 3, 3, 4, 4. Find the average number of parking tickets.
Compare your answer to **EXAMPLE 5.** What are the similarities in these problems? What are the differences?

3 AREA

The **area** of a flat surface is the number of square units needed to cover it. Area will be measured in square units, like square feet, square inches, or square meters.

A *square inch* is a square that measures one inch on each side. A *square foot* is a square that measures one foot on each side. You use square units to measure area because you can *cover* a surface with square units in much the same way that you can cover a floor with square tiles.

EXAMPLE 6 What is the area of a rectangular table top that measures 3 feet by 5 feet ?

To see how to do this, a picture would be helpful. Here is a picture of this table top covered with square feet:

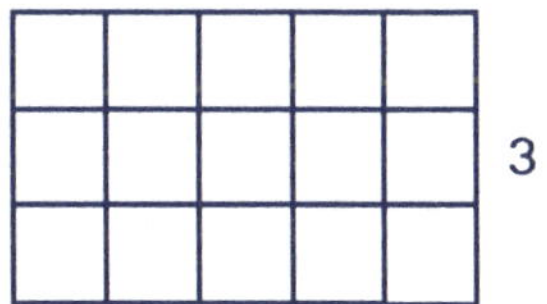

Since the length of the rectangle is 5 feet, five of these square feet will fit along the length. Since the width of the rectangle is 3 feet, three of these square feet will fit along the width. As you can see, 15 of these square feet exactly cover the table top, so its area is 15 square feet. This answer can also be obtained by simply *multiplying* the length of the rectangle by its width. This gives us a *formula* for the area of a rectangle.

Area of a Rectangle

To find the area (A) of a rectangle, we multiply the number that represents the length (l) by the number that represents the width (w). In short, we have the **formula** $A=l\cdot w$ (square units). The length and the width must both be in the same units (that is, both feet, both inches, etc.) before you find the area.

To use the formula for the area of a rectangle to solve this problem:
First, write the formula: $A=l\cdot w$

Substitute 5 ft for l and 3 ft for w: $A = (5\text{ ft})(3\text{ ft})$

$A = 15$ square feet

Answer: The area of the rectangular table top is 15 square feet.

♦ **You Try It** 5. The Ellis family's living room is 19 ft by 15 ft. What is the area?

♦ **Answers to You Try It** **1.** \$25 **2.** 1440 min. **3.** \$21,060 **4.** 3 **5.** 285 sq ft

SECTION 1.4 EXERCISES

For all problems , write all the steps out carefully and completely. Be sure your answer is written as a complete sentence.

1. A truck has the capacity to carry 4,200 pounds of weight. If 2,880 pounds are now in the truck, how many more pounds can the truck carry?

2. A storage tank can hold 7,500 gallons of oil. If 2,398 gallons are currently in the tank, how many gallons of oil should be ordered to fill the tank?

3. The area of the United States is 3,618,770 square miles. The area of Canada is 3,831,033 square miles.

a. What is the sum of the areas of the two countries?

b. Which country has more area?

c. What is the difference in the areas of the two countries?

4. In one recent year, the population of Canada was 27,483,000 and the U.S. population was 248,729,000.

a. Which country has the larger population?

b. What is the difference in the two populations?

c. What is the combined total of the two populations?

5. An empty truck has the capacity to haul 8,500 pounds. At one stop it picked up loads of 2,275 pounds, 920 pounds, and 3,650 pounds.

a. How many pounds were loaded onto the truck?

b. How many more pounds can the truck hold without exceeding its capacity?

c. Is there enough capacity left on the truck to carry two extra loads of 950 pounds and 1,340 pounds respectively?

6. A freight elevator has a capacity of 3,200 pounds. An attendant weighing 195 pounds is on the elevator along with a 435-pound refrigerator and a 1,460-pound generator.

a. What is the total weight on the elevator?

b. How much more weight can the elevator carry without exceeding its capacity?

c. Is there enough capacity left on the elevator to also carry a 760-pound condenser along with a 210-pound repairman?

7. David has a balance of $560 in his checking account. He wrote checks for $162, $87, $28, $63, and $104.

a. What is the total of the checks he wrote?

b. What is David's new balance?

8. Sarina has an equipment budget of $2,640. She furnished her office with a $760 sofa, two end tables for $180 each, a bookcase for $280, and a chair for $490.

a. How much has Sarina spent so far to furnish her office?

b. How much is still left in her equipment budget?

9. Audrey had a balance of $67 in her checkbook. She wrote checks for $32, $45, $17, and $53. She made deposits of $50 and $35. What is Audrey's new balance?

10. A hot air balloon is 450 feet above the ground. It drops 178 feet, then rises 310 feet. How high is the balloon?

11. Lauren makes $375 per week. How much will she make in 8 weeks?

12. George sold 12 air conditioners today. If each sold for $289, what were his total sales for the day?

13. Ed makes $23,556 a year. What does Ed make each week?

14. If 25 lamps cost a store owner $675, what is the cost of one lamp?

15. Irma's car gets 28 miles on a gallon of gas. How many miles will she get on a full tank of 18 gallons?

16. Tom's mortgage payment is $752 per month. How much does he pay each year for his mortgage?

17. A 30-foot camper gets 6 miles to a gallon of gas. How much gas is needed to drive 402 miles from Los Angeles to San Francisco?

18. A 26-foot motorboat gets 4 miles to a gallon of gas. How much gas is needed to travel 128 miles?

19. Diane can seed 990 square feet of lawn in one minute. Her favorite TV show is on in 45 minutes. Does she have time to seed her 1-acre lawn? 1 acre = 43,560 square feet.

20. A swimming pool requires 6,000 gallons of water. A hose pumps 8 gallons of water per minute. In how many minutes will the pool be filled?

21. You count 521 words on one page of a novel. Estimate how many words the 462-page novel contains.

22. A case of soda has 24 bottles. Each bottle has 16 ounces. Estimate how many ounces of soda are in the case.

23. Sandy makes $13 an hour. What will she make if she works 27 hrs?

24. Jim makes $9 an hour. What will he make if he works 40 hrs?

25. Each week, Joyce is paid $12 an hour for the first 40 hours, and $18 for each hour she works over 40. What will Joyce make if she works 62 hours this week?

26. Each week, Jayne makes $14 an hour for the first 40 hours, and $21 an hour for each hour over 40. What will her salary be if she works 43 hours this week?

27. A wedding caterer purchased 36 cases of regular soda at $7 per case and 28 cases of diet soda at $6 per case. What is the total cost?

28. The last performance of a play is sold out. If 40 box seats were sold at $28 each, 54 mezzanine seats were sold at $21, and 36 balcony seats were sold at $19, how much money did the theater collect from the last performance?

29. Susan, Diane, Nancy, and Joyce purchase a sailboat for $32,648. If they decide to evenly split the cost of the boat, how much does each woman pay?

30. Al, Bob, and Ted own a business. In the first year of operation, the business lost $4,659. If each man pays an equal share of this loss, how much will each man pay?

31. Bob bought a TV for $684. He put $120 down and paid the rest in 12 monthly payments. What was the monthly payment?

32. Jack bought a stereo for $2,597. He put $500 down and paid the rest in 9 equal monthly payments. Find the monthly payment.

33. Angela bought a car for $15,400. She put $3,400 down and paid the rest in 48 equal monthly payments. Find the monthly payment.

34. Hal bought a jade ring for $260. He put $40 down and paid the rest in 5 equal monthly payments. Find Hal's monthly payment.

2

35. Find the average of 4, 6, 7, 7, 9, and 15.

36. Find the average of 88, 70, 78, 90, 67, 76, 100, 90, 88, and 73.

37. Ron sold 12 vacuum cleaners in March, 9 in April, 13 in May, and 6 in June. What average number did he sell per month?

38. A baseball team scored 3, 6, 0, 1, 2, 3, 0, 4, and 8 runs over the last 9 games. Find the average number of runs scored per game.

39. Recently homes sold in Al's neighborhood for $124,000, $130,500, $121,250, $128,000, and $129,750. Find the average selling price.

40. A basketball team scored 77, 98, 102, 89, 85, 91, 112, 90, 96, 77, 79, and 84 points on a 12-game road trip. What is the average number of points scored per game?

3 *Find the area of each rectangle.*

41.

15 ft

10 ft

42.

4 m

6 m

43. Luis has a living room that measures 16 ft by 14 ft. What is the area?

44. Theresa has a rectangular vegetable garden that measures 8 ft by 12 ft. What is the area?

FIGURE 1.18

Treat Yourself To Sun and Fun!

Memberships As Low As $250 For Adults • $150 For Children • Children Under 5 FREE

Facilities Include:

- 42'x82'x30'x40' Outdoor Pool
- 42'x82' Outdoor Pool
- 875' Beach
- Filtered Kiddie Pool
- Parking for over 600 cars
- Swim Team/Lessons
- Furnished Sun Deck
- Basketball
- Sauna Room
- Hot-Cold Showers
- Oceanfront Bar
- Giant Hot Tub
- 2 Outdoor Tennis Courts
- Volleyball Court
- Snack Bar
- Cocktail Lounge
- Video Story Time
- Rainy Day Movies on our Big Screen

TRADE WINDS

B • E • A • C • H C • L • U • B

Reprinted with permission of Tradewinds, Inc.

45. *Use the advertisement in Figure 1.18 to answer the questions.*

   a. How much would a family of 5 (2 adults and 3 children of ages 2, 7, and 9) pay for membership?

   b. How much would a family of 5 (1 adult and 4 children of ages 6, 8, 10, 12) pay for membership?

   c. Do you think that the club should charge a flat fee for families? What should the charge be?

46. *Use Figure 1.19 to answer the questions.*

   a. Which two cities combined have more soccer players than New York?

   b. How many more soccer players are there in New York than there are in Boston?

FIGURE 1.19

47. *Use Figure 1.20 to answer the questions.*

a. In which of the counties listed was the average annual salary in 1991 closest to the state average?

b. Which pair of counties had average annual salaries in 1991 closest to each other?

c. Why do you think the answer to (b) is true?

FIGURE 1.20

Salary gains in 1991

In the private sector, workers in Monmouth and Ocean counties earned less on average than workers statewide last year and got smaller wage increases. Workers in Middlesex County, however, did better than the state average.

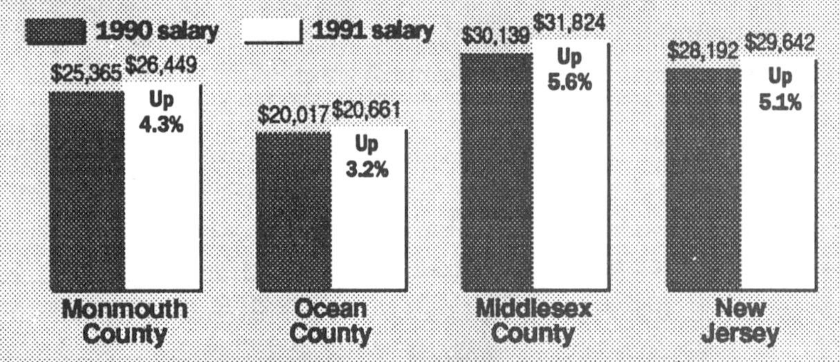

Change in average annual N.J. wages, by occupation

	1990*	1991	Increase
Construction	$34,389	$35,408	3.0%
Manufacturing	$34,223	$36,227	5.9%
Transportation/communications/public utilities	$33,881	$36,204	6.9%
Retail trade	$15,625	$16,251	4.0%
Finance/insurance/real estate	$33,364	$35,518	6.5%
Government	$29,891	$31,795	6.4%

SOURCE: New Jersey Department of Labor   Asbury Park Press graphic

FIGURE 1.21

USA SNAPSHOTS®

A look at statistics that shape our lives

Source: USA TODAY research by Patti Stang   By Bob Laird, USA TODAY

48. *Use Figure 1.21 to answer the questions.*

a. How many visitors would the White House have from Monday through Friday?

b. How many visitors would Graceland have from Monday through Friday?

c. Do you think either of these numbers is rounded? Explain your reasoning.

d. If it is open every day in July, how many visitors will Graceland have in July?

EXTEND YOUR THINKING ♦ ♦ ♦

♦ SOMETHING MORE

49. A pumpkin farmer planted 16 acres with pumpkins. He hopes to harvest about 420 pumpkins per acre. He can pack 12 pumpkins per crate for shipment to market. If he sells his pumpkins for $13 per crate, how much can the farmer expect to make?

50. A company employs 280 workers. Each employee works 8 hours each day, and is paid the exact same hourly wage. If the total payroll for one day is $20,160, what hourly wage are these workers paid?

WRITING TO LEARN ♦ ♦ ♦

51. You take five quizzes. One quiz score is 0. Explain why the score of 0 must be figured into your average.

52. Give a situation in which you would need to find the area.

53. Sketch an inch. Sketch a square inch. Describe the difference between them.

54. How is a square different than a rectangle. How would you find the area of a square?

♦ YOU BE THE JUDGE

55. Is it possible to completely submerge yourself in a lake whose average depth is 1 inch? Explain your decision.

1.5 FRACTIONS -- AN OVERVIEW

OBJECTIVES

1. Understand the idea of fraction
2. Use the Fundamental Principle of Fractions
3. Perform simple fraction operations

NEW VOCABULARY

numerator	denominator
proper fraction	improper fraction
mixed number	equivalent fractions
lowest terms	reducing a fraction
like terms	reciprocal

Basic Assumptions

We will be making some assumptions about you during this course. First, we assume that you are comfortable with the basic arithmetic of whole numbers. That doesn't mean that you never make any mistakes, but that you can generally carry out addition, subtraction, multiplication, and division. Second, we assume that you have at one time or another in your educational background worked with fractions, decimals, and percents. Again, that doesn't mean that you remember everything perfectly but that you will recognize the basic ideas when you see them. Because of these assumptions, we are going to focus on solving problems which may involve any of these types of numbers. For this reason, we will **briefly** review the basic ideas of and operations with fractions and then go on to concentrate on decimals. A working knowledge of fractions will enable you to use them if the need arises. As we proceed through our work, we will come back to fractional rules and investigate their origins. For now, we want to refresh your memory so that your "mathematical tool box" will have sufficient tools for you to be able to solve problems.

1 FRACTIONS

A fraction is a number that represents the quotient of two numbers, provided you do not divide by 0. For example, $\frac{2}{3}$ is a fraction that represents the quotient of 2 and 3. To say this a different way, the fraction $\frac{2}{3}$ is the same as the division $2 \div 3$. In the fraction $\frac{2}{3}$, the 3 is called the **denominator** to remind us of "denomination" or size, and the 2 is called the **numerator**, to remind us of "enumerate" or "count". This suggests that we have a whole that has been cut into **thirds** (or three equal pieces) and we have **taken** two of them. We can represent the fraction $\frac{2}{3}$ by a picture. Here are two possible pictures:

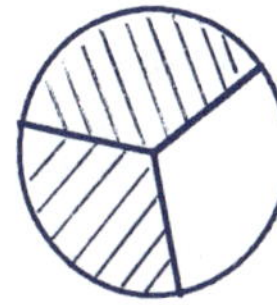

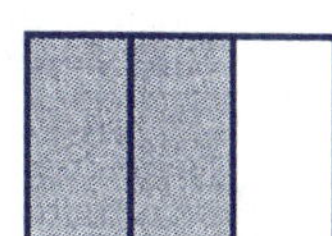

In each case, the whole (box or circle) has been cut into three equal parts and two of those parts has been "taken."

EXAMPLE 1 At a certain movie theater on a Saturday afternoon, there are 35 children and 20 adults. What fraction of the people attending the movie are children?

A fraction has two parts; the denominator tells us the total number of parts in the whole. In this example, the **total** number of parts in the whole is the total number of people attending the movie, so the denominator of the fraction is the **sum** of 35 children and 20 adults. The denominator is 55. The numerator is the number of parts we want. In this case we want the number of children, so the numerator is 35.

Answer: The fraction of children attending the movie is $\frac{35}{55}$.

OBSERVE When writing a fraction, the denominator cannot be zero. The fraction $\frac{2}{0}$ would mean cut the whole into zero parts and take 2 of them, which is meaningless. Zero can never be the denominator of a fraction.

Proper Fractions

If a fraction's numerator is **less** than its denominator, then the fraction is called a **proper** fraction, so $\frac{2}{3}$ is a proper fraction. Some other examples of proper fractions are $\frac{12}{50}, \frac{45}{92}, \frac{5}{17}$, and $\frac{1}{100}$.

♦ You Try It

1. Write 5 examples of proper fractions, different than the ones above.

2. Draw a picture, using either a box or a circle, to represent $\frac{3}{4}$.

3. Draw a picture, using either a box or a circle, to represent $\frac{5}{7}$.

4. In a kindergarten class, 9 out of 20 children are out with the chicken pox. Write the fraction of children with chicken pox. Write the fraction without chicken pox.

Improper Fractions

If a fraction's numerator is **greater than or equal to** its denominator, then the fraction is called **improper**. We should emphasize that this is simply the name of this type of fraction - there is really nothing "improper" about the fraction in the usual sense of the word. An improper fraction represents a number that is either a whole or larger than a whole. For example, the improper fraction $\frac{8}{5}$ cannot mean that <u>one</u> whole has been cut into 5 parts and 8 have been taken. Clearly, if the whole has been cut into 5 parts, each part is one-fifth and 5 of the parts uses up the whole so 8 fifths is one whole and 3 fifths more. A picture would look like this:

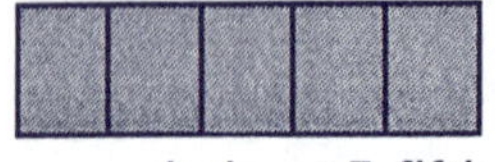
one whole or 5 fifths

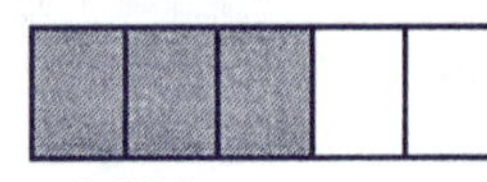
3 fifths

Another way to say this is that $\frac{8}{5}$ is the same as one whole and 3 fifths more or $1\frac{3}{5}$, which is called a **mixed number**. A procedure for writing an improper fraction as a mixed number is to divide the numerator, in this case 8, by the denominator, in this case, 5. The quotient, 1, is the whole number part of the mixed number and the remainder, 3, tells us how many fifths are left. If there is no remainder when we do this division, we have a whole number rather than a mixed number. The procedure looks like this:

$$\begin{array}{r} 1 \\ 5\overline{)8} \\ \underline{5} \\ 3 \end{array}$$

1 ⟵The quotient becomes the whole number

3 ⟵3 is the remainder, so $\frac{3}{5}$ is the fraction part of the mixed number.

Thus, $\frac{8}{5} = 1\frac{3}{5}$.

EXAMPLE 2 Write the improper fraction $\frac{25}{7}$ as a mixed number.

Divide the numerator by the denominator:

$$\begin{array}{r} 3 \\ 7\overline{)25} \\ \underline{21} \\ 4 \end{array}$$

3 ⟵The quotient becomes the whole number part.

4 ⟵The remainder becomes the numerator of the fraction part.

Answer: $\frac{25}{7} = 3\frac{4}{7}$

EXAMPLE 3 Write the improper fraction $\frac{8}{4}$ as a whole number.

Divide the numerator by the denominator.

Since $8 \div 4 = 2$, the improper fraction $\frac{8}{4}$ is the whole number 2.

Answer: $\frac{8}{4} = 2$

Of course, mixed numbers can also be represented as improper fractions. To change $1\frac{3}{5}$ to an improper fraction, look back at the division you did to get this mixed number in the first place. Essentially you reverse or "check" this division; that is, multiply the whole number by the denominator and then add the numerator. This will tell you how many fifths you have in the improper fraction.

$$1\frac{3}{5} = \frac{5 \cdot 1 + 3}{5} = \frac{8}{5}$$

EXAMPLE 4 Write the mixed number $4\frac{2}{9}$ as an improper fraction.

Multiply the whole number by the denominator and then add the numerator.

$$4\frac{2}{9} = \frac{9 \cdot 4 + 2}{9} = \frac{38}{9}$$

Answer: $4\frac{2}{9} = \frac{38}{9}$

♦ You Try It

5. Give 5 examples of improper fractions.

6. Draw a picture to represent the improper fraction $\frac{7}{3}$.

7. Draw a picture to represent the improper fraction $\frac{10}{5}$.

8. Write $\frac{12}{5}$ and $\frac{8}{3}$ as mixed numbers.

9. Write $4\frac{1}{3}$ and $3\frac{2}{5}$ as improper fractions.

---

2 EQUIVALENT FRACTIONS

Fractions have an interesting property: fractions provide different names for the same value. For example, the fraction $\frac{1}{2}$ means a whole that has been cut into 2 equal parts and one of those parts has been taken, like this:

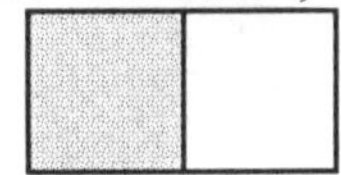

If that same whole had been cut into four parts and we take two, the picture would look like this:

This is the fraction $\frac{2}{4}$. Since both fractions represent the same size, we say that $\frac{1}{2}$ *is equivalent to* $\frac{2}{4}$, or $\frac{1}{2} = \frac{2}{4}$. In fact, $\frac{1}{2}$ can also be expressed as $\frac{3}{6}$, or $\frac{5}{10}$, or $\frac{60}{120}$, or any fraction whose numerator is half its denominator. One way to find new names for a fraction is to multiply the numerator and denominator of the fraction by the same number.

For example, $\frac{4}{5}$ is the same as $\frac{4\cdot 3}{5\cdot 3}$, which is the fraction $\frac{12}{15}$, so $\frac{4}{5}$ is equivalent to $\frac{12}{15}$, or $\frac{4}{5}=\frac{12}{15}$.

This is an important property of fractions, so we state it here:

Fundamental Principle of Fractions: If the numerator and denominator of a fraction are both multiplied by the same number (except zero), then the new fraction has the same value as the original one.

In symbols, the principle looks like this: $\frac{a}{b}=\frac{a\cdot c}{b\cdot c}$ if $c\neq 0$

EXAMPLE 5 $\frac{2}{3}=\frac{2\cdot 10}{3\cdot 10}=\frac{20}{30}$ Here the numerator and denominator have been multiplied by 10.

$\frac{2}{3}=\frac{2\cdot 4}{3\cdot 4}=\frac{8}{12}$ Here the numerator and denominator have been multiplied by 4.

$\frac{2}{3}=\frac{2\cdot 9}{3\cdot 9}=\frac{18}{27}$ Here the numerator and denominator have been multiplied by 9.

EXAMPLE 6 $\frac{8}{5}=\frac{8\cdot 6}{5\cdot 6}=\frac{48}{30}$

EXAMPLE 7 $\frac{8}{15}=\frac{8\cdot 2}{15\cdot 2}=\frac{16}{30}$

Every whole number can be written as a fraction whose denominator is 1.

EXAMPLE 8 $7=\frac{7}{1}$ Remember that $\frac{7}{1}$ means $7 \div 1$, which is 7.

$12=\frac{12}{1}$

$100=\frac{100}{1}$

Notice that whole numbers can be written with many different fraction names.

EXAMPLE 9 $4=\frac{4}{1}=\frac{4\cdot 3}{1\cdot 3}=\frac{12}{3}$ and $4=\frac{4}{1}=\frac{4\cdot 8}{1\cdot 8}=\frac{32}{8}$

$16=\frac{16}{1}=\frac{16\cdot 5}{1\cdot 5}=\frac{80}{5}$ and $16=\frac{16}{1}=\frac{16\cdot 2}{1\cdot 2}=\frac{32}{2}$

♦ You Try It

10. Write two other fractions that are equal to $\frac{3}{8}$.

11. Write three other fractions that are equivalent to $\frac{7}{2}$.

12. What can be said about the value of a fraction if its numerator and denominator are equal?

13. How can you write a whole number as a fraction?

14. Write 8 and 2 as fractions.

15. Write three other fractions that are equal to $\frac{4}{1}$.

16. What can you do to a fraction to produce a new name but the same value?

Reducing Fractions

You can also change the name of a fraction but keep its value by dividing the numerator and denominator of the fraction by the same (nonzero) number:

EXAMPLE 10 $\frac{12}{16}=\frac{12\div 4}{16\div 4}=\frac{3}{4}$ Here the numerator and denominator were divided by 4.

$\frac{60}{40}=\frac{60\div 10}{40\div 10}=\frac{6}{4}$ Here the numerator and denominator were divided by 10.

$\frac{100}{35}=\frac{100\div 5}{35\div 5}=\frac{20}{7}$ Here the numerator and denominator were divided by 5.

If the numerator and denominator of a fraction cannot be divided by the same number, then that fraction is said to be **reduced to lowest terms**.

In the examples above, $\frac{3}{4}$ and $\frac{20}{7}$ are reduced to lowest terms, whereas $\frac{6}{4}$ can be reduced to $\frac{3}{2}$ by dividing the numerator and denominator by 2.

♦**You Try It**

17. Reduce $\frac{12}{48}$ to lowest terms.

18. Reduce $\frac{75}{200}$ to lowest terms.

19. Can you reduce $\frac{4}{9}$? Why or why not?

20. Can you reduce $\frac{9}{5}$? Why or why not?

21. Explain how to reduce a fraction to lowest terms.

3 FRACTION OPERATIONS

There are four basic operations with fractions: addition, subtraction, multiplication, and division.

Addition and Subtraction

To add two quantities, they must be **like quantities** or **like terms**. For example, you can add 3 feet to 7 feet and get 10 feet, but you cannot add 3 feet to 7 pounds because they are not like quantities. With fractions, the denominator tells us the kind of quantity the fraction is; the denominator is the "denomination" of the fraction, just as feet is the denomination of 3 feet. For this reason, adding or subtracting fractions requires that all the fractions to be added must have the same denominator.

> To add (or subtract) fractions with the same denominator, keep the denominator and add (or subtract) the numerators.
>
> In symbols, $\frac{a}{c}+\frac{b}{c}=\frac{a+b}{c}$.

Then $\frac{3}{8}+\frac{2}{8}$ is 3 **eighths** added to 2 **eighths** and the sum is $\frac{5}{8}$. Here are some examples of fraction addition and subtraction:

EXAMPLE 11 $\frac{2}{7}+\frac{8}{7}=\frac{2+8}{7}=\frac{10}{7}$

EXAMPLE 12 $\frac{11}{12}-\frac{4}{12}=\frac{11-4}{12}=\frac{7}{12}$

EXAMPLE 13 $\frac{3}{5}+\frac{7}{10}$ Here the fractions do not have the same denominator so they cannot be added until they are rewritten. Careful inspection of these two fractions shows that both fractions can be written with a denominator of 10. (Although it is also possible to write both fractions with a denominator of 50 or 20 or 100 or many others, 10 is the smallest or least common denominator, often referred to as the LCD, and it is preferable to use the LCD in most cases.) We now rewrite $\frac{3}{5}$ as $\frac{3\cdot 2}{5\cdot 2}$ or $\frac{6}{10}$ and continue as in the first example:

$$\frac{3}{5}+\frac{7}{10}=\frac{6}{10}+\frac{7}{10}=\frac{13}{10}$$

EXAMPLE 14 $3+\frac{4}{5}=3\frac{4}{5}$ Adding a whole number and a proper fraction is, in fact, a mixed number, so no work is necessary for this addition.

♦ **You Try It**

22. Add $\frac{5}{14}$ and $\frac{6}{14}$.

23. Find the sum of $\frac{3}{5}$ and $\frac{9}{5}$.

24. Find the difference: $\frac{7}{9} - \frac{2}{9}$.

25. Find the sum: $6 + \frac{1}{4}$

26. Give an example of two fractions with the same denominator.

27. Give an example of two fractions with different denominators and find the common denominator.

28. Find the sum: $\frac{4}{5}+\frac{7}{10}$.

Multiplication

> To multiply fractions, multiply the numerators and multiply the denominators.
>
> In symbols, $\frac{a}{b}\cdot\frac{c}{d}=\frac{a\cdot c}{b\cdot d}$.

EXAMPLE 15 $\frac{2}{3}\cdot\frac{4}{7}=\frac{2\cdot 4}{3\cdot 7}=\frac{8}{21}$

Note: When finding the sum, difference, product, or quotient, we expect the result to be in lowest terms.

EXAMPLE 16 $\frac{5}{10}\cdot\frac{3}{10}=\frac{15}{100}$ which reduces to $\frac{3}{20}$ (Divide the numerator and denominator by 5.)

EXAMPLE 17 $\frac{1}{2}\cdot 9\cdot\frac{3}{4}=\frac{1}{2}\cdot\frac{9}{1}\cdot\frac{3}{4}=\frac{1\cdot 9\cdot 3}{2\cdot 1\cdot 4}=\frac{27}{8}$

Division To divide one fraction by another, the division problem is turned into a multiplication problem. To divide one fraction, $\frac{a}{b}$, by another fraction, $\frac{c}{d}$, multiply $\frac{a}{b}$ by the

reciprocal of $\frac{c}{d}$. The reciprocal of $\frac{c}{d}$ is $\frac{d}{c}$. For example, the reciprocal of $\frac{3}{4}$ is $\frac{4}{3}$ and the reciprocal of $\frac{1}{2}$ is $\frac{2}{1}$. The rule looks like this:

> To divide one fraction by another, multiply by the reciprocal of the second fraction.
>
> In symbols, $\frac{a}{b} \div \frac{c}{d} = \frac{a}{b} \cdot \frac{d}{c}$

Once the division has been rewritten as a multiplication, multiply the numerators and multiply the denominators.

EXAMPLE 18 $\frac{4}{5} \div \frac{3}{2} = \frac{4}{5} \cdot \frac{2}{3} = \frac{8}{15}$

↑

$\frac{2}{3}$ is the reciprocal of $\frac{3}{2}$

♦ You Try It

29. Explain how to multiply fractions.

30. What is the reciprocal of $\frac{4}{5}$?

31. What number has no reciprocal? Why?

32. Explain how to divide fractions.

33. Multiply $\frac{1}{4}$ by $\frac{3}{4}$.

34. Multiply 8 by $\frac{2}{9}$.

35. Divide $\frac{8}{3}$ by $\frac{1}{2}$.

♦ **Answers to You Try It** **1.** $\frac{7}{12}, \frac{3}{10}, \frac{7}{90}, \frac{1}{7}, \frac{8}{33}$ are a few proper fractions. **2.** $\frac{3}{4}$ 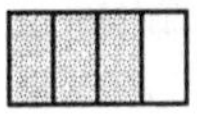 **3.** $\frac{5}{7}$ 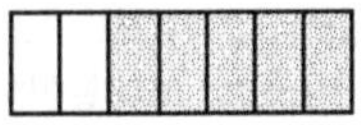 **4.** $\frac{9}{20}$ of the children have chicken pox. $\frac{11}{20}$ do not. **5.** $\frac{16}{3}, \frac{120}{7}, \frac{9}{8}, \frac{6}{6}, \frac{100}{100}$ are a few improper fractions. **6.** $\frac{7}{3}$  **7.** $\frac{10}{5}$ 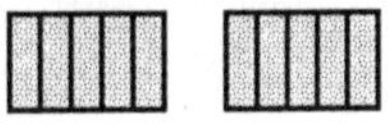 **8.** $\frac{12}{5} = 2\frac{2}{5}; \frac{8}{3} = 2\frac{2}{3}$ **9.** $4\frac{1}{3} = \frac{13}{3}; 3\frac{2}{5} = \frac{17}{5}$ **10.** $\frac{3}{8} = \frac{12}{32}$ and $\frac{3}{8} = \frac{18}{48}$ but there are many choices. **11.** $\frac{7}{2} = \frac{14}{4}, \frac{7}{2} = \frac{28}{8}$, and $\frac{7}{2} = \frac{35}{10}$ but there are many choices. **12.** The fraction is equal to 1. **13.** Write the whole number as the numerator and 1 as the denominator. **14.** $8 = \frac{8}{1}$ and $2 = \frac{2}{1}$ but there are many other choices. **15.** $\frac{4}{1} = \frac{8}{2}, \frac{4}{1} = \frac{20}{5}$, and $\frac{4}{1} = \frac{28}{7}$ but there are many choices. **16.** Multiply the numerator and denominator by the same number. **17.** $\frac{12}{48} = \frac{1}{4}$ **18.** $\frac{75}{200} = \frac{3}{8}$ **19.** $\frac{4}{9}$ is already reduced to lowest terms because the numerator and denominator cannot be divided by the same number. **20.** $\frac{9}{5}$ is already reduced to lowest terms because the numerator and denominator cannot be divided by the same number. **21.** Divide the numerator and denominator by the same number until that is no longer possible. **22.** $\frac{11}{14}$ **23.** $\frac{12}{5}$ **24.** $\frac{5}{9}$ **25.** $6\frac{1}{4}$ **26.** $\frac{3}{25}$ and $\frac{8}{25}$ have the same denominator, but there are many choices. **27.** $\frac{3}{4}$ and $\frac{5}{8}$ have different denominators. The common denominator is 8. **28.** $\frac{15}{10} = \frac{3}{2}$ **29.** Multiply the numerators and multiply the denominators. **30.** $\frac{5}{4}$ **31.** Zero, because zero cannot be the denominator of a fraction. **32.** Multiply by the reciprocal of the second fraction. **33.** $\frac{3}{16}$ **34.** $\frac{16}{9}$ **35.** $\frac{16}{3}$

SECTION 1.5 EXERCISES

1 *Draw a picture, using a circle or a box, to represent the fraction .*

1. $\frac{5}{8}$ **2.** $\frac{7}{4}$ **3.** $\frac{5}{2}$

4. At a garden center, there are 3 tables of red geraniums and 2 tables of white geraniums. What fraction of the geraniums is red? What fraction of the geraniums is white?

Write each improper fraction as a mixed number.

5. $\frac{9}{4}$ **6.** $\frac{25}{2}$ **7.** $\frac{50}{7}$ **8.** $\frac{31}{10}$

Write the following mixed numbers as improper fractions.

9. $4\frac{3}{10}$ **10.** $11\frac{1}{3}$ **11.** $8\frac{1}{8}$ **12.** $1\frac{1}{11}$

2 *Write two other fractions equal to each of the following fractions.*

13. $\frac{4}{5}$ **14.** $\frac{5}{3}$ **15.** $\frac{2}{9}$ **16.** $\frac{1}{9}$

Reduce each of the following fractions to lowest terms.

17. $\frac{9}{15}$ **18.** $\frac{6}{8}$ **19.** $\frac{21}{28}$ **20.** $\frac{12}{20}$

3 *Perform the operation indicated.*

21. $\frac{1}{8}+\frac{5}{8}$ **22.** $\frac{14}{15}-\frac{3}{15}$ **23.** $\frac{1}{2}+\frac{3}{4}$ **24.** $7+\frac{5}{8}$

25. $\frac{4}{5}\cdot\frac{3}{7}$ **26.** $6\times\frac{1}{5}$ **27.** $\frac{1}{2}\div\frac{2}{3}$ **28.** $\frac{2}{3}\div\frac{1}{2}$

29. $\frac{5}{7}+\frac{2}{7}$ **30.** $\frac{7}{12}-\frac{1}{12}$ **31.** $9+\frac{2}{5}$ **32.** $\frac{1}{2}\cdot\frac{3}{5}$

33. $\frac{5}{7} \times \frac{2}{3}$

34. $\frac{6}{7} \div \frac{2}{11}$

35. $\frac{1}{3} + \frac{1}{6}$

36. $\frac{5}{8} - \frac{1}{4}$

37. $11 + \frac{1}{2}$

38. $\frac{4}{9} \cdot \frac{2}{3}$

39. $\frac{4}{5} \div 2$

40. $\frac{4}{5} \div \frac{1}{2}$

EXTEND YOUR THINKING ♦♦♦

♦TROUBLESHOOT IT

Find and correct the error.

41. $\frac{3}{8} \div 5 = \frac{3}{8} \times \frac{5}{1} = \frac{15}{8}$

WRITING TO LEARN ♦♦♦

42. What is a proper fraction?

43. What is an improper fraction?

44. Write three examples of proper fractions.

45. Write three examples of improper fractions.

46. Explain how to write an improper fraction as a mixed number.

47. Explain how to write a mixed number as an improper fraction.

48. How can you produce a fraction whose value is the same as your original fraction but whose numerator and denominator are different?

49. How can you write a whole number as a fraction?

50. What does it mean to say that a fraction is reduced to lowest terms?

51. For which fraction operations is a common denominator required? Why?

52. Explain how to multiply fractions.

53. Explain how to divide fractions.

1.6 INTRODUCTION TO DECIMALS

OBJECTIVES

1. Understand decimal place value.
2. Write decimals as fractions.
3. Read and write decimal numbers.

NEW VOCABULARY

place value, tenths, hundredths, thousandths, rational number

1 DECIMAL PLACE VALUES

In Section 1.3 we discussed place value and how the position of a digit tells us its value. For example, in the whole number 5,892, the 2 is in the ones position, so we know there are 2 ones; the 9 is in the tens position, so we know there are 9 tens; the 8 is in the hundreds position, so we know there are 8 hundreds, and the 5 is in the thousands position, so we know there are 5 thousands. Place value is also important to indicate numbers between whole numbers.

A decimal point indicates the end of a whole number and the beginning of its fractional part. A decimal is another way of writing a fraction. In decimal numbers, as in whole numbers, the **position** of the digit tells us its value. For example, in the decimal number 0.476, the 4 is in the **tenths** position, so we know there are 4 tenths, the 7 is in the **hundredths** position, so we know there are 7 hundredths, and the 6 is in the thousandths position, so we know there are 6 thousandths. The number 0.476 is read "four hundred seventy-six thousandths". For decimal numbers, the place values are as follows:

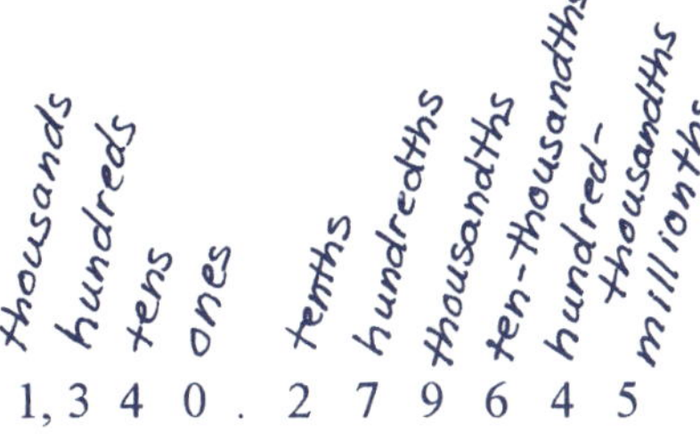

The decimal point is the symbol that separates any number into the whole number and the "fraction" or "part" of the whole number. When reading a number with a decimal point in it we should read the decimal point as **AND.** If we have no whole number part, we may use a zero on the left of the decimal point.

EXAMPLE 1

a. In the number 3.4076 the 6 is in the **ten-thousandths** place.

b. In the number 0.409 the 4 is in the **tenths** place.

c. In the number 11.3297 the 2 is in the **hundredths** place.

♦You Try It

Identify the place value of the digit 5 in each number.

1. 4.0519 **2.** 33.5860 **3.** 120.0035 **4.** 0.0259

2 DECIMALS AND FRACTIONS

In Section 1.2 we said that numbers between whole numbers could be indicated by the use of fractions or decimals. Decimals are preferred in many situations, especially those dealing with money. Decimals correspond to fractions whose denominators are 10, 100, 1,000, 10,000, etc. A fraction whose denominator is 10 is read "tenths", a fraction whose

denominator is 100 is read "hundredths", and a fraction whose denominator is 1,000 is read "thousandths", and these correspond to the first three places to the right of the decimal point. Observe the following patterns:

The **first** position to the right of the decimal point is **tenths**, which corresponds to a fraction denominator of **10**, which has **one** zero.

The **second** position to the right of the decimal point is **hundredths**, which corresponds to a fraction denominator of **100**, which has **two** zeros.

The **third** position to the right of the decimal point is **thousandths**, which corresponds to a fraction denominator of **1,000**, which has **three** zeros.

Using these patterns provides a guide for changing from decimal notation to fraction notation.

When changing from decimal notation to fraction notation:

1. The **whole number** part of a number is the same regardless of whether you use fraction notation or decimal notation.
2. A **decimal point** indicates the end of the whole number part and is translated as "and".
3. The **denominator** of the fraction is 10, 100, 1000, etc. The number of zeros is equal to the number of digits to the right of the decimal point.
4. The **numerator** of the fraction is the digits to the right of the decimal point.

EXAMPLE 2 The decimal number 12.3 is the same as 12 and $\frac{3}{10}$.

12.<u>3</u> has **one** decimal place so there is **one** zero after the 1 in the denominator, making the denominator 10.

EXAMPLE 3 The decimal number 0.43 is the same as 0 and $\frac{43}{100}$.

0.43 has **two** decimal places so there are **two** zeros after the 1 in the denominator, making the denominator 100.

EXAMPLE 4 The decimal number 7.019 is the same as 7 and $\frac{19}{1000}$.

7.019 has **three** decimal places so there are **three** zeros after the 1 in the denominator, making the denominator 1000.

As you can see from these examples, decimals and fractions are just different ways of writing the same number. Remember that the numbers 0, 1, 2, 3, 4, 5, . . . are called whole numbers. Fractions like $\frac{3}{10}$ and $\frac{13}{100}$ and decimals like 0.3 and 0.13 are not whole numbers. We call them **rational numbers**, which comes from the word "ratio", which means fraction. A rational number is any number that can be written as a fraction. Since all whole numbers can be written as fractions ($7 = \frac{7}{1}$, $4 = \frac{8}{2}$, etc.), whole numbers are also rational numbers.

♦ You Try It

Write each of the following as fractions or mixed numbers.

5. 0.087 **6.** 1.51 **7.** 0.1233
8. 25.7 **9.** 3.021 **10.** 0.9

The connection between fraction and decimals is one that you are familiar with from your experience with money. Consider \$0.25. You might say that it is twenty five cents, or you might say that it is a **quarter.** How do we relate \$0.25 to a quarter? We have a fraction with numerator of 25 and denominator of 100. (Remember **two** places, then **two** zeros.)

$\$0.25 = \frac{25}{100} = \frac{1}{4}$ You reduce the fraction to get "a quarter".

♦ You Try It

11. From your experience with money, what fraction is \$0.50?
12. From your experience with money, what fraction is \$0.75?

The fraction connection is very helpful in seeing an interesting decimal phenomenon. Zeros can be added or eliminated at the rightmost end of the decimal digits, as the need dictates.

$$32.9 = 32.90 = 32.900$$
$$471.88 = 471.880 = 471.88000$$
$$21.3000 = 21.3$$
$$11.000 = 11.00 = 11.0 = 11$$

Why can this be done? Consider $0.5 = \frac{5}{10}$. Also $0.50 = \frac{50}{100}$. Notice that $\frac{50}{100}$ reduces to $\frac{5}{10}$. (Divide the numerator and denominator by 10.) Since the two fractions are equal, we can conclude that the decimals are equal. Thus 0.5 = 0.50

In each of the cases listed above, if we reduce the fractions to lowest terms we see that the fractions represented by the decimal digits are all equal.

EXAMPLE 5 Write three other decimals that have the same value as 0.72.

0.720
0.7200 The zeros at the rightmost end of the decimal digits
0.72000 can be added on or eliminated, as needed.

♦ You Try It

13. Write three other decimals that have the same value as 0.1.
14. Write 23 in three other ways.
15. Are 2 and 2.00 equal? Explain.
16. Are 0.04 and 0.4 equal? Explain.

3 READING DECIMALS

EXAMPLE 6 What is the proper way to read 0.17 ?

First, it is important to realize that the 0 in this number tells us that the whole number part of the number is 0. Writing 0.17 is the same as writing .17, but writing the 0 emphasizes the decimal point, so we will use it throughout the text. You will notice that most scientific calculators also display a whole number part of 0 even if you don't enter it.

To read a decimal number, there are two steps:

1. Read the number as if there were no decimal point. For 0.17 read "seventeen".

2. Read the rightmost position occupied by the number. In 0.17, there are two decimal places (think: two places, two zeros, so hundredths) The rightmost position is hundredths.

Putting these two together, the number 0.17 is read as "seventeen hundredths".

EXAMPLE 7 The number 0.075 is read "seventy-five thousandths". (Three places, three zeros, thousandths.)

EXAMPLE 8 The number 11.05 is read " eleven **and** five hundredths." Note that the decimal point is read "and" to emphasize the place where the whole number ends and the fractional part begins.

EXAMPLE 9 The number 401.041 is read "four hundred one **and** forty one thousandths.

TIP: Some students find it helpful to first write the decimal number as a fraction or mixed number when they are translating. They know that the denominators will be 10, 100, 1000, etc and they also know that the denominator for the problem has the same number of zeros as there are decimal digits.

In the examples above, $0.17 = \frac{17}{100}$. 2 decimal places; 2 zeros in the denominator

It is now easy to see that it is 17 hundredths.

$0.\underline{075} = \frac{75}{1000}$ 3 decimal places; 3 zeros in the denominator

75 <u>thousandths.</u>

♦ You Try It

Write the following decimal numbers in words.

17. 60.6 **18.** 6.06 **19.** 0.6

20. 0.66 **21.** 0.006 **22.** 42.37

23. What is the value of the second position to the right of the decimal point?

24. What is the value of the first position to the right of the decimal point?

When you translate from words to decimals look for the **AND.** It separates the whole number part from the fractional part. Look at the last words in the written number and it will indicate how many decimal places are needed for your decimal digits. You may need zeros to fill in places. It may be helpful to place blanks for all of the decimal places you need. Fill in the number so that it is on the right end (right justified). Fill in any blanks with zeros.

EXAMPLE 10 Write "thirty and three hundredths " in decimal notation.

1. Look for the AND to place your decimal point. Write the whole number that is specified before the AND, here "thirty" as 30.
2. Read the place value for the decimal part. Here it is hundredths. Hundredths implies 2 decimal places. Use blanks to indicate this.

   30. ___ ___ In the blanks, you must fit "three". How do you handle two blanks and a "3" ? The three must be at the right end so you lead with a zero.

The answer is 30.03

OBSERVE 30.03 is the same as $30\frac{3}{100}$. The number of zeros in the denominator is the same as the number of blanks we filled in above.

EXAMPLE 11 Write "thirty-three hundredths" in decimal notation.

There is no AND so this number does not have both a whole number and decimal part. The last words will let you know if it is a decimal number. If it ends in "ths", we have a decimal number.

"thirty three hundredths" is a decimal number with 2 decimal places.

We can write the whole number part as zero or we can leave it out.

0 . ___ ___ Fill in the blanks with 33 and we get 0.33 for the answer.

Again, if we write this number as a fraction, you'll see the connection between the zeros in the denominator and the number of blanks. $0.33 = \frac{33}{100}$ = thirty-three hundredths

---

♦ You Try It

Write the following in decimal notation.

25. eight tenths

26. sixty-six hundredths

27. sixty and six hundredths

28. eighty-five thousandths

29. eighty and five thousandths

EXAMPLE 12 Decimals and fractions both appear when writing checks. To write a check for $27.52, payable to Ace Hardware, the entire amount, in decimal form, is written in the upper right location. Then on the longer line, the dollar amount is written in words, while the decimal part is written as a fraction. Here is a sample:

Ann Clark
5 Pell Street
Monmouth, NJ 07785

125

Jan 25 1994

Pay To Ace Hardware $ 27.52

Twenty-seven and $\frac{52}{100}$ DOLLARS

MONMOUTH STATE BANK

For ______ Ann Clark

♦**You Try It** **30.** Write the following check out to New Jersey Natural Gas for the amount of \$48.74.

Ann Clark
5 Pell Street
Monmouth, NJ 07785

125

________19__

Pay To ______________________ $

______________________ DOLLARS

MONMOUTH STATE BANK

For__________ ______________

♦**Answers to You Try It** **1.** hundredths **2.** tenths **3.** ten-thousandths **4.** thousandths **5.** $\frac{87}{1000}$ **6.** $1\frac{51}{100}$ **7.** $\frac{1233}{10000}$ **8.** $25\frac{7}{10}$ **9.** $3\frac{21}{1000}$ **10.** $\frac{9}{10}$ **11.** $\frac{1}{2}$ **12.** $\frac{3}{4}$ **13.** 0.10, 0.100, 0.1000 **14.** 23.0, 23.00, 23.000 **15.** Yes **16.** No **17.** sixty and six tenths **18.** six and six hundredths **19.** six tenths **20.** sixty-six hundredths **21.** six thousandths **22.** forty-two and thirty-seven hundredths **23.** hundredths **24.** tenths **25.** 0.8 **26.** 0.66 **27.** 60.06 **28.** 0.085 **29.** 80.005 **30.** The amount should be written: Forty-eight and $\frac{74}{100}$.

SECTION 1.6 EXERCISES

1 *Identify the place value of the digit 8 in each number.*

1. 1.2865 **2.** 0.0081 **3.** 80.17 **4.** 12.583

Identify the place value of the digit 2 in each number.

5. 6.21 **6.** 0.00425 **7.** 203.05 **8.** 8.9236

2 *Write each of the following decimal numbers as a fraction or mixed number.*

9. 0.21 **10.** 2.1 **11.** 2.01 **12.** 0.201

13. 4.061 **14.** 40.01 **15.** 100.1 **16.** 11.11

17. 300.01 **18.** 0.003 **19.** 21.001 **20.** 0.7

3 *Write each of the following in words.*

21. 67.6 **22.** 6.76 **23.** 0.676 **24.** 60.06

25. 20.02 **26.** 0.09 **27.** 90.9 **28.** 100.01

Write each of the following in decimal notation.

29. forty-two thousandths **30.** forty and two thousandths

31. one hundred and three thousandths **32.** one hundred three thousandths

33. eighty-six thousandths **34.** eighty-six thousand

35. eleven thousandths **36.** eleven thousand

37. two hundred three thousandths **38.** two hundred and three thousandths

39. two hundred three thousand **40.** sixty-six and six hundredths

---

SKILLSFOCUS (Section 1.3) *Estimate the answer by first rounding each number to the indicated place.*

41. 4,226 + 8,745 thousands **42.** 2,804 – 481 hundreds

43. 5,799 – 836 hundreds **44.** 15,204 + 6,888 + 1,103 thousands

EXTEND YOUR THINKING ♦ ♦ ♦

WRITING TO LEARN ♦ ♦ ♦

45. Write a number that has a 6 in the thousands place.

46. Write a number that has a 6 in the thousandths place.

47. Discuss the location of the decimal point in a whole number. Give an example.

1.7 ROUNDING AND COMPARING DECIMALS

OBJECTIVES

1 Round decimal numbers.

2 Use rounding to estimate.

3 Compare decimal numbers

1 ROUNDING DECIMALS

Rules for Rounding

1. Locate the digit in the place you are rounding to.
2. Look at the digit to the right of that.
3. If the digit to the right is 5 or more, round **up**; if the digit to the right is less than 5, round **down**.
4. If you are rounding to a whole number position, change all digits to the right of the rounding position to zeros and drop any decimal digits. If you are rounding to a decimal position, drop all digits to the right of the rounding position.The **last** digit of the number must be **in** the rounding position.

EXAMPLE 1 Round 82.5678 to the nearest hundredth.

Using the rule for rounding, the 6 is in the hundredths place and the digit to its right is a 7. Since this 7 is larger than 5, we round **up** to 82.57. Notice that we drop off the decimal digits to the right of the hundredths place (they do not become zeros).

Answer: 82.5678, rounded to the nearest hundredth, is 82.57.

EXAMPLE 2 Round 14.795 to the nearest hundredth.

5 or more

1 4 . 7 9 5

hundredths

Using the rule for rounding, the 9 is in the hundredths place and the digit to its right is a 5. Since this is 5 or more, we round **up** to 14.80. Notice that the digit in the hundredths place becomes a zero because rounding up means adding 1, so 79 hundredths becomes 80 hundredths.

Answer: 14.795, rounded to the nearest hundredth, is 14.80.

EXAMPLE 3 Round 21.8997 to the nearest thousandth.

5 or more

2 1 . 8 9 9 7

thousandths

Using the rule for rounding, the 9 is in the thousandths place and the digit to its right is a 7. Since this is 5 or more, we round **up**. This time, adding 1 to the digit in the thousandths place forces a carry to the hundredths place as well. The rounded answer will be 21.900.

Answer: 21.8997, rounded to the nearest thousandth, is 21.900.

♦ **You Try It**

1. Round 1,657 to the nearest hundred.
2. Round 52.365 to the nearest whole number.
3. Round 28,945.03675 to the nearest thousand.
4. Round 28,945.03675 to the nearest thousandth.
5. Round 99.9999 to the nearest tenth.
6. Round 99.9999 to the nearest hundredth.
7. Round 99.9999 to the nearest hundred.
8. Round 0.2 to the nearest whole number.

There are times when you must report an exact value but there are also times when an approximation is not only adequate, it is preferable. If the MacDonald's sign told **exactly** how many hamburgers they sold, would it be effective advertising? Imagine the sign at MacDonald's saying, "More than 11,234,567,891 hamburgers sold", instead of, "More than 11 billion sold". Imagine having to change the sign every time they sell a hamburger! Consider the following situations and decide if an exact answer or an approximation is preferable. Make sure you can explain your choice.

Situation	**Exact or Approximate**
population of the US	________________
your salary	________________
number of cars on the Garden State Parkway in one day	________________
your grade on your first test	________________
the money a friend owes you	________________

The last example can lead to an interesting examination of the advantages and disadvantages of rounding. Suppose your friend owes you $145.11. He decides to pay you back the amount to the nearest hundred dollars. You would like him to pay you back the amount to the nearest ten dollars. Explain what happens under each plan.

Round to the nearest hundred dollars.________________________

Round to the nearest ten dollars.________________________

As you can see, one method is to your advantage and the other method is to your friend's advantage. Which is which?________________________

____

Rounding and estimating have advantages and disadvantages. Fill in the table below with some of your ideas.

Advantages	Disadvantages

2 ESTIMATING Rounding can be a very useful tool for estimating the answer to a calculation. It enables you to get an idea of the answer by doing mental calculations. Before they put calculators on shopping carts, some shoppers would estimate their grocery bills by rounding the prices of the items that they were buying to the nearest dollar. This way they could be relatively sure they were not spending more money than they had.

EXAMPLE 4 Suppose that you are buying the following items and you have $10 with you. Do you have enough money? Do the problem by rounding to the nearest dollar. Check your answer by using a calculator.

		Rounded
peanut butter	$1.89	$2
bread	$1.39	$1
soda	$ .99	$1
TV dinner	$3.89	$4
ice cream	$2.49	$2

The rounded total is $10 but the exact total is $10.65. Why do you think the rounded total is smaller than the exact total? ______________________

____

Can you give some advice to someone who will use estimates in his or her calculations?

____

____

♦ You Try It *Estimate the sum or difference by rounding to the indicated value. Check your answer by using a calculator.*

9. 6.825 – 3.479 Round to the nearest tenth.
Estimated difference ________ – ________ =
Exact difference ______________

10. 92,132 + 34,582 + 28,904 Round to the nearest thousand.
Estimated sum ______ + ______ + ______ =
Exact sum ______________

11. 345.675 – 259.299 Round to the nearest whole number.

Estimated difference _________ – _________ =

Exact difference ______________________

12. Carol is preparing 4000 letters for mailing. She has already finished 2,155. Estimate, to the nearest hundred, the number of letters she has left to do.

---

3 COMPARING DECIMALS

We have seen that the symbols < (less than) and > (greater than) can be used to compare two numbers. For example, since 12 *is greater than* 7, we write 12 > 7, and since 0 *is less than* 28, we write 0 < 28. Also, since 9 *is between* 3 and 10, we can write 3 < 9 < 10. We now turn our attention to comparing decimals to see which is larger.

If two decimals have the same number of decimal places, it is very easy to compare them.

EXAMPLE 5 Compare 0.29 and 0.92.

Since 0.29 is 29 hundredths and 0.92 is 92 hundredths, they are both hundredths and 0.92 has more hundredths than 0.29, so 0.29 is less than 0.92.

Answer: 0.29 < 0.92

In this last example, we could have decided which number was larger based only on the tenths position. In decimals, the further a place is to the right, the smaller the value of the place. For example, one thousandth is a smaller value than one tenth. Think of fractions and eating $\frac{1}{1000}$ of a pizza (still hungry?) or $\frac{1}{10}$ of a pizza. For this reason, if a decimal number has larger digits in the equivalent beginning decimal places, it is a larger number than a decimal number that has smaller digits in its beginning decimal places. So in Example 5, since 0.92 has a 9 in the tenths position, while 0.29 only has a 2 in the tenths position, 0.92 is larger than 0.29.

> When comparing decimal numbers, compare the digits in corresponding positions, working from left to right. A larger digit indicates the larger decimal.

EXAMPLE 6 Compare 0.29 and 0.029.

Sometimes it is easier to compare if we line up the numbers vertically:

0.29
0.029

and then compare position by position. Right away, in the tenths position, the first number has a 2, while the second number has a 0. Since 2 is larger than 0, 0.29 is larger than 0.029.

Answer: 0.29 > 0.029

Often when people see a decimal with a large number of digits, they assume that it is a large number. Look at this next example.

EXAMPLE 7 Compare 0.12345678 and 0.6.

Again, lining the numbers up vertically

0.12345678
0.6

we see that the first number has a 1 in the tenths position, while the second number has a 6. That means that 0.12345678 is less than 0.6, even though it has more digits. **The number of digits in a decimal number is not an indication of the magnitude (size) of the number.**

Note also that both of these numbers are between 0 and 1. 0.12345678 is very close to 0, while 0.6 is just to the right of .5 (one half).

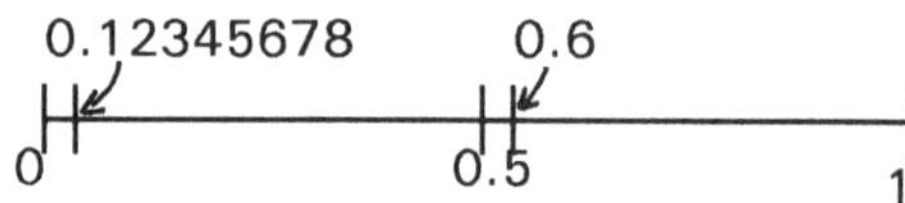

Answer: 0.12345678 < 0.6

EXAMPLE 8 Compare 32.01 and 3.201.

Write the numbers vertically, lining up the decimal points:

32.01
 3.201

Here, the location of the decimal point is of critical importance. Each of these numbers has a whole number part. In 32.01, 32 is the whole number part, so 32.01 is between 32 and 33. In 3.201, 3 is the whole number part, so 3.201 is between 3 and 4. From looking at the whole number parts, we can see that 32.01 is larger than 3.201.

Answer: 32.01 > 3.201

♦You Try It

13. True or False: When comparing two numbers, the one with the most digits is the larger number. Explain your answer.

14. True or False: 0.603421598 is a very large number. Explain your answer.

Compare the numbers in each pair. Write your answer using < or >.

15. 0.456 0.546 **16.** 0.317 0.0317

17. 0.222 0.2 **18.** 0.000009 0.001

19. 25.08 25.1 **20.** 7.356 73.56

21. Which of the following numbers does **not** equal 0.86 ?

0.860, 0.86, 0.086

In each pair of numbers, which number is larger?

22. 500,000,000 500,000,001
23. 0.40000001 0.4
24. 0.10000004 0.4

EXAMPLE 9 Write these numbers in order from smallest to largest: 0.402, 0.420, 4.02, 0.042, 4.20

Write the numbers vertically, lining up the decimal points, and compare positions from left to right.

0.402 ← second
0.420 ← third
4.02 ← fourth
0.042 ← first
4.20 ← fifth

In order from smallest to largest, we have 0.042, 0.402, 0.420, 4.02, 4.20

♦ You Try It

25. Write these numbers in order from smallest to largest: 5.1, 5.001, 5.01

26. Write these numbers in order from smallest to largest: 0.107, 0.170, 1.07, 1.70, 0.017

♦ Answers to You Try It **1.** 1700 **2.** 52 **3.** 29,000 **4.** 28,945.037 **5.** 100.0 **6.** 100.00 **7.** 100 **8.** 0 **9.** Estimate $6.8 - 3.5 = 3.3$ Exact 3.346 **10.** Estimate $92{,}000 + 35{,}000 + 29{,}000 = 156{,}000$ Exact 155,618 **11.** Estimate $346 - 259 = 87$ Exact 86.376 **12.** Carol has *about* $4{,}000 - 2{,}200 = 1{,}800$ letters left. **13.** False. The number of digits is not important; the place value of the digits determines how large the number is. **14.** False, 0.603421598 is less than 1 **15.** $0.456 < 0.546$ **16.** $0.317 > 0.0317$ **17.** $0.222 > 0.2$ **18.** $0.000009 < .001$ **19.** $25.08 < 25.1$ **20.** $7.356 < 73.56$ **21.** 0.086 **22.** 500,000,001 **23.** 0.40000001 **24.** 0.4 **25.** 5.001 , 5.01, 5.1 **26.** 0.017, 0.107, 0.170, 1.07, 1.70

SECTION 1.7 EXERCISES

1 *Round as indicated.*

1. 4.72 to tenths

2. 2.091 to tenths

3. 264.99 to tens

4. 568.09 to tens

5. 4.4821 to hundredths

6. 4,612.517 to hundreds

7. 76.416 to ones

8. 2.7021 to ones

9. 90.0267 to thousandths

10. 9.8261 to thousandths

11. 450.95 to hundreds

12. 2,789.419 to hundredths

13. 6.488 to tenths

14. 50.05 to tenths

15. 27.089 to ones

16. 101.50 to ones

17. 509.5 to hundreds

18. 900.0561 to thousandths

19. 3.0089 to thousandths

20. 56.0293 to hundredths

21. 0.254 to tenths

22. 0.062 to tenths

23. 683.811 to ones

24. 0.49 to ones

25. 38.99 to hundreds

26. 55.015 to hundredths

27. 0.0664 to thousandths

28. 0.07623 to hundredths

29. 427.03 to tenths

30. 58.394 to tenths

31. 2.00475 to ten-thousandths

32. 31.7746 to thousandths

33. 0.00057123 to millionths

34. 0.00000588 to millionths

35. 4.97 to tenths

36. 397.67 to tenths

37. 0.0068 to hundredths

38. 69.996 to hundredths

39. 599.9995 to thousandths

40. 54.994 to tenths

41. Round $4,572.0472 to the nearest cent.

42. Round $32.59245 to the nearest cent.

43. Round $0.495 to the nearest cent.

44. Round $0.0749 to the nearest cent.

2

45. Estimate 49.95 + 7.50 + 18 + 2.4 by rounding to tens.

46. Estimate $121.89 + $88.16 + $275.56 + $65.70 by rounding to hundreds.

47. Estimate 1,465.7 + 531.87 + 1,013.7 + 87.03 by rounding to hundreds.

48. Estimate $15.26 + $7.32 + $11.68 + $0.45 by rounding to ones.

49. Estimate 45.7 – 28.09 by rounding to tens.

50. Estimate 6.409 – 3.18 by rounding to ones.

51. Estimate $560.38 – $256.90 by rounding to hundreds.

52. Estimate $0.80 – $0.39 by rounding to tenths.

3 *Compare the numbers in each pair. Write your answer using $<$ or $>$.*

53. 0.8   0.089

54. 0.04   0.015

55. 0.0971   0.103

56. 0.78   0.8

57. 0.00091   0.002

58. 5.016   5.0201

Write the numbers in order from smallest to largest.

59. 0.603, 0.6104, 0.6

60. 9.03, 9.105, 9.0084

61. 3.03, 3.0063, 3.026

62. 0.044, 0.0404, 0.042

63. 0.254, 0.2042, 0.26, 0.205

64. 0.082, 0.09, 0.0815, 0.085

65. 4.03, 4.1, 4.0069, 4

66. 1.53, 1.053, 1.503, 1.5

67. 0.02, 0.0061, 0.019, 0.1

68. 5.071, 5.1, 5.0708, 5.07

69. 7.07, 7.018, 7.2, 7.09, 7.089

70. 23.05, 23.108, 23.009, 23.3, 23.0802

SKILLSFOCUS (Section 1.4)

71. Connie bought 3 shirts at $15 each and 2 belts at $11 each. She gave the cashier $70. How much change did she get?

72. a. Find the area of a rectangular lot that is 155 ft by 102 ft.

b. A house on this lot occupies 37 ft by 25 ft. What is the area covered by this house?

73. If your quiz grades are 87, 75, 93, 68, and 82, what is your quiz average?

EXTEND YOUR THINKING ♦ ♦ ♦

WRITING TO LEARN ♦ ♦ ♦

74. a. Between which two consecutive whole numbers is 4.57 ? Write this as an inequality using mathematical symbols.

b. Illustrate this inequality on a number line.

c. Which whole number is 4.57 closer to? Round 4.57 to the nearest whole number.

75. Write three numbers greater than 1.42 but less than 1.43.

76. When would you prefer to have each of the following numbers rounded to the given position?

a. your math test average rounded to the nearest ten points

b. your age rounded to the nearest whole number

c. the cost of a new car rounded to the nearest thousand dollars.

1.8 ADDING AND SUBTRACTING DECIMALS; APPLICATIONS

OBJECTIVES

1. Add and subtract decimals
2. Solve application problems using addition and subtraction of decimals
3. Find the perimeter of a figure.

NEW VOCABULARY

Perimeter

1 ADDITION AND SUBTRACTION OF DECIMALS

> To add or subtract decimals, line up the decimal points and add or subtract in the same way as whole numbers. In this way, digits in the same position are added or subtracted.

EXAMPLE 1 Add 3.42 + 4.9 + 80.672.

$$\begin{array}{r} 3.420 \\ 4.900 \\ \underline{80.672} \\ 88.992 \end{array}$$

NOTE: Since these numbers all have different numbers of decimal places, you might wish to add zeros to the right end so that they all have three decimal places. Remember that adding zeros to the right end of a decimal number does not change its value.

EXAMPLE 2 Find the sum of 4.09, 812, and 12.8.

Remember that 812 is a whole number so its decimal point is understood to be to the right: 812 = 812.0

$$\begin{array}{r} 4.09 \\ 812.00 \\ \underline{12.80} \\ 828.89 \end{array}$$

EXAMPLE 3 Find the difference of 16.92 and 8.33.

$$\begin{array}{r} 16.92 \\ \underline{-\ 8.33} \\ 8.59 \end{array}$$

You can check this subtraction by adding the answer, 8.59, to 8.33. The sum should be 16.92.

EXAMPLE 4 Find the difference of 50.27 and 30.105.

Here, you must add a zero at the end of 50.27 in order to complete the subtraction.

$$\begin{array}{r} 50.270 \\ \underline{-\ 30.105} \\ 20.165 \end{array}$$

♦You Try It

1. How do you add decimal numbers?
2. How do you subtract decimal numbers?
3. Does every whole number have a decimal point? If so, where is it?
4. Find the sum of 4.82, 3.56, and 1.25.
5. Find the sum of 2.05, 8.119, and 0.096.
6. Find the sum of 22.17, 312, 18.2, and 10.75.
7. Find the difference of 16.28 and 4.09.
8. Find the difference of 32.7 and 15.25.
9. Subtract 48 − 3.65.

2 PROBLEMS LEADING TO ADDITION AND SUBTRACTION

As we discussed earlier, solving problems requires a strategy. Here are the steps we suggested in Section 1.4.

Steps for Problem Solving

1. **READ** carefully. Identify what each of the numbers in the problem measures. (dollars? miles? chickens?)

2. **WRITE**, in your own words, the question that is being asked. Often it is helpful to include a sketch.
   It is also important to note the units for the answer. Again, do we want dollars, miles, or chickens?

3. **ANALYZE** how you can use the information in the problem to answer the question.
   - Should you use all or only some of the values stated in the problem?
   - Must you do one or more arithmetic operations (addition, subtraction, multiplication, or division) to answer the question?
   - Make an outline of the steps you need to follow to solve the problem.

4. **CHECK** your answer to see if it makes sense. Use rounding to estimate the answer.

EXAMPLE 5 Lauren had a balance of $105.75 in her checking account. She wrote checks for the following amounts: $23.60, $5.75, $84.85, and $2.60. Did she have enough money in her checking account to cover the checks? If so, how much is left in her account? If not, by how much is she overdrawn?

READ, WRITE, ANALYZE

Analysis: We know Lauren had $105.75 to begin with.
She wrote four checks.
To see if she had enough money to cover the four checks, we need to know the TOTAL of these checks.
Once we know the total, we can compare it to her original balance.

Outline: 1. Find the total of the checks.
2. Compare the total of the checks to the original amount.

Solution: 1. Total of the checks = \$23.60 + \$5.75 + \$84.85 + \$2.60
= \$116.80

2. Compare to the balance: \$116.80 > \$105.75

Since the total of the checks is larger than the original balance, Lauren's account is now **overdrawn**.

3. Amount overdrawn = total of checks − original balance
= \$116.80 − \$105.75
= \$ 11.05

Check: To check this, we use rounding to estimate the sums and do some mental computations. The amounts of the checks are *about* \$24 + \$6 +\$85 + \$3, which is about \$30 +\$88 or \$118, so it seems that Lauren has overdrawn her account by ***about*** \$120 – \$105 or \$15. These are simply rough checks on the arithmetic. It is possible that you might approximate differently but the outcome would probably be the same or close. Remember that the object is to see if the answer is **reasonable.**

Answer: Lauren's account is overdrawn by \$11.05.

EXAMPLE 6 Normal body temperature is 98.6°F. When Ted had the flu, his temperature was 102°F. What was his rise in temperature?

READ, WRITE, ANALYZE

Analysis: We know the normal temperature and the temperature during the flu and we need to find the **rise**, which in this case means the **difference**.

Outline: Subtract 98.6 from 102

Solution: rise in temperature = 102°F – 98.6°F
= 3.4°F

Work:

$$\begin{array}{r} 102.0 \\ -\ \ 98.6 \\ \hline 3.4 \end{array}$$

Remember that 102 is a whole number so its decimal point is to the right of the 2 and can be followed by as many zeros as necessary.

Check: Ted's temperature did go up and 99 from 102 is 3 so the answer is **reasonable.**

Answer: Ted's temperature rose 3.4°F.

♦ You Try It

Solve each of the following problems in your notebook. Follow the problem solving strategy in the examples. Be sure to provide an outline, a complete solution, and a complete answer.

10. On June 14, Mary's savings account had a balance of \$5,285.90. She withdrew \$350.25 to pay off a loan on June 20, deposited \$125.89 on June 25 and deposited \$100 on June 29. During the month of June, her account earned \$35.68 in interest. What was her new balance on July 1?

11. Joyce earns \$234.25 a week and her brother Tom earns \$175.80 a week. What is their combined salary? How much more does Joyce earn than Tom?

12. A building contractor paid bills of \$131.25, \$278.16, and \$78.29 for materials and labor on a certain job. He is being paid \$1500 for the job. What is his profit?

13. Ellen loves to shop at Marshall's. One day, she found a sweater for \$24.50, jeans for \$21.95, and sneakers for \$33.75. If she gave the cashier a \$100 bill, what was her change?

3 PERIMETER

The **perimeter** of an object is the distance around it. To find the perimeter of a straight sided object, add the lengths of all the sides. The perimeter is measured in the same units as the sides.

EXAMPLE 7 A farmer has a field with an irregular shape, as indicated below. He wants to enclose this field with fence and has 245.5 meters (m) of fencing on hand. Does he have enough? If not, how much more will he need?

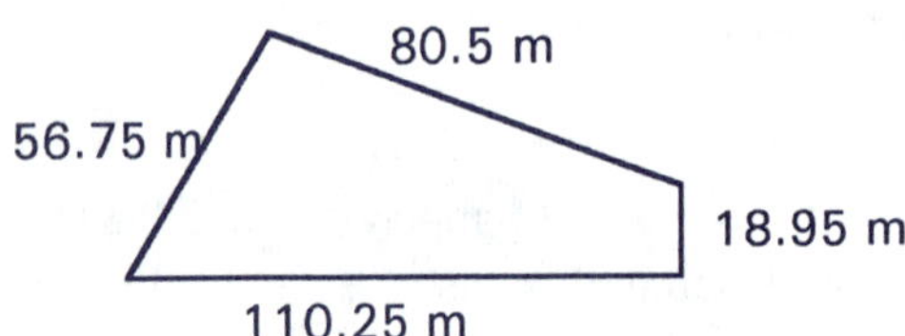

READ, WRITE, ANALYZE

Analysis: Although the shape of the field is irregular, the amount of fence needed is the total distance around the field. The distance around an object is its **perimeter** so we must find the perimeter of the field by finding the total distance around the field. Since we know the lengths of all the sides of the field, we can add them to find the perimeter, which will also be the amount of fence needed. We can then see if the farmer has enough fencing material on hand.

Outline:
1. Find the perimeter of the field by adding the lengths of all the sides.
2. Compare the perimeter to the amount of fence on hand.
3. If the farmer does not have enough fence, subtract to find the amount of fence he must buy.

Solution:
1. Perimeter = 56.75 m + 80.5 m + 18.95 m + 110.25 m
   = 266.45 m
2. Since 245.5 m < 266.45 m, the farmer does not have enough fence.
3. Amount needed = 266.45 m – 245.5 m = 20.95 m

Check: The distance around is *approximately* 60 m + 80 m + 20 m + 110 m, which is 270 m, which supports the answer that the farmer does not have enough fence. He needs to buy **about** 270 – 250 or 20 more meters of fence, which is very close to the answer we got.

Answer: The farmer will need to buy 20.95 m of fence in order to enclose the field.
NOTE: It is probably not practical to buy 20.95 m of fence, so the farmer will probably buy 21 m.

EXAMPLE 8 Find the perimeter

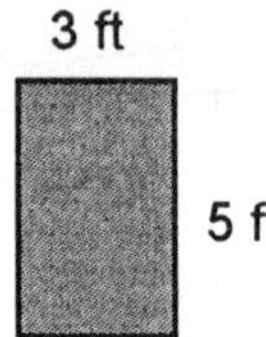

This figure is a rectangle. Notice that only two sides are labeled with their measurements. In a rectangle, *opposite sides have equal lengths*. The side opposite the 5 ft side is also 5 ft and the side opposite the 3 ft side is also 3 ft. The rectangle can be drawn with all four sides labeled.

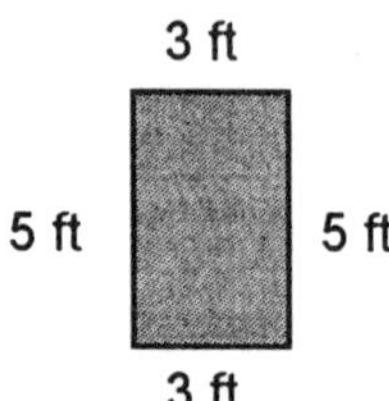

Now we can find the perimeter by adding the lengths of the sides:

Perimeter = 5 ft + 3 ft + 5 ft + 3 ft
Perimeter = 16 ft

In Section 2.8, we will use a formula for finding the perimeter of a rectangle.

♦ You Try It

Find the perimeter.

14.

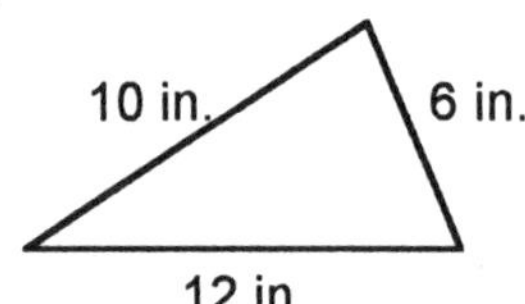

15.

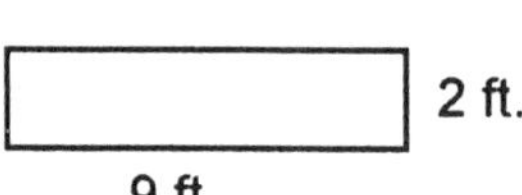

16. Louise has a vegetable garden that is a 20 ft by 10 ft rectangle. How many feet of fence would she need for the garden?

♦ Answers to You Try It **1.** Line up the decimal points and add as usual. **2.** Line up the decimal points and subtract as usual, adding zeros to the top number if needed. **3.** Every whole number has a decimal point to the right of the ones place. **4.** 9.63 **5.** 10.265 **6.** 363.12 **7.** 12.19 **8.** 17.45 **9.** 44.35 **10.** Mary's new balance on July 1 was $5,197.22. **11.** Joyce and Tom's combined salary is $410.05. Joyce earns $58.45 a week more than Tom. **12.** The builder's profit is $1,012.30. **13.** Ellen's total bill was $80.20. Her change was $19.80. **14.** Perimeter = 28 in. **15.** Perimeter = 22 ft. **16.** Louise needs 60 ft. of fence.

SECTION 1.8 EXERCISES

1 *Add or subtract.*

1. $4.62 + 9.7$ **2.** $54.02 + 17.304$ **3.** $89.3 + 607.14$

4. $7.051 + 24.7926$ **5.** $0.014 + 0.5 + 0.58$ **6.** $0.0034 + 0.6 + 0.882$

7. $12 + 0.068 + 2$ **8.** $8 + 0.09 + 1.637$ **9.** $60 + 0.06 + 0.882$

10. $1 + 0.02 + 0.003$ **11.** $582 + 6.85 + 24.377 + 0.0088$ **12.** $15.905 + 40 + 6.085$

13. $300.56 + 0.8 + 34.0056 + 0.0008 + 70 + 200.003$ **14.** $0.345 + 0.0390 + 0.4 + 3 + 0.007 + 0.00306$

15. $4.6 - 3.2$ **16.** $12.8 - 7.9$ **17.** $25.07 - 15.42$

18. $5.7 - 2.61$ **19.** $23.9 - 14.87$ **20.** $46.239 - 30.4$

21. $6.937 - 2.61$ **22.** $0.104 - 0.02$ **23.** $25.66 - 8.912$

24. $7.34 - 2.0056$ **25.** $0.01 - 0.009$ **26.** $0.035 - 0.008$

27. $7 - 3.8$ **28.** $29 - 13.4$ **29.** $7 - 0.472$

30. $10 - 0.59$ **31.** $4.97 - 3$ **32.** $45.904 - 20$

33. 45 feet + 7.25 feet + 0.5 foot + 9.375 feet **34.** \$35 + \$4.65 + \$0.36 + \$9.40 + \$3 + \$0.04

Use your calculator to compute the sum or difference.

35. 0.5 inch + 3.25 inches + 1.875 inches + 4 inches + 3.75 inches

36. 100 meters + 45.03 meters + 0.478 meter + 30.3 meters + 16 meters + 0.73 meter

37. $380.6 - 45.9$ **38.** $12{,}600 - 56.89$ **39.** $100.00 - 39.95$

---

40. Subtract 0.3 from 5. **41.** Subtract 0.06 from 1.

---

2 *Solve each problem by using the problem solving strategies discussed in this section.*

42. Sharon purchased a coffee table for \$175.42, a sofa for \$856.92, and a love seat for \$587.03. What total did she spend?

43. When shopping for his son's birthday party, Bill spent \$6.75 on a cake, \$0.79 on balloons, \$12.84 on party favors, and hired a clown for \$40. How much did Bill spend?

44. A $7.36 lunch check is paid for with a $10 bill. What is the change?

45. A $29.67 sweater is paid for with a $50 bill. What is the change?

46. Phil has a temperature of 102.3°F. If a normal temperature is 98.6°F, how many degrees above normal is Phil's temperature?

47. Casey's batting average was 0.273 last year. This year it is 0.305. How much higher is his batting average this year?

48. Pieces of wire 5.3 centimeters , 3.75 centimeters, and 4.0 centimeters are cut from a wire 17.5 centimeters long. How much wire is left?

49. This morning the checking account balance was $341.73. During the day checks were written for $30, $17.54, $104.32, and $15.68. What was the balance at the end of the day?

50. Four athletes run a 1 mile relay. Their times for each leg of the relay are 50.23 seconds, 49.8 seconds, 51.68 seconds, and 47.37 seconds.

a. What was the team's time for the relay?

b. By how much did they beat the old record of 200.73 seconds?

51. Anna runs 140 miles per week when training for a marathon. This week she has run 18.2, 17, 17.4, 16.9, 20.2, and 19.0 miles.

a. How many miles did Anna run this week:?

b. How many more miles must she run to make 140 miles?

52. What must be subtracted from 20.013 to get 7.92?

53. What do you subtract from $347.08 to get $280.45?

54. What do you add to 128.07 to get 157.391?

55. What do you add to $5,206.17 to get $8,000?

56. Zack has $45.16. He needs $162.50 to purchase a TV.

a. Estimate how much more money he needs by rounding to tens.

b. What is the actual answer?

57. Evan needs $12,451.04 to buy a car. He has $2,578.57.

a. Estimate how much money he must borrow to purchase the car by rounding to thousands.

b. What is the actual answer?

58. Helen's gross pay is $2,153.18. Her deductions are $88.34 for state tax, $221.13 for federal tax, $124.05 for FICA, $46.87 for medical insurance, and a $328.22 deduction for a car loan. What is Helen's net pay?

59. Alan's gross pay is $1,658.33. His deductions are $66.54 for state tax, $150.67 for federal tax, $97.04 for FICA, $23.11 for health insurance, and a $50 deduction for a savings plan. What is Alan's net pay?

60. Dean's checking account balance was $670.45 this morning. Today he wrote checks for $26.12, $248.90, $170, and $15.78. He made deposits of $35.56 and $100.

a. Estimate his new balance.

b. What is his actual new balance?

61. Jayne has $400.00 in her checking account. She writes checks for $62.34, $45.16, $65.40, $12, and $6.17. She makes deposits of $120.45, $14.25, and $20.

a. Estimate her new balance.

b. What is her actual new balance?

62. On vacation, Vince purchased 12.6 gallons of gas for $14.60, 13.7 gallons for $15.34, 11.9 gallons for $13.80, and 14.1 gallons for $16.
  a. How many gallons fo gas did Vince purchase on the trip?
  b. What was the total cost of this gas?

63. Alicia bought 1.47 pounds of roast beef for $8.34, 0.78 pounds of salami for $2.88, and 2.23 pounds of ham for $11.23.
  a. How much lunch meat did Alicia buy?
  b. What was the total cost?

64. Sal purchased a radio for $69.95, a TV for $329.95, a stereo for $479.95, and four tapes for $10.29 each.
  a. Estimate the total spent by rounding to hundreds.
  b. Estimate the total spent by rounding to tens.
  c. Which estimate is more accurate? Why?
  d. How much did Sal actually spend?

65. Erica spent $275 on skis, $89.95 on boots, $32.50 on poles, and $28 for a lift ticket.
  a. Estimate the total spent by rounding to hundreds.
  b. Estimate the total spent by rounding to tens.
  c. Which estimate is more accurate? Why?
  d. How much did Erica actually spend?

Use Figure 1.22 to answer 66 - 69.

66. Which of the pieces of money is the most expensive to make?

67. Which of the pieces of money is the least expensive to make?

68. How much more does it cost to make a dollar bill than to make a dime?

69. How much more does it cost to make a quarter than to make a nickel?

FIGURE 1.22

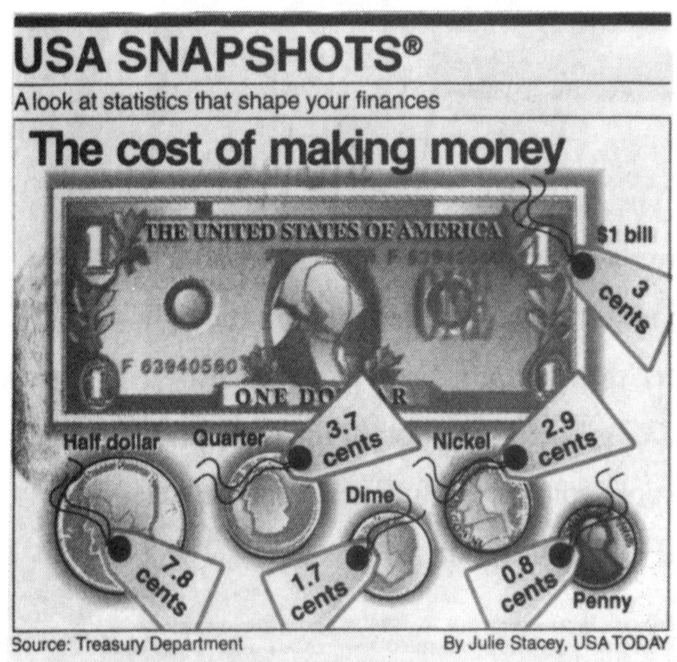

3 *Find the perimeter of each figure.*

70.

4 ft
6 ft
7 ft
3 ft

71.

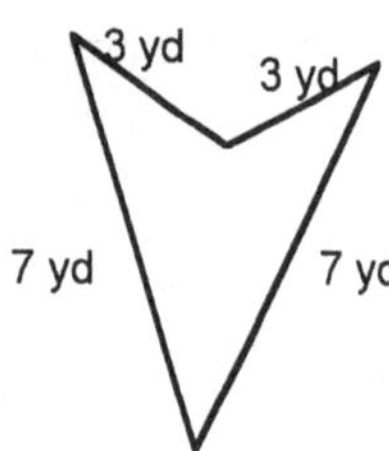

72.

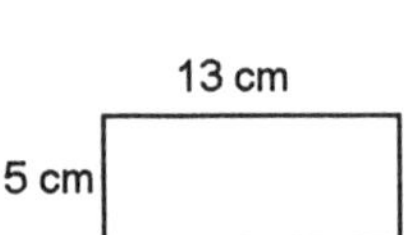

73.

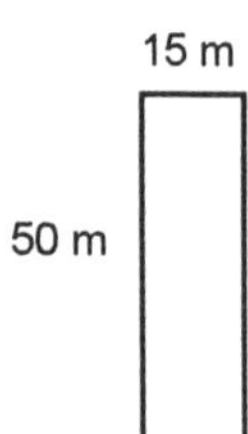

74.

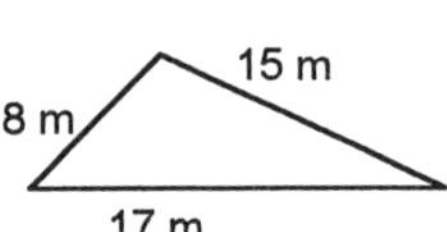

75.

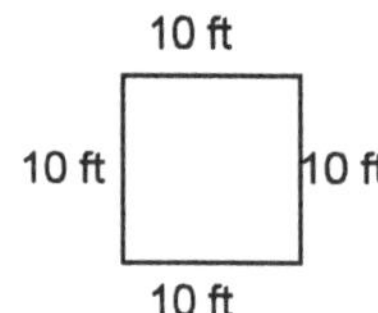

76.

2 m
2 m
2 m
3 m
3 m

77. Find the perimeter of a 20 ft by 40 ft rectangular garden.

78. Find the perimeter of a rectangular room 14 ft 3 in long by 10 ft 8 in wide.

79. Find the perimeter of a square room with 8.37 m on each side.

SKILLSFOCUS (Section 1.6) *Write each decimal as a fraction or mixed number.*

80. 0.73 **81.** 0.109 **82.** 4.3

EXTEND YOUR THINKING ♦ ♦ ♦

♦SOMETHING MORE

83. A large hole is drilled in the center of a board. What is the diameter of the hole? (The diameter is the distance across a circle through the center, as shown in the figure.)

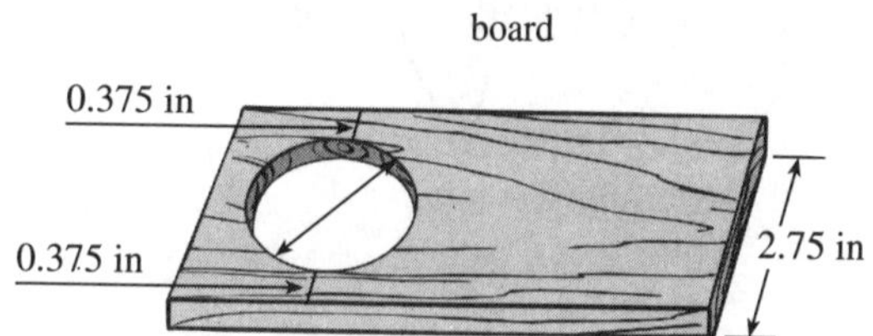

84. Find the perimeter. (Hint: You will need to determine the missing lengths first.)

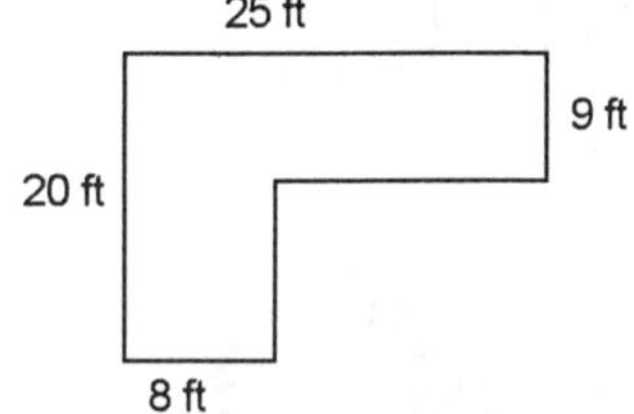

85. A square garden measures 12 ft on a side.

a. What is the perimeter of the garden?

b. Suppose you double the length of each side of the garden. What is the new perimeter?

WRITING TO LEARN ♦ ♦ ♦

86. Write a word problem whose answer is 2.71. The word problem must be about a grocery store purchase. It must contain three numbers, one addition, and one subtraction.

♦ YOU BE THE JUDGE

87. Don said the figure shown contains 5 rectangles. Fran said it had more than 10. Who is right? Explain your decision.

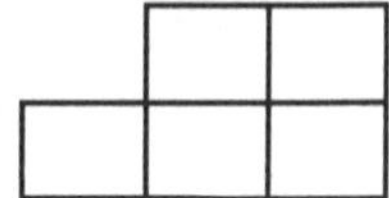

CHAPTER 1 REVIEW

VOCABULARY AND MATCHING

New words and phrases introduced in this chapter are shown in the left column. Match each term on the left with the word or sentence on the right that best describes it.

A. 8 more than 2	_____ 111.1
B. 8 is more than 2	_____ ab
C. variable	_____ 4.0, 4.00
D. area	_____ numerator is smaller than the denominator
E. average of 4, 5, and 6	_____ inches
F. proper fraction	_____ a symbol used to represent a number
G. lowest terms	_____ when adding or subtracting fractions
H. forty-two hundredths	_____ number of square units to cover an object
I. forty and two hundredths	_____ 5
J. product of a and b	_____ 0.42
K. quotient of a and b	_____ $\frac{4}{1}, \frac{8}{2}$
L. reciprocal of $\frac{2}{3}$	_____ square feet
M. when you need like denominators	_____ 6
N. 111.111 rounded to the nearest tenth	_____ $a \div b$
O. 111.111 rounded to the nearest ten	_____ when numerator and denominator have no divisors in common
P. ways to write 4 as a fraction	_____ $\frac{3}{2}$
Q. ways to write 4 as a decimal number	_____ $8 > 2$
R. xy when $x = 2$ and $y = 3$	_____ 40.02
S. units to measure area	_____ 110
T. units to measure perimeter	_____ $2 + 8$

REVIEW EXERCISES

1.1 Numbers in the News

Use Figure 1.23 to answer questions 1 - 4.

FIGURE 1.23
Death Rates per 100,000 from Selected Causes, 1900 and 1990

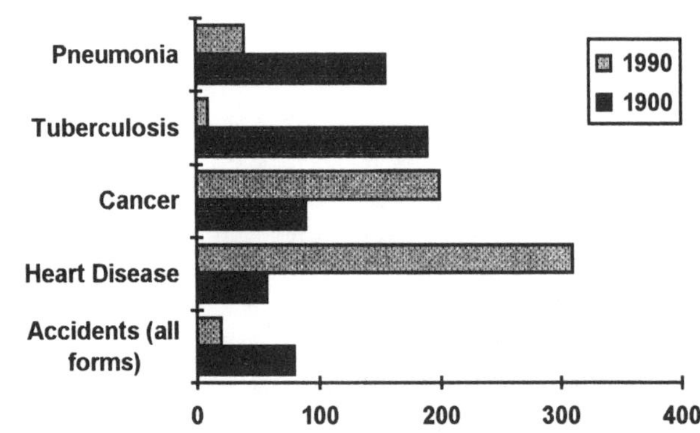

1. **a.** About how many people per 100,000 died from heart disease in 1900?
   b. In 1990?
   c. About how many times more people died from heart disease in 1990 compared to 1900?

2. Which cause of death was almost eliminated from 1900 to 1990?

3. Which causes resulted in more deaths per 100,000 in 1990 than 1900?

4. In 1900, which disease was closest to causing the same number of deaths as pneumonia?

Use Figure 1.24 in problems 5 - 9.

FIGURE 1.24
Average Price of 1 Share of XYZ Stock for Each of Six Months, March through August

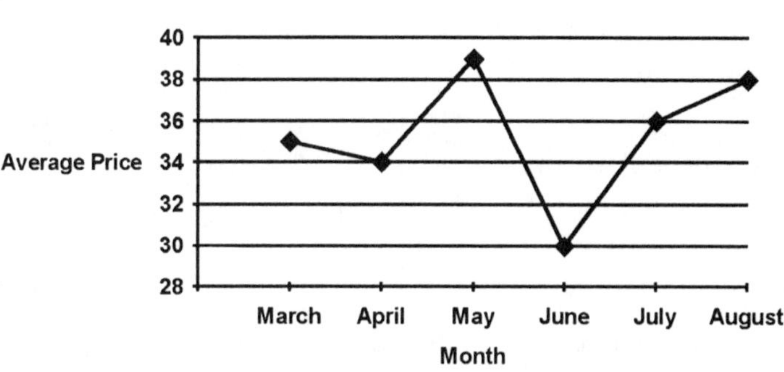

5. What was the average cost of one share of XYZ stock in May?

6. Between which two consecutive months did the value of one share rise the least?

7. **a.** Between which two successive months did the average value of one share drop the most?
   b. How much was this drop?

8. Estimate the average price per share for each of the six months shown.

9. Teresa purchased 240 shares of XYZ in June and sold them in July. Using the average price per share, estimate how much profit she made.

Use Figure 1.25 to answer questions 12 - 15.

FIGURE 1.25

Smoking rate falling fast

By Mike Snider
USA TODAY

Smoking is declining faster in the USA now than at any time since tracking began in 1955, a government study reports today.

But a Cancer Society study says a fifth of the developed world's people will die of smoking-related causes.

Social pressure, health risks and cost led to the USA decline, says Gary Giovino of the Centers for Disease Control. "You can't smoke *anywhere* anymore."

The prevalence of smoking has gradually declined from 40% in 1965 to 25.5% in 1990. It dropped about 0.5% annually since 1966, and 1.1% a year since 1987.

Half of USA adults — about 90 million — smoked at one time. Now: 45.8 million.

More results from the survey of 40,666:

▶ The same percentage of blacks and whites smoke — 26% — but fewer young blacks do: 17% of blacks age 20-24 vs. 28% of whites.

▶ Smoking is highest among those age 25-44.

▶ About 24 million men smoke; 22 million women.

The international study on smoking deaths also finds:

▶ Deaths almost doubled from the mid-'60s to more than 2 million this year.

▶ Women's deaths will overtake men's by 2010.

▶ In 1995, about half of the 510,000 women who die will be from the USA.

10. Which of the numbers presented in the article do you find most interesting? Why?

11. How many people were in the survey?

12. What fraction of USA adults smoked at one time?

13. What age group has the highest amount of smokers?

14. What percent of USA adults smoked in 1965?

15. What percent of USA adults smoked in 1990?

1.2 Rediscovering Mathematics: Numbers and Operations

Use the correct symbol, < or >, between the pair of numbers to write a true statement.

16. 0 12 **17.** 72.31 72 **18.** $\frac{1}{3}$ 1 **19.** $14\frac{1}{4}$ 15

Rewrite the English phrase using symbols. If you use a variable, define it.

20. 6 less than 11

21. 6 is less than 11

22. the product of 7 and a number

23. the quotient of 7 and a number

Evaluate the following expressions when $x = 3$ and $y = 4$.

24. $2 + x$ **25.** $y - 2$ **26.** $\frac{12}{y}$ **27.** $3x$

Using the given values for the variables, determine whether the equation is true or false.

28. $x - 16 = 4$ if $x = 20$

29. $2y = t + 1$ if $y = 3$ and $t = 4$

30. $\frac{48}{n} = p$ if $n = 9$ and $p = 6$

31. $\frac{a}{8} = x$ if $a = 56$ and $x = 7$

1.3 Rounding and Estimating

Round the following numbers to the indicated place.

32. 345 to the nearest hundred

33. 12,345 to the nearest thousand

34. 765,123 to the nearest ten thousand

35. 12,345 to the nearest ten

1.4 Problem Solving

36. If there are 28 teams in the National Football League and 47 players are on each team, how many players are there in the NFL?

37. Nina's grades on her tests are 87, 76 and 92. What is her average grade?

38. Juan's living room is 8 ft. by 12 ft. If the carpeting he wants costs \$7 per square foot, how much will he pay for carpeting?

1.5 Fractions - An Overview

Draw a picture, using a circle or a box, to represent the following.

39. $\frac{5}{3}$

40. $\frac{5}{8}$

Write each of the following fractions in two different ways.

41. $\frac{1}{4}$

42. $\frac{6}{1}$

Write as a mixed number.

43. $\frac{11}{2}$

44. $\frac{5}{3}$

Write as an improper fraction.

45. $4\frac{1}{7}$

46. $11\frac{1}{2}$

Find the sum.

47. $\frac{2}{13}+\frac{4}{13}+\frac{6}{13}$

48. $\frac{1}{3}+\frac{2}{9}$

Find the difference.

49. $\frac{10}{11}-\frac{3}{11}$

50. $\frac{3}{4}-\frac{3}{8}$

Find the product.

51. $\frac{2}{3} \times \frac{7}{11}$

52. $\frac{3}{5} \times 5$

Find the quotient.

53. $\frac{3}{7} \div \frac{1}{2}$

54. $\frac{3}{5} \div 5$

1.6 Introduction to Decimals

Write the following decimal numbers in words.

55. 34.06

56. 0.001

Write the following decimal numbers as mixed numbers or fractions.

57. 34.06

58. 0.001

Write the following in decimal digits.

59. three thousandths

60. three hundred and one tenth

1.7 Rounding and Comparing Decimal Digits

Round each of the following numbers to the place indicated.

61. 123.456 to the nearest tenth

62. 0.654 to the nearest hundredth

Write the following numbers in order from lowest to highest.

63. 0.02, 2.002, 0.2, 0.2002

64. 50.01, 51, 50.1, 50.001

1.8 Adding & Subtracting Decimals; Applications

Perform the operation indicated.

65. $35.02 + 12 + 0.675$

66. $12 - 2.156$

67. Dion pays $2.58 for a hamburger, $1.29 for a large drink and $0.99 for fries. How much change will he receive from a $5 bill?

68. Emma was making $5.96 an hour and receives a raise of $1.09 an hour. What is her new hourly pay rate?

More Problem Solving

MORTGAGE PAYMENTS

Nancy and Dave are purchasing a home. They plan to borrow \$100,000 at 10.875% interest. Their monthly payments for a 20-year and a 30-year mortgage are shown here.

Term	*Monthly Payment*
20 years	\$1,023.69
30 years	\$ 942.89

A mortgage payment consists of principal plus interest. Each time the bank receives your payment, they first deduct the interest you owe on the loan. They then apply the rest toward reducing your principal balance. The couple wants to know how much interest they will save with the 20-year mortgage compared to the 30-year mortgage.

With the 30-year loan, you pay \$942.89 per month for 12 months a year for 30 years, or \$942.89 × 12 × 30 over 30 years. The total interest paid with the 30-year mortgage is this product less the \$100,000 borrowed.

$$\begin{array}{c}\text{total interest for}\\ \text{30-year loan}\end{array} = \$942.89 \times 12 \times 30 - \$100{,}000$$

Next, calculate the interest for the 20-year loan, and compare. This problem will be solved completely in this chapter.

In this chapter you will continue your study of decimal numbers. You will learn the rules to multiply and divide decimals, which will allow you to solve practical problems like the one here. You will also use estimation to help you check the reasonableness your answers.

2.1 PROBLEM SOLVING AND MULTIPLICATION

OBJECTIVES

1 Multiply decimals and estimate answers.

2 Solve applications involving multiplication of decimals

NEW VOCABULARY

product factor

1 MULTIPLICATION OF DECIMALS

When doing multiplications, the numbers being multiplied are called *factors*. The answer to the multiplication is called the *product*. In the multiplication $5 \times 7 = 35$, 5 and 7 are the *factors* and 35 is the *product*.

To multiply decimals, there are two steps:
1. Multiply the two factors, ignoring the decimal points.
2. Count up the number of **decimal** places in the two factors. This is how many decimal places there should be in the product.

EXAMPLE 1 Find the product of 0.3 and 0.02.

First, multiply the two factors, ignoring the decimal points:

$$3 \times 2 = 6.$$

Now locate the decimal point: 0.3 has one decimal place and 0.02 has two decimal places, so the product will have 3 decimal places:

$$0.3 \times 0.02 = 0.006$$

That is, 3 tenths times 2 hundredths is 6 thousandths.
Notice that this agrees with the corresponding fraction multiplication.

$0.3 = \frac{3}{10}$ and $0.02 = \frac{2}{100}$ so that $0.3 \times 0.02 = \frac{3}{10} \cdot \frac{2}{100} = \frac{6}{1000} = 0.006$.

Counting the decimal places corresponds to multiplying denominators.

Answer: $0.3 \times 0.02 = 0.006$

EXAMPLE 2 Multiply 5×0.72.

$5 \times 0.72 = 3.60$ Notice that there are no decimal places in 5. There are 2 decimal places in 0.72. There are 2 decimal places in the answer.

Answer: $5 \times 0.72 = 3.60$ The ending zero can be eliminated so the answer can be written as 3.6.

EXAMPLE 3 Multiply 0.008×1.04.

$0.008 \times 1.04 = 0.00832$

3 places + 2 places = 5 places

Answer: $0.008 \times 1.04 = 0.00832$

♦ You Try It

1. If you multiply 1.56 by 25.367, how many decimal places will there be in the product?

2. Multiply 0.05 by 0.4.

3. Multiply 1.5 by 7.

4. Multiply 4.25×0.06

5. Multiply 1.005×0.003.

How Big Should the Answer Be?

Many people think that multiplication always produces answers **(products)** larger than either of the original numbers **(factors)**. This is only true when both factors are larger than 1. For example, $5 \times 8 = 40$, which is larger than either 5 or 8. Also $2.5 \times 9.3 = 23.25$, which is larger than either 2.5 or 9.3. However, if one or both of the factors is between 0 and 1, then the results are quite different. Try these problems:

1. 4×0.5 ____________

2. 0.02×8 ____________

3. 0.001×50 ____________

In each of these problems, one of the **factors** is larger than 1 and the other is between 0 and 1. What can you say about the size of the **product** compared to the size of the two **factors?**

____

Now try these problems.

4. 0.02×0.6 ____________

5. 0.22×0.4 ____________

6. 0.05×0.3 ____________

In each of these problems, both of the factors are between 0 and 1. What can you say about the size of the product compared to the size of the two factors? ____________

____

We can make some general conclusions about the size of the product based on the size of the factors.

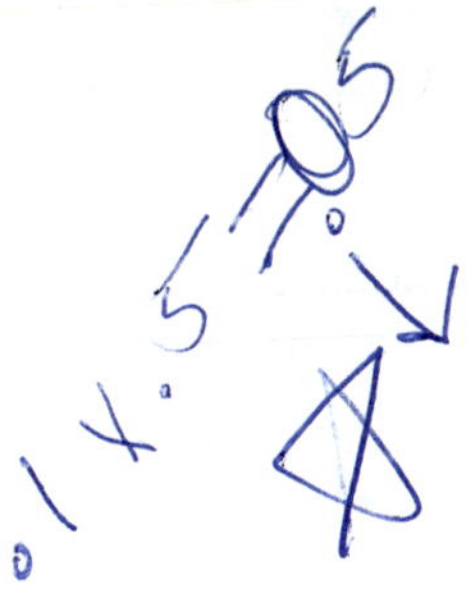

1. If the two numbers being multiplied (factors) are larger than 1, the product is larger than either factor.
2. If one of the factors is larger than 1 and the other is between 0 and 1, the product is between the two original numbers.
3. If both factors are between 0 and 1, the product is smaller than either of the original numbers, but still between 0 and 1.

♦ You Try It

*Decide how big the product should be: larger than either factor, in between the two factors, or smaller than either factor. Do **not** do the multiplication.*

6. 75×96 **7.** 4.2×3.05 **8.** 0.42×3.05

9. 0.42×0.305 **10.** 6.3×0.04 **11.** 0.035×0.112

12. 2.001×1.02

Estimating Products

We can combine the ideas of size of answers with rounding to be able to estimate answers to decimal multiplications. Remember a good estimate is a way to sharpen your mathematical intuition. You should always be sure that you are providing a reasonable answer to a problem.

EXAMPLE 4 Estimate the answer to 5.34×2.79.

First, see how big the product should be. Since both factors are larger than 1, the answer should be larger than either factor, so the product should be larger than 5.

Second, try rounding both numbers to the nearest whole number. 5.34 rounds to 5 and 2.79 rounds to 3, so the product is approximately 5×3 or 15. Notice that the exact answer is 14.8986 so the approximation is a good one.

Answer: 5.34×2.79 is approximately 15.

EXAMPLE 5 Estimate the answer to 4.821×0.602.

Size: Since 4.821 is larger than 1 and 0.602 is between 0 and 1, the product will be between the two factors, so 4.821×0.602 will be between 0.602 and 4.821.

Round: Since 0.602 is between 0 and 1, it would be best to round to the nearest **tenth** to approximate the product. 0.602 rounds to 0.6 and 4.821 rounds to 4.8, so 0.602×4.821 is approximately 0.6×4.8, which is 2.88. Note that the exact answer is 2.902242, so the approximation is a good one.

Answer: 4.821×0.602 must be between 0.602 and 4.821 and is approximately 2.88.

EXAMPLE 6 Estimate the answer to 0.234×5.112.

Size: Since 0.234 is between 0 and 1 and 5.112 is larger than 1, the product will be between 0.234 and 5.112.

Round: Again, since 0.234 is between 0 and 1, approximate by rounding to the nearest tenth. 0.234 rounds to 0.2 and 5.112 rounds to 5.1, so 0.234×5.112 is approximately 0.2×5.1 or 1.02.

Answer: 0.234×5.112 must be between 0 and 5 and is approximately 1.02.

EXAMPLE 7 Estimate the answer to 0.435×0.229

Size: Both numbers are between 0 and 1, so the product will be smaller than either factor, but still between 0 and 1.

Round: If we round both numbers to the nearest tenth, we get 0.4×0.2 or 0.08. The exact answer is .099615.

Answer: 0.435×0.229 is smaller than either number, must be between 0 and 1, and is approximately 0.08.

♦You Try It *Use the ideas of size and rounding to estimate the answers to each multiplication. If both numbers are larger than 1, round to the nearest whole number. If either number is between 0 and 1, round to the nearest tenth.*

5.8 × .7

13. 6.89×9.21	**14.** 3.025×6.4	**15.** 5.77×0.722
16. 0.35×2.001	**17.** 0.442×0.976	**18.** 0.403×0.601

2 APPLICATIONS

Many problems require the use of multiplication in the solution. We will investigate several examples and we will use the problem-solving strategy developed in earlier sections as well as the estimation techniques we have just discussed.

EXAMPLE 8 Maria is buying her son Miguel new clothes for school. She chooses 3 pairs of pants at \$22.99 each and 3 shirts, 2 of which cost \$15.98 each and one which costs \$20.55. What is the total cost of her selections? If she has \$150.00, does she have enough to cover these items? If so, what is her change?

A general rule when buying multiple amounts of anything is that

Total cost = Number of items × cost of one item

READ, WRITE, ANALYZE

Outline the problem; that is, give a list of steps needed to solve the problem.
1. Find the cost of the pants by multiplying 3 times cost of one pair of pants.
2. Find the cost of the shirts - be careful because there are two different prices.
3. Find the total.
4. Find the change.

Estimate the total cost: the pants are *about* \$23 each, so $3 \times 23 = \$69$, 2 shirts at *about* \$16 is \$32, and a third shirt at *about* \$21, so \$69 + \$32 + \$21 is *about* \$70 + \$30 +20 or approximately \$120, so it seems that Maria will have enough money.

Solution:

3 pairs of pants at $22.99 each	=	$68.97
2 shirts at $15.98 each	=	$31.96
1 shirt at $20.55	=	$20.55
Total	=	$ 121.48

Since $150 > $121.48, Maria does have enough to cover the cost of these items. Her change will be
$150.00 – $121.48 = $28.52

Answer: The total cost of Maria's purchases is $121.48. Her change will be $28.52.

CALCULATOR To do these computations on your calculator, use the following keystrokes:
3 [×] 22.99 [+] 2 [×] 15.98 [+] 20.55 [=] to get the total, which should be 121.48
To find the change, use these keystrokes:
150 [−] 121.48 [=] and the display should read 28.52.

EXAMPLE 9 Gerry has bought a new car but to do so he took out a loan of $11,500. His loan payment each month is $265.94, part of which goes toward paying off the loan and the rest is interest. The loan period is 4 years (or 48 months). What is the amount Gerry is paying in interest?

Outline

1. We need to know the total amount of the loan payments. This can be found by multiplying the monthly payment, $265.94 by the number of months of the loan period, 48.
2. Since the amount of the loan was $11,500, the interest must be the difference between this number and the total from step 1.

Estimate

1. Round the monthly payment to the nearest hundred dollars. ________________
2. Round the number of months to the nearest ten. ________________
3. Use the rounded answers to estimate the total amount of the loan payments.

   ____
4. Use your rounded answer from #3 to estimate the amount of interest.

   ____

Because the monthly payment and the number of months both rounded **up**, we got a rather large **overestimate** of the amount owed. Can you think of any way to avoid this problem?

____

____

Solution

1. Amount paid back = $265.94 per month × 48 months = $12,765.12.
2. Interest = $12,765.12 – $11,500 = $1265.12

Answer: Gerry is paying $1265.12 in interest.

EXAMPLE 10 Anna's hourly rate of pay is \$12.50 per hour. If she works overtime, she gets time and a half. What is her overtime rate of pay?

Analyze: "Time and a half " means the overtime rate of pay is $1\frac{1}{2}$ or 1.5 times the regular rate of pay. Since Anna's regular rate of pay is \$12.50 per hour, her overtime rate will be 1.5 × \$12.50 per hour = \$18.75 per hour

Answer: Anna's overtime rate of pay is \$18.75 per hour.

EXAMPLE 11 Suppose Anna earns \$12.50 per hour for her regular 40 hour work week and she gets time and a half for working overtime. How much would she earn for a week in which she worked 48 hours?

Outline

1. We need to know how much Anna will earn for her regular work week.
2. We need to know what Anna's overtime rate of pay is.
3. We need to know how many overtime hours Anna worked.
4. We need to know how much Anna will earn for her overtime hours.
5. We can then find Anna's total earnings.

Estimate

1. Regular pay is approximately \$13 × 40 = \$520
2. Overtime pay is approximately \$20 × 8 = \$160
3. Total is approximately \$500 + \$200 or \$700.

Solution

1. Since the rate of pay for the regular 40 hour week is \$12.50 per hour, Anna's pay for the regular work week is:
   \$12.50 per hour × 40 hours = \$500.00
2. We calculated Anna's overtime rate in the previous example: \$18.75 per hour.
3. Anna worked 48 hours, 40 of which are her regular work week, so she worked 8 hours overtime.
4. Anna worked 8 hours overtime at \$18.75 per hour, so her overtime earnings are:
   \$18.75 per hour × 8 hours = \$150.00
5. Anna's total earnings for the week is the sum of her regular pay, \$500, and her overtime pay, \$150:
   \$500.00 + \$150.00 = \$650.00

Answer: Anna earned \$500 for her regular work week plus \$150 in overtime pay, for a total of \$650.00 for the week.

EXAMPLE 12 Find the area of a rectangle which is 3.75 meters long and 2.05 meters wide. Round your answer to the nearest hundredth.

Outline

1. Draw a picture of the rectangle. A picture of the rectangle would look something like this:

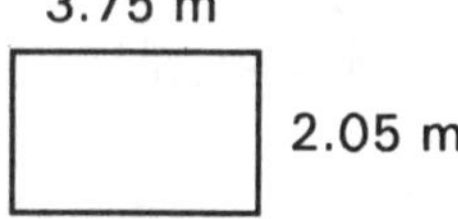

2. Use the formula for the area of a rectangle.
Remember that the formula for the area of a rectangle is $A = lw$. In this rectangle, the length, l, is 3.75 meters, and the width, w, is 2.05 meters. Substituting into the formula, we get:

$$A = lw$$
$$A = 3.75 \text{ m} \times 2.05 \text{ m}$$

Estimate: Since both factors are larger than 1, the product will be larger than either factor. Rounding gives approximately $4 \text{ m} \times 2 \text{ m}$ or 8 square meters.

Solution

$A = lw$
$A = (3.75\text{m})(2.05\text{m})$
$A = 7.6875$ sq. m.
rounded: $A \doteq 7.69$ sq. m.

Here is the decimal multiplication:

3.75	2 decimal places
× 2.05	2 decimal places
1875	
7500	
7.6875	4 decimal places in the answer

Answer: The area of the rectangle is approximately 7.69 square meters.

EXAMPLE 13 A car rental company charges \$14.95 per day plus \$.22 per mile. If you rent a car for 3 days and 550 miles, how much will it cost?

Outline
1. Find the cost of 3 days at \$14.95 per day.
2. Find the cost of 550 miles at \$.22 per mile.
3. Find the total cost.

Estimate
1. 3 days at approximately \$15 per day = \$45
2. 600 miles at \$.20 per mile = \$120.00
3. Total cost is approximately \$165

Solution

1. 3 days at \$14.95 per day = 3 days × \$14.95 per day
= \$44.85

Work:
14.95
× 3
44.85

2. 550 miles at \$.22 per mile = 550 miles × \$.22 per mile
= \$121.00

Work:
550
× .22
1100
1100
121.00

3. Total cost = cost of days plus cost of miles
= \$44.85 + \$121.00
= \$165.85

Work:
44.85
121.00
165.85

Answer: The total cost of renting the car is \$165.85.

◆**You Try It** *For each problem, provide an outline and estimate the answer before doing calculations. Write your answer in a complete sentence.*

19. A volleyball coach is ordering supplies for a new season. She orders one new net for \$45.27, four new volleyballs at \$18.75 each, and 2 dozen towels at \$26.00 per dozen. Find the total cost of her order.

20. For a long distance call, the telephone company charges \$0.35 for the first minute and \$0.27 for each additional minute. How much would a 14 minute long distance call cost?

21. For her new suit, Kathy needs 4.7 yards of material for the jacket and 2.3 yards for the skirt. If the fabric sells for \$8.99 per yard, how much will the outfit cost?

EXAMPLE 14 **Chapter Problem** Read the problem at the beginning of this chapter.

30-year loan:

\$942.89 × 12 × 30 =	\$339,440.40	total paid over 30 years
	− 100,000.00	borrowed
	239,440.40	interest paid to borrow \$100,000 for 30 years

20-year loan:

\$1,023.69 × 12 × 20 =	\$245,685.60	total paid over 20 years
	− 100,000.00	borrowed
	145,685.60	interest paid to borrow \$100,000 for 20 years

Savings with a 20-year loan:

You pay:		You make:	You save:	
\$1,023.69	(20-yr)	10 years	\$239,440.40	(30-yr)
−942.89	(30-yr)	fewer	− 145,685.60	(20-yr)
\$80.80	more per month	payments	\$93,754.80	in interest ■

OBSERVE The \$80.80 more per month goes directly towards reducing your principal balance. None of it is used to pay interest on your mortgage loan.

▶ **You Try It** **22.** A couple plan to borrow \$60,000 at 12.5% interest. Their monthly payments for a 20-year and a 30-year mortgage are \$681.68 and \$640.35, respectively. How much interest is saved with the 20-year mortgage when compared to the 30-year mortgage?

♦**Answers to You Try It** **1.** five **2.** 0.020 **3.** 10.5 **4.** 0.2550 **5.** 0.003015 **6.** larger than either **7.** larger than either **8.** between the two **9.** smaller than either **10.** between the two **11.** smaller than either **12.** larger than either **13.** $6.89 \times 9.21 \doteq 7 \times 9 = 63$ Answer is larger than either factor. **14.** $3.025 \times 6.4 \doteq 3 \times 6 = 18$. Answer is larger than either factor. **15.** $5.77 \times 0.722 \doteq 5.8 \times 0.7 = 4.06$ Answer is between the two factors. **16.** $0.35 \times 2.001 \doteq 0.4 \times 2.0 = 0.8$ Answer is between the two factors. **17.** $0.442 \times 0.976 \doteq 0.4 \times 1.0 = 0.4$ Answer is smaller than either factor. **18.** $0.403 \times 0.601 \doteq 0.4 \times 0.6 = 0.24$ Answer is smaller than either factor.
19. The total cost of her order was $172.27. **20.** A 14-minute long distance call would cost $3.86. **21.** The outfit will cost $62.93. **22.** $66,922.80 is saved with the 20-year mortgage.

SECTION 2.1 EXERCISES

1 *Identify the factors and the product in each problem.*

1. $7.2(3.1) = 22.32$

2. $8.12 \times 0.5 = 4.06$

3. $11.1 \times 3 = 33.3$

4. $0.23(0.4) = 0.092$

5. $0.001 \times 0.02 = 0.00002$

6. $0.007 \times 4.2 = 0.0294$

7. $8.5(450) = 3{,}825$

8. $23.7(105.62) = 2{,}503.194$

9. $99.2 \times 81.3 = 8{,}064.96$

*Decide how big the product must be: larger than either factor, in between the two factors, or smaller than either factor. Do **not** do the multiplication.*

10. 82×44

11. 0.61×2.36

12. 0.61×0.236

13. 6.1×0.236

14. $5(7.3)$

15. $50(0.73)$

16. 0.5×7.3

17. 0.5×0.73

Use the ideas of size and rounding to estimate the answer to each multiplication.

- *If both numbers are larger than 1, round to the nearest whole number.*
- *If either number is between 0 and 1, round to the nearest tenth.*

18. 8.82×2.49

19. 3.358×8.601

20. 5.512×0.64

21. $0.580(0.911)$

22. 19×3.75

23. $320(0.125)$

24. 4.32×11.13

25. $0.752(0.073)$

Multiply. Do not use a calculator.

26. 0.4×0.7

27. 0.9×0.2

28. 4.5×0.5

29. 2.7×0.6

30. 2.83×6.5

31. 7.9×1.42

32. 8.05×180.3

33. 45.82×36.14

34. 0.04×0.6

35. 0.8×0.009

36. 0.007×0.0002

37. 0.01×0.001

38. 58×0.5

39. 31×0.62

40. 0.056×210

41. 0.0054×720

42. $2{,}000 \times 0.009$

43. 800×0.0125

44. 312.5×0.01

45. 39.3×0.0001

46. 0.08×125

47. 0.004×250

48. 10×2.34

49. 100×4.036

---

2 *Solve each application problem.*

50. Gas costs \$1.179 per gallon. What is the cost to fill a 14.6 gallon tank?

51. Diesel fuel costs \$1.349 per gallon. What is the cost to a trucker who purchases 120 gallons?

52. You purchase 11 dress shirts selling for \$21.69 each.

a. Estimate the total cost.

b. What is the exact cost?

53. Jayne worked 28.5 hours and made $9.48 per hour.
a. Estimate her total wage.
b. What are her exact earnings?

54. A doctor needs 200 injections of an antibiotic. One injection uses 3.8 milligrams of antibiotic. The cost of 1 milligram is $0.06. Find the total cost.

55. Alice makes $8.09 per hour. She works 40 hours per week. How much will she make during the 13 weeks of summer?

56. A farm worker gets $3.15 an hour plus $0.92 for every bushel of peaches he picks in a day. If he works 10 hours and picks 52 bushels, how much does he make for the day?

57. A seamstress makes $5.42 an hour plus $0.14 for each piece of clothing she stitches. How much will she make if she works 7 hours and stitches 136 pieces of clothing?

58. Kate purchased 0.78 pound of ham at $3.29 per pound. What did she pay?

59. Ed bought 1.26 pounds of cheese at $3.79 per pound. What did he pay?

60. The charge for a long-distance phone call is $0.67 for the first minute and $0.48 for each additional minute. What is the cost of a 17-minute call?

61. Al called his sister long-distance. He paid $0.45 for the first minute and $0.38 for each additional minute. What was the charge for his half hour phone call?

62. Arnold makes $10.68 an hour plus time and a half for each hour worked over 40 hours each week. What does he earn for working
a. a 34.5-hour week?
b. a 40-hour week?
c. a 51-hour week?

63. Suzanne makes $14.72 an hour plus time and a half for each hour worked over 37.5 hours each week. What does she earn for working
a. a 16-hour week?
b. a 37.5-hour week?
c. a 52.5-hour week?

64. Dale paid for a TV on the credit plan. He paid $42.16 per month for 16 months. The selling price for the TV was $499.95.
a. What total did he pay for the TV?
b. What was the total interest charge?

65. Lucy paid $237.45 per month for 36 months to pay off a $6,200 car loan.
a. What total did she pay on the loan?
b. What was the total interest charge?

66. Gasoline costs $1.269 per gallon using a credit card. It costs $1.209 per gallon if you pay cash. How much money do you save if you pay cash for 16.3 gallons of gas?

67. A trucker purchases 150 gallons of gasoline. The credit card price is $1.189 per gallon. The cash price is $1.147 per gallon. How much is saved by paying cash?

68. Renting a truck costs $23 a day plus $0.12 per mile. What do you owe if you use the truck for 3 days and drive 1,673 miles?

69. You rent a car for 7 days and drive 512 miles. What do you owe if you pay $16 per day, $0.15 per mile, and $1.50 per day for insurance?

70. Estimate the cost of 62 radios at $39.95 each.

71. Estimate the cost of 207 ounces of gold at $317.62 per ounce.

72. Find the area of a rectangle that is 1.2 meters long and 3.4 meters wide.

73. A kitchen floor measures 12 ft by 9 ft. What is the cost to tile this floor at $0.39 per tile? Each tile is 1 sq ft.

74. A basement measures 24 ft by 36 ft. What is the cost to tile the floor if the tiles cost $0.62 each? Each tile is 1 sq ft.

75. A living room is 8 yd by 6 yd. What will be the cost to carpet this room at $23.99 per square yard?

76. A dining room is 4 yd by 3 yd. What will be the cost to carpet this room at $21.50 per sq yd?

SKILLSFOCUS (Section 1.2) *Using the given values for the variables, decide if the equation is true or false.*

77. $x - 4 = 16$ if $x = 12$ **78.** $3.5 + y = 8.9 - y$ if $y = 2.7$ **79.** $8t = 32.24$ if $t = 4.03$

EXTEND YOUR THINKING ♦ ♦ ♦

♦ SOMETHING MORE

80. Susan makes $14.58 per hour if she works 40 hours or less per week. She makes time and a half for each hour worked over 40, up to 20 hours. She makes double time for each hour worked over 60. If Susan worked 64.75 hours in one week, what will she make?

Use Figure 2.1 to answer questions 81 - 82.

FIGURE 2.1

81. What is the difference in the area of a typical 1981 vegetable garden and a typical 1991 vegetable garden?

82. Which of the following gardens has an area closest to 547 sq. ft.?

a.

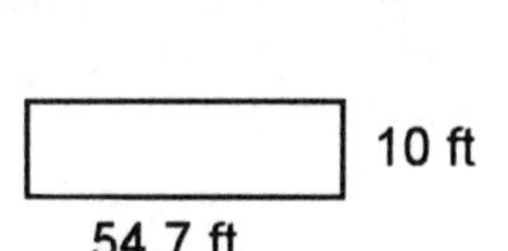

b.

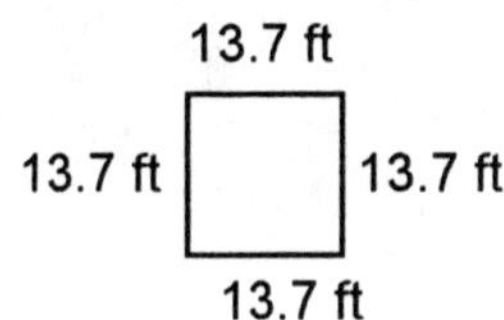

c.

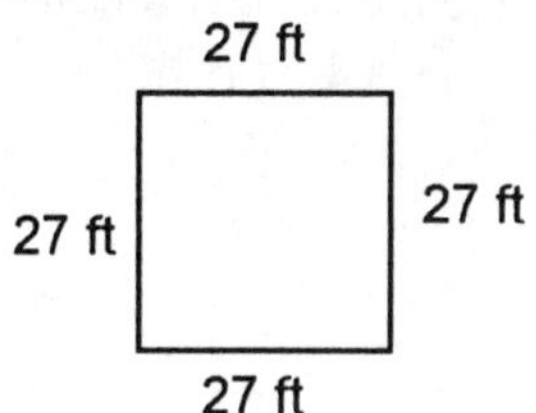

83. If the vegetable garden is to have an area of 241 sq. ft., what could its length and width be?

WRITING TO LEARN ♦♦♦

84. Explain how to multiply two decimal numbers. Illustrate using your own example.

2.2 EXPONENTS

OBJECTIVE

Evaluate exponents.

NEW VOCABULARY

term, base, exponent, power, square, cube

TERMS AND FACTORS

When doing additions or subtractions, the numbers being added or subtracted are called *terms*. For example, in the expression $5+7$, the terms are 5 and 7. In the expression $x+y+z$ the terms are x, y, and z.

Recall from Section 2.1 that when doing multiplications, the numbers being multiplied are called *factors*.

For example, in the expression $8 \cdot 9$, the factors are 8 and 9. In the expression $6x$, the factors are 6 and x.

EXAMPLE 1 In the expression $2.7-3.4+x$, there are **three terms**: 2.7, 3.4, and x.

EXAMPLE 2 In the expression (5.6)(600)(95.2), there are **three factors**: 5.6, 600, and 95.2.

EXAMPLE 3 In the expression $\frac{1}{2}bh$, there are **three factors**: $\frac{1}{2}$, b, and h.

♦You Try It

1. Explain the difference between a term and a factor.
2. In the expression $9.1+3.6+11$, how many terms are there and what are they?
3. In the expression $\frac{5}{8}xyz$, how many factors are there and what are they?

EXPONENTS

Sometimes one number is used as a factor several times in a multiplication. For example, suppose you were devising a new lottery in which each ticket consisted of a 6-digit number and each digit could be chosen from the numbers 1 through 9. Since there are 9 choices for the first digit and 9 for the second digit, etc., the total number of possible choices of numbers for the lottery would be $9 \cdot 9 \cdot 9 \cdot 9 \cdot 9 \cdot 9$ or 531,441. Notice that $9 \cdot 9 \cdot 9 \cdot 9 \cdot 9 \cdot 9$ is a product in which 9 is used as a factor six times. In mathematics we use a shorthand notation for this type of product. It is called exponential notation, and we write $9 \cdot 9 \cdot 9 \cdot 9 \cdot 9 \cdot 9$ as 9^6. The 9 is called the base and the 6 is called the exponent or power. 9^6 is read "nine to the sixth power."

> The expression a^b is read "a to the bth power" and it means a product in which the base, a, is used as a factor b times.

EXAMPLE 4 What is the value of 4^3 ?

4^3, which is read "4 to the third power", means $4 \cdot 4 \cdot 4$ or 64.
Thus, $4^3 = 64$.

Some exponents are more common than others. The exponents 2 and 3 are the most common exponents and so they have special names. When 2 is used as an exponent, the base is said to be **squared.** That is, 3^2 can be read "3 squared" or it can be read "3 to the second power." When 3 is used as an exponent, the base is said to be **cubed**. That is, 5^3 can be read "5 cubed" or it can be read "5 to the third power".

EXAMPLE 5 What is the value of 8 squared?

"Squared" means an exponent of 2, so $8^2 = 8 \cdot 8 = 64$

EXAMPLE 6 List the squares of the first twelve whole numbers.

Complete the following chart.

Number	Square
0	0
1	1
2	4
3	
4	
5	
6	
7	49
8	
9	
10	
11	

EXAMPLE 7 What is the value of 5 cubed?

$5^3 = 5 \cdot 5 \cdot 5 = 125$

EXAMPLE 8 List the first eight powers of 2.

That means find the value of $2^1, 2^2, 2^3, 2^4, 2^5, 2^6, 2^7$, and 2^8

$2^1 = 2$, $2^2 = 4$, $2^3 = 8$, $2^4 = 16$, $2^5 = 32$, $2^6 = 64$, $2^7 = 128$, and $2^8 = 256$

OBSERVE Each power of 2 is 2 times the previous power of 2. For example, $2^7 = 2 \cdot 2^6$.

EXAMPLE 9 List the first six powers of 10.

$10^1 = 10$, $10^2 = 100$, $10^3 = 1{,}000$, $10^4 = 10{,}000$, $10^5 = 100{,}000$, $10^6 = 1{,}000{,}000$

Powers of Ten

Notice something special about the powers of 10: the exponent and the number of zeros are the same. This is always true, and that means that you never really have to calculate the value of a power of ten, since the exponent will tell you how many zeros to write after the 1. For example, 10^{12} will be a 1 followed by 12 zeros: $10^{12} = 1{,}000{,}000{,}000{,}000$ or one trillion.

Also notice that each power of 10 is 10 times bigger than the previous power of 10. For example, $10^5 = 10 \cdot 10^4$.

EXAMPLE 10 Evaluate $(0.3)^2$.

$(0.3)^2 = 0.3 \times 0.3 = 0.09$

EXAMPLE 11 Evaluate $(0.07)^2$

$(0.07)^2 = 0.07 \times 0.07 = 0.0049$

EXAMPLE 12 Evaluate $\left(\frac{2}{3}\right)^2$.

$$\left(\frac{2}{3}\right)^2 = \frac{2}{3} \times \frac{2}{3} = \frac{4}{9}$$

♦ You Try It

4. In your own words, explain what an exponent is.

5. Find the value of 3^5 .

6. Evaluate 7^4.

7. List the squares of the numbers from 12 to 20.

8. List the cubes of the first ten whole numbers.

9. List the first six powers of 3.

10. Evaluate 0^8.

11. What do you think will be the value of 0^{67} ? Explain your answer.

12. Find 1^7.

13. What do you think will be the value of 1^{20} ? Explain your answer.

14. Are these two expressions the same? Why or why not?

a. 3^4 **b.** 4^3

15. What power of 10 is one million?

16. What power of 10 is one thousand?

17. Evaluate $(0.2)^2$.

18. Evaluate $(0.09)^2$.

19. Evaluate $\left(\frac{4}{5}\right)^2$.

Exponents on Your Calculator

There are two keys on your calculator that make it very easy to evaluate exponential expressions. The $\boxed{x^2}$ key is used for finding the square of a number. To use it, type in the number you wish to square and then press the $\boxed{x^2}$ key. The square of the number will be displayed. This key allows you to square a number with the fewest keystrokes. The $\boxed{y^x}$ key is used for finding other powers. This is the exponentiation key and y is the base and x is the exponent. To use it, type in the base of the expression you wish to evaluate, then press the $\boxed{y^x}$ key, then type in the exponent, and finally press $\boxed{=}$.

EXAMPLE 13 Use your calculator to evaluate 15^2.

Press 15 and then press the $\boxed{x^2}$ key.

Keystrokes: 15 $\boxed{x^2}$

The display should read 225.

Answer: $15^2 = 225$

EXAMPLE 14 Use your calculator to evaluate 8^7.

Press 8 first, then press the $\boxed{y^x}$ key, and then enter the exponent, 7, and finally press $\boxed{=}$.

Keystrokes: 8 $\boxed{y^x}$ 7 $\boxed{=}$

The display should read 2,097,152.

Answer: $8^7 = 2{,}097{,}152$.

EXAMPLE 15 Use your calculator to evaluate $(1.29)^3$.

Keystokes: 1.29 $\boxed{y^x}$ 3 $\boxed{=}$.

The display should read 2.146689.

Answer: $(1.29)^3 = 2.146689$.

◆ **You Try It**

20. Use your calculator to evaluate 75^2.

21. Use your calculator to find the square of 10.4.

22. Use your calculator to evaluate 8^9.

23. Use your calculator to evaluate 9^8.

24. Is $8^9 = 9^8$?

25. Use your calculator to find the cube of 8.38.

26. Use your calculator to evaluate $(4.7)^2$

◆ **Answers to You Try It** **1.** Terms are added; factors are multiplied. **2.** 3 terms: 9.1, 3.6, 11 **3.** 4 factors: $\frac{5}{8}$, x, y, z **5.** 243 **6.** 2401 **7.** $12^2 = 144$, $13^2 = 169$, $14^2 = 196$, $15^2 = 225$, $16^2 = 256$, $17^2 = 289$, $18^2 = 324$, $19^2 = 361$, $20^2 = 400$ **8.** $0^3 = 0$, $1^3 = 1$, $2^3 = 8$, $3^3 = 27$, $4^3 = 64$, $5^3 = 125$, $6^3 = 216$, $7^3 = 343$, $8^3 = 512$, $9^3 = 729$ **9.** $3^1 = 3, 3^2 = 9, 3^3 = 27, 3^4 = 81, 3^5 = 243, 3^6 = 729$ **10.** 0 **11.** 0; When 0 is a factor, the product will always be 0. **12.** 1 **13.** 1; When you multiply one times one, the product is always one. **14.** $3^4 = 81$ and $4^3 = 64$. In the first case 3 is the base. In the second case 3 is the exponent. **15.** 10^6 **16.** 10^3 **17.** 0.04 **18.** 0.0081 **19.** $\frac{16}{25}$ **20.** 5,625 **21.** 108.16 **22.** 134,217,728 **23.** 43,046,721 **24.** No **25.** 588.480472 **26.** 22.09

SECTION 2.2 EXERCISES

1. Identify the factors in the product $8 \cdot 7$.

2. Identify the factors in the product $9x$.

3. Identify the factors in the product $5xy$

4. Identify the terms in the expression $5 + 9$.

5. Identify the terms in the expression $7 + x - t$.

6. Identify the terms in the expression $2a + 3b - 11c$.

Identify the base and the exponent. Do not evaluate.

7. 4^3 **8.** 3^4 **9.** 11^2 **10.** 2^{11} **11.** $(0.5)^3$

12. 5^3 **13.** $(5.5)^3$ **14.** 8^1 **15.** 1^{17} **16.** 8^8

Write the expression in exponential form.

17. $6 \times 6 \times 6 \times 6$ **18.** $2 \times 2 \times 2 \times 2 \times 2 \times 2$ **19.** $7 \times 7 \times 7$

Write the English phrase using symbols.

20. seven to the fourth power

21. ten to the sixth power

22. eight cubed

23. twelve squared

Evalaute each expression without a calculator.

24. 2^3 **25.** 3^2 **26.** 2^4 **27.** 5^3 **28.** 15^2

29. 7^3 **30.** 13^2 **31.** 20^2 **32.** 2^2 **33.** $(0.2)^2$

34. $(0.02)^2$ **35.** $(0.002)^2$ **36.** 3^4 **37.** 4^3 **38.** $(0.5)^3$

39. 3^5 **40.** $(0.3)^2$ **41.** $(0.3)^3$ **42.** $(0.3)^4$ **43.** $\left(\frac{3}{10}\right)^2$

44. $\left(\frac{3}{10}\right)^3$ **45.** $\left(\frac{1}{2}\right)^3$ **46.** $\left(\frac{1}{3}\right)^2$ **47.** $\left(\frac{2}{5}\right)^2$ **48.** $\left(\frac{2}{3}\right)^4$

Use your calculator to evaluate each expression.

49. 43^2 **50.** $(7.7)^2$ **51.** 16^5 **52.** $(0.24)^3$ **53.** 41^4

54. 6^9 **55.** 3^{15} **56.** $(1.76)^3$ **57.** $(45.7)^3$ **58.** $3{,}149^2$

SKILLSFOCUS (Section 1.2)

59. Between what two consecutive whole numbers is 3.01?

60. Write three numbers between 1.2 and 1.3.

EXTEND YOUR THINKING ♦♦♦

♦TROUBLESHOOT IT

Find and correct the error.

61. $2^3 = 2 \times 3 = 6$

62. $4^3 = 4 \times 4 \times 3 = 48$

WRITING TO LEARN ♦♦♦

63. What is a *factor*?

64. What is a *term*?

65. Give an example of an expression with four terms.

66. Consider the two expressions:

a. abc **b.** $a + b + c$

How are they alike? How are they different? Use the words *term* and *factor* in your explanation.

67. Consider the two expressions:

a. $3 \cdot 4$ **b.** 3^4

How are they alike? How are they different? Include the words *product*, *exponent*, and *factor* in your explanation.

68. Consider the two expressions:

a. 11^5 **b.** 10^5 .

How are they alike? How are they different? Use the words *base* and *exponent* in your explanation.

69. Explain the difference between the $\boxed{x^2}$ key and the $\boxed{y^x}$ key on your calculator. Could you use either key to evaluate 32^2 ? Explain the keystrokes.

2.3 LARGE NUMBERS and POWERS OF TEN

OBJECTIVES

1. Write large numbers in standard form.
2. Multiply by powers of 10.
3. Change dollars to cents.

NEW VOCABULARY

million billion
exponential form
standard form

1 LARGE NUMBERS

How big is a thousand? How big is a million? Most of us have not really had hands-on experience with very large numbers, so we are sometimes very casual about numbers with lots of zeros on the end. To get an idea how big a million is, think about how long it would take for one million seconds to pass by. Since there are 60 seconds in each minute, 1 million ÷ 60 is approximately 16,667 minutes. Since there are 60 minutes in one hour, 16,667 min ÷ 60 min/hr is approximately 278 hours. Since there are 24 hours in a day, 278 hr ÷ 24 hr/day is approximately 12 days.

Even though this seems like a very long time, if you did the same analysis for one billion, you might be surprised at the result. Remember that 1 billion = 1,000,000,000, which is the same as one thousand million. This time, 1 billion seconds ÷60 seconds per minute ÷ 60 minutes per hour ÷ 24 hours per day ÷ 365 days per year is approximately 32 years!

Exponents can be used as a shorthand way of representing very large numbers. The powers of 10 are helpful when working with numbers in the millions or billions. Remember from the last section that a special pattern develops when different powers are applied to a base of 10.

$$10^2 = 10 \times 10 = 100 = \text{one hundred}$$

We have placed equal signs between these expressions since they are all different ways of saying the same thing.

> 10^2 is called EXPONENTIAL FORM.
>
> 100 is called STANDARD FORM.

Complete the following chart by filling in the missing items on each line.

EXPONENTIAL	STANDARD	WORDS
10^3	____________	____________
____________	1,000,000	____________
____________	____________	one billion

These amounts, a thousand, a million, and even a billion, are of particular importance since they are frequently mentioned in news reports. Do you notice a pattern with the exponents in the charts? Can you offer an explanation for the pattern?

____

You can also see that: 1 billion is 1 thousand million.

1 million is 1 thousand thousand.

EXAMPLE 1 What is the difference between one billion and one million?

one billion = 1,000,000,000 or 10^9
one million = 1,000,000 or 10^6

The difference is 1,000,000,000 – 1,000,000 = 999,000,000.

Answer: The difference between one billion and one million is 999 million.

EXAMPLE 2 If you had a bank account of 1 million dollars and you withdrew 1 thousand dollars every month, how many months would your savings last?

one million = 1,000,000
one thousand = 1,000

1,000,000 ÷ 1,000 = 1,000 months. We mentioned above that a million is a thousand thousand.

1,000 months ÷ 12 ≐ 83.3 years

Answer: The money would last approximately 83.3 years.

♦ You Try It

1. Which is larger, 10^5 or 10^4 ? How do you know?
2. How much larger is one million than one thousand?
3. If you owed $10,000 on your car loan and paid off $100, how much would be left to pay?
4. How long would it take (in minutes) for one thousand seconds to pass by?
5. If you had one billion dollars and spent one million dollars a year, how long would it take to spend all the money?

---

2 MULTIPLICATION OF DECIMALS BY POWERS OF TEN

Here are the first six powers of 10:

EXPONENTIAL	STANDARD	WORDS
10^1	10	ten
10^2	100	one hundred
10^3	1,000	one thousand
10^4	10,000	ten thousand
10^5	100,000	one hundred thousand
10^6	1,000,000	one million

We see from this list that the number of zeros in any power of 10 is equal to the power, which makes it easy to write any power of 10 in standard form. For example, 10^8 would be a 1 followed by 8 zeros, or 100,000,000 or one hundred million. Our goal in this section is to see what happens when a decimal number is multiplied by a power of 10 and then generalize our discoveries to create an easy-to-use rule. Try these problems.

1. Use your calculator to do the following multiplications.

a. 3.429×10 ______________________

b. 5.06×10 ______________________

c. 20.339×10 ______________________

d. 120.656×10 ______________________

2. Based on your answers to problem 1, complete the following sentence:

When a number is multiplied by 10, the decimal point moves ____________

____

3. Use your calculator to do the following multiplications.

a. 3.429×100 ______________________

b. 5.06×100 ______________________

c. 20.339×100 ______________________

d. 120.656×100 ______________________

4. Based on your answers to problem 3, complete the following sentence:

When a number is multiplied by 100, the decimal point moves ____________

____

5. Based on your answers to problems 2 and 4, what happens when a number is multiplied by 1,000?

____

____

6. What happens when a number is multiplied by 10,000?

____

____

We have, then, a general (and very easy) rule for multiplying a number by a power of 10:

> To multiply a number by a power of 10, move the decimal point to the right as many places as there are zeros in the power of 10.

EXAMPLE 3 Suppose you own a store and you decide to purchase 100 calculators at \$5.00 each. A clerk at the company selling you the calculators makes an error and puts an extra zero on the number of calculators in your order, so you will be receiving 1,000 calculators. How much should your proper bill be? How much is your incorrect bill? What is the amount of overcharge?

Your correct bill is 100 calculators x \$5 per calculator = \$500.
The incorrect bill is for 1,000 calculators x \$5 per calculator = \$5,000.
The amount of overcharge is \$5,000 – \$500 = \$4,500. Notice that the overcharge, \$4,500, is **nine times** the size of your correct bill, \$500 ! One extra zero at the end of a number can really change the size of a number.

♦You Try It *Use this easy method to do each of the following problems*:

6. 14.015×100

7. $2.119 \times 1{,}000$

8. 0.0125×10^5

9. 6.5×10^4

10. A builder bought 100 lighting fixtures from an electrical supplier at \$22.56 each. What was the total cost of the order?

11. A discount electronics store received a shipment of 1,000 calculators. They paid \$5.79 each for the calculators and they will sell them for \$11.50 each. If all the calculators are sold, what will be the total profit? (Note: profit is the difference between your total sales and your cost.)

Remember that one thousand (1,000) is 10^3, one million (1,000,000) is 10^6, and one billion (1,000,000,000) is 10^9. Often you will see newspaper articles that refer to a number as 2.3 million. What number is that? Well,

2.3 million is the same as 2.3×1 million

That is, 2.3 million $= 2.3 \times 1$ million
$= 2.3 \times 10^6$
$= 2{,}300{,}000$ Move the decimal point 6 places to the right

Answer: 2.3 million is two million, three hundred thousand.

EXAMPLE 4 An author recently signed a book deal for 1.8 million dollars. Write this number in standard notation.

\$1.8 million $= \$1.8 \times 1$ million
$= \$1.8 \times 10^6$
$= \$1{,}800{,}000$ Move the decimal point 6 places to the right.

Answer: \$1.8 million is one million, eight hundred thousand dollars.

♦ **You Try It**

12. The population of the United States is approximately 255.6 million people. Write this number in standard notation.

13. In 1992, the United States spent $840 billion on health care. Write this number in standard notation.

3 DOLLARS AND CENTS

Changing from dollars to cents is an exercise in multiplying by a power of 10. $1 is 100 cents. We read $.25 and 25¢ the same way, 25 cents, but the monetary unit is different in each. When we write 25¢, we mean 25 cents or 25 pennies. When we write $.25, we really mean 25 **hundredths** of a dollar, since there are 100 cents in a dollar. Sometimes dollar notation is more useful and sometimes cents notation is better, so we need to be able to convert from one to the other. For now, we will focus on changing from dollars to cents.

EXAMPLE 5 Write $1.25 in cents.

$1.25 can be thought of as 1.25 dollars or 1.25 × 1 dollar, so

$$\$1.25 = 1.25 \times 1 \text{ dollar}$$

and since 1 dollar = 100 cents

$$\begin{aligned} \$1.25 &= 1.25 \times 100 \text{ cents} \\ &= 125¢ \quad \text{Move the decimal point two places to the right} \end{aligned}$$

Answer: $1.25 = 125¢

EXAMPLE 6 Sometimes retailers are charged a unit price like $.024 each for pencils. Write $.024 in cents.

$$\begin{aligned} \$.024 &= .024 \times 1 \text{ dollar} \\ &= .024 \times 100¢ \\ &= 2.4¢ \end{aligned}$$

Answer: The retailer is charged about $2\frac{1}{2}$¢ each for the pencils.

♦ **You Try It**

14. Write $0.014 in cents.

15. Gasoline costs $\$1.24\frac{9}{10}$ per gallon. Write this amount in cents.

♦ **Answers to You Try It** **1.** $10^5 > 10^4$, $100{,}000 > 10{,}000$ **2.** 999,000 **3.** 9,900 **4.** 16.67 minutes **5.** 1000 years **6.** 1401.5 **7.** 2119 **8.** 1250 **9.** 65,000 **10.** The total cost of the order was \$2,256.00. **11.** The profit will be \$5,710.00.

12. 255,600,000 **13.** \$840,000,000,000 **14.** 1.4¢ **15.** $124\frac{9}{10}$¢

SECTION 2.3 EXERCISES

1

1. The national debt is \$3.24 trillion. Write this number in standard notation.

2. It is reported that the movie *Jurassic Park* cost \$75 million to make. Write this number in standard notation.

3. One year, a local town submitted a \$9.54 million referendum to fund school repairs and improvements. The voters rejected the proposal and the following year, the school board submitted a \$3.49 million referendum for the same purpose. This proposal was approved by the voters. What is the difference between the two proposals? Write this amount in standard notation.

4. What is the largest power of 10 that can be displayed in your calculator in standard notation? Write this number in words.

5. A *google* is a 1 followed by 100 zeros. Write a google as a power of 10.

6. Suppose you had \$1 million and you spend \$100 every day. How many days would it take to spend all of the \$1 million? How many years is this?

2

7. Write the numbers represented by the first six powers of ten. Write these numbers in words.

Do each of the following multiplications without using a calculator and without doing any long multiplication.

8. 16.05×100

9. $1.3546 \times 1{,}000$

10. 0.01766×10^2

11. $3.56(10^3)$

12. $3 \times 10{,}000$

13. 0.0003598×10^4

14. $11.1(10)$

15. 4.2×10^3

16. 4.2×10^4

17. 4.2×10^5

18. $\$3.95 \times 10{,}000$

19. $\$14.36 \times 100{,}000$

---

20. An electronics store buyer purchased 100 VCRs at \$159.35 each. What is the total cost?

21. A political organization photocopied 10,000 flyers to be distributed at a rally. The flyers cost \$0.015 each. What is the total cost?

22. Find the area of a rectangle whose length is 52.3 cm. and whose width is 100 cm.

23. Find the area of a rectangular piece of land that measures 1,000 ft. by 827 ft.

3

24. Write \$3.77 in cents.

25. Photocopying at a local printing center costs \$0.035 per page. Write this amount in cents.

26. A grocery store pays \$0.455 per can for canned fruit. Write this amount in cents.

27. In a grocery store, the unit price of an item is listed at \$0.235 per quart. Write this amount in cents.

SKILLSFOCUS (Section 1.2) *Write using symbols.*

28. the difference of 25 and 13

29. the quotient of 45 and 7

30. twice the sum of 8 and 3

EXTEND YOUR THINKING ♦ ♦ ♦

WRITING TO LEARN ♦ ♦ ♦

31. Explain in your own words how to multiply a number by a power of 10.

Use Figure 2.2 to answer questions 32 - 33.

FIGURE 2.2

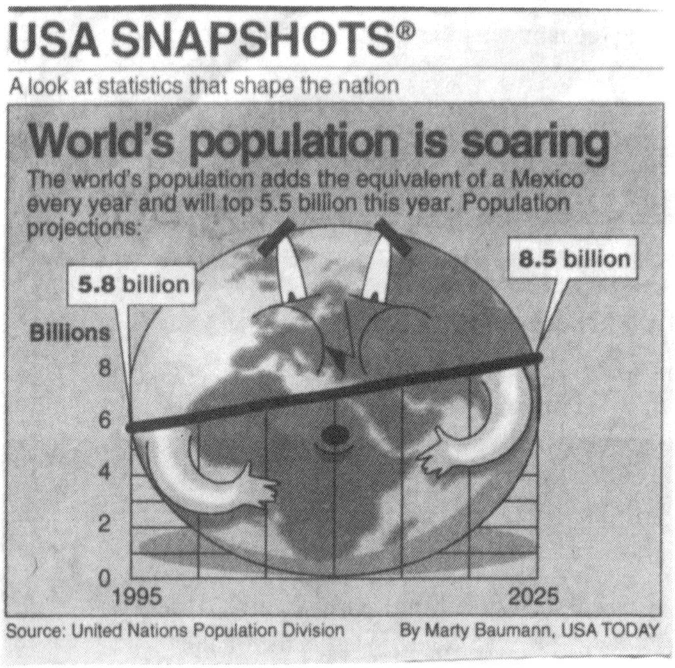

USA SNAPSHOTS®

A look at statistics that shape the nation

World's population is soaring

The world's population adds the equivalent of a Mexico every year and will top 5.5 billion this year. Population projections:

5.8 billion

8.5 billion

Billions

8

6

4

2

0

1995

2025

Source: United Nations Population Division

By Marty Baumann, USA TODAY

32. What is the projected increase in the world's population between 1995 and 2025?

33. Write the increase in standard notation.

Use Figure 2.3 to answer questions 34 - 35.

FIGURE 2.3

34. Write both the number of sesame seeds and the national debt in standard notation.

35. If there are 250 million people in the U.S., how much would each need to contribute to pay off the national debt?

2.4 SQUARE ROOTS

OBJECTIVE
Understand the use of the square root symbol and evaluate square roots.

NEW VOCABULARY
radical sign
square root

SQUARE ROOTS Although there are four basic arithmetic operations (addition, subtraction, multiplication, and division), two of these are the primary ones - addition and multiplication. Subtraction and division can be thought of as operations that "undo" addition and multiplication.

For example, if you start at a number, say 5, and then add 2, you arrive at a sum of 7. How can you get back to 5 from 7? You can "undo" the addition by subtracting 2 from 7.

If $5 + 2 = 7$ then $7 - 2 = 5$. Now you are back to the number you started with.

If you start at 5 and multiply by 2, you arrive at a product of 10. To "undo" this multiplication, you can divide 10 by 2.

If $5 \cdot 2 = 10$ then $10 \div 2 = 5$. Now you are back to the number you started with.

When we studied exponents, using a power of 2 was called "squaring". Squaring a number is also an operation and it is reasonable to ask whether there is another operation that can "undo" squaring. First, let's review squaring. To square a number means to use it as a factor two times. Here is a table of the squares of the first ten whole numbers:

Number	Square
0	0
1	1
2	4
3	9
4	16
5	25
6	36
7	49
8	64
9	81

Suppose we square 4 to get 16. To get back to 4, we will need to "undo" the squaring, or "unsquare" the 16. This means we will need to find the number whose square is 16. Thus,

If $4^2 = 16$ then the "unsquare" of $16 = 4$. Now you are back to the number you started with.

In the same way, if $3^2 = 9$ then the "unsquare" of $9 = 3$. Of course, we can't continue to call this new operation "unsquaring", just as subtraction isn't called "unadding". The operation of "unsquaring" requires its own name and also its own symbol. The name we give to "*unsquaring*" is called "*finding the square root*", and the symbol is $\sqrt{}$, which is called a *radical sign*. Now we can write the above statements correctly.

If $4^2 = 16$ then $\sqrt{16} = 4$ Read this: the square root of 16 is 4.

If $3^2 = 9$ then $\sqrt{9} = 3$ Read this: the square root of 9 is 3.

EXAMPLE 1 Evaluate $\sqrt{100}$.

$\sqrt{100} = 10$ because $10^2 = 100$

EXAMPLE 2 Evaluate $\sqrt{0.04}$.

$\sqrt{0.04} = 0.2$ because $(0.2)^2 = 0.04$.

EXAMPLE 3 Evaluate $6\sqrt{4}$.

Notice that there is no operation symbol between the 6 and the radical sign. The **only** operation that can be written without a symbol is multiplication, so $6\sqrt{4}$ is read " 6 **times** the square root of 4 ". Before we can do the multiplication, we must find the square root of 4. Here are the steps:

$$6\sqrt{4} = 6 \cdot 2$$
$$= 12$$

Answer: $6\sqrt{4} = 12$

♦ You Try It

1. How do you read the equation $\sqrt{100} = 10$?

2. What are you looking for when you are asked to find $\sqrt{49}$?

3. Evaluate $\sqrt{49}$. **4.** Evaluate $\sqrt{25}$.

5. Evaluate $\sqrt{0.09}$. **6.** Evaluate $\sqrt{0.0016}$.

7. Evaluate $7\sqrt{64}$. **8.** Evaluate $\sqrt{49} + \sqrt{4}$.

Square Roots on Your Calculator

Look on your calculator. You may see a key with the radical symbol, [√] or you may see the [x^2] key with $\sqrt{x}$ above it. These two possibilities require different keystrokes to evaluate a square root.

EXAMPLE 4 Evaluate $\sqrt{2{,}025}$.

Here are the keystrokes for each type of calculator:

If you have a calculator with a [√] key, the keystrokes are: 2025 [√]

The display should read 45. This means that $\sqrt{2{,}025} = 45$.

If you have a calculator with $\sqrt{x}$ above the [x^2] key, you will need to use the [2nd] key and the [x^2] key in order to access the square root function. Here are the keystrokes for finding $\sqrt{2{,}025}$: 2025 [2nd] [x^2]

The display should read 45. Once again, this means that $\sqrt{2{,}025} = 45$.

EXAMPLE 5 Evaluate $\sqrt{0.0121}$.

If you have a calculator with a [$\sqrt{\ }$] key, the keystrokes are: .0121 [$\sqrt{\ }$]

The display should read 0.11. This means that $\sqrt{0.0121} = 0.11$.

If you have a calculator with $\sqrt{x}$ above the [x^2] key, here are the keystrokes:

.0121 [2nd] [x^2]

The display should read 0.11. Once again, this means that $\sqrt{0.0121} = 0.11$.

Answer: $\sqrt{0.0121} = 0.11$

EXAMPLE 6 Evaluate $\sqrt{441} + \sqrt{256}$.

Here are the keystrokes for each type of calculator:

If you have a calculator with a [$\sqrt{\ }$] key, the keystrokes are:

441 [$\sqrt{\ }$] [+] 256 [$\sqrt{\ }$] [=]

The display should read 37.

If you have a calculator with $\sqrt{x}$ above the [x^2] key, here are the keystrokes:

441 [2nd] [x^2] [+] 256 [2nd] [x^2] [=]

The display should read 37.

Answer: $\sqrt{441} + \sqrt{256} = 37$

♦ You Try It

Find each of these square roots on your calculator.

9. $\sqrt{196}$ **10.** $\sqrt{625}$ **11.** $\sqrt{0.0529}$

12. $\sqrt{900} + \sqrt{576}$ **13.** $3\sqrt{484}$

♦ **Answers to You Try It** **1.** The square root of 100 is 10. **2.** A number whose square is 49. **3.** 7 **4.** 5 **5.** 0.3 **6.** 0.04 **7.** 56 **8.** 9 **9.** 14 **10.** 25 **11.** 0.23 **12.** 54 **13.** 66

SECTION 2.4 EXERCISES

1. Write the equation $\sqrt{36} = 6$ in words.

2. Write the equation $\sqrt{81} = 9$ in words.

Evaluate each square root without using your calculator.

3. $\sqrt{16}$ **4.** $\sqrt{25}$ **5.** $\sqrt{4}$ **6.** $\sqrt{0}$

7. $\sqrt{100}$ **8.** $\sqrt{1}$ **9.** $\sqrt{49}$ **10.** $\sqrt{9}$

11. $\sqrt{144}$ **12.** $\sqrt{64}$ **13.** $\sqrt{0.01}$ **14.** $\sqrt{0.25}$

15. $\sqrt{0.0064}$ **16.** $\sqrt{0.0001}$ **17.** $5\sqrt{36}$ **18.** $10\sqrt{81}$

19. $8\sqrt{25}$ **20.** $2\sqrt{121}$ **21.** $\sqrt{4}+\sqrt{49}$ **22.** $\sqrt{36}+\sqrt{64}$

23. $\sqrt{100}+\sqrt{25}$ **24.** $\sqrt{9}+\sqrt{16}$

Use your calculator to find each square root.

25. $\sqrt{324}$ **26.** $\sqrt{1,156}$ **27.** $\sqrt{4,356}$ **28.** $\sqrt{441}$

29. $\sqrt{529}$ **30.** $\sqrt{576}$ **31.** $\sqrt{0.0196}$ **32.** $\sqrt{0.0484}$

33. $\sqrt{10.24}$ **34.** $\sqrt{17.64}$ **35.** $6\sqrt{289}$ **36.** $4.7\sqrt{676}$

37. $\sqrt{784}+\sqrt{625}$ **38.** $\sqrt{841}+\sqrt{0.0036}$

SKILLSFOCUS (Section 2.1) *Multiply without using your calculator.*

39. 0.3×0.3 **40.** 0.03×0.03 **41.** 1.4×1.4 **42.** 0.14×0.14

Use Figure 2.4 to answer questions 43 - 44.

FIGURE 2.4

Biggest Hispanic-owned firms in '91

COMPANY LOCATION	TYPE OF BUSINESS CHIEF EXECUTIVE	'91 REVENUE IN MILLIONS
Bacardi Imports Miami, Fla.	**Rum importer/dist.** Juan Grau	**$580**
Burt on Broadway Englewood, Colo.	**Auto sales, service** Lloyd Chavez	**$423**
Goya Foods Secaucus, N.J.	**Hispanic foods** Joseph Unanue	**$410**
Sedano's Supermarkets Miami	**Supermarkets** Manuel Herran	**$221**
Int'l. Bancshares Laredo, Texas	**Financial services** Dennis Nixon	**$162**
Troy Ford Troy, Mich.	**Auto sales, service** Irma Elder	**$154**
Pizza Management San Antonio, Texas	**Restaurants** Arturo Torres	**$150**
Cal-State Lumber Sales San Ysidro, Calif.	**Wood products** Benjamin Acevedo	**$147**
Handy Andy Supermarkets San Antonio, Texas	**Supermarkets** A. Jimmy Jimenez	**$143**
Ancira Enterprises San Antonio, Texas	**Auto sales, service** Ernesto Ancira Jr.	**$132**

Source: *Hispanic Business* magazine

43. What is the difference in revenue between the first company listed and the last company listed on the chart?

44. Do the totals of Goya and Handy Andy together exceed Baccardi Imports?

EXTEND YOUR THINKING ♦♦♦

♦ TROUBLESHOOT IT

Find and correct the error.

45. $\sqrt{36} = 6^2$ **46.** $\sqrt{16} = \sqrt{4} = 2$ **47.** $\sqrt{0.0004} = 0.0002$

WRITING TO LEARN ♦♦♦

48. Explain in your own words what square roots are.

49. Explain how to check the answer to a square root.

2.5 PROBLEM SOLVING AND DIVISION

OBJECTIVES

1. Perform decimal divisions.
2. Solve applications involving division.
3. Divide by powers of 10.
4. Change cents to dollars.

NEW VOCABULARY

divisor	quotient
dividend	per

1 DIVISION OF DECIMALS

Every division problem involves three numbers. Each one has a role and a name. In the equation $16 \div 8 = 2$, 16 is called the **dividend**, 8 is called the **divisor**, and 2 is called the **quotient**.

$$\text{divisor}\overset{\text{quotient}}{\overline{\smash{)}\text{dividend}}} \qquad \text{dividend} \div \text{divisor} = \text{quotient}$$

> In order to carry out divisions involving decimal numbers, the divisor must be a whole number.

In a division problem the order of the numbers matter. Remember in Section 1.2 we said that if the order of the numbers in an operation matters, then the operation (here, division) is not *commutative*. The divisor is the number doing the dividing. In the expression $16 \div 8$, 8 is the divisor. In the expression $\frac{35}{5}$, 5 is the divisor, and in the expression $4\overline{)24}$, 4 is the divisor. Since the divisor must be a whole number in order to carry out the division, there are two possible types of division problems: either (1) the divisor is a whole number or (2) the divisor is a decimal. We will treat these two cases separately.

> **Case 1: Decimal Division when the Divisor is a Whole Number**
>
> To divide a decimal by a whole number, set the problem up for long division, move the decimal point directly up from the dividend to the quotient, and divided as you would whole numbers. If necessary, use zeros as placeholders.

EXAMPLE 1 Divide 14.65 by 5.

The division problem is $14.65 \div 5$ or $5\overline{)14.65}$

$$\begin{array}{r} 2.93 \\ 5\overline{)14.65} \\ \underline{10} \\ 46 \\ \underline{45} \\ 15 \\ \underline{15} \end{array}$$

Notice that the divisor, 5, is a whole number, so the decimal point moves directly up to the quotient.

Answer: $14.65 \div 5 = 2.93$ Remember that you can check the division by multiplying the quotient, 2.93, by the divisor, 5. The answer should be the dividend, 14.65.

EXAMPLE 2 Find the quotient of 0.02 and 4.

The division problem is $0.02 \div 4$ or $4\overline{)0.02}$

4 is the divisor and it is a whole number.

Don't forget this zero in the quotient. 4 divides into 0, 0 times. Write 0 above the 0 in the quotient.

4 divides into 2, 0 times. Write 0 above the 2 in the quotient.

$$\begin{array}{r} 0.005 \\ 4\overline{)0.020} \\ \underline{20} \\ 0 \end{array}$$

← Remember that you can add zeros to the end of a decimal number without changing its value.

Answer: $0.02 \div 4 = 0.005$

Case 2: Decimal Division when the Divisor is a Decimal

To divide by a decimal number, move the decimal point in **both** the divisor **and** the dividend to the right as many places as it takes for the divisor to become a whole number.

EXAMPLE 3 Divide 1.25 by 0.5.

Here, the division problem is $1.25 \div 0.5$. Since 0.5 is the divisor and it is a decimal number, the division cannot be completed in this form. To see why we can move the decimal point as stated in the rule in the above box, we will write the division problem in fraction form.

$$1.25 \div 0.5 = \frac{1.25}{0.5}$$

Recall that the Fundamental Property of Fractions says that we can multiply the numerator and denominator of a fraction by the same number without changing the value of the fraction. Since our goal is to have a whole number as denominator, we can accomplish this by multiplying the numerator and denominator of this fraction (that is, the dividend and divisor of the division problem) by 10:

Multiply numerator and denominator by 10.

$$1.25 \div 0.5 = \frac{1.25}{0.5} = \frac{1.25 \times 10}{0.5 \times 10} = \frac{12.5}{5} = 12.5 \div 5$$

The divisor is now a whole number.

Notice that we were able to rewrite the division problem as a new one in which the denominator is a whole number. We usually write the process as follows:

$$1.25 \div 0.5 = 1.2.5 \div 0.5. = 12.5 \div 5$$

Move the decimal point **one** place to the right in **both** numbers.

Now we can divide as in the previous examples:

$$\begin{array}{r} 2.5 \\ 5\overline{)12.5} \\ \underline{10} \\ 2\,5 \\ \underline{2\,5} \end{array}$$

Answer: $1.25 \div 0.5 = 2.5$

EXAMPLE 4 Divide $0.064 \div 0.08$.

Here, we need to move the decimal point two places (in **both** the divisor and the dividend) in order to make the divisor, 0.08, a whole number:

$$0.064 \div 0.08 = 0.06.4 \div 0.08. = 6.4 \div 8 = 0.8$$

Answer: $0.064 \div 0.08 = 0.8$

EXAMPLE 5 Find the quotient: $3 \div 0.3$.

Here we need to move the decimal point one place. Remember that a whole number has a decimal point located at the right end.

$$3 \div 0.3 = 3.0. \div 0.3. = 30 \div 3 = 10$$

Answer: $3 \div 0.3 = 10$

♦ You Try It

1. In the division problem $22 \div 11$, which number is the divisor?
2. In the division problem $4.8 \div 1.6$, which number is the divisor?
3. In the division problem $\frac{1.6}{4.8}$, which number is the divisor?
4. True or false: The divisor is always the smaller number. Explain your answer.
5. True or false: The divisor is determined by its position in the division problem.
6. How do you decide whether you need to move the decimal point in a division problem?
7. Find the quotient of 0.45 and 5.
8. Find the quotient: $0.45 \div 0.5$.
9. Divide $4.5 \div 0.5$.
10. Divide $45 \div 0.5$.
11. Divide $1.2 \div 0.04$.

Divisions on Your Calculator

Suppose you use your calculator to do the division in the last problem. Look carefully at your calculator to see what buttons it has for division. The only button on your calculator for division is [÷] (divided **by**). There is no [$\overline{)}$] (into) key. That means if we want to divide 1.25 **by** 0.5, we must use the notation $1.25 \div 0.5$, **not** $0.5\overline{)1.25}$. Therefore, on calculator, the dividend, 1.25 , must be keyed in first, then press the [÷] key, then key in the divisor, 0.5. The keystrokes look like this:

1.25 [÷] .5 [=]

The display should read 2.5.

The general rule for dividing on a calculator is:

> When dividing using a calculator, the **dividend** must be keyed in first, then the [÷] key, then the **divisor**, then [=]. The keystrokes for $a \div b$ or $\frac{a}{b}$ are a [÷] b [=]

♦ You Try It

12. Write down the calculator keystrokes necessary for the division in **You Try It** #7 above.
Do the problem on your calculator and write down the answer in the display.

13. Write down the calculator keystrokes necessary for the division in **You Try It** #8 above.
Do the problem on your calculator and write the answer in the display.

14. Write down the calculator keystrokes necessary for the division in **You Try It** #9 above.
Do the problem on your calculator and write the answer in the display.

15. Write down the calculator keystrokes necessary for the division in **You Try It** #10 above.
Do the problem on your calculator and write the answer in the display.

16. Write down the calculator keystrokes necessary for the division in **You Try It** #11 above.
Do the problem on your calculator and write the answer in the display.

How Big Should the Answer be?

In order to be able to estimate the reasonableness of the answer to a division, we will look at the outcomes of division by examining the relative sizes of the dividend and divisor. There are three possibilities: the dividend and divisor are equal, the divisor is smaller than the dividend, or the divisor is larger than the dividend.

If the dividend and the divisor are **equal**, then the quotient is 1. For example, $16 \div 16 = 1$ and $0.12 \div 0.12 = 1$.

If the divisor is **smaller** than the dividend, the quotient will **be larger than 1.** For example $27 \div 9 = 3$ because if you divide 27 into 9 equal pieces, each piece is exactly 3. Another way to look at it is that exactly three 9's fit into 27.

If the divisor is **larger** than the dividend, the quotient will be **smaller than 1.** In this case, you are dividing a smaller number by a larger number. Suppose we divide 2 by 5. We write this as $2 \div 5$, which could mean, "take $2 and divide it equally among 5 people". Clearly, each person will get less than a whole dollar, so the answer should be less than one.

We can summarize these observations as follows:

Determining the Size of the Answer to a Division Problem

1. If the divisor and dividend are **equal**, the quotient is **1**. That is, if a number is divided by itself, the answer is 1.
2. If the divisor is **smaller** than the dividend, the quotient will be **larger than 1**. That is, if a number is divided by a number smaller than itself, the answer is larger than 1.
3. If the divisor is **larger** than the dividend, the quotient will be **smaller than 1**. That is, if a number is divided by a number larger than itself, the answer is smaller than 1.

EXAMPLE 6

1. In the division $15 \div 6$, the divisor is 6 and it is **smaller** than the dividend 15, so the quotient is **larger than 1**.
2. In the division $2 \div 12$, the divisor is 12 and it is **larger** than the dividend 2, so the quotient is **smaller than 1**.
3. In the division $2.02 \div 2.02$, the divisor is 2.02 and it is **equal** to the dividend 2.02, so the quotient is **equal to 1**.
4. In the division $45 \div 0.5$, the divisor is 0.5 and it is **smaller** than the dividend 45, so the quotient is **larger than 1**.
5. In the division $0.45 \div 0.5$, the divisor is 0.5 and it is **larger** than the dividend 0.45, so the quotient is **smaller than 1**.
6. In the division $5 \div \frac{1}{2}$, the divisor is $\frac{1}{2}$ and it is **smaller** than the dividend 5, so the quotient is **larger than 1**.

♦ You Try It

For the following problems, decide if the answer will be smaller than, equal to, or larger than 1. Do not carry out the division.

17. $78 \div 12$ **18.** $12 \div 78$ **19.** $12 \div 12$

20. $6 \div \frac{1}{3}$ **21.** $\frac{1}{3} \div 6$ **22.** $4\frac{1}{2} \div 4\frac{1}{2}$

2 APPLICATIONS

You know that there are two ways to write a division problem. For example $8 \div 4$ and $4\overline{)8}$ both mean the same thing. However, the two ways to write a division problem come from two different ways to think about division. For example, $8 \div 4$ means taking 8 and dividing it into 4 pieces whereas $4\overline{)8}$ means seeing how many 4's there are in 8. We will examine these two types of division problems so that you will have a better understanding of the kinds of problems for which division is the appropriate operation.

Type 1 Division Problem

In this type of problem, you begin with a total amount of something (pounds, miles, dollars) that you want to *divide into parts* (for each hamburger, each hour, each person).

$$$$$ ✂ $ $ $ $ $

Example: Consider 500 pounds of ground beef to make 200 meatloaves. How many pounds are in each meatloaf?

$500 \div 200$

Example: Consider 400 miles in a trip and 8 hours to travel. How many miles do you travel in each hour?

$400 \div 8$

Example: Consider $1,000 to be shared by 5 people. How many dollars are there for each person?

$1{,}000 \div 5$

Sometimes the phrasing of the problem points you to this type of division problem. Instead of the underlined phrases above, you might see the word *PER*. *PER* comes from Latin and means "for every" or "for each", so mathematically, *PER* means "divided by."

pounds	*PER*	meatloaf	means	pounds	÷	meatloaves
miles	*PER*	hour	means	miles	÷	hours
dollars	*PER*	person	means	dollars	÷	persons

Remember mathematics is a language and the word *PER* means a division needs to be done or was done already. If you are asked to **find** miles per hour or dollars per person, then you need to divide. On the other hand, if you are **given** miles per hour or dollars per person and asked to find something else, the division has already been done. If you do need to divide, remember to divide in the proper order. The units mentioned just before *PER* must go before the ÷ symbol.

EXAMPLE 7 Ann walked 3.6 miles in 1.5 hours. At what rate in miles per hour did she walk?

Take the total number of miles and divide by the number of hours.

miles *PER* hour (Often written as mph)
miles ÷ hour
$3.6 \div 1.5 = 2.4$

Work: $3.6 \div 1.5 = 36 \div 15$ (move each decimal point one place right)

Work:
$$\begin{array}{r} 2.4 \\ 15\overline{)36.0} \\ \underline{30} \\ 6\,0 \\ \underline{6\,0} \end{array}$$

Answer: Ann walked at a rate of 2.4 miles per hour.

> In general, "quantity a" **PER** "quantity b" means "quantity a" ÷ "quantity b" and can be written
>
> $$\frac{\text{quantity } a}{\text{quantity } b} \text{ or } \frac{a}{b}.$$

EXAMPLE 8 When Harriet started an exercise program she weighed 139.4 pounds. After 12 weeks, she weighed 132.9 pounds. What was her average weight loss per week?

Analyze
We know she weighed 139.4 pounds at the start and 132.9 pounds later.
We know that 12 weeks have gone by.
We need to find the weight loss **per** week, which means total weight lost ÷ number of weeks.
This means that we first need to find the total weight lost. This will be a subtraction:

139.4 pounds − 132.9 pounds = 6.5 pounds

Finally, we can divide 6.5 pounds ÷ 12 weeks

Estimate In the division 6.5 ÷ 12, the divisor is 12 and it is **larger** than the dividend 6, so the quotient will be **smaller than 1**, so she has lost **less than 1** pound per week.

Solution 6.5 pounds ÷ 12 weeks = 0.541666... pound per week

Work

```
       0.541666
   12)6.500000
      6 0
        50
        48
         20
         12
          80
          72
           80
           72
            80
            72
             8
```

Notice that the 6's in the quotient will keep repeating because the remainder, 8, keeps repeating.

Calculator keystrokes 6.5 [÷] 12 [=]
Display 0.5416667 Notice that the calculator rounded the last digit displayed.

Answer: In this case, it makes sense to round the answer to the nearest tenth of a pound, so Harriet lost approximately 0.5 pound per week. That is, she lost about half a pound each week.

Type 2 Division Problem

Begin with the total amount of something again. This time you want to know how many *pieces of a certain size* it can be divided into. The phrasing of these problems is often, "How many ______ 's fit into _________ ?

For example, how many □ 's fit into ▭ ?

Example: Consider a water cooler with 128 gallons of water. How many 4 gallon bottles can be filled from the water cooler?

Example: Consider 264 hours and knowing that there are 24 hours in a day, how many days are there in 264 hours?

Example: Consider a trip of 1260 miles. If you travel at a speed of 60 mph, how many hours will the trip take?

You need to picture the smaller quantity fitting into the larger quantity. These problems use the idea of division as *going into*.

How many 4 gallon bottles *go into* 128 gallons? $4\overline{)128}$

How many 24 hour days *go into* 264 hours? $24\overline{)264}$

How many 60 mph *go into* 1260 miles? $60\overline{)1260}$

EXAMPLE 9 Cal drove 450 miles at a steady rate of 50 miles per hour. How long did he drive?

Analyze Here we know total miles = 450 miles and we know that the rate is 50 miles per hour, so we can divide miles by miles per hour to find hours:

450 miles ÷ 50 miles per hour

Another way to think about this is: Every 50 miles uses 1 hour of time. If Cal drove 450 miles, each 50 miles used up 1 hour of time. We need to find out how many 50's are in 450, or how many 50 mph *go into* 450 miles. This is one of the meanings of division. A picture might help:

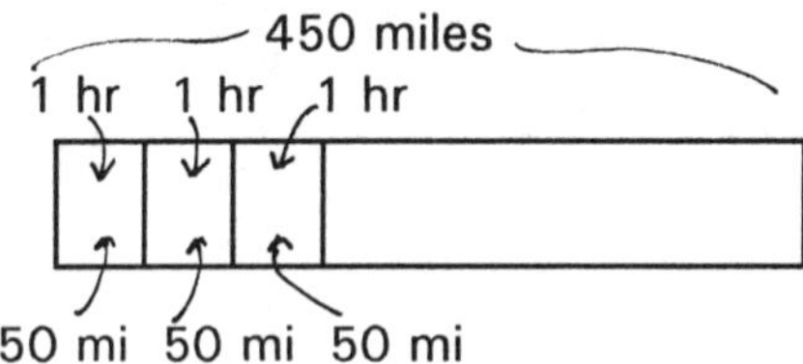

We need to find out how many of these 50-mile segments are in 450 miles.

Estimate No rounding is necessary for this since 450 ÷ 50 = 9.

Solution 450 miles ÷ 50 miles per hour = 9 hours

Answer: Cal drove 9 hours.

EXAMPLE 10 A shirt pattern requires 2.25 yards of fabric. If a tailor has 66.75 yards of fabric available, how many shirts can be made?

Analyze
We know that the total amount of fabric available is 66.75 yards.
We know that one shirt requires 2.25 yards. That is, the pattern requires 2.25 yards *per* shirt.
We want to know how many 2.25-yard shirts can be made from 66.75 yards. Picture the total number of yards as if it were laid out on a table:

66.75 yards

Every 2.25 yards is one shirt, so we can take the total yardage and cut it up into shirt-size pieces:

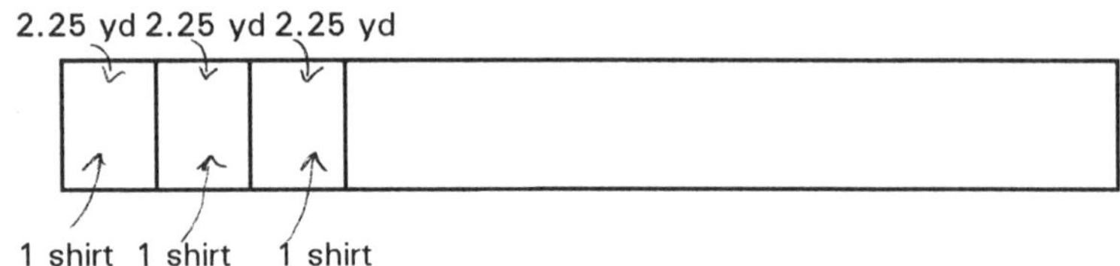

From this picture, we see that we need to ask "how many 2.25 yards are there in 66.75 yards?" This is definitely a division question and we need to divide the total, 66.75 yards, by the yards per shirt, 2.25.

$$66.75 \text{ yards} \div 2.25 \text{ yards per shirt}$$

Estimate: 66.75 is *about* 67 and 2.25 is *about* 2, so 67 ÷ 2 is *about* 33, so the tailor should be able to make about 33 shirts.

Solution
66.75 yards ÷ 2.25 yards per shirt = 29.6666.... shirts

Work

```
        29.666
  225)6675.000
      450
      2175
      2025
       150 0
       135 0
        15 00
        13 50
         1 500
         1 350
           150
```

Notice that the 6's in the quotient will keep repeating because the remainder, 150 keeps repeating.

Calculator keystrokes: 66.75 [÷] 2.25 [=]
Display 29.666667 Notice that the calculator rounded the last displayed digit.

Answer: Before we can answer this question, it is important to realize that **for this problem**, only a whole number answer makes sense. A tailor will not make part of a shirt. Even though 29.66667 rounds to 30, there is not enough material for 30 shirts, so the answer must be 29 shirts. Thus, the tailor can make 29 shirts.

◆ **You Try It**

23.. April is planning a 480 mile trip. Usually on long trips her car gets about 20 miles to the gallon. How many gallons will she need for the trip? If her tank holds 12 gallons, how many tanks of gas will she need?

24. New Jersey Transit services 73,800 passengers on a typical workday. If it runs 123 trains, what is the average number of passengers per train should there be?

3 DIVISION BY POWERS OF TEN

Remember that 10, 100, 1,000, 10,000, 100,000, etc., are powers of ten, and that when a decimal is *multiplied* by a power of 10, the decimal point moves to the *right* as many places as there are zeros in the power of 10. Now let's see what happens if you divide a decimal by a power of 10. (You may have guessed the answer.)

1. Use your calculator to do the following divisions.

a. $3429.6 \div 10$ ________________

b. $\frac{17.06}{10}$ ________________

c. $893.1 \div 10$ ________________

2. Based on your answers to problem 1, complete the following sentence:

When a number is divided by 10, the decimal point moves ________

____

3. Use your calculator to do the following divisions.

a. $\frac{3429.6}{100}$ ________________

b. $17.06 \div 100$ ________________

c. $\frac{893.1}{100}$ ________________

4. Based on your answers to problem 3, complete the following sentence:

When a number is divided by 100, the decimal point moves ________

____

5. Based on your answers to problems 2 and 4, what happens when a number is divided by 1,000 ?

____

____

6. What happens when a number is divided by 10,000 ?

____

____

We have, then, another general and easy-to-use rule involving powers of 10:

To **divide** a number by a power of 10, move the decimal point to the **left** as many places as there are zeros in the power of 10.

♦ **You Try It**

Use the easy method to do each division.

25.. $61.29 \div 100$ **26.** $\frac{185.4}{10}$ **27.** $\frac{12.5}{1000}$

28. $6 \div 100$ **29.** $0.9 \div 10$ **30.** $\frac{16}{1000}$

31. A sewer project costing \$44,350 will be paid for by 100 families. What will be the cost per family?

32. A shipment of 1000 calculators cost a store \$5,938.80. Find the cost per calculator to the store.

4 DOLLARS AND CENTS

In an earlier section we discussed how to write a money amount that is given in dollars in cents. Here is an example:

EXAMPLE 11 $\$4.27 = 4.27$ dollars
$= 4.27 \times 1$ dollar
$= 4.27 \times 100$ cents
$= 427$ cents

Thus, $\$4.27 = 427¢$.

Now we will use a similar approach for writing an amount given in cents in dollars. Since there are 100 cents in 1 dollar, 1 cent is one-hundredth of a dollar, or $\frac{1}{100}$ dollar.

EXAMPLE 12 Write $67¢$ in dollars.

$67¢ = 67$ cents
$= 67 \times 1$ cent

$= 67 \times \frac{1}{100}$ dollar   Remember that to multiply fractions, we multiply numerators and multiply denominators.

$= \frac{67}{100}$ dollar   Now divide by 100, which means moving the decimal point two places to the left.

$= .67$ dollar

$= \$.67$ Write the dollar sign in front of the decimal.

Thus, 67¢ = \$0.67.

Although it is doubtful that you really needed that many steps to write 67¢ in dollars, it is useful to go through the steps once to see the reasoning behind the general rule.

To change cents to dollars, we need to divide by 100 and that means moving the decimal point two places to the left.

EXAMPLE 13 On a grocery store shelf, the unit price of an item is listed as 12.2 cents per ounce. Write this amount in dollars.

To change cents to dollars, divide by 100 (move the decimal point **two** places to the **left**):

$$12.2¢ \text{ per ounce} = \frac{12.2}{100} \text{ dollars per ounce} = \$0.122 \text{ per ounce}$$

Notice that \$.122 is not an amount of money that anyone can actually pay. Unit prices are often given to more than two decimal places so that we can compare items which have very close unit prices.

♦ You Try It

33. Write 13.5¢ in dollars.

34. Find the cost, in dollars, of 3 pounds of grapes at 89¢ per pound.

35. An 11-ounce can of fruit costs \$0.79. Find the unit price, which is the price per ounce. Express your answer in cents per ounce. Round your answer to the nearest tenth of a cents per ounce.

♦ Answers to You Try It **1.** 11 is the divisor **2.** 1.6 is the divisor **3.** 4.8 is the divisor **4.** false; Look at problem 3. **5.** true **6.** If the divisor is a decimal, move the decimal point in both the divisor and the dividend **7.** 0.09 **8.** 0.9 **9.** 9 **10.** 90 **11.** 30 **12.** .45 [÷] 5 [=] Display: 0.09 **13.** .45 [÷] .5 [=] Display: 0.9 **14.** 4.5 [÷] .5 [=] Display: 9 **15.** 45 [÷] .5 [=] Display: 90 **16.** 1.2 [÷] .04 [=] Display: 30 **17.** larger than 1 **18.** smaller than 1 **19.** equal to 1 **20.** larger than 1 **21.** smaller than 1 **22.** equal to 1 **23.** April will need 24 gallons of gas, which is 2 tanks. **24.** There are 600 passengers per train. **25.** 0.6129 **26.** 18.54 **27.** 0.0125 **28.** 0.06 **29.** 0.09 **30.** 0.016 **31.** The sewer project will cost each family \$443.50 **32.** Each calculator cost \$5.9388 or about \$5.94. **33.** \$0.135 **34.** \$2.67 **35.** The fruit costs approximately 7.2 cents per oz.

SECTION 2.5 EXERCISES

1 *Label the divisor, dividend and quotient.*

1. $3\overline{)180}$ with quotient 60

2. $31.05 \div 1.035 = 30$

3. $\dfrac{4.5}{0.5} = 9$

4. $45.87 \div 0.03 = 1529$

Specify whether the quotient will be larger than 1, equal to 1, or smaller than 1. Do NOT do the division.

5. $1{,}134 \div 5.678$

6. $5.678 \div 1{,}134$

7. $\dfrac{1.134}{1.134}$

8. $\dfrac{47}{0.001}$

9. $2\overline{)0.47}$

10. $2 \div 0.47$

11. $4\frac{1}{2} \div \frac{1}{2}$

12. $\frac{1}{2} \div \frac{1}{4}$

Divide.

13. $4.8 \div 6$

14. $6.4 \div 16$

15. $12.6 \div 9$

16. $41.3 \div 7$

17. $32.4 \div 15$

18. $214.24 \div 26$

19. $1020.5 \div 13$

20. $45.144 \div 18$

21. $3.6 \div 0.9$

22. $2.8 \div 0.4$

23. $7.2 \div 0.6$

24. $8.5 \div 0.5$

25. $0.56 \div 0.8$

26. $0.35 \div 0.7$

27. $44 \div 1.1$

28. $2.25 \div 1.5$

29. $0.112 \div 8$

30. $0.4347 \div 9$

31. $2.56 \div 0.4$

32. $72.4 \div 0.2$

33. $7 \div 8$

34. $3 \div 5$

35. $36 \div 50$

36. $12 \div 75$

37. $0.6 \div 12.5$

38. $33.659 \div 6.94$

39. $1.3 \div 13$

40. $0.13 \div 13$

Divide. Round to the indicated place.

41. $1.3 \div 7$ thousandths

42. $4.7 \div 6$ hundredths

43. $4.8 \div 0.7$ hundredths

44. $8.3 \div 0.6$ thousandths

45. $24.6 \div 5$ tenths

46. $45.08 \div 12$ ten-thousandths

47. $12.7 \div 0.41$ tenths

48. $16.4 \div 1.18$ hundredths

49. $2 \div 9$ hundredths

50. $4 \div 7$ thousandths

Divide. Use your calculator.

51. $24.288 \div 4.8$

52. $33.659 \div 6.94$

53. $395.415 \div 5.05$

54. $0.04293 \div 0.81$

Divide. Use your calculator and round to the indicated place.

55. $128.56 \div 32$ hundredths

56. $1{,}835.6 \div 120$ thousandths

57. $3.094 \div 10.75$ thousandths

58. $0.95 \div 12.7$ ten-thousandths

2 *Solve each application problem. Round dollar amounts to the nearest cent.*

59. Tom makes $42.08 for working 8 hours. What does he make per hour?

60. Amy makes $16 an hour. How many hours must she work to make $500?

61. If you earn $28,462.75 per year, how much do you make per week? (Assume you work a 52-week year.)

62. How many gallons of gas can you buy for $8.30 if gas costs $1.269 per gallon? Round to the nearest tenth of a gallon.

63. If 450 bricks cost $180, what is the cost of one brick?

64. A case of 144 pens cost $122.40. What is the cost of one pen?

65. Jeff travels 364 miles on 15.8 gallons of gas. How many miles per gallon did he get, rounded to the nearest tenth?

66. Amelia drove from Baltimore to San Francisco, a distance of 2,875 miles. She used a total of 165.2 gallons of gas. How many miles per gallon did she get, rounded to the nearest tenth?

67. You pay $3,284.16 in 24 equal monthly payments. Find the payment.

68. You owe $12,683.56, which you plan to pay in 48 equal monthly payments. How large is each payment?

69. A package of three bars of soap sells for $0.99. A package of twelve bars costs $3.60.
   a. What is the price per bar in the three-pack?
   b. What is the price per bar in the twelve-pack?
   c. What is the savings per bar with the twelve-pack?

70. A box with 6 quarts of oil costs $8.04. A box of 24 quarts costs $30.96.
   a. What is the price per quart in the 6-pack?
   b. What is the price per quart in the 24-pack?
   c. What is the savings per quart in the 24-pack?

71. Mailing a package costs $0.28 for the first ounce and $0.17 for each additional ounce.
   a. What will it cost to mail a 32-ounce package?
   b. How heavy a package can you mail for $6.91?

72. Pencils cost $0.12 each for the first 150, $0.10 each for the next 150, and $0.07 each for every pencil thereafter.
   a. What will it cost to buy 500 pencils?
   b. How many pencils can you purchase for $52.25?

73. You need 9,000 bricks to build a new home. A bricklayer lays 45 bricks per hour at a cost of $21 per hour. The price of a brick is $0.36. Mortar costs $10 per 500 bricks. What will it cost to brick the house?

74. The U.S. national debt was $907.7 billion in 1980. The population of the United States was 223 million in 1980. What was the debt per person in 1980? Round to the nearest ten dollars.

75. Your club has $126.19 to purchase toys for needy youngsters.
   a. If each toy costs $8.62, how many toys can be purchased?
   b. What change is left over?

76. A double-decker hamburger costs $1.69.
   a. How many can you buy for $14?
   b. What change is left over?

77. What is the average of the eight test scores 78, 90, 65, 80, 98, 91, 100, and 85? Round to the nearest tenth.

78. Five spiders of the same species weighed 0.020, 0.023, 0.0185, 0.019, and 0.027 ounces. Find the average weight, rounded to thousandths of an ounce.

3 *Divide by the power of ten.*

79. $12.6 \div 10$

80. $351.8 \div 10$

81. $417.5 \div 1{,}000$

82. $4{,}503.34 \div 1{,}000$

83. $69.301 \div 100$

84. $318.7 \div 100$

85. $5{,}930.16 \div 10{,}000$

86. $2.7 \div 100$

87. $6.08 \div 1{,}000$

88. $6 \div 10$

89. $0.0028 \div 1{,}000$

90. $0.039 \div 10$

91. Divide $256 by 100.

92. Divide 2,480 pounds by 1,000.

93. Ten identical toys cost a total of $176. How much does one toy cost?

94. Suppose 12,560 acres of land is evenly divided. among 100 farmers. How much land does each farmer get?

95. A lottery jackpot of $12,503,410 is evenly divided among 1,000 ticket holders. How much does each get?

96. Suppose 10,000 pencils cost $900. How much does one pencil cost?

4 *Write each amount in dollars.*

97. 49¢

98. 3¢

99. 2.5¢

100. 187¢

101. Find the cost, in dollars, of four cans of beans at 79¢ per can.

102. Find the cost, in dollars, of seven pencils at 14.5¢ each.

SKILLSFOCUS (Sections 2.2 and 2.4) *Evaluate without a calculator.*

103. 6^3

104. 15^2

105. $\sqrt{81}$

106. $\sqrt{49}$

EXTEND YOUR THINKING ♦ ♦ ♦

Use Figure 2.5 to answer questions 107 - 109.

FIGURE 2.5

USA SNAPSHOTS®

A look at statistics that shape the nation

How much we spend on Mother's Day for...

(in millions)

$329 Candy

$225 Cards

$76 Phone calls

$61 Sending flowers

Source: The Mother's Day Council, Ayer Public Relations

By Gary Visgaitis, USA TODAY

107. What is the total amount spent on the Mother's Day items listed? Write the answer in standard notation.

108. If there are 250 million people in the country, what is the Mother's Day spending PER person?

109. Is the answer to problem 108 meaningful? Explain.

Use Figure 2.6 to answer questions 110 - 111.

110. If you had to send 100 2-page letters across the USA during the day on a Tuesday, how much would you save by faxing? Write your answer in dollars.

111. If you could wait until Saturday to fax the letters, how much cheaper would it be than mailing the letters? Write your answer in dollars.

FIGURE 2.6

Use Figure 2.7 to answer questions 112 - 114.

112. Which ballpark is the most expensive to attend?

113. Which ballpark is the least expensive to attend?

114. To the nearest cent, what is the price per person at Shea Stadium?

FIGURE 2.7

On the tab

How much it costs a family of four to attend a baseball game at the 28 major league parks (includes four tickets, two beers, four hot dogs, four soft drinks, two baseball caps, two programs and parking):

Team	Ballpark	Price
Atlanta Braves	Atlanta Fulton County Stadium	$97.06
Baltimore Orioles	Oriole Park at Camden Yards	$102.96
Boston Red Sox	Fenway Park	$104.18
California Angels	Anaheim Stadium	$84.60
Chicago Cubs	Wrigley Field	$103.94
Chicago White Sox	Comiskey Park	$97.81
Cincinnati Reds	Riverfront Stadium	$77.31
Cleveland Indians	Municipal Stadium	$87.30
Colorado Rockies	Mile High Stadium	$81.12
Detroit Tigers	Tiger Stadium	$92.17
Florida Marlins	Joe Robbie Stadium	$98.83
Houston Astros	Astrodome	$80.52
Kansas City Royals	Royals Stadium	$86.29
Los Angeles Dodgers	Dodger Stadium	$88.60
Milwaukee Brewers	County Stadium	$91.69
Minnesota Twins	Metrodome	$82.63
Montreal Expos	Olympic Stadium	$86.14
New York Mets	Shea Stadium	$86.46
New York Yankees	Yankee Stadium	$113.43
Oakland Athletics	Oakland Stadium	$99.97
Philadelphia Phillies	Veterans Stadium	$82.81
Pittsburgh Pirates	Three Rivers Stadium	$87.70
St. Louis Cardinals	Busch Stadium	$78.54
San Diego Padres	Jack Murphy Stadium	$85.31
San Francisco Giants	Candlestick Park	$87.23
Seattle Mariners	Kingdome	$85.41
Texas Rangers	Arlington Stadium	$92.73
Toronto Blue Jays	SkyDome	$116.00

Source: Team Marketing Report

By Sam Ward, USA TODAY

WRITING TO LEARN ♦ ♦ ♦

115. True or false: In a division problem, the divisor is always the larger number. Explain your answer.

116. True or false: Whenever a division problem involves at least one decimal, you need to move the decimal point. Explain your answer.

117. True or false: In a division problem involving decimals, the answer is always a decimal. Explain your answer.

118. In the division problem $7.5 \div 25$, do you need to move the decimal point? Why or why not?

♦YOU BE THE JUDGE

119. Theresa says the following divisions are related. Is she correct? You may use a calculator to prove your point. Explain why you get the answers you get.

$$0.075\overline{)0.4086} \qquad 75\overline{)408.6} \qquad 0.75\overline{)4.086} \qquad 7.5\overline{)40.86}$$

120. A turkey is priced at \$13.67 and weighs 8.92 pounds. The price per pound is listed at \$1.39.

a. Are these numbers correct? Explain your decision.

b. If not, can you determine which number is wrong? Explain.

2.6 PROBLEM SOLVING USING ORDER OF OPERATIONS

OBJECTIVES

1. Use the rule for Order of Operations to evaluate expressions.
2. Translate English phrases to mathematical expressions and vice versa.
3. Use the distributive property to rewrite expressions.

NEW VOCABULARY

grouping symbols
distributive property
common factor

1 ORDER OF OPERATIONS

One of the most amazing things in the study of mathematics is that the rules and strategies are universally used. You can "talk mathematics" on a beach in the south of France and it will be the same as on the Jersey shore. Mathematics is truly a universal language. Every student of mathematics must read mathematical statements in the same way. Consider the following expression:

$$2+2\cdot 2$$

Write down what you think is the value of this expression. _______

Everyone who does this problem MUST get the same answer. For mathematics to be a universal language, mathematical statements must have a universal interpretation. Did you choose one of the two possibilities below?

1. The answer is **6**. It was obtained by multiplying first:
$$2+2\cdot 2=2+4=6$$

2. The answer is **8**. It was obtained by working from left to right.
$$2+2\cdot 2=4\cdot 2=8$$

It is important to know that only one of the above choices is correct and that there must be a reason for your choice of the order in which to do the operations listed in a problem. It is not an arbitrary order. At this point in the history of mathematics, it is not up for discussion or choice. We are using the order that was established centuries ago.

The correct choice is Answer #1. There is an Order of Operations agreement or rule that everyone uses when evaluating expressions. Before we write down this rule, look back at the original problem. Is there any symbol or symbols that could have been inserted into the problem to make the order of operations more apparent to any reader?

Mathematicians use **parentheses** to group numbers together and to make expressions easier to read. If we had written $2+2\cdot 2$ as $2+(2\cdot 2)$, your eyes would have focused on the parentheses first and you would probably have done the operation inside them first. Other grouping symbols can be used. The most common examples are brackets, [ ] or braces, { }. These are usually chosen if parentheses are used elsewhere in the problem.

The rule for **Order of Operations** is:

A. Work within parentheses or other grouping symbols first.
B. Do the operations in this order:
1. Evaluate exponents and square roots.
2. Multiply and divide from left to right.
3. Add and subtract from left to right.

Note that this rule says that exponents and square roots are to be completed *before* multiplications or additions, and multiplications and divisions must be completed *before* additions and subtractions.

EXAMPLE 1 Evaluate $3+7(5-2)$

Copy the whole expression: $3+7(5-2)$

Parentheses first: $=3+7(3)$

At this point, it is very important to read this expression carefully. There are three numbers left and 2 operations still to be done. **The parenthesis between 7 and 3 indicates multiplication.** Multiplication must be done before the addition. $=3+21$

$=24$

OBSERVE In Example 1, it is **very** tempting to add the 3 and the 7 together at the beginning. However, as we saw in the solution, the 7 is not "free" to be added to the 3 because it is connected to the $5-2$ by parentheses which indicated multiplication. Since multiplication is done *before* addition, adding the 3 must wait until the multiplication of the 7 and the answer to the parentheses has been completed.

EXAMPLE 2 Evaluate $3\sqrt{25}+2(18-10)$

Copy the whole expression: $3\sqrt{25}+2(18-10)$

Parentheses first: $=3\sqrt{25}+2(8)$

Square root: $=3\cdot5+2(8)$

Multiply: $=15+16$

Add: $=31$

TIP: Read the problem to yourself as you choose which operation to perform next. Between every two numbers there will be an operation. Sometimes students forget that a parenthesis between two numbers can indicate multiplication and that **no** operation symbol means multiplication.

$3\sqrt{25}+2(8)$ is read "3 TIMES the square root of 25 plus 2 TIMES 8". Remember to multiply before you add.

EXAMPLE 3 Evaluate $2^3+16\div 4\cdot 2-(7-3)$

Copy the whole expression:	$2^3+16\div 4\cdot 2-(7-3)$	
Parentheses first:	$=2^3+16\div 4\cdot 2-4$	Once the subtraction is done, there is no need for the parentheses.
Exponents:	$=8+16\div 4\cdot 2-4$	
Multiply and divide left to right:	$=8+8-4$	Notice the division is done first because we work left to right
Add and subtract left to right:	$=16-4$	Notice the addition is done first because we work left to right.
The answer is:	$=12$	

ORDER OF OPERATIONS ON YOUR CALCULATOR

Most scientific calculators are programmed to follow the rule for order of operations. This means that if you want to use your calculator to compute $3+4\cdot 5$, you would use the following keystrokes 3 [+] 4 [×] 5 [=] . Your calculator "knows" to multiply first so the display will read 23, which is the correct answer.

EXAMPLE 4 Use your calculator to evaluate 4(12 – 8)

Here we will need to use the parenthesis keys. Here are the keystrokes:

4 [×] [(] 12 [−] 8 [)] [=]

The display should read 16.

EXAMPLE 5 Use your calculator to check the answer to Example 3.

You need to evaluate $2^3+16\div 4\cdot 2-(7-3)$. The keystrokes are:

2 [y^x] 3 [+] 16 [÷] 4 [×] 2 [−] [(] 7 [−] 3 [)] [=]

The display should read 12.

♦ You Try It

Use the rule for order of operations to evaluate each expression. Use your calculator to check each answer.

1. $6\times 12-(30+5)\div 7-(8-4)$

2. $(3+8)^2$

3. 3^2+8^2

4. $5\sqrt{36}+6\cdot 2^4-(1+3)^3$

PROBLEM SOLVING USING ORDER OF OPERATIONS

Many of the problems we have solved in this chapter required more than one mathematical operation. Usually you thought about the problem, made an outline of the steps involved, and then did one operation at a time. It was important for you to decide the *order* in which each step was to be done. Think about the *order* of the following six statements.

1. You hand the clerk a \$20 bill.
2. You order a hamburger for \$1.50, a soda for \$.90, and fries for \$.99.
3. The clerk totals your bill.
4. The clerk subtracts your bill from \$20.
5. You enter the store.
6. You receive your change.

Clearly these are not in the correct order. The correct order of the statements is

_____, _____, _____, _____, _____, _____.

Again, ORDER MATTERS and we must establish the order in which we perform mathematical operations. To do the arithmetic of the problem in the above 6 statements, we need to add \$1.50, \$.90, and \$.99 and then subtract the total from \$20. To write this as **one** mathematical statement, parentheses are helpful:

$$\$20 - (\$1.50 + \$.90 + \$.99)$$

OBSERVE A common error in this situation is to write the subtraction backwards: (\$1.50 + \$.90 + \$.99) – \$20. This says " subtract the \$20 from the cost of the food", which is clearly not the correct order. Remember that subtraction is not *commutative*; the order in which you subtract two numbers makes a difference in the answer.

In the next few examples, we will take problems that require more than one operation and write the necessary arithmetic as a single mathematical statement. For now, the object is not to solve the problems, but to practice writing the necessary operations as a single expression.

EXAMPLE 6 Tony bought some new clothes for school. He bought 4 pairs of jeans at \$42 each, 5 shirts at \$28 each, one jacket for \$95, and 7 pairs of socks at \$4 per pair. How much did he spend on clothes?

OUTLINE:

1. We need the cost of 4 pairs of jeans at \$42 each: $4 \cdot 42$
2. We need the cost of 5 shirts at \$28 each: $5 \cdot 28$
3. We need the cost of 1 jacket at \$95: $1 \cdot 95$
4. We need the cost of 7 pairs of socks at \$4 per pair: $7 \cdot 4$
5. We need to add all these products together.

The operations needed to find the answer to this problem can be written as a single mathematical expression in which we write the sum of all the products listed above:

$$4 \cdot 42 + 5 \cdot 28 + 1 \cdot 95 + 7 \cdot 4$$

Notice that if the expression is written in this way, the multiplications **must** be done first, according to the rule for order of operations, which is exactly what we had intended!

EXAMPLE 7 Suppose Tony (see Example 6) had a clothing allowance of $500. How much money does he have left after his purchases?

OUTLINE:
1. We have already written an expression for his total purchases.
2. We need to subtract the total *from* $500.

$$500 - (4 \cdot 42 + 5 \cdot 28 + 1 \cdot 95 + 7 \cdot 4)$$

Parentheses are **essential** here. They tell you to complete the multiplications and additions before subtracting. Without the parentheses, only $4 \cdot 42$ would be subtracted from 500. The rest would be added.

♦ You Try It For each problem, write the operations needed to find the answer as a **single mathematical expression.** If necessary, outline the problem first.

5. A student pays $576 a month for rent and $194 a month for food during 9 months at college. What is the total cost?

6. A local car dealership lists the price of a new car at $16,435. They advertise a giant price break of $1250 on this model. Suppose that you wanted to buy the car for cash, but had saved only $7890 towards the price of the car. How much more will you need in order to pay for the car?

7. Leah has just received a bill for her car insurance. She can pay the whole amount, $876 at once, or she can pay in three installments. Under the three-payment plan, she would pay $300 now and then pay two other payments of $295 each. How much will she pay under the three-payment plan? How much is the company charging for this service?

---

2 TRANSLATIONS

As we move ahead in mathematics, it becomes more and more important to be able to move back and forth between the language of mathematics and the English language by translating mathematical expressions into English and English phrases into mathematics. We did some of this in Section 1.2 and now we will work with expressions that involve more than one operation. Grouping symbols can be used to highlight one operation and then you can apply a second operation to the result. You must follow order of operations in your translation. Remember to use the words, **sum, difference, product,** or **quotient** as appropriate.

EXAMPLE 8 Translate the mathematical expression $4(9-2)$ into English.

Mathematical expression:	$4(9-2)$
English phrase:	4 times the difference of 9 and 2

Note that the parentheses tell us to find the difference first and then multiply by 4. Writing "4 times the difference ..." tells us to find the difference first.

EXAMPLE 9 Translate the mathematical expression $4 \cdot 9 - 2$ into English.

Mathematical expression: $4 \cdot 9 - 2$
English phrase: Subtract 2 from the product of 4 and 9.

Order of Operations tells us to multiply first and then subtract 2. Writing "Subtract 2 from the product ..." tells us to find the product first.

Alternate translation: Since the answer to $4 \cdot 9 - 2$ is smaller than $4 \cdot 9$, we can write

2 less than the product of 4 and 9

EXAMPLE 10 Translate the mathematical expression $2x - 9$ into English.

Mathematical expression: $2x - 9$
English phrase: 9 subtracted from twice a number
Alternate translations: 9 less than twice a number
9 subtracted from double a number
The difference of twice a number and 9

EXAMPLE 11 Translate the mathematical sentence $8 < \frac{x}{2}$ into English.

Mathematical sentence: $8 < \frac{x}{2}$
English sentence: 8 is less than the quotient of a number and 2.
Alternate translations: 8 is less than a number divided by 2.
8 is less than half a number.

♦ You Try It Translate each mathematical expression or sentence into English.

8. $3(8 + 11)$ **9.** $3 \cdot 8 + 11$ **10.** $\frac{14}{5} + 6$

11. $\frac{14}{5+6}$ **12.** $3(7 - 4)$ **13.** $3 \cdot 7 - 4$

14. $4 - 3 \cdot 7$ **15.** $5y + 11$ **16.** $5y - 11$

17. $4 < 3x$ **18.** $t + 6 < 8$

Now we'll translate from English phrases to mathematical expressions. Once again, we'll be using more than 2 numbers and we must be careful to use grouping symbols when they are necessary.

Carefully consider these two expression $2(8 + 4)$ and $2(8) + 4$. In the first case, $2(8 + 4)$ we multiply the sum of 8 and 4 by 2. In the second case, we multiply only the 8 by 2 and add 4 to the product. The grouping symbols highlight a particular operation.

EXAMPLE 12 Translate the English phrase "twice the difference of 8 and 4" into mathematics.

English phrase: twice the difference of 8 and 4

Mathematical expression: $2(8-4)$

Since the subtraction must be done first, the parentheses are essential.

EXAMPLE 13 Translate the English phrase " 3 less than the product of 9 and 11" into mathematics.

English phrase: 3 less than the product of 9 and 11

Mathematical expression: $9 \cdot 11 - 3$

We could have used parentheses to say that $9 \cdot 11$ is done first: $(9 \cdot 11) - 3$. However, the parentheses are not necessary since the multiplication will be done first without them according to the rule for Order of Operations.

OBSERVE A common error is to write this expression $3 - 9 \cdot 11$ which writes the subtraction in the wrong order. "3 less than a number" means "take 3 away **from** a number" so we write the number **first** and then minus 3.

EXAMPLE 14 Translate the English sentence "3 is less than the product of 9 and 11" into mathematics.

Compare this *sentence* closely with the *phrase* of the previous example. They appear very similar but the presence of the verb "is" in the sentence means we have a mathematical sentence. The complete verb is " is less than " which translates $<$.

English sentence: 3 is less than the product of 9 and 11

Mathematical sentence: $3 < 9 \cdot 11$

EXAMPLE 15 Translate the English phrase " double your age and divide by 7 " into mathematics.

In this phrase, "your age" needs to be identified by a variable. We will use a to represent "your age".

English phrase: double your age and divide by 7

Mathematical phrase: $2a \div 7$

Alternate expression: $\frac{2a}{7}$

♦You Try It Translate the following English phrases or sentences into mathematics. If you use a variable, define it.

19. 10 times the sum of 4 and 8
20. twice 5, divided by 3
21. 3 times 9 plus twice the square of 9
22. 8 less than the product of 7 and 2
23. 8 is less than the product of 7 and 2
24. add 3 to a number and multiply the sum by 7
25. 5 is added to the product of 6 and some number

3 THE DISTRIBUTIVE PROPERTY

EXAMPLE 16 Suppose you own a clothing store and decide to have a two-day sale on shirts. The sale price of every shirt in the store is \$15. On the first day of the sale you sold only 12 shirts because you didn't advertise, but on the second day, word spread and you sold 45 shirts. How much money did you take in from the sale?

There are two possible ways to find the answer to this problem.

Method 1

Outline

1. Find the total number of shirts sold. $12 + 45$
2. Multiply by \$15.

We can write this as a single expression: $15(12 + 45)$
Using the rule for order of operations, we get: $= 15(57)$
$= 855$

Answer: You made \$855 from the two-day sale.

Method 2

Outline

1. Find the money earned from the first day. $15 \cdot 12$
2. Find the money earned from the second day. $15 \cdot 45$
3. Add the sales from the two days together.

We can write this as a single expression: $15 \cdot 12 + 15 \cdot 45$
Using the rule for order of operations, we get: $= 180 + 675$
$= 855$

Answer: You made \$855 from the two-day sale.

Both methods give the same answer, so we can conclude that the two expressions, $15(12 + 45)$ and $15 \cdot 12 + 15 \cdot 45$ are equal.

$$15(12 + 45) = 15 \cdot 12 + 15 \cdot 45$$

This is an example of a very useful property of numbers, called the *distributive property of multiplication over addition.* This property says that if you want to find the product of one number times the sum of two other numbers, you will get the same answer by multiplying that number times each of the numbers that were added. It is a property that lets us change the way an expression is written without changing its value.

The left side of the equation

$$15(12 + 45) = 15 \cdot 12 + 15 \cdot 45$$

is an expression in which addition is done first and multiplication last. The right side of the equation is an expression in which multiplication is done first and addition last. The distributive property lets us write a new expression with the same value as the old one but with a different order of operations. This is a very valuable tool in algebra.

Distributive Property of Multiplication over Addition

$$a(b+c) = a \cdot b + a \cdot c$$

EXAMPLE 17 Use the distributive property to rewrite the expression $5(8 + 7)$.

To use the distributive property to rewrite this expression, multiply each number inside the parentheses by the 5 on the outside:

$$5(8 + 7) = 5 \cdot 8 + 5 \cdot 7$$

You should check that both sides of this equation have the same value.

EXAMPLE 18 Use the distributive property to rewrite the expression $12(x + 2)$.

To use the distributive property to rewrite this expression, multiply each number inside the parentheses by the 12 on the outside:

$$12(x + 2) = 12 \cdot x + 12 \cdot 2$$

We do not need the multiplication symbol between the 12 and the x:

$$12(x + 2) = 12x + 12 \cdot 2$$

EXAMPLE 19 Use the distributive property to rewrite the expression $3 \cdot 6 + 3 \cdot 9$.

This time we have the expression in which the multiplication is done first. We want to write the expression in which the addition is done first. To do this, notice that **3** is a factor of both terms in this expression:

3 is called a **common factor**

$$\mathbf{3} \cdot 6 + \mathbf{3} \cdot 9$$

This common factor, 3, will be on the outside of the parentheses and the other two numbers, 6 and 9, will be added inside the parentheses:

$$\mathbf{3} \cdot 6 + \mathbf{3} \cdot 9 = \mathbf{3}(6 + 9)$$

♦ **You Try It**

Use the distributive property to rewrite each expression. Do not evaluate.

26. $5(4 + 10)$ **27.** $8(3 + 7)$ **28.** $7(y + 9)$ **29.** $6(11 + t)$

30. $7 \cdot 2 + 7 \cdot 5$ **31.** $4 \cdot 8 + 4 \cdot 6$ **32.** $12 \cdot 2 + 12 \cdot 10$ **33.** $6x + 6y$

◆ **Answers to You Try It** **1.** 63 **2.** 121 **3.** 73 **4.** 62 **5.** $9(576 + 194) =$ 6,930 The total cost of food and rent is \$6,930 **6.** $16{,}435 - 1{,}250 - 7{,}890 = 7{,}295$ You will need \$7,295 more to buy the car. **7.** $(300 + 2 \cdot 295) = 890$ Leah will pay \$890 under the three-payment plan. $(300 + 2 \cdot 295) - 876 = 14$ The service charge is \$14. **8.** triple the sum of 8 and 11 or 3 times the sum of 8 and 11 or the product of 3 and the sum of 8 and 11 **9.** 11 more than the product of 3 and 8 or 11 added to the product of 3 and 8 **10.** the quotient of 14 and 5 added to 6 or 6 more than the quotient of 14 and 5 **11.** the quotient of 14 and the sum of 5 and 6 or 14 divided by the sum of 5 and 6 **12.** 3 times the difference of 7 and 4 or the product of 3 and the difference of 7 and 4 or triple the difference of 7 and 4 **13.** 4 subtracted from the product of 3 and 7 or 4 less than the product of 3 and 7 **14.** the product of 3 and 7 subtracted from 4 or the difference of 4 and the product of 3 and 7 **15.** the sum of 5 times a number and 11 or 11 added to the product of 5 and a number **16.** 11 subtracted from 5 times a number or 11 subtracted from the product of 5 and a number or 11 less than the product of 5 and a number **17.** 4 is less than the product of 3 and a number. **18.** The sum of a number and 6 is less than 8 **19.** $10(4+8)$ **20.** $\frac{2(5)}{3}$ or $2 \cdot 5 \div 3$ **21.** $3 \cdot 9 + 2 \cdot 9^2$ **22.** $7(2) - 8$ **23.** $8 < 7 \cdot 2$ **24.** $7(x+3)$, where x is the number **25.** $6t + 5$, where t is the number **26.** $5 \cdot 4 + 5 \cdot 10$ **27.** $8 \cdot 3 + 8 \cdot 7$ **28.** $7y + 7 \cdot 9$ **29.** $6 \cdot 11 + 6t$ **30.** $7(2 + 5)$ **31.** $4(8 + 6)$ **32.** $12(2 + 10)$ **33.** $6(x + y)$

SECTION 2.6 EXERCISES

1 *Use the rule for order of operations to evaluate the expression. Use your calculator to check each answer.*

1. $8+6\cdot 5$

2. $8+6\cdot 5^2$

3. $8+6\sqrt{49}$

4. $12\div 2\cdot 3$

5. $12\div(2\cdot 3)$

6. $12\cdot 2\div 3$

7. $4^2-5\sqrt{9}$

8. $3(2.4)-0.5(8.6-3.9)$

9. $(4+2)(4-2)$

10. $4+2(4-2)$

11. $4+2\cdot 4-2$

12. $5\cdot 6\div 3\cdot 2\div 10$

13. $6(3)-1^8+3^3$

14. $\left(6+\sqrt{16}\right)\left(6-\sqrt{16}\right)$

15. $4\sqrt{36}-2\left(4^2-2^3\right)$

16. $\frac{3^3}{9}+2\cdot 3-15\div 5$

17. $(1.1)^2+3\left(\sqrt{0.04}-0.1(0.2)\right)$

18. $4^2-2[27\div 3-2(6-6\div 3)]$

For the following problems, three similar expressions are given. Two of them give the same answer. The only difference is in the placement of parentheses. Look at the expressions carefully and see if you can tell which two give the same answer before you evaluate any of them. Then evaluate each expression and see if you were right.

19. a. $2+3\cdot 5$ **b.** $(2+3)5$ **c.** $2+(3\cdot 5)$

20. a. $4\cdot(6-3)$ **b.** $(4\cdot 6)-3$ **c.** $4\cdot 6-3$

21. a. $\left(\frac{20}{4}\right)5$ **b.** $\frac{20}{4}\cdot 5$ **c.** $\frac{20}{(4\cdot 5)}$

22. a. $60\div 30\cdot 2$ **b.** $60\div(30\cdot 2)$ **c.** $\left(\frac{60}{30}\right)\cdot 2$

23. a. $(2\cdot 5)^2$ **b.** $2\cdot 5^2$ **c.** $2(5)^2$

---

*Write the operations needed to find the answer as a **single mathematical expression**. If necessary, outline the problem first. After you write one expression to solve the problem, solve the problem. Write your answer in a complete sentence.*

24. At the bookstore, Mike bought one textbook at \$65, five notebooks at \$3.59 each, and three pens at \$2.35 each. What is his total bill?

25. A car rental company charges \$14.95 per day plus \$.22 per mile. If you rent a car for 3 days and drive 550 miles, how much will it cost?

26. Allen bought 3 paint brushes for \$1.75 each and 7 tubes of oil paint for \$2.49 each. How much change did he get from \$25?

27. *Time* magazine costs \$2.50 per issue at the newsstand. A yearly subscription of 52 issues costs \$57.25. How much is saved in a year by paying for a yearly subscription?

28. At a recent Macy's White Sale, Jennifer bought 2 full-size sheet sets at \$34.99 a set, 2 pillows at \$14.99 each, and 2 pillow cases at \$10.00 each. How much change does she get if she gives the cashier \$200 for the purchases?

29. Samira has \$1,000 to spend on stocks. She buys 25 shares of one stock at \$21 per share and 30 shares of another stock at \$14 per share. How much money does she have left?

2 *Translate each mathematical expression or sentence into English. Do not evaluate the expressions.*

30. $5(7+2)$

31. $5 \cdot 7 + 2$

32. $\frac{16}{4} + 9$

33. $\frac{16}{4+9}$

34. $2(12-5)$

35. $2 \cdot 12 - 5$

36. $5 - 2 \cdot 12$

37. $2x + 3$

38. $2x - 3$

39. $2(x+3)$

40. $2(x-3)$

41. $\frac{y+2}{9}$

42. $\frac{y}{9} + 2$

43. $5t < 12$

44. $5t - 12$

45. $6 < \frac{n}{13}$

46. $\frac{n}{13} - 6$

Translate each phrase or sentence into mathematics. If you use a variable, define it. Do not evaluate any expressions.

47. Four times the sum of 5 and 7.

48. Three more than the product of 8 and 2.

49. Nine added to the quotient of 80 and 5.

50. Twelve divided by the sum of 4 and 6.

51. Ten times the difference of 25 and 7.

52. Seven less than the product of 10 and 25.

53. Seven is less than the product of 10 and 25.

54. Nine more than the quotient of 42 and 6.

55. Nine is more than the quotient of 42 and 6.

56. Three subtracted from twice 11.

57. Eight times the sum of a number and 15.

58. Twice the sum of ten and the number of tests.

59. The difference of four times five and three times two

60. Ten less than the product of three and the number of calculators.

61. Five more than the quotient of x and y.

62. Five is more than the quotient of x and y.

3 *Use the distributive property of multiplication over addition to rewrite the expression. Do not evaluate.*

63. $9(4 + 5)$ **64.** $3(11 + 2)$ **65.** $6(8 + 16)$ **66.** $20(13 + 7)$

67. $4(x + 2)$ **68.** $16(5 + t)$ **69.** $25(n + 20)$ **70.** $3(x + y)$

71. $8 \cdot 4 + 8 \cdot 2$ **72.** $5 \cdot 3 + 5 \cdot 8$ **73.** $40 \cdot 15 + 40 \cdot 12$ **74.** $6 \cdot 25 + 6 \cdot 9$

75. $3x + 3 \cdot 4$ **76.** $12t + 12k$ **77.** $16t + 16a$ **78.** $9 \cdot 7 + 9y$

SKILLSFOCUS (Section 2.3) *Multiply or divide by the power of ten.*

79. 7.31×10^3 **80.** 0.0112×10^2 **81.** $\frac{3{,}487}{10^3}$ **82.** $223{,}407 \div 10^5$

EXTEND YOUR THINKING ♦ ♦ ♦

SOMETHING MORE

83. Determine if each equation is true or false. Use the rule for order of operations to evaluate each side of the equation separately.

a. $\sqrt{(25 + 144)} = \sqrt{25} + \sqrt{144}$

b. $\sqrt{(169 - 25)} = \sqrt{169} - \sqrt{25}$

c. $\sqrt{(16 \cdot 25)} = \sqrt{16} \cdot \sqrt{25}$

d. $\sqrt{\frac{36}{9}} = \frac{\sqrt{36}}{\sqrt{9}}$

84. Make the equation true by inserting the correct operational symbols on the left side of each equal sign.

a. 8 8 8 8 = 3

b. 2 2 2 2 = 2

c. 6 6 6 6 = 13

d. 4 4 4 4 = 4

e. 1 1 1 = 0

f. 9 9 9 9 9 = 0

g. 2 1 7 = 7

h. 3 2 4 1 = 25

i. 1 0 0 1 = 1

j. 4 3 2 1 = 5

TROUBLESHOOT IT

Find and correct the error.

85. $7+2\cdot 3=9\cdot 3=27$

86. $3\cdot 6-3=3\cdot 3=9$

87. $10\div 5\cdot 2=10\div 10=1$

88. $20\div 10\div 2=20\div 5=4$

89. $10\cdot 10\div 10\cdot 10=100\div 100=1$

90. $8+3-4+2=11-6=5$

91. $2\cdot 3^2=6^2=36$

92. $3\cdot 2+4\cdot 7=6+4\cdot 7=10\cdot 7=70$

93. $2+2\cdot 2-2=4\cdot 2-2=4\cdot 0=0$

94. $5(6+2)=5\cdot(6+2)=30+2=32$

95. $5+2\sqrt{16}-7=5+2\sqrt{9}=5+2\cdot 3=5+6=11$

96. $4+6\div 2=10\div 2=5$

WRITING TO LEARN ♦ ♦ ♦

97. Write an expression for your age. It must include an addition, a power, a root and one other operation. Then evaluate it.

98. Consider these two expressions:

a. $(4+3)^2$ **b.** 4^2+3^2

How are they alike?
How are they different?
Evaluate each expression.

99. Consider these two expressions:

a. $(4\times 3)^2$ **b.** 4×3^2

How are they alike?
How are they different?
Evaluate each expression.

100. Write the rule for Order of Operations.

2.7 USING FORMULAS

OBJECTIVE
Perform calculations using formulas.

NEW VOCABULARY
volume
cubic units
right triangle
hypotenuse
Pythagorean Theorem

FORMULAS

Sometimes an equation uses variables to express a relationship among some quantities. Remember that we studied the formula for the area of a rectangle. This formula, $A = lw$, expresses a relationship that is **true all the time**. It says that the area of **any** rectangle is **always** found by multiplying the length of the rectangle by its width.

> A **formula** is an equation that uses variables to express a relationship that is **true** all the time.

EXAMPLE 1 Find the area of a rectangle with length 2.5 cm and width 1.7 cm.

Write the formula: $A = lw$

Substitute 2.5 for l and 1.7 for w: $A = (2.5\text{ cm})(1.7\text{ cm})$

$A = 4.25$ sq cm   Remember that area is measured in square units

Answer: The area of the rectangle is 4.25 sq cm.

EXAMPLE 2 Ted is purchasing new carpet for his living room, which measures 4 yd by 5 yd. The carpet costs $18.99 per square yard. How much will Ted pay for the carpet?

In this problem there is no mention of finding the area of the room. You need to know that area is a measure of the number of square units needed to *cover* a surface and since carpet *covers* the surface of the floor, the area of the floor must be found to determine the number of square yards of carpet needed.

OUTLINE
1. Find the area of the room in square yards.
2. Find the cost.

SOLUTION
1. To find the area, use the formula: $A = lw$

$A = (4\text{ yd})(5\text{ yd})$

$A = 20$ sq yd

2. To find the cost, multiply area by cost per square yard:

Cost = 20 sq yd × $18.99 per sq yd

Cost = $379.80

Answer: Ted will pay $379.80 to carpet his living room.

Another commonly used formula is one used for finding the distance an object travels if you know its rate of speed and the length of time it traveled. This formula is distance = rate × time; using variables we could write $d = rt$.

$d = rt$, where d is the distance traveled, r is the rate of speed, and t is the time traveled.

EXAMPLE 3 Use the formula $d = rt$ to find the distance you would travel if you drove at 48 miles per hour for 6 hours.

Write the formula:	$d = rt$
Substituting 48 for r and 6 for t:	$d = 48 \times 6$
	$d = 288$ miles

Answer: If you drove for 6 hours at 48 miles per hour, you would cover 288 miles.

Perimeter

There are other important formulas with which you should be familiar. Remember that in Chapter 1 we said that the *perimeter* of an object is the distance around it. One of the most common geometric shapes is a rectangle. Here is a rectangle with length l and width w:

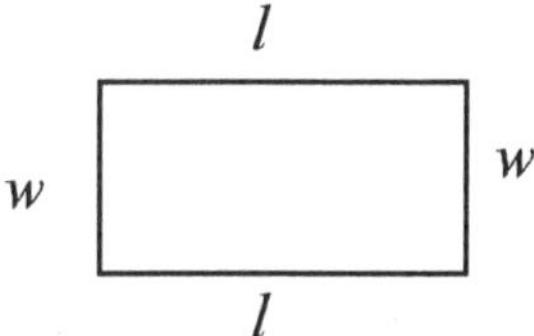

To find the perimeter of the rectangle, we need to add the lengths of all the sides. Of course, two of the sides measure l units and the other two sides measure w units, so we should add two l's and two w's. In the symbols of mathematics, we would write that as $2l + 2w$. This gives us a **formula** for the perimeter of a rectangle.

The *perimeter* of a rectangle whose length is l and whose width is w is $P = 2l + 2w$. The perimeter is measured in the same units as the sides of the rectangle.

EXAMPLE 4 Find the perimeter of a rectangle with length 15 feet and width 12 feet.

Write the formula:	$P = 2l + 2w$	
Substitute 15 for l and 12 for w:	$P = 2(15\text{ ft}) + 2(12\text{ ft})$	Multiply before you add.
	$P = 30\text{ ft} + 24\text{ ft}$	Notice that these are both **feet** and can be added.
	$P = 54\text{ ft}$	

Answer: The perimeter is 54 feet.

EXAMPLE 5 Glenda is digging a vegetable garden and wants to enclose it with a wire fence. The rectangular garden is 18 feet long and 6 feet wide. The wire fence costs \$3.50 per foot. How much will Glenda's fence cost?

In this problem there is no mention of finding the perimeter of the garden. You need to know that perimeter measures the distance *around* an object and since a fence *goes around* the garden, the perimeter of the garden must be found to determine the number of feet of fence needed.

OUTLINE
1. Find the perimeter.
2. Find the cost.

SOLUTION
1. To find the perimeter, use the formula:

$$P = 2l + 2w$$
$$P = 2(18 \text{ ft}) + 2(6 \text{ ft})$$
$$P = 36 \text{ ft} + 12 \text{ ft}$$
$$P = 48 \text{ ft}$$

2. To find the cost, multiply perimeter by cost per foot:

$$\text{Cost} = 48 \text{ ft} \times \$3.50 \text{ per ft}$$
$$\text{Cost} = \$168$$

Answer: The fence for the garden will cost $168.

Squares

A square is a rectangle whose length and width are equal. We could use the formula for the perimeter of a rectangle to find the perimeter of a square, but we could also write a new formula for the perimeter of a square.

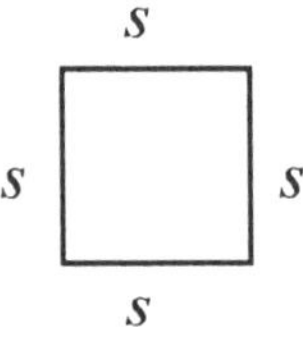

In a square, all four sides have the same length. If the side of a square is s units, then to find the perimeter we add $s + s + s + s$, which is the same as $4 \times s$ or $4s$. This gives us a formula.

The formula for the perimeter of a square with side s units is $P = 4s$.

EXAMPLE 6 Find the perimeter of a square with a side of 8 inches.

Write the formula: $P = 4s$

Substitute 8 for s: $P = 4(8 \text{ in})$

$P = 32 \text{ in}$

Answer: The perimeter is 32 in.

Volume

Another formula from geometry concerns volume. Remember that to find the **perimeter** of an object you are measuring length, and length is measured in plain units, like inches, centimeters, yards, etc. Area **covers** and if you need to cover a flat surface (like a floor), you cannot use units of length. **Area** is measured in square units, like square inches or square centimeters. Think of covering a floor with tiles. **Volume** does not measure length and does not cover, but instead fills up the space within a three-dimensional shape, such as a box. We cannot fill up a space with plain units, nor with square units, since they have no height. To fill, we need **cubic units**; that is, we fill up the space with little cubes 1 unit on each side: Sugar **cubes** and ice **cubes** are called cubes because they are three dimensional objects. They fill up your sugar bowl or your ice cube tray. To be a real cube the length, width and height of the object must all have the same measurement (which is not usually the case with either sugar or ice cubes.)

> **Volume** is the number of cubic units needed to **fill up** the space within a three-dimensional solid.

We will find a formula for the volume of a box, sometimes called a rectangular solid. Suppose the box is l units long, w units wide, and h units high.

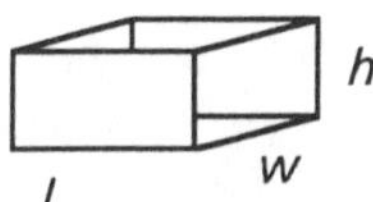

You need to picture this box filled with little cubes 1 unit on each side:

It takes $l \times w$ of these cubes to cover the bottom of the box and then there are h layers, so the total number of cubes needed to fill the box is $l \times w \times h$.

> The formula for the volume of a rectangular solid is $V = lwh$ cubic units.

EXAMPLE 7 Find the volume of a box that is 3.2 meters long, 1.7 meters wide, and 2.6 meters high.

Use the formula: $V = lwh$

Substitute 3.2 for l, 1.7 for w, and 2.6 for h: $V = (3.2\text{ m})(1.7\text{ m})(2.6\text{ m})$

$V = 14.144$ cubic meters

Answer: The volume of the box is 14.144 cubic meters.

EXAMPLE 8 Pete has a swimming pool that is 24 ft long, 15 ft wide, and 5 ft deep. How much water is needed to fill the pool?

In this problem there is no mention of finding the volume of the pool. You need to know that volume *fills up* and since a pool gets *filled up* with water, the volume of the pool must be found to determine the number of cubic feet of water needed.

Find the volume by using the formula: $V = lwh$

$V = (24\text{ ft})(15\text{ ft})(5\text{ ft})$

$V = 1{,}800$ cubic feet

Answer: The pool requires 1,800 cubic feet of water.

♦ **You Try It**

For each problem, choose an appropriate formula. For practice, do the calculations without a calculator.

1. Find the perimeter of a rectangle with a width of 12 centimeters (cm) and a length of 8 cm.
2. Paul has a new dog and wants to fence in a rectangular portion of the backyard 8 yards long and 6 yards wide for it. How much fence will he need? If the fence costs $8.29 per yard, how much will the fence cost?
3. Suppose you took a high speed train that averages 75 miles per hour. If your trip lasts 2.5 hours, how far will you travel?
4. How would you find the perimeter of a triangle?
5. Find the perimeter of a triangle with sides 11 ft, 8 ft, and 7 ft.
6. Describe a situation in which you would need to find the perimeter of a square, rectangle, or triangle.
7. You want to put a chair rail around the wall of your dining room. The room is 9 feet by 14 feet with one 4 foot doorway. The rail you are looking at costs $7.50 per foot. How much will it cost to put the rail around your dining room?
8. Find the area of a rectangle that is 12.6 meters by 8.2 meters.
9. Now you decide to carpet your dining room (see problem 7). How much carpet will you need?
10. Find the volume of a rectangular solid that is 2.4 m long, 1.1 m wide, and 0.4 m high.
11. Fran has built a sandbox in the backyard for her son. It is 3 feet square and 1 foot deep. How much sand will fill the sandbox?

The Pythagorean Theorem

The last geometry formula for this section concerns a special type of triangle. As you know, a triangle is a straight-sided figure with three sides and three angles. If two of those sides are perpendicular, the angle formed is called a **right angle**. For example, a right angle occurs when the wall is perpendicular to the floor. A right angle measures 90°. A right angle looks like one of these:

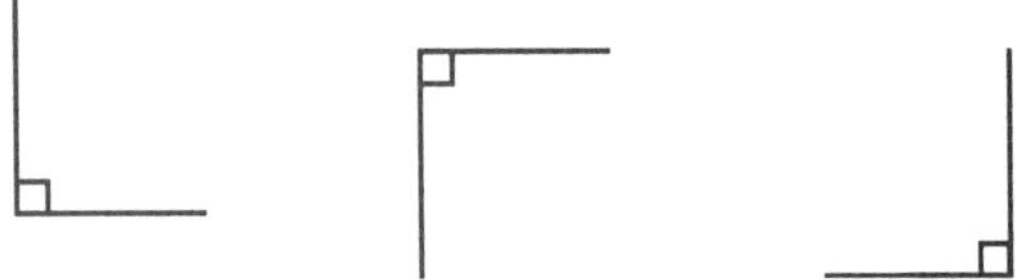

If a triangle has a right angle it is called a **right triangle**. Here are some examples of right triangles:

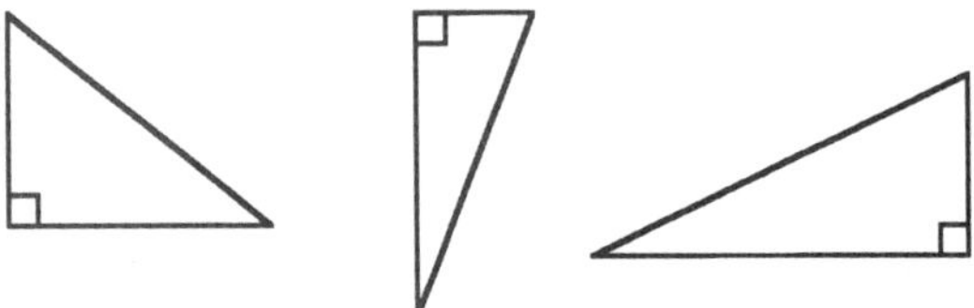

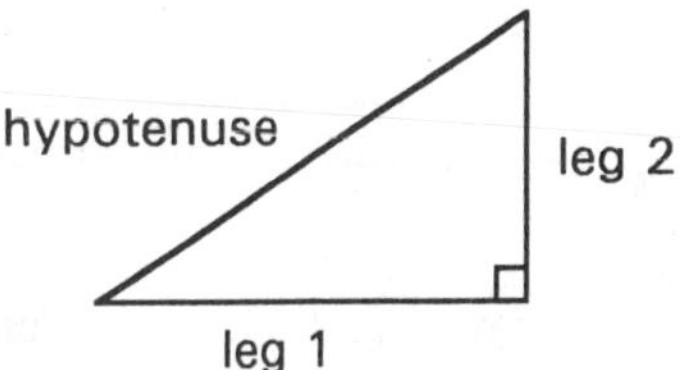

There is an amazing relationship among the sides of a right triangle. This relationship has been known for thousands of years and is named after a Greek philosopher and mathematician named Pythagoras. The relationship is called the **Pythagorean Theorem**.

The Pythagorean Theorem

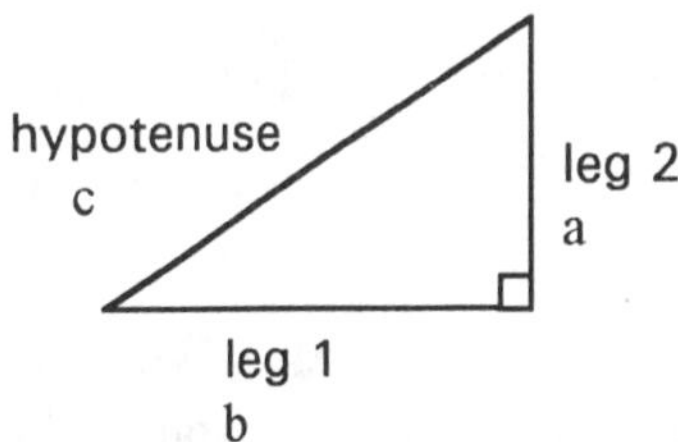

In any right triangle, the sum of the squares of the legs is equal to the square of the hypotenuse. That is, in any right triangle,

$$(\text{leg }1)^2 + (\text{leg }2)^2 = (\text{hypotenuse})^2$$

If the legs are called a and b and the hypotenuse is called c, then

$$c^2 = a^2 + b^2$$

We can use the Pythagorean Theorem to find the length of the third side of a right triangle if we know the lengths of the other two sides. For now, we will use it to find the length of the hypotenuse if we know the lengths of the two legs.

EXAMPLE 9 Find the hypotenuse of a right triangle whose legs measure 3 inches and 4 inches.

First, draw a picture and call the length of the hypotenuse c.

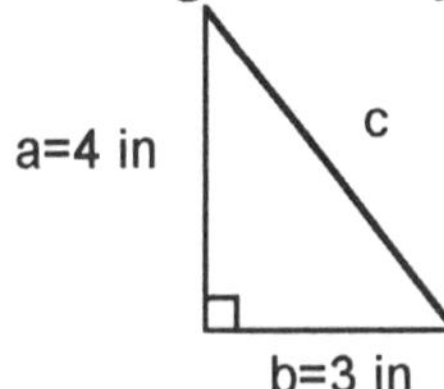

Next, write the Pythagorean Theorem: $c^2 = a^2 + b^2$

Substitute 4 in for one leg, a, and 3 in for the other leg, b: $c^2 = (4)^2 + (3)^2$ Remember to square first.

$c^2 = 16 + 9$

$c^2 = 25$

$$c^2 = 16 + 9$$
$$c^2 = 25$$

Notice what this equation says. The **square** of c is 25. We want to know the value of c and we know the square. To find the number whose **square** is 25 means we want the **square root** of 25. Thus if $c^2 = 25$, then c is the square root of 25. We write this:

$$c^2 = 25$$
$$c = \sqrt{25}$$
$$c = 5 \qquad \text{Recall that } \sqrt{25} = 5.$$

Answer: The length of the hypotenuse is 5 inches.

EXAMPLE 10 Find the hypotenuse of the right triangle whose legs measure 12 cm and 5 cm.

Draw a picture and label the legs and hypotenuse:

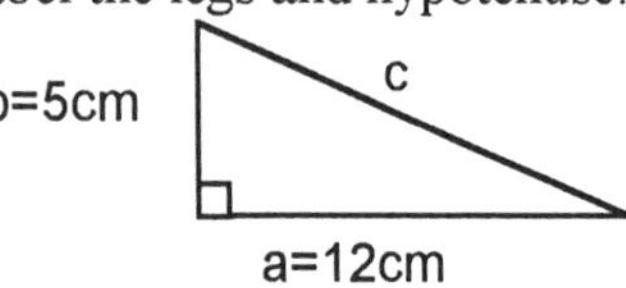

Next, write the Pythagorean Theorem: $c^2 = a^2 + b^2$

Substitute 12 for a and 5 for b: $c^2 = (12)^2 + (5)^2$

$c^2 = 144 + 25$ Square first.

$c^2 = 169$ This says that the square of c is 169.

$c = \sqrt{169}$ Take the square root of 169 to find c.

$c = 13$ $\sqrt{169} = 13$

Calculator keystrokes: 12 $\boxed{x^2}$ $\boxed{+}$ 5 $\boxed{x^2}$ $\boxed{=}$ $\boxed{\sqrt{\ }}$

Note that the calculator "knows" to square before adding.

Answer: The length of the hypotenuse is 13 cm.

♦ **You Try It** For these problems you may use your calculator. However, be sure to write the Pythagorean Theorem and show all steps of the solution.

12. Find the length of the hypotenuse of this right triangle:

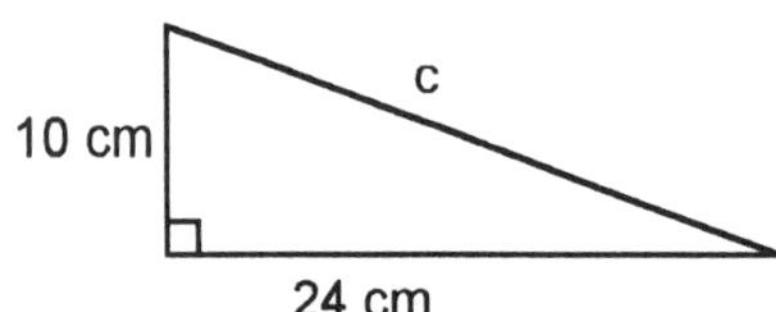

13. Find the hypotenuse of the right triangle whose legs measure 9 ft. and 12 ft.

14. Find the hypotenuse of the right triangle whose legs measure 40 cm and 9 cm.

SOMETHING MORE

Did the Scarecrow Get a Dud? *The Wizard of Oz* (1939), starring Judy Garland, is a movie classic. The movie centers around four characters each in need of something they feel only the wizard can give. Dorothy wants to return to her Kansas home, the Lion wants courage, the Tin Man a heart, and the Scarecrow wants a brain. Eventually, each receives his or her wish.

The Scarecrow, proud of his newly acquired brain, exercises it by reciting the Pythagorean Theorem.

Unfortunately, the Scarecrow's rendition of this famous theorem had three errors. He said:

> "The sum of the square roots of any two sides of an isosceles triangle equals the square root of the remaining side."

Can you find the three errors? Reread the Scarecrow's statement and try, before looking at the answers.

Admittedly, poetic license reigns in the making of a movie. But was this poetic license, or did the Scarecrow really get a dud? File this one under cinematic trivia. The three errors are:

1. It is not square roots, but squares.
2. It is not isosceles triangle, but right triangle.
3. It is not any two sides, but the two sides touching the right angle (the legs).

♦**Answers to You Try It** **1.** 40 cm **2.** 28 yd and $232.12 **3.** 187.5 miles **4.** Add the lengths of the 3 sides of the triangle. **5.** 26 ft. **6.** Picture frames, molding or any edging require perimeter. **7.** 42 ft. at a cost of $315 **8.** 103.32 sq m **9.** 126 sq ft **10.** 1.056 cu m **11.** 9 cu ft **12.** 26 cm **13.** 15 ft **14.** 41 cm

SECTION 2.7 EXERCISES

Find the perimeter of each rectangle. Write down the appropriate formula.

1.

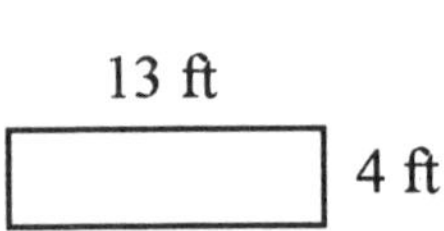

2.

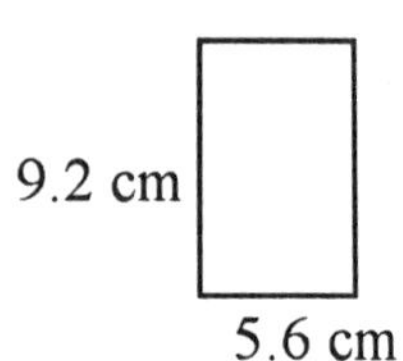

3.

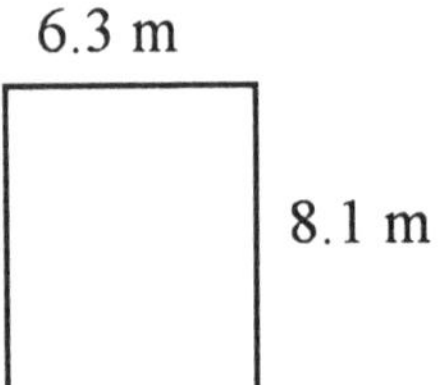

4.

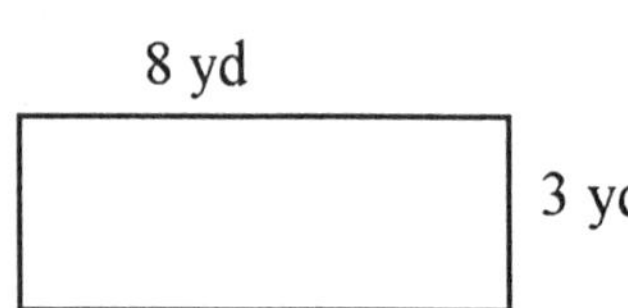

Find the perimeter of each square. Write down the appropriate formula.

5.

6.

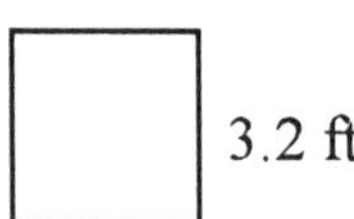

7.

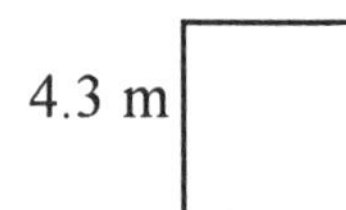

8.

Find the perimeter. Sketch the figure and write down the appropriate formula.

9. A rectangle 16 ft long and 3.6 ft wide.

10. A rectangle 4.27 mm wide by 7.45 mm long.

11. A square that measures 8 in on each side.

12. A square whose side is 5 yd.

13. A 15 in by 8 in rectangle.

14. A 17 in square.

Find the volume of each rectangular solid. Write down the formula.

15.

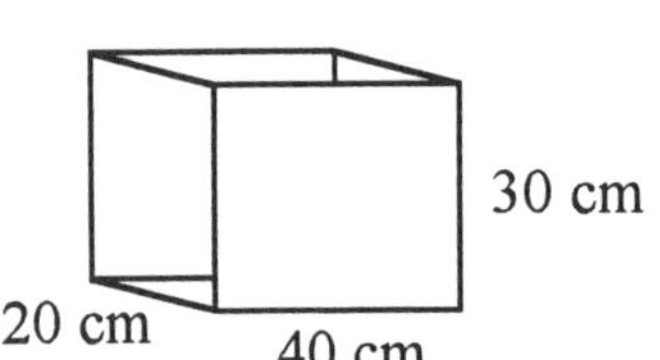

16.

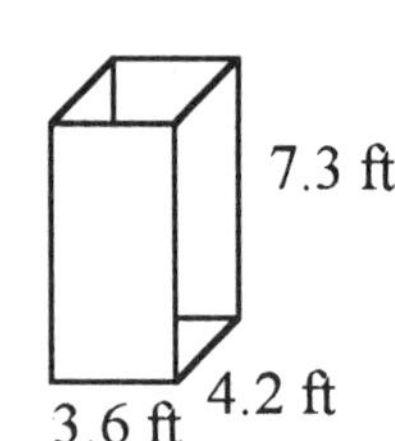

Find the volume. Write down the formula.

17. A rectangular solid that is 9 ft long, 3 ft wide, and 6.5 ft high.

18. A rectangular solid that is 16 cm long, 12 cm wide, and 22 cm high.

19. A cube that is 1 foot on each edge.

20. A cube that is 2 ft on each edge.

Decide whether you need to find perimeter, area, or volume to solve each problem. Use the appropriate formula.

21. A farmer wants to fence a rectangular plot of land that is 400 meters long and 32 meters wide. If the fencing material costs $2.95 per meter, find the cost of fencing in this plot.

22. A painting that is 2 ft by 1.5 ft is to be framed with a frame that costs $12 per foot. How much will it cost to frame the picture?

23. After some construction, a homeowner wants to seed a rectangular patch of her lawn. The area she wants to seed is 36 ft by 48 ft. Small bags of grass seed cover 1,000 square feet. How many bags will she need?

24. Maria's dining room is 4 yards by 3.5 yards.

a. What is the area of her dining room?

b. If Maria buys carpeting for $23.99 per square yard, how much will it cost to carpet the dining room?

c. If Maria buys baseboard molding at $3.50 per yard, how much will it cost for baseboard molding for the dining room? (You can ignore doorways.)

25. Deana wants to build a concrete walkway. It will be 30 feet long, 5 feet wide and 0.5 feet deep.

a. How many cubic feet of concrete will be needed for the job?

b. If the concrete costs $10 per cubic foot, how much will she have to pay?

26. An addition is being built onto a house. The hole for the foundation is 22 ft by 36 ft by 10 ft. How much dirt was removed to make this hole?

Use the Pythagorean Theorem to find the length of the hypotenuse.

27.

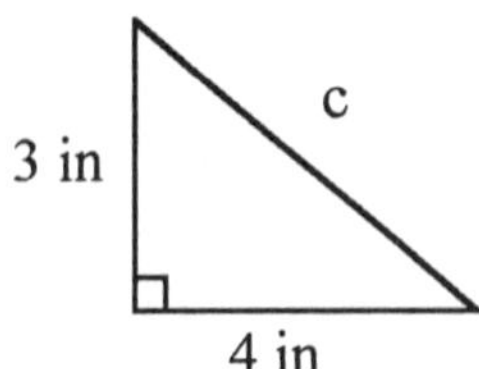

28.

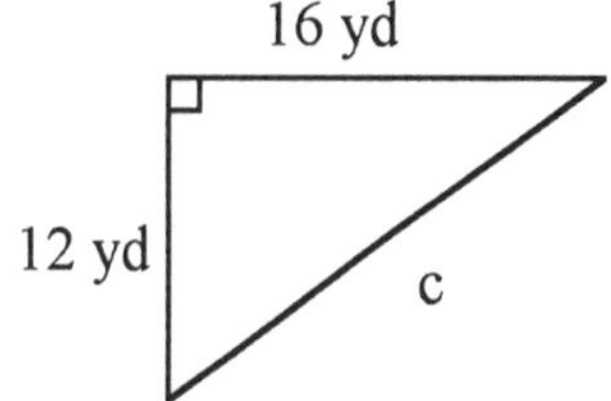

29.

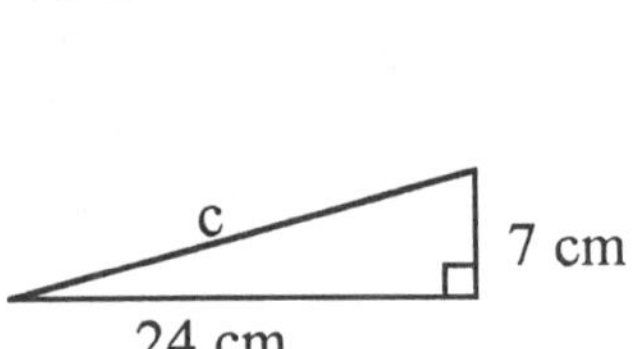

30.

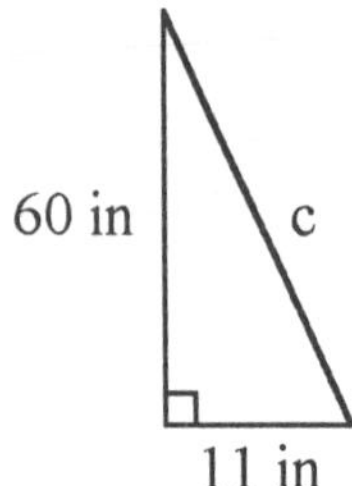

For each triangle, a) find the length of the hypotenuse, and b) find the perimeter of the triangle.

31.

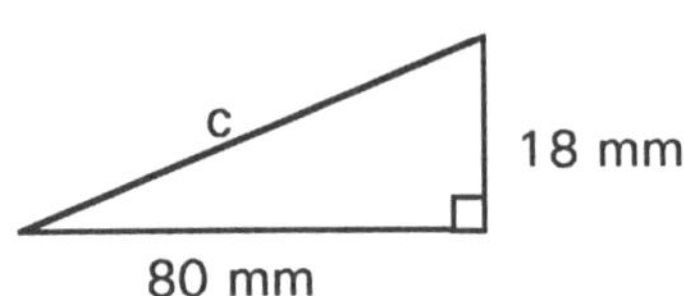

32.

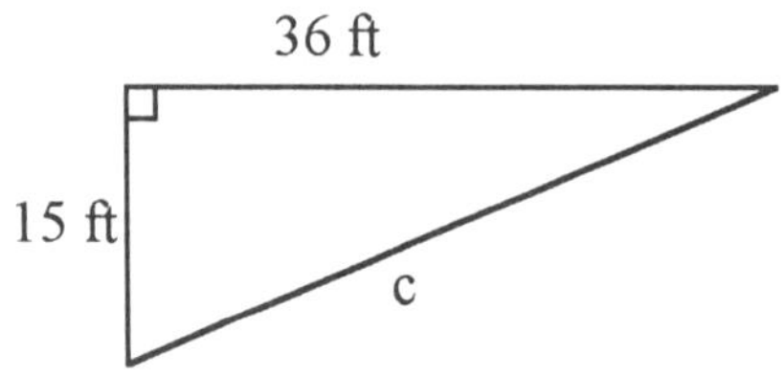

SKILLSFOCUS (Section 2.2) *Evaluate each expression.*

33. 7^3

34. $(0.03)^2$

35. $(0.003)^2$

36. 2^6

Use Figure 2.8 to answer questions 37 - 39.

37. What is the total number of Pulitzers won by the five newspapers?

38. The New York Times has how many more awards than the paper with the second highest number of awards?

39. Compare the combined total of the number of Pulitzer Prizes won by the L.A. Times and the Philadelphia Inquirer to the number of Pulitzer Prizes won by the Associated Press. Use one of these symbols: < or > .

FIGURE 2.8

Use Figure 2.9 to answer questions 40 - 42.

40. How many people were killed in accidents in the U.S. in 1992?

41. What is the leading cause of accidental deaths?

42. List some of the other top causes for accidental deaths.

43. How does the 1992 accidental death rate compare to that of 1922?

FIGURE 2.9

Accidental death rate falls

By Lori Sharn
USA TODAY

Accidents killed 84,000 people in 1992, the USA's lowest toll in 70 years, the National Safety Council announced Wednesday.

A poor economy and successful safety programs contributed to the decline, says the council's Alan Hoskin.

"More people out of work . . . reduces work-related deaths," Hoskin says, and results in less driving.

The council estimates:

▶ Accidental deaths dropped 5% from '91.

▶ Motor vehicles are the leading cause, with 40,100 deaths in '92, an 8% drop.

▶ Deaths in the workplace fell 7% to 9,200.

▶ Though not yet computed for '92, falls are generally the second leading cause.

▶ Other top causes: poisonings - including drug overdoses - drownings, fires.

The '92 accidental death rate of about 33 per 100,000 people — higher than some countries — is less than half of 1922's. Since then, the population more than doubled.

Considering how the population and economy have grown, says Hoskin, "that's a remarkable achievement."

EXTEND YOUR THINKING♦ ♦ ♦

♦SOMETHING MORE

44. Suppose you have a 6 ft by 9 ft rectangular room.
  a. If you double the length of each side, does the perimeter also double?
  b. If you double the length of each side does the area also double?

WRITING TO LEARN♦ ♦ ♦

45. Write the formula for the perimeter of a rectangle. What kind of units are used to measure perimeter?

46. Write the formula for area of a rectangle. What kind of units are used to measure area?

47. Write the formula for the volume of a rectangular solid. What kinds of units are used to measure volume?

48. Explain, in your own words, the difference between perimeter, area and volume.

49. What is a right triangle?

50. What does the Pythagorean Theorem say?

2.8 FRACTIONS AND DECIMALS: THE NEED FOR BOTH

OBJECTIVES

1. Change fractions to decimals.
2. Solve applications with both fractions and decimals.
3. Evaluate square roots of numbers that are not perfect squares.

NEW VOCABULARY

terminating decimal
repeating decimal

BACKGROUND At the beginning of this course, we reviewed the kinds of numbers we encounter in our daily lives. Some people might think that in a calculator and computer-based world the need for fractions is eliminated. However, fractions are rooted in our culture in ways that are hard to change. For example, tailors and dressmakers buy fabric in increments of eighths of a yard, carpenters measure wood in fractions of an inch, many tools are measured in fractions of an inch, and we buy cold cuts in quarter or half pound amounts. Even our money is often described using fractions: a quarter is 1/4 of a dollar. The origin of our use of fractions has to do with the history of our measuring system; the units that were developed, like feet, yards, miles, ounces, pounds, etc., did not have a consistent conversion factor. For example, there are 12 inches to the foot, but 3 feet in a yard, and 16 ounces in a pound. To convert from one unit to another, we may need to divide by 12 or by 3, which does not always lend itself to "nice" decimals. Fractions, then, became the numbers of choice. In the metric system, however, conversion from one unit to another is accomplished by multiplying or dividing by a power of 10. As we saw earlier in this chapter, multiplying or dividing by a power of 10 simply moves the decimal point in the number. In the metric system decimals are preferred because they are easy to work with. Since it is highly unlikely that we will give up our methods of measuring in the near future, a working knowledge of fractions is required, and furthermore, we will need to be able to change easily from fraction to decimal and decimal to fraction in order to deal with problems that involve both kinds of numbers.

1 CHANGING FRACTIONS TO DECIMALS

In Section 1.6 we saw how decimals could be written as fractions with denominators that were powers of 10. Fractions can be written as decimals, giving us different names for the same number. The form we use depends only on which is most appropriate in a given situation. Some fraction to decimal conversions are obvious from our experience with money. For example:

$\frac{1}{2} = 0.50$ is obvious since one-half dollar is the same as 50 cents

$\frac{1}{4} = 0.25$ is obvious since one-quarter of a dollar is the same as 25 cents

$\frac{3}{4} = 0.75$ is obvious since three-quarters of a dollar is the same as 75 cents

However, most decimal forms of fractions are less obvious. To see how to find these, recall that the bar in a fraction means "per" or "divided by". Thus

$\frac{5}{8}$ means 5 divided by 8 or $5 \div 8$

$\frac{7}{20}$ means 7 divided by 20 or $7 \div 20$

$\frac{11}{5}$ means 11 divided by 5 or $11 \div 5$

To find the decimal form of these fractions, we need to complete the division, supplying a decimal point at the end of the numerator and adding zeros as necessary.

EXAMPLE 1 Write $\frac{5}{8}$ in decimal form.

$\frac{5}{8} = 5 \div 8 = 0.625$

Remember: The denominator is the divisor.

Work:

$$\begin{array}{r} 0.625 \\ 8\overline{)5.000} \\ \underline{4\,8} \\ 20 \\ \underline{16} \\ 40 \\ \underline{40} \end{array}$$

When the division comes to a stop with a remainder of 0, the decimal is called a terminating decimal. The decimal 0.625 is a terminating decimal.

Calculator keystrokes: 5 [÷] 8 [=]

Answer: $\frac{5}{8} = 0.625$

EXAMPLE 2 Write $\frac{7}{20}$ in decimal form.

$\frac{7}{20} = 7 \div 20 = 0.35$

Work:

$$\begin{array}{r} 0.35 \\ 20\overline{)7.00} \\ \underline{6\,0} \\ 1\,00 \\ \underline{1\,00} \end{array}$$

Calculator keystrokes: 7 [÷] 20 [=]

Answer: $\frac{7}{20} = 0.35$ (Again, 0.35 is a terminating decimal.)

EXAMPLE 3 Write $\frac{11}{5}$ in decimal form.

$\frac{11}{5} = 11 \div 5 = 2.2$

Work:

$$\begin{array}{r} 2.2 \\ 5\overline{)11.0} \\ \underline{10} \\ 1\,0 \\ \underline{1\,0} \end{array}$$

Calculator keystrokes: 11 [÷] 5 [=]

Answer: $\frac{11}{5} = 2.2$

> **To Write a Fraction as a Decimal**
>
> To write the fraction $\frac{a}{b}$ as a decimal, divide the numerator by the denominator. That is,
>
> $\frac{a}{b} = a \div b$ or $b\overline{)a}$.

♦ You Try It Write each fraction in decimal form.

1. $\frac{3}{8}$ **2.** $\frac{3}{5}$ **3.** $\frac{4}{25}$ **4.** $\frac{5}{16}$

5. Explain how to write a fraction in decimal form.

Mixed numbers can be put into decimal form by leaving the whole number part as it is given and changing the fraction part to a decimal number.

EXAMPLE 4 Write $5\frac{3}{4}$ in decimal form.

The 5 is the same in either fraction or decimal form. $\frac{3}{4} = .75$

so $5\frac{3}{4} = 5.75$.

Repeating Decimals

EXAMPLE 5 Write $\frac{2}{3}$ in decimal form.

$\frac{2}{3} = 2 \div 3 = 0.6666...$

Work:

$$\begin{array}{r} 0.6666 \\ 3\overline{)2.0000} \\ \underline{1\,8} \\ 20 \\ \underline{18} \\ 20 \\ \underline{18} \\ 20 \\ \underline{18} \\ 2 \end{array}$$

It seems clear that this pattern will go on forever.

This example is different. The decimal form of $\frac{2}{3}$ is called a repeating decimal because the 6 repeats forever in the quotient. We sometimes write a repeating decimal with a bar over the digits that repeat:

$\frac{2}{3} = 2 \div 3 = 0.\overline{6}$ The bar over the 6 indicates that the 6 repeats forever.

However, when actually using the decimal form of $\frac{2}{3}$ in doing arithmetic, it is usually necessary to round it.

0.6666... rounded to the nearest tenth is 0.7.

Which is larger, 0.6666... or 0.7 ? Writing the numbers vertically, we have

0.6666...

0.7

Comparing place by place, we see that in the tenths place, 7 is larger than 6, so 0.7 is larger than 0.6666.... . This means that if 0.7 is used as a rounded version of $\frac{2}{3}$, we are using a slightly larger number, so the answer we get using 0.7 will be different from, but close to, the answer we get when using $\frac{2}{3}$.

0.6666.... rounded to the nearest hundredth is 0.67. Here, 0.67 is larger than 0.6666.... .

0.6666.... rounded to the nearest thousandth is 0.667, which is again larger than 0.6666....

There are two important things to notice in all this:

1. No matter what place we round to, the rounded version of $\frac{2}{3}$ will be larger than the actual repeating decimal.

2. Any rounded decimal we use will be an **approximation** to $\frac{2}{3}$ and not the exact value.

The **only** way to use the **exact** value of $\frac{2}{3}$ is to use the fraction form.

When working with fractions whose decimal form repeats, you must decide how accurate an answer you need. If you must have the exact value, the only way to get it is to work with the fraction itself. If you choose to work with the decimal form, your answer will become more accurate if you round to a position further to the right of the decimal point.

EXAMPLE 6 Find the decimal form of $\frac{8}{11}$. Round the decimal to the nearest: a) tenth, b) hundredth, c) thousandth, d) ten-thousandth. In each case decide whether the decimal is larger or smaller than the exact value.

$\frac{8}{11} = 8 \div 11 = 0.\overline{72}$ Work:

$$\begin{array}{r} 0.727272 \\ 11\overline{)8.000000} \\ \underline{7\ 7} \\ 30 \\ \underline{22} \\ 80 \\ \underline{77} \\ 3 \end{array}$$

a. $\frac{8}{11}$ rounded to the nearest tenth is 0.7 and $0.7 < 0.727272...$ so 0.7 is *smaller* than the exact value

b. $\frac{8}{11}$ rounded to the nearest hundredth is 0.73 and $0.73 > 0.727272...$ so 0.73 is *larger* than the exact value

c. $\frac{8}{11}$ rounded to the nearest thousandth is 0.727 and $0.727 < 0.727272...$ so 0.727 is *smaller* than the exact value

d. $\frac{8}{11}$ rounded to the nearest ten-thousandth is 0.7273 and $0.7273 > 0.727272...$ so 0.7273 is *larger* than the exact value

♦ **You Try It**

Write each of the following fractions in decimal form. Do the division by hand. If the decimal is repeating, round it to the nearest a) tenth, b) hundredth, c) thousandth, d) ten-thousandth. In each case, decide whether the decimal is larger or smaller than the exact value.

6. $\frac{1}{3}$ **7.** $\frac{5}{9}$ **8.** $\frac{7}{12}$ **9.** $\frac{9}{16}$ **10.** $\frac{5}{11}$

11. Consider the statement $\frac{5}{6} = 0.83$. Is it true or false? Explain your answer.

12. What is the exact value of $\frac{4}{9}$?

2 PROBLEM SOLVING

There are many instances in daily life of problems that involve both fractions and decimals. We will consider just a few of these at this time.

EXAMPLE 7 Find the cost of $\frac{3}{4}$ pound of turkey breast at $5.25 per pound.

This is an example of a situation in which we use a fraction to describe the amount of turkey breast we wish to purchase and a decimal to describe the cost per pound. Regardless of the kinds of numbers involved in a situation, we first need to analyze the problem to see what needs to be done to arrive at the answer. Once we have determined what needs to be done, we can focus on the method for doing it.

ANALYZE

Since we know the cost per pound and the number of pounds and we wish to know total cost, we need to multiply since

dollars per pound × pounds = total dollars

In this case we will multiply $5.25 per pound × $\frac{3}{4}$ pound

ESTIMATE

$5.25 per pound × $\frac{3}{4}$ pound $\doteq 5 \times 1 = 5$. The turkey should cost *about* $5.

SOLUTION

If we are using a calculator, using decimal form would be the easiest way to do the multiplication. We know that $\frac{3}{4} = 0.75$, so the multiplication becomes

$5.25 per pound × 0.75 pound = $3.9375

Since this is money, we must round to the nearest cent (penny), which is the hundredths position, so

$5.25 per pound × 0.75 pound ≐ $3.94

Calculator keystrokes: 5.25 [×] .75 [=]

ANSWER: $\frac{3}{4}$ pound of turkey breast at $5.25 per pound will cost $3.94.

EXAMPLE 8 A quilt maker has chosen several fabrics for a new quilt. She will use a total of $3\frac{5}{8}$ yards of the fabric that costs \$4.99 per yard and $2\frac{1}{2}$ yards of the fabric that cost \$5.75 per yard. What is the total cost of her purchases?

ANALYZE

We need to find the cost of $3\frac{5}{8}$ yards at \$4.99 per yard and the cost of $2\frac{1}{2}$ yards at \$5.75 per yard. Each of these is a multiplication since

$$\text{dollars per yard} \times \text{yards} = \text{total dollars}$$

We need to find $3\frac{5}{8}$ yards × \$4.99 per yard and $2\frac{1}{2}$ yards × \$5.75 per yard

and then add. We can write this as one mathematical expression as follows:

$$3\frac{5}{8} \text{ yards} \times \$4.99 \text{ per yard} + 2\frac{1}{2} \text{ yards} \times \$5.75 \text{ per yard}$$

so the computation we need to do is

$$3\frac{5}{8} \times 4.99 + 2\frac{1}{2} \times 5.75$$

ESTIMATE

$$3\frac{5}{8} \times 4.99 + 2\frac{1}{2} \times 5.75 \doteq 4 \times 5 + 3 \times 6 = 20 + 18 = 38$$

Notice that all the numbers were rounded up, so the estimate is an *overestimate*.

SOLUTION

We can write the mixed numbers in decimal form to do the calculations. The whole number part stays the same and the fraction part changes to decimal.

$\frac{5}{8} = 0.625$ and $\frac{1}{2} = 0.5$ so the problem becomes $3.625 \times 4.99 + 2.5 \times 5.75$

Calculator keystrokes: 3.625 [×] 4.99 [+] 2.5 [×] 5.75 [=]

Remember that your calculator will do the multiplication first.

The display reads 32.46375. Since the answer is money, we round to the nearest penny and get \$32.46.

ALTERNATE SOLUTION

Instead of taking the time to write the mixed numbers in decimal form, we can write the mixed numbers as improper fractions and then go directly to the calculator:

$$3\frac{5}{8} \times 4.99 + 2\frac{1}{2} \times 5.75 = \frac{29}{8} \times 4.99 + \frac{5}{2} \times 5.75$$

Calculator keystrokes: Remember that the fraction bar means division and the calculator will do the division first, thus changing the fraction to a decimal.

29 [÷] 8 [×] 4.99 [+] 5 [÷] 2 [×] 5.75 [=]

Once again, the display is 32.46375.

ANSWER: The total cost of the fabric is \$32.46.

EXAMPLE 9 Carol rides her bike to school every day. The distance from home to school is $1\frac{1}{3}$ miles. Every Friday she rides home for lunch and then rides back to school again. What is the total distance she rides on Friday?

ANALYZE

Every Friday, Carol rides to school ($1\frac{1}{3}$ miles), home for lunch ($1\frac{1}{3}$ miles), back to school ($1\frac{1}{3}$ miles), and finally home again ($1\frac{1}{3}$ miles). So she rides $1\frac{1}{3}$ miles four times. That means we need to find the product

$$4 \times 1\frac{1}{3} \text{ miles}$$

SOLUTION

We now have to decide whether to use fraction or decimal. We have already seen that $\frac{1}{3}$ gives a repeating decimal, so if we use the decimal form of $\frac{1}{3}$, we will get only an approximate answer. The only way to get the exact answer is to use fractions.

FRACTION SOLUTION

$$4 \times 1\frac{1}{3} = \frac{4}{1} \times \frac{4}{3}$$

Notice that we wrote 4 with a denominator of 1 and changed the mixed number to an improper fraction.

$$= \frac{4 \cdot 4}{1 \cdot 3}$$

Remember that to multiply fractions, we multiply numerators and multiply denominators.

$$= \frac{16}{3}$$

$$= 5\frac{1}{3}$$

Change $\frac{16}{3}$ to a mixed number.

DECIMAL SOLUTION

To get the most accurate answer possible using decimals, use your calculator, write the mixed number as an improper fraction. and enter the fraction as a division:

$$4 \times 1\frac{1}{3} = 4 \times \frac{4}{3}$$

Calculator keystrokes: 4 [×] 4 [÷] 3 [=]

The display reads 5.3333333.

ANSWER: On Fridays, Carol rides a distance of $5\frac{1}{3}$ miles. (Notice that this answer is the same as 5.3333.... miles).

OBSERVE $1\frac{1}{3} = 1.33333333...$ The decimal representation has 3's repeating forever. If we try to do the calculation $4 \times 1\frac{1}{3}$ by rounding the decimal first and multiplying 4×1.3, we get only 5.2, which is much smaller than the correct answer. Even if we use 1.33 as an approximation to $1\frac{1}{3}$, we get $4 \times 1.33 = 5.32$, which is still smaller than the correct answer. The only way to get the exact answer is to use the fraction itself and the best decimal approximation comes from entering the fraction into your calculator as a division.

Having a working knowledge of both fractions and decimals and an understanding of how they relate to one another are essential parts of our mathematical "tool box." These tools enable us to make the best choice of solutions to any problem and give us needed flexibility. Most mathematicians look for the quickest, easiest way to solve a problem. As we develop skills with fractions in this unit, try working some of the problems by changing to decimals. This will give you an idea of which situations lend themselves to fractions and which work better with decimals. Only then will you have enough background to make an intelligent choice.

♦ **You Try It**

For these problems, decide what operation needs to be done to answer the question. You may use a calculator to do the computations, but show all your steps.

13. Boiled ham sells for \$6.99 per pound. How much will $2\frac{1}{4}$ pounds cost?

14. To make a bat costume for Halloween, you need $3\frac{3}{8}$ yards of black fabric that costs \$2.98 per yard. How much will the fabric cost?

15. There are 15 books in a carton. Each book weighs about $1\frac{3}{4}$ pounds and the carton itself weighs about $\frac{1}{4}$ pound. What is the total weight of the filled carton?

Try the next problem by first working with the fraction form of the number and doing the calculations by hand. Then do the calculations on your calculator, entering the fraction as a division. Compare your answers.

16. There are 15 books in a carton. Each book weighs about $1\frac{2}{3}$ pounds. How much do all the books weigh?

---

3 SQUARE ROOTS

In the beginning of this section, we discovered that when fractions are converted to decimal form, the decimals generated are of only two types, terminating or repeating. A terminating decimal is one in which eventually all the digits are zeros. A repeating decimal is one in which a digit or a sequence of digits repeats forever. It is reasonable to ask whether there are decimals that do neither of these things; that is, they do not terminate and they do not have a repeating sequence of digits. Actually, it is quite easy to simply make up such a decimal. Here is one example: 0.1010010001000010000001.... Although there is a pattern, the decimal does not terminate nor does it repeat a specific sequence of digits. Do such decimals come about in any other way? In several other sections, we studied the concept of square roots. Notice that $1^2 = 1$ and $2^2 = 4$ so that $\sqrt{1} = 1$ and $\sqrt{4} = 2$. The answers, 1 and 2 are consecutive, but the numbers we started with, 1 and 4, are not. What if we want to take the square root of a number between 1 and 4? Let's find the square root of 2.

Use your calculator to find $\sqrt{2}$. Note: If your calculator has a [√] key, the keystrokes are 2 [√]. If your calculator has $\sqrt{x}$ above the [x^2] key, the keystrokes are 2 [2nd] [x^2]. (The keystrokes on your calculator may be different. Consult your calculator manual or ask your teacher if you need help.)

The answer in the display might have surprised you, but look more carefully. Remember that $\sqrt{1} = 1$ and $\sqrt{4} = 2$, so you should expect that $\sqrt{2}$ should have an answer somewhere between 1 and 2. In fact, the answer you got is just under 1.5. You may think that the exact value of $\sqrt{2}$ is what you see in the display on your calculator. However, remember that your calculator only displays a small number of digits. The unusual thing about this answer is that the decimal representation of $\sqrt{2}$ never ends and it is not a repeating decimal. There are two important things to realize because of this:

- The only **exact** representation of $\sqrt{2}$ is $\sqrt{2}$.
- If we need to use a decimal representation for $\sqrt{2}$ we must round to some decimal place, and it will be an **approximation**, not an exact value.

EXAMPLE 10 Approximate $\sqrt{2}$ to the nearest thousandth.

smaller than 5

$$\sqrt{2} = 1.41\underline{4}2\,1\,3\,5\,6\,2\ldots$$

thousandths

Answer: $\sqrt{2} \doteq 1.414$

EXAMPLE 11 Approximate $\sqrt{7}$ to the nearest ten-thousandth.

5 or more

$$\sqrt{7} = 2.645\underline{7}5\,1\,3\,1\,1\ldots$$

ten-thousandths

Answer: $\sqrt{7} \doteq 2.6458$

EXAMPLE 12 Approximate $5\sqrt{10}$ to the nearest thousandth.

There are two operations in this expression, square root and multiplication. Since your calculator "knows" order of operations, we can enter the expression as we see it and the calculator will evaluate the square root first and then multiply. The rounding should **not** be done until the very end.

Calculator keystrokes: 5 [×] 10 [2nd] [x^2] [=]

The display reads: 15.811388. Rounding to the nearest thousandth, we get 15.811.

Answer: $5\sqrt{10} \doteq 15.811$.

It is useful to be able to decide the approximate value of a square root without having to find the actual decimal. To do this, we need to be familiar with numbers we call perfect squares. These are numbers that are the squares of whole numbers. Here are the first five perfect squares -- remember that they come from squaring the first five whole numbers: 0, 1, 4, 9, 16.

List the next ten perfect squares: ______________________________

____

Knowing these perfect squares and their square roots gives you the ability to estimate other square roots.

EXAMPLE 13 $\sqrt{20}$ is between what two consecutive whole numbers?

From your list above, we can see that 20 is not a perfect square. That means that its decimal representation is nonterminating and nonrepeating. However, we can get an idea of the value of $\sqrt{20}$ by looking for the perfect squares closest to 20.

What is the perfect square closest to 20 but less than 20 ? ________________

What is the perfect square closest to 20 but greater than 20 ? ________________

Since $16 < 20 < 25$, their square roots will be in the same order; that is, the square root of 20 is between the square root of 16 and the square root of 25.
Since $16 < 20 < 25$, then
$\sqrt{16} < \sqrt{20} < \sqrt{25}$ but $\sqrt{16} = 4$ and $\sqrt{25} = 5$ so now we know that $4 < \sqrt{20} < 5$.
That means that $\sqrt{20}$ is between 4 and 5. That estimate may be all we need to know about $\sqrt{20}$.

Answer: $\sqrt{20}$ is between 4 and 5.

EXAMPLE 14 $\sqrt{40}$ is between which two consecutive whole numbers?

Since 40 is between the perfect squares 36 and 49, we know that

$36 < 40 < 49$ so that

$\sqrt{36} < \sqrt{40} < \sqrt{49}$ which means that

$6 < \sqrt{40} < 7$

Answer: $\sqrt{40}$ is between 6 and 7.

EXAMPLE 15 Find, to the nearest tenth, the hypotenuse of a right triangle whose legs measure 8 ft and 17 ft.

This problem involves a right triangle in which the legs are given and we are asked to find the hypotenuse. We will use the Pythagorean Theorem.

First, draw a picture, label the legs and label the hypotenuse c:

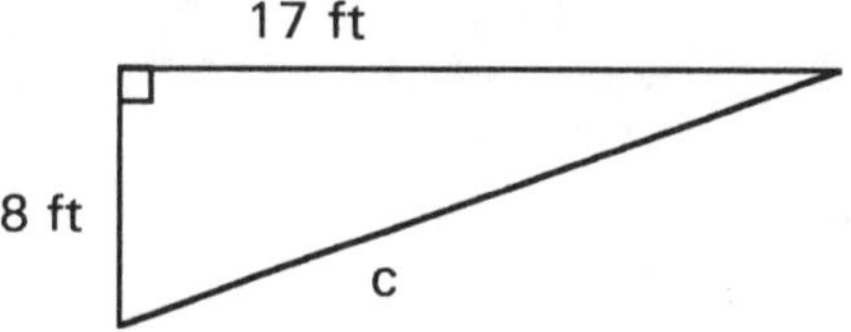

Next, write the Pythagorean Theorem: $c^2 = a^2 + b^2$

Now substitute 8 ft and 17 ft for the legs, a and b: $c^2 = (8)^2 + (17)^2$

$c^2 = 64 + 289$ Evaluate exponents first.

$c^2 = 353$

$c = \sqrt{353}$

$c = 18.78829423$ Evaluate the square root.

$c \doteq 18.8$ ft

Answer: The hypotenuse is approximately 18.8 ft.

♦ **You Try It**

17. a. $\sqrt{73}$ is between which two consecutive whole numbers?

b. Write down your calculator's display for $\sqrt{73}$

18. a. $\sqrt{14}$ is between which two consecutive whole numbers?

b. Write down your calculator's display for $\sqrt{14}$

19. Approximate $\sqrt{22}$ to the nearest thousandth.

20. Approximate $\sqrt{80}$ to the nearest hundred-thousandth.

21. $\sqrt{75}$ is between which two consecutive whole numbers?

22. $\sqrt{56}$ is between which two consecutive whole numbers?

23. What is the exact value of $\sqrt{39}$?

24. Find, to the nearest tenth, the hypotenuse of a right triangle whose legs measure 9 cm and 22 cm.

25. Find, to the nearest hundredth, the hypotenuse of a right triangle, whose legs measure 2.3 in. and 6.1 in.

♦**Answers to You Try It** **1.** 0.375 **2.** 0.6 **3.** 0.16 **4.** 0.3125 **5.** Divide the denominator into the numerator. **6. a.** 0.3 **b.** 0.33 **c.** 0.333 **d.** 0.3333 **7. a.** 0.6 **b.** 0.56 **c.** 0.556 **d.** 0.5556 **8. a.** 0.6 **b.** 0.58 **c.** 0.583 **d.** 0.5833 **9. a.** 0.6 **b.** 0.56 **c.** 0.563 **d.** 0.5625 **10. a.** 0.5 **b.** 0.45 **c.** 0.455 **d.** 0.4545 **11.** $\frac{5}{6} = 0.8\overline{3}$ **12.** $\frac{4}{9} = 0.\overline{4}$ **13.** $15.73 **14.** $10.06 **15.** 26.5 lbs **16.** 25 lbs **17. a.** 8 and 9 **b.** 8.5440037 **18. a.** 3 and 4 **b.** 3.7416574 **19.** 4.690 **20.** 8.94427 **21.** 8 and 9 **22.** 7 and 8 **23.** $\sqrt{39}$ **24.** 23.8 **25.** 6.52

SECTION 2.8 EXERCISES

1 *Change the fraction or mixed number to an exact decimal number.*

1. $\frac{4}{5}$ **2.** $7\frac{1}{8}$ **3.** $\frac{9}{20}$ **4.** $5\frac{1}{16}$

5. $\frac{8}{100}$ **6.** $\frac{355}{1000}$ **7.** $3\frac{12}{25}$ **8.** $\frac{11}{40}$

9. $\frac{19}{4}$ **10.** $\frac{23}{5}$ **11.** $\frac{7}{8}$ **12.** $\frac{3}{16}$

13. $\frac{3}{20}$ **14.** $4\frac{5}{8}$ **15.** $2\frac{1}{16}$ **16.** $\frac{3}{50}$

Change each fraction or mixed number to an approximate decimal number. Round to the indicated place.

17. $\frac{1}{9}$ hundredths **18.** $2\frac{7}{9}$ thousandths

19. $\frac{1}{12}$ tenths **20.** $4\frac{5}{33}$ ten-thousandths

21. $\frac{7}{22}$ hundredths **22.** $6\frac{1}{37}$ tenths

23. $1\frac{2}{3}$ hundredths **24.** $\frac{7}{18}$ tenths

25. $2\frac{2}{7}$ thousandths **26.** $1\frac{5}{6}$ ten-thousandths

27. $4\frac{1}{6}$ hundredths **28.** $\frac{1}{7}$ hundred-thousandths

29. $\frac{4}{15}$ thousandths **30.** $\frac{1}{11}$ thousandths

31. $\frac{2}{11}$ thousandths

32. $\frac{4}{11}$ thousandths

2 *For the following problems, decide what operation needs to be done. Outline the problem and estimate the answer before doing any operations. You may use a calculator to do the computations, but show all your steps. Be sure to write your answer in a complete sentence.*

33. A carpet store advertises Stainmaster carpet at a sale price of \$20.99 per square yard. Suppose you decide to carpet your living room and you need $30\frac{1}{4}$ square yards. How much will the carpet cost?

34. Jay works $37\frac{1}{2}$ hours per week and earns \$6.25 per hour. What is his weekly salary?

35. If a gallon of gas costs \$1.16, how much will $12\frac{3}{4}$ gallons cost?

36. Peggy bought 100 shares of stock at $\$18\frac{3}{16}$ per share. What was her total cost?

37. Find the cost of 1,000 manila folders at $3\frac{1}{2}$¢ each.

38. For Halloween, Carol wants to make witch costumes for her three children. If each costume requires $3\frac{1}{2}$ yards of fabric and the fabric costs \$2.95 per yard, how much will all three costumes cost?

39. Jennifer bought $1\frac{1}{4}$ pounds of Swiss cheese and $\frac{3}{4}$ pound of turkey breast. The cheese costs \$3.99 per pound and the turkey costs \$5.99 per pound. What is the total cost of Jennifer's purchase?

40. Ed went to the deli and bought $1\frac{1}{2}$ pounds of potato salad at \$1.20 per pound, 2 pounds of ham at \$2.99 per pound, and 6 rolls at 35¢ each. How much change will he get if he gives the cashier \$10.00?

3 *Use your calculator to find the decimal value of each square root. Round your answer to the indicated place.*

41. $\sqrt{19}$ thousandths

42. $\sqrt{412}$ ten-thousandths

43. $\sqrt{44}$ tenths

44. $\sqrt{61}$ hundredths

45. $\sqrt{687}$ ten-thousandths

46. $\sqrt{83}$ thousandths

47. $\sqrt{1{,}312}$ hundredths

48. $\sqrt{204}$ ten-thousandths

Use your calculator and the rule for order of operations to evaluate each expression. Round your answer to the indicated place.

49. $3\sqrt{6}$ hundredths

50. $12\sqrt{2}$ thousandths

51. $5\sqrt{5}$ tenths

52. $\frac{\sqrt{14}}{6}$ hundredths

53. $\frac{\sqrt{2}}{2}$ thousandths

54. $9\sqrt{20}+3\sqrt{5}$ hundredths

55. $\frac{1}{\sqrt{2}}$ thousandths

56. $\frac{\sqrt{3}+\sqrt{2}}{7}$ tenths

Each of the following square roots lies between two consecutive whole numbers. Find these two whole numbers. Do not use your calculator.

57. $\sqrt{18}$

58. $\sqrt{52}$

59. $\sqrt{43}$

60. $\sqrt{90}$

61. $\sqrt{20}$

62. $\sqrt{30}$

63. $\sqrt{12}$

64. $\sqrt{78}$

---

65. Use the Pythagorean Theorem to find the hypotenuse of a right triangle whose legs measure 14 in. and 24 in.. Round your answer to the nearest tenth of an inch.

66. A right triangle has legs that measure 15.2 cm. and 8.7 cm.
    a. Find the hypotenuse of the triangle to the nearest tenth of a centimeter.
    b. Find the perimeter of this triangle.

67. a. Find, to the nearest hundredth, the hypotenuse of this triangle:

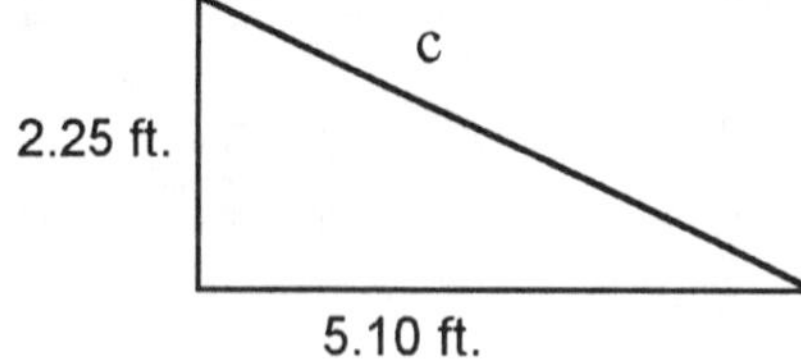

b. Find the perimeter of this triangle.

EXTEND YOUR THINKING ♦ ♦ ♦

WRITING TO LEARN ♦ ♦ ♦

68. Consider the statement: $\frac{4}{9} = 0.4$. Is it true or false? Explain.

69. What is the exact value of $\frac{6}{11}$?

70. Read the following problem:

> A recipe calls for $1\frac{2}{3}$ cups of flour. If you want to double the recipe, how much flour should you use?

a. What operation is necessary to solve this problem?
b. Would the answer be more useful in fraction form or in decimal form? Why?
c. Use your choice to solve the problem.

71. Give an example of a problem situation where the most useful form of the answer would be decimal form.

72. Give an example of a problem situation where the most useful form of the answer would be fraction form.

73. **a.** What is a perfect square?
b. Give three examples of perfect squares.

74. Is 18 a perfect square? How do you know?

75. Consider the statement $\sqrt{34} = 5.83$. Is this statement true or false? Explain.

76. Consider the statement "The exact value of $\sqrt{34}$ is 5.8309519." Is this statement true or false? Explain.

77. **a.** In the expression $\sqrt{14+8}$, there are two mathematical operations. What are they and which one is performed first?

b. Evaluate $\sqrt{14+8}$ to the nearest hundredth.

c. In the expression $\sqrt{14}+\sqrt{8}$, what are the mathematical operations and which one is performed first?

d. Evaluate $\sqrt{14}+\sqrt{8}$ to the nearest hundredth.

e. Is $\sqrt{14+8}$ equal to $\sqrt{14}+\sqrt{8}$?

78. State the Pythagorean Theorem.

CHAPTER 2 REVIEW

VOCABULARY AND MATCHING

New words and phrases introduced in this chapter are shown in the left column. Match each term on the left with the word or sentence on the right that best desscribes it.

A. term
B. factor
C. 5 cubed
D. squaring
E. perfect square
F. one million
G. one billion
H. divide by 1000
I. multiply by 1000
J. \$0.07
K. 1.2¢
L. square root
M. perimeter
N. volume of a box
O. Pythagorean Theorem
P. repeating decimal
Q. terminating decimal
R. double a number
S. change a fraction to a decimal

_____ a whole number multiplied by itself
_____ distance around an object
_____ 7¢
_____ \$0.012
_____ Move the decimal point 3 places left.
_____ Multiply a number by 2.
_____ 125
_____ Divide denominator into numerator.
_____ number that is added
_____ $c^2 = a^2 + b^2$
_____ 1,000,000
_____ Move the decimal point 3 places right.
_____ a decimal value you must approximate
_____ multiplying a number by itself
_____ 10^9
_____ number that is multiplied
_____ a number that when squared gives the value
_____ $V = lwh$
_____ a decimal value you can find exactly

REVIEW EXERCISES

2.1 Problem Solving and Multiplication

Do the following multiplications without a calculator.

1. 32.4×0.04 **2.** 0.004×1.02 **3.** 88×2.04 **4.** 3.6×9

Use the ideas of size and rounding to estimate the answers to problems 5 - 8. DO NOT do the actual multiplication. SHOW how you estimated the answer.

5. 6.78×9.87 **6.** 11.23×0.992 **7.** 0.042×8.01 **8.** 0.85×2.04

9. A rectangular parking lot measures 300.2 ft. by 211.1 ft. Find its area. (Include a sketch.)

10. Emilio buys 5 shirts for \$29.95 each and 2 pairs of jeans for \$37.95.
a. Estimate the total cost. (Show how you did this.)
b. Figure the actual total cost.

2.2 Exponents

11. Write an expression with 3 factors.

12. Write an expression with 3 terms.

Evaluate the following expressions without a calculator.

13. 3^2 **14.** 2^3 **15.** 7^2 **16.** 1^8 **17.** 10^2 **18.** 10^3

19. Use a calculator to evaluate 15^3.

20. Use a calculator to evaluate $(4.3)^2$.

2.3 Large Numbers and Powers of 10

21. What is the difference between a billion and a million?

Multiply without a calculator.

22. 81.235×100 **23.** 0.0953×1000 **24.** 54×1000 **25.** 89×10^3

26. Tasha is making 1000 copies of a flyer at the copy shop where they charge $0.015 for each copy. How much will she pay?

2.4 Square Roots

Evaluate without a calculator.

27. $\sqrt{25}$ **28.** $\sqrt{81}$ **29.** $\sqrt{0.04}$ **30.** $3\sqrt{100}$

31. Use a calculator to evaluate $\sqrt{529}$.

32. Use a calculator to evaluate $3\sqrt{2.89}$

2.5 Problem Solving and Division

Divide without a calculator.

33. $5.6 \div 7$ **34.** $5.6 \div 0.7$ **35.** $56 \div 0.7$ **36.** $13.4 \div 1000$

37. $47 \div 100$ **38.** $\dfrac{0.23}{10}$

Decide if the quotient will be greater than, less than, or equal to 1. Do not evaluate.

39. $15 \div 31$ **40.** $31 \div 15$ **41.** $0.54 \div 0.54$

42. Write 11.2¢ in dollars.

43. Write $1.21 in cents.

44. Ten members of an investment club are buying stock for $12,345. What is each member's share?

2.6 Problem Solving Using Order of Operations

Use the rule for order of operations to evaluate each expression.

45. $12 + 2 \times 4$

46. $13 - 2(7 - 5)$

47. $5\sqrt{64} + 9\sqrt{81}$

48. $\frac{18}{2} + 5 \times 3 - 2^3$

49. $(8 - 3)(8 + 3)$

50. $3\sqrt{25} - 2(16 - 4^2)$

Translate into English. Do not evaluate.

51. $2(32 + 5)$

52. $\frac{45}{9} + 6$

53. $5(2) - 4$

54. $\frac{x - 3}{2}$

Translate into mathematics. Do not evaluate.

55. double the sum of 25 and a number.

56. 5 is added to the product of 11 and 8.

57. 4 less than the quotient of 32 and 2.

58. 4 is less than the quotient of 32 and 2.

Use the distributive property of multiplication over addition to rewrite each expression. Do not evaluate.

59. $3(6 + 11)$

60. $4(x + t)$

61. $3 \cdot 5 + 3 \cdot 7$

62. $18a + 18b$

2.7 Using Formulas

63. Find the perimeter of a rectangle with length 4.5 in. and width 1.7 in.

64. Find the volume of a box that is 7.1cm by 5.5 cm by 1.9 cm.

65. Find the hypotenuse of a right triangle with legs of 21 m and 28 m.

2.8 Fractions and Decimals: The Need for Both

Without a calculator, change each fraction to a decimal.

66. $\frac{7}{8}$

67. $\frac{1}{3}$

68. $\frac{13}{20}$

69. Which is larger $\frac{4}{9}$ or 0.4 ?

70. Which is larger $\frac{2}{3}$ or $.67$?

71. Use your calculator to evaluate $\sqrt{123}$. Round to the nearest hundredth.

72. Find the cost of $1\frac{1}{4}$ pounds of salami at \$4.99 a pound.

Multiplication and Division of Fractions

CONSTRUCTION

A truck can deliver $6\frac{1}{4}$ tons of stone in one trip. How many trips are needed to deliver 85 tons of stone? To solve this problem, find how many $6\frac{1}{4}$-ton loads are contained in 85 tons. This is written

$$85 \div 6\frac{1}{4}.$$

In this chapter you will see how to solve this problem. You will also learn how to correctly interpret the answer.

Fractions are used in recipe books, in carpentry, with tools to mark wrench and drill bit sizes, in stock quotations on the financial page of the newspaper, and in other areas where a part of a whole is used. You will learn how the operations applied to whole numbers in the last two chapters can also be applied to fractions. You will solve a variety of real world applications so you can see for yourself the role fractions play in everyday life.

3.1 PRIME NUMBERS AND PRIME FACTORING

OBJECTIVES

1. Find all the divisors of a counting number.
2. Determine if a number is prime or composite.
3. Use the divisibility rules.
4. Prime factor a counting number.

NEW VOCABULARY

divisor	2-3-5-7 rule
divisible	prime factor
prime number	prime factorization
composite number	

REVIEW OF BASIC FRACTION CONCEPTS

In Section 1.5 we did some work with the basic ideas and operations with fractions. In this chapter we will investigate these ideas and operations in greater depth and focus especially on applications of fractions. Before we go on, let's review some of the basic concepts of fractions.

> The **fraction** $\frac{a}{b}$ means $a \div b$. The top number a is called the *numerator* and the bottom number b is the denominator. The denominator of a fraction cannot be zero.
>
> The fraction $\frac{a}{b}$ tells us to divide the whole into b equal parts and take a of them.

EXAMPLE 1 In the diagram below, a box has been divided into **eight** equal pieces. **Five** of these are shaded and **three** are not shaded.

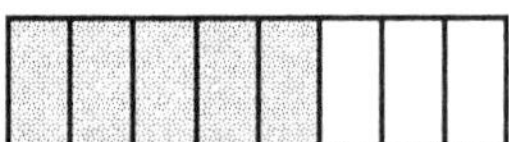

The fraction of the box that is shaded is $\frac{5}{8}$. The fraction of the box that is not shaded is $\frac{3}{8}$.

OBSERVE The fraction of the box that **is** shaded, $\frac{5}{8}$, plus the fraction of the box that is **not** shaded, $\frac{3}{8}$, gives us the **whole** box, or $\frac{8}{8}$.

EXAMPLE 2 Of the 15 students in a school band, 10 are female.

a. What fraction is female?
b. What fraction is male?

a. The **whole** in this situation is 15 students and the fraction we want is

$$\frac{\text{female band members}}{\text{total band members}} = \frac{10}{15}$$ Notice that this fraction reduces to $\frac{2}{3}$.

Answer: $\frac{2}{3}$ of the band is female.

b. If 10 band members are female and there are 15 students in the band, then there must be 15 – 10 or 5 male band members. The fraction we want is

$$\frac{\text{male band members}}{\text{total band members}} = \frac{5}{15}$$ Notice that this fraction reduces to $\frac{1}{3}$.

Answer: $\frac{1}{3}$ of the band is male.

OBSERVE We could have used the answer to part **a** to answer part **b.** If $\frac{2}{3}$ of the band is female and the **whole** band is $\frac{3}{3}$, then $\frac{3}{3} - \frac{2}{3}$ or $\frac{1}{3}$ of the band is male.

> A **proper fraction** is a fraction in which the numerator is less than the denominator.
> An **improper fraction** is a fraction in which the numerator is greater than or equal to the denominator.
> If the numerator and denominator of a fraction are ***equal,*** then fraction is equal to 1.
> If the numerator is greater than the denominator, the improper fraction can also be expressed as a **mixed number,** which is the sum of a whole number and a proper fraction.

EXAMPLE 3

a. $\frac{11}{3} = 3\frac{2}{3}$

To write $\frac{11}{3}$ as a mixed number, divide the denominator into the numerator. 3 divides into 11 three times wiith a remainder of 2. The 3 becomes the whole number part of the mixed number and the 2 is the numerator of the fraction part. The denominator stays the same.

b. $4\frac{3}{5} = \frac{5 \cdot 4 + 3}{5} = \frac{23}{5}$

To write $4\frac{3}{5}$ as an improper fraction, multiply the denominator by the whole number and add the numerator. This number becomes the numerator of the improper fraction and the denominator stays the same.

♦ You Try It

a. What fraction of each figure is shaded? **b.** What fraction is not shaded?

1. 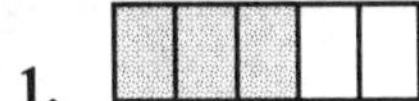

2. 

3. In a group of 28 people, 7 had blue eyes.

a. What fraction of the people had blue eyes? Express your answer in reduced form.

b. What fraction of the people did not have blue eyes? Express your answer in reduced form.

Write each improper fraction as a mixed number.

4. $\frac{7}{4}$

5. $\frac{9}{2}$

Write each mixed number as an improper fraction.

6. $2\frac{3}{8}$

7. $1\frac{2}{9}$

In the rest of this section we will develop skills that will be useful in our study of fractions.

1 Finding the Divisors of a Counting Number

The numbers 3 and 4 are called factors of 12 because $3 \cdot 4 = 12$. The word *factor* is used when referring to a product.

Since $12 \div 3$ has a zero remainder, 3 is called a **divisor** of 12. 4 is also a divisor of 12. 5 is not a divisor of 12 because $12 \div 5$ has a remainder of 2. The word *divisor* is used when referring to a quotient.

Divisors are found in pairs. If a number is a divisor for a counting number, its quotient is also a divisor.

EXAMPLE 4 The number 3 is a divisor of 12 because $12 \div 3 = 4$. Then the quotient 4 is also a divisor of 12 ($12 \div 4 = 3$). ■

Since 3 and 4 are divisors of 12, 12 is said to be **divisible** by 3 and divisible by 4.

The words *factor* and *divisor* are closely related. A factor of a number is also a divisor of the number. A divisor of a number is also a factor of the number.

EXAMPLE 5 The numbers 3 and 5 are factors of 15 because $3 \cdot 5 = 15$. 3 and 5 are also divisors of 15 because

$$15 \div 3 = 5 \qquad \text{and} \qquad 15 \div 5 = 3. \quad ■$$

CAUTION The word *divisor* can be used in two contexts. In the division $5\overline{)12}$, 5 is called *the divisor*. However, 5 is not *a divisor of* 12 because $12 \div 5$ gives a remainder of 2.

To find all the divisors of a counting number n

1. Divide n by 1, 2, 3, 4, and so on.
2. When you get a zero remainder, both that number and its quotient are divisors of n.
   When you get a nonzero remainder, that number is not a divisor of n. Skip it. Test the next number.
3. Stop when the list of divisors begins to repeat (or when the square of the number being tested exceeds n).

EXAMPLE 6 Find all of the divisors of 12.

Since $12 \div 1 = 12$, 1 and 12 are divisors.

Since $12 \div 2 = 6$, 2 and 6 are divisors.

Since $12 \div 3 = 4$, 3 and 4 are divisors.

Since $12 \div 4 = 3$, stop because the set of divisors is beginning to repeat. The divisors 4 and 3 were found in the previous step.

The divisors of 12 are 1, 2, 3, 4, 6, and 12. ■

▶ You Try It 8. Find all the divisors of 21.

EXAMPLE 7 Find all the divisors of 15.

$15 \div 1 = 15$, 1 and 15 are divisors.
$15 \div 2 = 7$ R 1, 2 is not a divisor. The remainder is not zero.
$15 \div 3 = 5$, 3 and 5 are divisors.
$15 \div 4 = 3$ R 3, 4 is not a divisor.
$15 \div 5 = 3$, stop. The divisors are repeating.

The divisors of 15 are 1, 3, 5, and 15. ■

Every counting number (except 1) has at least two divisors, 1 and the number itself. The number 1 is the smallest divisor of every counting number. The largest divisor is the number itself.

EXAMPLE 8 Find all the divisors of 7.

$7 \div 1 = 7$, 1 and 7 are divisors.
$7 \div 2 = 3$ R 1, 2 is not a divisor.
$7 \div 3$, stop. Since $3^2 > 7$, 7 has no more divisors.

The divisors of 7 are 1 and 7. ■

▶ You Try It 9. Find all the divisors of 30. 10. Find all the divisors of 13.

2 Prime and Composite Numbers

The number 7 has exactly two divisors, 1 and 7. Therefore, 7 is called a prime number. A **prime number** is any counting number greater than 1 having exactly two divisors, 1 and the number itself. The smallest prime number is 2. There is no largest prime. The first fifteen prime numbers are shown below.

2, 3, 5, 7, 11, 13, 17, 19, 23, 29, 31, 37, 41, 43, 47, . . .

EXAMPLE 9 **a.** 3 is a prime number. Its only divisors are 1 and 3.
b. 17 is a prime number. Its only divisors are 1 and 17.
c. 53 is a prime number. 1 and 53 are its only divisors. ■

If a counting number larger than 1 is not prime, it is called a **composite number**. Every composite number has three or more divisors: 1, itself, and at least one other number in between. The smallest composite number is 4. It has three divisors: 1 and 4, as well as 2. The first fifteen composite numbers are shown below.

4, 6, 8, 9, 10, 12, 14, 15, 16, 18, 20, 21, 22, 24, 25, . . .

EXAMPLE 10 **a.** 15 is a composite number. It has four divisors: 1, 3, 5, and 15.
b. 9 is composite. It has three divisors: 1, 3, and 9.
c. 32 is composite. Its divisors are 1, 2, 4, 8, 16, and 32. ■

The number 1 is neither prime nor composite. It has exactly one divisor, itself.

3 Divisibility Rules

Is 111 a prime number? No, because 3 is a divisor of 111. In fact, $111 \div 3 = 37$. To help you determine if larger numbers are prime or composite, use the following divisibility rules. These rules will be helpful in Chapters 3 and 4 when reducing fractions.

2 is a divisor of a number if its ones digit is even: 0, 2, 4, 6, or 8.

EXAMPLE 11 4, 26, 50, 288, and 5,732 are all divisible by 2 because each number has an even ones digit. ■

3 is a divisor of a number if 3 is a divisor of the sum of its digits.

EXAMPLE 12

a. 3 is a divisor of 15 because 3 is a divisor of the sum of its digits $1 + 5 = 6$.
b. 3 is a divisor of 174 because 3 is a divisor of the sum of its digits $1 + 7 + 4 = 12$.
c. 3 is not a divisor of 53 because 3 is not a divisor of $5 + 3 = 8$. ■

5 is a divisor of any number if its ones digit is 0 or 5.

EXAMPLE 13 10, 25, 40, 700, and 8,245 are all divisible by 5 because each number has 0 or 5 in the ones place. ■

7 is a divisor of a number if 7 is a divisor of the difference between *the number left when the ones digit is crossed out* and *2 times the ones digit.*

EXAMPLE 14

a. 7 is a divisor of 35 because 7 is a divisor of the difference between $3\not{5}$ and $2 \cdot 5 = 2 \cdot 5 - 3 = 10 - 3 = 7$.
b. 7 is a divisor of 476 because 7 is a divisor of the difference between $47\not{6}$ and $2 \cdot 6 = 47 - 2 \cdot 6 = 35$.
c. 7 is not a divisor of 185 because 7 is not a divisor of $18 - 2 \cdot 5 = 8$.
d. 7 is a divisor of 63 because 7 is a divisor of $6 - 2 \cdot 3 = 0$. ■

The **2-3-5-7 rule** is useful for finding divisors of counting numbers. Simply test the primes 2, 3, 5, and 7 first for divisibility, before trying larger primes.

▶ **You Try It** Determine if the number is divisible by 2, 3, 5, 7, or none of these.

11. 28 **12.** 39 **13.** 60 **14.** 157

---

4 Prime Factoring

To **prime factor** a counting number, write it as a product of prime numbers only. Prime factoring will be used in Chapter 4 to help you add and subtract fractions.

EXAMPLE 15 6 is prime factored as $2 \cdot 3$ because $6 = 2 \cdot 3$, and 2 and 3 are both prime numbers. ■

The expression $2 \cdot 3$ is called the **prime factorization** of 6. Every counting number greater than 1 can be prime factored in only one way, though the order of the factors may differ.

EXAMPLE 16 30 is prime factored as $2 \cdot 3 \cdot 5$ because $30 = 2 \cdot 3 \cdot 5$, and 2, 3, and 5 are prime numbers.

CAUTION 30 is not prime factored as $5 \cdot 6$ because even though $30 = 5 \cdot 6$, 6 is not prime.

■

Prime factor a counting number using the method of **successive division by primes.**

1. Divide the number by the primes 2, 3, 5, 7, 11, and so on, starting with the smallest prime, 2. Only primes giving a zero remainder are used as factors.
2. Stop when the quotient is a prime number.
3. The prime factorization is the product of all the prime divisors and the final prime quotient.

EXAMPLE 17 Prime factor 20.

First, divide 20 by the smallest prime, 2. $20 \div 2 = 10$. Write the quotient 10 above 20.

$$2\,\overline{)\,20}\quad\text{quotient } 10$$

Since 10 is even, divide by 2 again. $10 \div 2 = 5$. Write the quotient 5 above 10. Since the quotient 5 is prime, you are done.

$$\begin{array}{r|l} & 5 \\ 2 & 10 \\ 2 & 20 \end{array}$$

Multiply the prime divisors (2 and 2), and the final quotient (5), to get the prime factorization of 20.

20 is prime factored as: $20 = 2 \cdot 2 \cdot 5 = 2^2 \cdot 5$. ■

EXAMPLE 18 Prime factor 105.

2 is not a divisor of 105, because 105 ends in an odd digit. Try the next prime, 3. 3 is a divisor of 105 because 3 divides $1 + 0 + 5 = 6$. $105 \div 3 = 35$.

$$3\,\overline{)\,105}\quad\text{quotient } 35$$

3 does not divide the quotient 35. Try the next prime, 5. 5 divides 35. $35 \div 5 = 7$. Since the quotient 7 is prime, you are done.

$$\begin{array}{r|l} & 7 \\ 5 & 35 \\ 3 & 105 \end{array}$$

$105 = 3 \cdot 5 \cdot 7$

The prime factorization of $105 = 3 \cdot 5 \cdot 7$. ■

▶ **You Try It** Prime factor **15.** 45 **16.** 72 **17.** 140

EXAMPLE 19 Prime factor 54.

$$\begin{array}{r|l} & 3 \\ 3 & 9 \\ 3 & 27 \\ 2 & 54 \end{array}$$

The prime factorization of 54 is $54 = 2 \cdot 3 \cdot 3 \cdot 3 = 2 \cdot 3^3$. ■

▶ **You Try It** Prime factor **18.** 84 **19.** 270 **20.** 127

♦**Answers to You Try It** **1. a.** $\frac{3}{5}$ **b.** $\frac{2}{5}$ **2. a.** $\frac{9}{10}$ **b.** $\frac{1}{10}$ **3. a.** $\frac{1}{4}$ **b.** $\frac{3}{4}$ **4.** $1\frac{3}{4}$ **5.** $4\frac{1}{2}$ **6.** $\frac{19}{8}$ **7.** $\frac{11}{9}$ **8.** 1, 21, 3, 7 **9.** 1, 30, 2, 15, 3, 10, 5, 6 **10.** 1, 13 **11.** 2, 7 **12.** 3 **13.** 2, 3, 5 **14.** none **15.** $3^2 \cdot 5$ **16.** $2^3 \cdot 3^2$ **17.** $2^2 \cdot 5 \cdot 7$ **18.** $2^2 \cdot 3 \cdot 7$ **19.** $2 \cdot 3^3 \cdot 5$ **20.** prime

Oct. 12 1999

SECTION 3.1 EXERCISES

1 2 3 *Find all of the divisors for the given number. Use the 2-3-5-7 rule as a guide. Identify each number as prime or composite.*

1. 6 **2.** 9 **3.** 16 **4.** 10

5. 14 **6.** 18 **7.** 13 **8.** 29

9. 26 **10.** 30 **11.** 33 **12.** 42

13. 80 **14.** 100 **15.** 24 **16.** 125

17. 79 **18.** 47 **19.** 142 **20.** 181

21. 51 **22.** 39 **23.** 59 **24.** 75

25. What is the smallest odd prime number?

26. What is the only even prime number?

27. What is the smallest two-digit prime number?

28. What is the smallest prime number?

29. Find the largest two-digit prime number.

30. Find the largest one-digit prime number.

31. What is the smallest composite number?

32. What is the smallest odd composite number?

33. Find the smallest two-digit composite number.

34. Find the largest two-digit composite number.

35. What is the largest one-digit composite number?

36. (True/False) Every composite number is divisible by 2.

37. (True/False) All prime numbers are odd.

38. (True/False) A number cannot be both prime and composite.

4 *Prime factor each number below. Use the 2-3-5-7 rule as a guide. If a number is prime, write prime.*

39. 4 **40.** 9 **41.** 13 **42.** 29 **43.** 14

44. 18 **45.** 16 **46.** 40 **47.** 33 **48.** 10

49. 80 **50.** 100 **51.** 81 **52.** 42 **53.** 144

54. 225 **55.** 51 **56.** 26 **57.** 66 **58.** 71

59. 77 **60.** 666 **61.** 210 **62.** 39

SKILLSFOCUS (Section 1.8) *Perform the operation indicated.*

63. $8 + 8.8 + 0.88$ **64.** $12 - 3.456$ **65.** $54.1 - 34.56$ **66.** $120 + 12.1 + 1.21 + 0.121$

EXTEND YOUR THINKING ▶▶▶

▶ SOMETHING MORE

67. Perfect Numbers A perfect number is a counting number equal to the sum of all of its divisors less than itself. For example, 6 is a perfect number. The divisors of 6 are 1, 2, 3, and 6, and $1 + 2 + 3 = 6$. Which of the following are perfect numbers?

a. 20 **b.** 28 **c.** 100 **d.** 12 **e.** 496

Consider the following two sets of numbers:

$$\begin{array}{cccccccc} 1 & 2 & 4 & 8 & 16 & 32 & \ldots \\ 1 & 3 & 7 & 15 & 31 & 63 & \ldots \end{array}$$

The top row of numbers corresponds to powers of 2: $2^0 = 1$, $2^1 = 2$, $2^2 = 4$, $2^3 = 8$, and so on. Each number in the bottom row is the sum of the number above it with all the numbers above and to the left. For example, $15 = 8 + 4 + 2 + 1$. The ancient mathematician Euclid (c.300 B.C.) claimed that whenever the lower number is a prime number, the product of it and the number above is a perfect number. For example, since 3 in the lower row is prime, then $3 \cdot 2 = 6$ is perfect. Was Euclid right for the other prime numbers shown in the bottom row?

▶ TROUBLESHOOT IT

Find and correct the error.

68. Prime factor 48.

$$\begin{array}{r|l} & 6 \\ \hline 2 & 12 \\ \hline 2 & 24 \\ \hline 2 & 48 \end{array}$$

Therefore, $48 = 2^3 \cdot 6$.

WRITING TO LEARN ▶▶▶

69. Explain the difference between a prime and a composite number.

70. Explain in words how to check a number for divisibility by 3.

▶ YOU BE THE JUDGE

71. Dan wanted to impress his younger brother Kile. He claimed that in the ten expressions that follow,

$$\begin{array}{ccccc} 2^2 - 1 & 2^3 - 1 & 2^4 - 1 & 2^5 - 1 & 2^6 - 1 \\ 2^7 - 1 & 2^8 - 1 & 2^9 - 1 & 2^{10} - 1 & 2^{11} - 1 \end{array}$$

if the power on 2 is a prime number, the expression simplifies to a prime. For example, 3 is prime, and $2^3 - 1 = 8 - 1 = 7$ is prime. If the power on 2 is a composite number, it simplifies to a composite number. For example, 4 is composite, and $2^4 - 1 = 16 - 1 = 15$ is composite. Did Kile have reason to be impressed?

72. Jan claims that the divisibility rules for 2, 3, 5, and 7, can be used to check divisibility by the numbers 4, 6, 8, 9, or 10. Do you agree with Jan? Explain your decision.

73. Rob claims if a number is divisible by 3, then any number you get by arranging its digits in a different order is also divisible by 3. Is he correct? Explain your decision.

3.2 REDUCING FRACTIONS

OBJECTIVES

1 Define equivalent fractions and lowest terms form.
2 Reduce fractions by inspection.
3 Reduce fractions by prime factoring.
4 Reduce mixed numbers and improper fractions.

NEW VOCABULARY

equivalent fractions
lowest terms
reducing
reducing by inspection
reducing by prime factoring
greatest common factor

1 Equivalent Fractions and Lowest Terms Form

A fraction can be written in many ways. For example, the fraction one half can be written in each of the following ways.

$$\frac{1}{2}, \quad \frac{2}{4}, \quad \frac{3}{6}, \quad \frac{4}{8}, \quad \frac{5}{10}, \quad \frac{6}{12}, \ldots$$

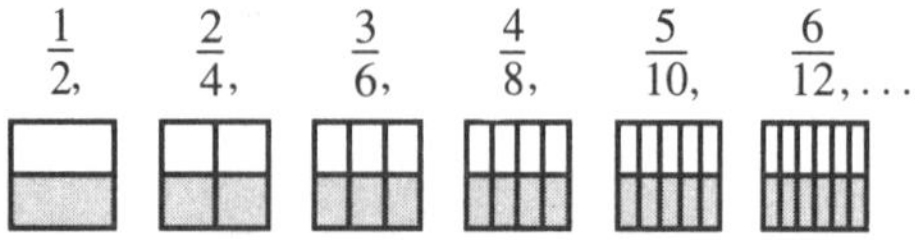

All of these fractions are equal. In each case, half of the figure is shaded. Fractions that are equal are called **equivalent fractions**. In this example, the fraction $\frac{1}{2}$ is in **lowest terms** form. This means both terms in the fraction are the smallest numbers that can be used to make one half.

> A fraction is in lowest terms form if the only counting number that divides exactly into *both* numerator and denominator is the number 1.

The process of writing a fraction in lowest terms form is called **reducing**. In this text all fractional answers will be reduced to lowest terms.

2 Reducing Fractions by Inspection

> To reduce a fraction to lowest terms by **inspection**
>
> **1.** Divide both numerator and denominator by any counting number that you can see is a divisor of both. Use the divisibility rules for primes to help you find divisors.
> **2.** Keep performing these divisions until the only number that divides both numerator and denominator exactly is 1.

EXAMPLE 1 Reduce $\frac{6}{12}$ to lowest terms by inspection.

A number that divides 6 and 12 exactly is the number 6.

$$\frac{6}{12} = \frac{6 \div 6}{12 \div 6} = \frac{1}{2} \quad \text{(lowest terms)}$$

The fraction $\frac{1}{2}$ is in lowest terms form because the only number that divides both the numerator 1 and the denominator 2 exactly is the number 1. ■

Fractions that can be reduced to the same lowest terms answer are equivalent. The fractions $\frac{1}{2}$ and $\frac{6}{12}$ are equivalent fractions. Reducing a fraction to lowest terms means finding an equivalent fraction with the smallest possible terms.

EXAMPLE 2 Reduce $\frac{15}{24}$ to lowest terms by inspection.

3 is a divisor of both 15 and 24. Reduce using 3.

$$\frac{15}{24} = \frac{15 \div 3}{24 \div 3} = \frac{5}{8}$$

$\frac{5}{8}$ is in lowest terms form because the only number that divides both 5 and 8 exactly is 1. ■

▶ **You Try It** Reduce to lowest terms by inspection.

1. $\frac{8}{24}$ **2.** $\frac{12}{30}$ **3.** $\frac{32}{42}$

EXAMPLE 3 Reduce $\frac{36}{60}$ to lowest terms using inspection.

$$\frac{36}{60} = \frac{36 \div 2}{60 \div 2} = \frac{18}{30} = \frac{18 \div 6}{30 \div 6} = \frac{3}{5} \quad \text{(lowest terms)}$$

reduce using 2 reduce using 6 ■

The reduction in Example 3 may also be written vertically.

$$\frac{\overset{\overset{3}{\cancel{18}}}{\cancel{36}}}{\underset{\underset{5}{\cancel{30}}}{\cancel{60}}} = \frac{3}{5}$$

First reduce 36 and 60 using 2.
Then reduce 18 and 30 using 6.

EXAMPLE 4 Reduce $\frac{63}{147}$ to lowest terms by inspection.

Use the divisibility rules to find a divisor for 63 and 147.
3 is a divisor of 63 because 3 divides $6 + 3 = 9$.
3 is a divisor of 147 because 3 divides $1 + 4 + 7 = 12$.

$$\frac{63}{147} = \frac{63 \div 3}{147 \div 3} = \frac{21}{49} = \frac{21 \div 7}{49 \div 7} = \frac{3}{7} \quad \text{(lowest terms)}$$

reduce using 3 reduce using 7 ■

▶ **You Try It** Reduce to lowest terms by inspection.

4. $\frac{42}{63}$ **5.** $\frac{231}{264}$

---

3 Reducing Fractions by Prime Factoring

To reduce fractions to lowest terms using **prime factoring**

1. Prime factor both numerator and denominator.
2. Reduce numerator and denominator by all factors common to both prime factorizations.
3. Simplify for the lowest terms answer.

EXAMPLE 5 Reduce $\frac{6}{10}$ to lowest terms using prime factoring.

6 is prime factored as $6 = 2 \times 3$.
10 is prime factored as $10 = 2 \times 5$.

$$\frac{6}{10} = \frac{2 \times 3}{2 \times 5}$$

2 is the only factor common to both numerator and denominator.

$$\frac{6}{10} = \frac{\overset{1}{\cancel{2}} \times 3}{\underset{1}{\cancel{2}} \times 5} = \frac{1 \times 3}{1 \times 5} = \frac{3}{5}$$

Reduce both numerator and denominator by the common factor 2. Since $2 \div 2 = 1$, write 1 in place of each 2. ■

EXAMPLE 6 Reduce $\frac{30}{75}$ to lowest terms using prime factoring.

30 is prime factored as $30 = 2 \times 3 \times 5$.
75 is prime factored as $75 = 3 \times 5 \times 5$.

$$\frac{30}{75} = \frac{2 \times 3 \times 5}{3 \times 5 \times 5}$$

3 and 5 are common factors of both numerator and denominator.

$$\frac{30}{75} = \frac{2 \times \overset{1}{\cancel{3}} \times \overset{1}{\cancel{5}}}{\underset{1}{\cancel{3}} \times \underset{1}{\cancel{5}} \times 5} = \frac{2}{5}$$

Reduce both numerator and denominator by 3, then by 5. Replace all reduced factors by 1. ■

EXAMPLE 7 Reduce $\frac{60}{96}$ to lowest terms using prime factoring.

60 is prime factored as $60 = 2 \times 2 \times 3 \times 5$.
96 is prime factored as $96 = 2 \times 2 \times 2 \times 2 \times 2 \times 3$.

$$\frac{60}{96} = \frac{\overset{1}{\cancel{2}} \times \overset{1}{\cancel{2}} \times \overset{1}{\cancel{3}} \times 5}{2 \times 2 \times 2 \times \underset{1}{\cancel{2}} \times \underset{1}{\cancel{2}} \times \underset{1}{\cancel{3}}}$$

Reduce numerator and denominator by the common factors 2, 2, and 3.

$$= \frac{5}{2 \times 2 \times 2} = \frac{5}{8}$$ ■

OBSERVE In Example 7, the factors common to both 60 and 96 are 2, 2, and 3. The product of these common factors is called the **greatest common factor,** or **GCF.**

GCF of 60 and 96 = $2 \times 2 \times 3 = 12$

The GCF is the number that will reduce the fraction to lowest terms in one step.

$$\frac{60}{96} = \frac{60 \div 12}{96 \div 12} = \frac{5}{8}$$

lowest terms in one step

▶ You Try It Reduce to lowest terms using prime factoring.

6. $\frac{15}{18}$ **7.** $\frac{40}{64}$ **8.** $\frac{72}{120}$

4 Reducing Mixed Numbers and Improper Fractions

Reduce a mixed number by reducing its fractional part.

EXAMPLE 8 Reduce the mixed number $5\frac{8}{12}$.

$$5\frac{8}{12} = 5 + \frac{8}{12} = 5 + \frac{\overset{2}{\cancel{8}}}{\underset{3}{\cancel{12}}} = 5 + \frac{2}{3} = 5\frac{2}{3} \quad \left(\text{In short, } 5\frac{\overset{2}{\cancel{8}}}{\underset{3}{\cancel{12}}} = 5\frac{2}{3}\right)$$ ■

As seen in Example 8, to reduce a mixed number, just reduce the proper fraction. Example 9 shows how to reduce an improper fraction.

EXAMPLE 9 Reduce $\frac{35}{30}$ to lowest terms.

$$\frac{35}{30} = 1\frac{5}{30} = 1\frac{\overset{1}{\cancel{5}}}{\underset{6}{\cancel{30}}} = 1\frac{1}{6}$$

Change the improper fraction to a mixed number.

Reduce the proper fraction using 5. ■

▶ You Try It Reduce to lowest terms. **9.** $3\frac{21}{24}$ **10.** $\frac{40}{25}$

EXAMPLE 10 Simplify $5\frac{24}{9}$.

$$5\frac{24}{9} = 5 + \frac{24}{9} = 5 + 2\frac{6}{9} = (5 + 2) + \frac{6}{9} = 7 + \frac{\overset{2}{\cancel{6}}}{\underset{3}{\cancel{9}}} = 7\frac{2}{3}$$

This mixed number contains an improper fraction.

Change the improper fraction to a mixed number.

Add the whole numbers.

Reduce. ■

▶ You Try It Simplify. **11.** $4\frac{15}{6}$

▶ Answers to You Try It 1. $\frac{1}{3}$ 2. $\frac{2}{5}$ 3. $\frac{16}{21}$ 4. $\frac{2}{3}$ 5. $\frac{7}{8}$ 6. $\frac{5}{6}$ 7. $\frac{5}{8}$ 8. $\frac{3}{5}$ 9. $3\frac{7}{8}$ 10. $1\frac{3}{5}$ 11. $6\frac{1}{2}$

SECTION 3.2 EXERCISES

1 **2** *Reduce to lowest terms using inspection.*

1. $\frac{6}{8}$
2. $\frac{4}{8}$
3. $\frac{6}{9}$
4. $\frac{5}{10}$
5. $\frac{6}{14}$
6. $\frac{14}{21}$
7. $\frac{15}{20}$
8. $\frac{8}{24}$
9. $\frac{12}{20}$
10. $\frac{7}{28}$
11. $\frac{12}{16}$
12. $\frac{21}{24}$
13. $\frac{9}{36}$
14. $\frac{10}{30}$

3 *Reduce to lowest terms using prime factoring* ***when necessary.***

15. $\frac{21}{35}$
16. $\frac{25}{30}$
17. $\frac{32}{40}$
18. $\frac{24}{42}$
19. $\frac{33}{77}$
20. $\frac{42}{56}$
21. $\frac{24}{32}$
22. $\frac{9}{33}$
23. $\frac{45}{75}$
24. $\frac{49}{84}$
25. $\frac{52}{65}$
26. $\frac{60}{72}$
27. $\frac{48}{120}$
28. $\frac{25}{35}$

Reduce to lowest terms using any method.

29. $\frac{24}{64}$
30. $\frac{18}{90}$
31. $\frac{34}{44}$
32. $\frac{28}{49}$
33. $\frac{57}{78}$
34. $\frac{54}{81}$
35. $\frac{15}{35}$
36. $\frac{12}{80}$
37. $\frac{60}{100}$
38. $\frac{120}{150}$
39. $\frac{90}{270}$
40. $\frac{300}{480}$
41. $\frac{75}{225}$
42. $\frac{135}{270}$

43. What fractional part of an hour is 15 minutes?

44. What fractional part of a day is 18 hours?

45. A person making $30,000 a year pays $5,000 in taxes. What fractional part of the salary pays for taxes?

46. A person making $1,680 a month in wages pays $630 a month for rent. What fraction of the monthly salary is rent?

47. Out of 12,000 students attending a university, 8,400 are women. What fraction are women?

48. 64,000 out of 80,000 inhabitants live in villages. What fraction live in villages?

49. In a high school of 371 students, 318 plan to attend college.
    a. What fraction plan to attend college?
    b. What fraction do not plan to attend college?

50. Out of a sample of 710 people, 639 are holding full-time jobs.
    a. What fraction of the sample is this?
    b. What fraction are not holding full-time jobs?

51. Which fraction does not reduce to $\frac{1}{3}$?

$$\frac{3}{9}, \frac{14}{42}, \frac{12}{36}, \frac{18}{48}, \text{ or } \frac{13}{39}$$

52. Which fraction does reduce to $\frac{3}{8}$?

$$\frac{15}{24}, \frac{16}{32}, \frac{15}{40}, \frac{24}{72}, \text{ or } \frac{36}{48}$$

4 *Express each number as a mixed number reduced to lowest terms.*

53. $4\frac{32}{40}$

54. $2\frac{18}{24}$

55. $5\frac{15}{21}$

56. $7\frac{12}{28}$

57. $\frac{33}{27}$

58. $\frac{25}{20}$

59. $\frac{57}{33}$

60. $\frac{108}{48}$

61. $\frac{75}{40}$

62. $\frac{720}{300}$

63. $6\frac{15}{10}$

64. $10\frac{24}{18}$

65. $7\frac{38}{9}$

66. $12\frac{27}{13}$

SKILLSFOCUS (Section 1.6) *Write each of the following in decimal notation.*

67. fifty and five hundredths

68. fifty five hundredths

69. forty six thousandths

70. forty six thousand

EXTEND YOUR THINKING ▶▶▶▶

▶ TROUBLESHOOT IT

Find and correct the error.

71. $\frac{36}{54} = \frac{36 \div 6}{54 \div 6} = \frac{6}{8} = \frac{3}{4}$

72. $\frac{48}{72} = \frac{\overset{1}{\cancel{2}} \times \overset{1}{\cancel{2}} \times 2 \times 2 \times \overset{1}{\cancel{3}}}{\underset{1}{\cancel{2}} \times \underset{1}{\cancel{2}} \times 3 \times 3 \times \underset{1}{\cancel{3}}} = \frac{4}{9}$

WRITING TO LEARN ▶▶▶▶

73. Explain why $\frac{3}{8}$ is reduced to lowest terms, but $\frac{6}{16}$ is not.

▶ YOU BE THE JUDGE

74. Jim claims that $\frac{10}{16}, \frac{25}{40}$, and $\frac{15}{24}$, are three different ways to write $\frac{5}{8}$. Is he correct? Explain how you arrived at your decision.

3.3 MULTIPLYING FRACTIONS

OBJECTIVES

1. Multiply proper fractions without reducing.
2. Multiply proper fractions with reducing.
3. Multiply mixed numbers.
4. Evaluate fractions raised to powers.

1 Multiplying Proper Fractions

EXAMPLE 1

$$\frac{1}{3} \times \frac{2}{5} = \frac{1 \times 2}{3 \times 5} = \frac{2}{15} \quad \blacksquare$$

> To multiply two proper fractions
>
> 1. Multiply the two numerators for the new numerator.
> 2. Multiply the two denominators for the new denominator.
> 3. Reduce your answer, if possible.

EXAMPLE 2

$$\frac{3}{4} \times \frac{5}{7} = \frac{3 \times 5}{4 \times 7} = \frac{15}{28} \quad \blacksquare$$

EXAMPLE 3

$$\frac{5}{9} \times \frac{2}{3} \times \frac{4}{7} = \frac{5 \times 2 \times 4}{9 \times 3 \times 7} = \frac{40}{189} \quad \blacksquare$$

▶ **You Try It** 1. $\frac{2}{3} \times \frac{4}{9}$ 2. $\frac{7}{8} \times \frac{3}{5}$ 3. $\frac{5}{6} \times \frac{1}{2} \times \frac{7}{9}$

2 Multiplying Proper Fractions with Reducing

In the next example, the answer must be reduced to lowest terms.

EXAMPLE 4

$$\frac{3}{4} \times \frac{2}{5} = \frac{3 \times 2}{4 \times 5} = \frac{6}{20} = \frac{\cancel{6}^{\,3}}{\cancel{20}_{\,10}} = \frac{3}{10}$$

Reduce 6 and 20 using 2. ■

In Example 4, you reduced after multiplying. You can reduce fractions before multiplying. If a numerator and a denominator have a common factor, reduce both using the factor. In Example 4, the numerator 2 and the denominator 4 have a factor of 2 in common. Reduce 2 and 4 using 2 before you multiply.

$$\frac{3}{\cancel{4}_{\,2}} \times \frac{\cancel{2}^{\,1}}{5} = \frac{3 \times 1}{2 \times 5} = \frac{3}{10} \quad \text{(same lowest terms answer as in Example 4)}$$

The advantage to reducing before you multiply is that you avoid reducing a fraction with a very large numerator or denominator.

EXAMPLE 5 Multiply $\frac{7}{8} \times \frac{36}{45}$.

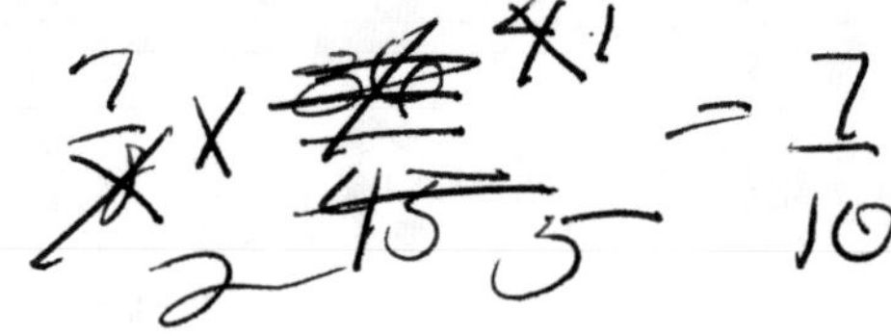

Solution #1: Multiply first, then reduce.

$$\frac{7}{8} \times \frac{36}{45} = \frac{7 \times 36}{8 \times 45} = \frac{\overset{\overset{\overset{7}{\cancel{63}}}{\cancel{126}}}{\cancel{252}}}{\underset{\underset{\underset{10}{\cancel{90}}}{\cancel{180}}}{\cancel{360}}} = \frac{7}{10}$$

Reduce 252 and 360 using 2 (/).
Reduce 126 and 180 using 2 (—).
Reduce 63 and 90 using 9 (\).

Solution #2: Now solve the same problem by reducing first, then multiplying. Reduce any numerator with any denominator that share a common factor.

Reduce 36 and 45 using 9 (/).
Reduce 8 and 4 using 4 (—).

$$\frac{7}{\underset{2}{\cancel{8}}} \times \frac{\overset{\overset{1}{\cancel{4}}}{\cancel{36}}}{\underset{5}{\cancel{45}}} = \frac{7 \times 1}{2 \times 5} = \frac{7}{10}$$ ■

The answer is the same in each case. By reducing before multiplying (see Solution #2), the numbers you reduced and multiplied were smaller. In Solution #2, you may reduce *vertically*, as with 36 and 45, or *diagonally* as with 4 and 8. The order in which the terms are reduced will not change the answer.

> To reduce before multiplying fractions
>
> 1. If a numerator has a factor in common with a denominator, reduce both by the common factor.
> 2. Repeat this process until the only factor any numerator has in common with any denominator is 1.
> 3. Multiply across. Your answer will be in lowest terms.

EXAMPLE 6

$$\frac{10}{15} \times \frac{7}{24} = \frac{\overset{\overset{1}{\cancel{2}}}{\cancel{10}}}{\underset{3}{\cancel{15}}} \times \frac{7}{\underset{12}{\cancel{24}}} = \frac{1 \times 7}{3 \times 12} = \frac{7}{36}$$

Reduce 10 and 15 using 5 (/).
Reduce 2 and 24 using 2 (—). ■

CAUTION You may not reduce 15 and 24 using 3, because 15 and 24 are both denominators.

▶ **You Try It** Reduce, then multiply.

4. $\frac{5}{8} \times \frac{4}{7}$

5. $\frac{10}{12} \times \frac{2}{5}$

EXAMPLE 7

$$\frac{15}{21} \times \frac{28}{35} = \frac{\overset{1}{\cancel{\overset{\cancel{5}}{\cancel{15}}}}}{\underset{7}{\cancel{21}}} \times \frac{\overset{4}{\cancel{28}}}{\underset{1}{\cancel{\underset{\cancel{5}}{\cancel{35}}}}} = \frac{1 \times 4}{7 \times 1} = \frac{4}{7}$$

Reduce 28 and 35 using 7 (/).
Reduce 15 and 21 using 3 (—).
Reduce 5 and 5 using 5 (\). ■

CAUTION Always check to see if your answer can still be reduced. If so, you did not reduce completely before multiplying across.

EXAMPLE 8

$$\frac{5}{9} \times \frac{16}{35} \times \frac{3}{40} = \frac{\overset{1}{\cancel{5}}}{\underset{3}{\cancel{9}}} \times \frac{\overset{2}{\cancel{16}}}{35} \times \frac{\overset{1}{\cancel{3}}}{\underset{1}{\cancel{\underset{\cancel{5}}{\cancel{40}}}}} = \frac{1 \times 2 \times 1}{3 \times 35 \times 1} = \frac{2}{105}$$

Reduce 3 and 9 using 3 (/).
Reduce 16 and 40 using 8 (—).
Reduce 5 and 5 (\). ■

▶ **You Try It** 6. $\frac{28}{42} \times \frac{12}{16}$ 7. $\frac{5}{9} \times \frac{3}{10}$ 8. $\frac{3}{8} \times \frac{6}{9} \times \frac{8}{15}$

3 Multiplying Mixed Numbers

To multiply mixed numbers, first write each mixed number as an improper fraction.

EXAMPLE 9

$$3\frac{1}{2} \times \frac{5}{6} = \frac{7}{2} \times \frac{5}{6} = \frac{7 \times 5}{2 \times 6} = \frac{35}{12} = 2\frac{11}{12}$$

Rewrite $3\frac{1}{2}$ as $\frac{7}{2}$. Rewrite $\frac{35}{12}$ as $2\frac{11}{12}$. ■

> To multiply two or more mixed numbers
>
> 1. Change each mixed number to an improper fraction.
> 2. Reduce, if possible.
> 3. Multiply across.

EXAMPLE 10

$$2\frac{4}{5} \times 4\frac{2}{7} = \frac{\overset{2}{\cancel{14}}}{\underset{1}{\cancel{5}}} \times \frac{\overset{6}{\cancel{30}}}{\underset{1}{\cancel{7}}} = \frac{2 \times 6}{1 \times 1} = \frac{12}{1} = 12$$

Reduce 14 and 7 using 7 (/).
Reduce 30 and 5 using 5 (—). ■

▶ **You Try It** 9. $7\frac{1}{3} \times \frac{2}{5}$ 10. $4\frac{3}{5} \times 3\frac{1}{3}$

To multiply a fraction by a whole number, first write the whole number as an improper fraction with a denominator of 1.

EXAMPLE 11

$$6 \times \frac{3}{5} = \frac{6}{1} \times \frac{3}{5} = \frac{6 \times 3}{1 \times 5} = \frac{18}{5} = 3\frac{3}{5}$$

Rewrite 6 as $\frac{6}{1}$. ■

EXAMPLE 12

$$3\frac{9}{10} \times \frac{5}{12} \times 5\frac{1}{3} = \frac{\overset{13}{\cancel{39}}}{\underset{\underset{1}{\cancel{2}}}{\cancel{10}}} \times \frac{\overset{1}{\cancel{5}}}{\underset{3}{\cancel{12}}} \times \frac{\overset{\overset{2}{\cancel{4}}}{\cancel{16}}}{\underset{1}{\cancel{3}}} = \frac{13 \times 1 \times 2}{1 \times 3 \times 1} = \frac{26}{3} = 8\frac{2}{3}$$

Reduce 39 and 3 using 3 (/).
Reduce 12 and 16 using 4 (—).
Reduce 10 and 5 using 5 (\).
Reduce 2 and 4 using 2 (//). ■

▶ **You Try It**

11. $8 \times \frac{5}{6}$

12. $2\frac{5}{8} \times \frac{3}{7} \times 1\frac{1}{9}$

4 Evaluating a Fraction Raised to a Power

Recall from Section 2.2 that a power on a number means repeated multiplication of that number. When a fraction is raised to a power, write the fraction the number of times indicated by the power, then multiply.

EXAMPLE 13

$$\left(\frac{3}{5}\right)^2 = \frac{3}{5} \times \frac{3}{5} = \frac{9}{25}$$ ■

EXAMPLE 14

$$\left(2\frac{1}{4}\right)^2 = 2\frac{1}{4} \times 2\frac{1}{4} = \frac{9}{4} \times \frac{9}{4} = \frac{81}{16} \text{ or } 5\frac{1}{16}$$ ■

EXAMPLE 15

$$\left(\frac{1}{2}\right)^4 = \frac{1}{2} \times \frac{1}{2} \times \frac{1}{2} \times \frac{1}{2} = \frac{1}{16}$$ ■

▶ **You Try It**

Evaluate.

13. $\left(\frac{3}{5}\right)^2$

14. $\left(3\frac{1}{2}\right)^2$

15. $\left(\frac{5}{4}\right)^3$

▶ **Answers to You Try It** 1. $\frac{8}{27}$ 2. $\frac{21}{40}$ 3. $\frac{35}{108}$ 4. $\frac{5}{14}$ 5. $\frac{1}{3}$ 6. $\frac{1}{2}$ 7. $\frac{1}{6}$ 8. $\frac{2}{15}$ 9. $\frac{44}{15}$ 10. $15\frac{1}{3}$ 11. $6\frac{2}{3}$ 12. $1\frac{1}{4}$ 1 $\frac{9}{25}$ 14. $12\frac{1}{4}$ 15. $1\frac{61}{64}$

SECTION 3.3 EXERCISES

1 **2** *Multiply and express all answers in lowest terms.*

1. $\frac{3}{5} \times \frac{2}{3}$

2. $\frac{3}{4} \times \frac{1}{2}$

3. $\frac{4}{7} \times \frac{8}{9}$

4. $\frac{5}{6} \times \frac{7}{8}$

5. $\frac{4}{7} \times \frac{3}{4}$

6. $\frac{5}{6} \times \frac{1}{5}$

7. $\frac{3}{4} \times \frac{4}{7}$

8. $\frac{3}{5} \times \frac{4}{9}$

9. $\frac{7}{10} \times \frac{4}{5}$

10. $\frac{8}{9} \times \frac{5}{12}$

11. $\frac{6}{11} \times \frac{11}{6}$

12. $\frac{7}{10} \times \frac{10}{7}$

13. $\frac{3}{5} \times \frac{25}{36}$

14. $\frac{9}{10} \times \frac{2}{3}$

15. $\frac{15}{28} \times \frac{35}{70}$

16. $\frac{24}{60} \times \frac{30}{36}$

17. $\frac{27}{45} \times \frac{6}{15}$

18. $\frac{24}{32} \times \frac{12}{16}$

19. $\frac{60}{90} \times \frac{20}{30}$

20. $\frac{25}{63} \times \frac{56}{35}$

21. $\frac{5}{6} \times \frac{2}{15} \times \frac{20}{24}$

22. $\frac{4}{5} \times \frac{3}{8} \times \frac{6}{9}$

23. $\frac{50}{75} \times \frac{16}{28} \times \frac{21}{24}$

24. $\frac{18}{42} \times \frac{36}{45} \times \frac{55}{64}$

3 *Multiply and express each answer in lowest terms.*

25. $\frac{5}{6} \times 4\frac{1}{2}$

26. $\frac{5}{8} \times 3\frac{1}{5}$

27. $5\frac{1}{2} \times \frac{4}{5}$

28. $6\frac{3}{4} \times 1\frac{1}{7}$

29. $2\frac{1}{4} \times 3\frac{1}{2}$

30. $4\frac{2}{3} \times 1\frac{3}{5}$

31. $2\frac{1}{3} \times 4\frac{1}{2}$

32. $6\frac{1}{8} \times 3\frac{1}{7}$

33. $5\frac{1}{9} \times 2\frac{2}{3}$

34. $3\frac{3}{4} \times 1\frac{1}{10}$

35. $4 \times 3\frac{1}{5}$

36. $8 \times 2\frac{1}{4}$

37. $5\frac{1}{4} \times \frac{6}{7}$

38. $2\frac{2}{9} \times \frac{4}{5}$

39. $2\frac{2}{3} \times 6$

40. $4\frac{2}{5} \times 3$

41. $\frac{2}{3} \times 10 \times 4\frac{1}{5}$

42. $3 \times 4\frac{1}{5} \times \frac{5}{7}$

43. $4\frac{1}{2} \times 3\frac{2}{3} \times 5\frac{5}{9}$

44. $1\frac{1}{4} \times 4\frac{1}{3} \times 2\frac{1}{2}$

45. $4\frac{1}{5} \times 2\frac{1}{7} \times 3\frac{2}{3}$

46. $4\frac{1}{12} \times \frac{3}{70} \times \frac{8}{14}$

47. What is 2 times $\frac{1}{2}$?

48. What is $5\frac{5}{6}$ times $\frac{12}{35}$?

4 *Evaluate each power.*

49. $\left(\frac{3}{5}\right)^2$ **50.** $\left(\frac{1}{4}\right)^2$ **51.** $\left(2\frac{5}{6}\right)^2$ **52.** $\left(4\frac{1}{2}\right)^2$

53. $\left(\frac{6}{7}\right)^3$ **54.** $\left(\frac{2}{5}\right)^3$ **55.** $\left(\frac{3}{10}\right)^4$ **56.** $\left(\frac{1}{3}\right)^4$

57. $\left(\frac{5}{3}\right)^4$ **58.** $\left(\frac{5}{2}\right)^3$ **59.** $\left(3\frac{1}{3}\right)^2$ **60.** $\left(6\frac{1}{2}\right)^3$

61. What is $\frac{7}{8}$ squared?

62. What is $\frac{3}{4}$ cubed?

SKILLSFOCUS (Section 2.4) *Evaluate:*

63. $\sqrt{36}$ **64.** $\sqrt{81}$ **65.** $\sqrt{1}$ **66.** $\sqrt{100}$

EXTEND YOUR THINKING ▶▶▶▶

▶ TROUBLESHOOT IT

Find and correct the error.

67. $\frac{3}{8} \times \frac{1}{4} = \frac{3}{\cancel{8}_{2}} \times \frac{1}{\cancel{4}_{1}} = \frac{3 \times 1}{2 \times 1} = \frac{3}{2}$

68. $2\frac{1}{2} \times 5\frac{3}{4} = (2 \times 5) + \left(\frac{1}{2} \times \frac{3}{4}\right) = 10 + \frac{3}{8} = 10\frac{3}{8}$

69. $3\frac{1}{4} \times \frac{2}{5} = 3\frac{1}{\cancel{4}_{2}} \times \frac{\cancel{2}^{1}}{5} = 3\frac{1}{2} \times \frac{1}{5} = \frac{7}{2} \times \frac{1}{5} = \frac{7}{10}$

WRITING TO LEARN ▶▶▶▶

70. Explain each step you would use to multiply a mixed number times a whole number. Demonstrate using your own example.

71. Explain the advantage to reducing before multiplying fractions. Demonstrate your point using your own example.

▶ YOU BE THE JUDGE

72. Ron claims $\left(\frac{24}{40}\right)^3$ can only be evaluated by reducing within parentheses first, then cubing. Barb says no, you must write $\frac{24}{40}$ times itself 3 times first. Then you reduce and multiply. Settle the dispute.

3.4 DIVIDING FRACTIONS

OBJECTIVES

1. Find the reciprocal of a fraction.
2. Divide fractions.
3. Divide mixed numbers.

NEW VOCABULARY

reciprocal

1 The Reciprocal of a Fraction

In order to divide fractions, you need to understand reciprocals. The **reciprocal** of a fraction is found by interchanging the numerator with the denominator.

EXAMPLE 1 **a.** The reciprocal of $\frac{2}{3}$ is $\frac{3}{2}$. **b.** The reciprocal of $\frac{7}{12}$ is $\frac{12}{7}$.

c. The reciprocal $\frac{1}{4}$ is $\frac{4}{1}$, or 4. ■

To find the reciprocal of a counting number, first write the counting number with a denominator of 1.

EXAMPLE 2 **a.** Since $6 = \frac{6}{1}$, the reciprocal of 6 is $\frac{1}{6}$. **b.** The reciprocal of $25 = \frac{25}{1}$ is $\frac{1}{25}$. ■

CAUTION Since the reciprocal of $0 = \frac{0}{1}$ is $\frac{1}{0}$, 0 has no reciprocal. $\frac{1}{0}$ is division by 0, which is not possible.

▶ **You Try It** Find the reciprocal.

1. $\frac{4}{7}$ **2.** $\frac{9}{10}$ **3.** $\frac{1}{8}$ **4.** 2 **5.** 1 **6.** 18

To find the reciprocal of a mixed number, first write the mixed number as an improper fraction.

EXAMPLE 3 **a.** Since $3\frac{1}{2} = \frac{7}{2}$, the reciprocal of $3\frac{1}{2}$ is $\frac{2}{7}$.

b. Since $2\frac{7}{9} = \frac{25}{9}$, the reciprocal of $2\frac{7}{9}$ is $\frac{9}{25}$. ■

▶ **You Try It** Find the reciprocal.

7. $2\frac{1}{5}$ **8.** $5\frac{3}{8}$

The product of a fraction and its reciprocal is 1.

EXAMPLE 4 a. The reciprocal of $\frac{2}{5}$ is $\frac{5}{2}$, and $\frac{2}{5} \times \frac{5}{2} = \frac{10}{10} = 1$.

b. The reciprocal of $4 = \frac{4}{1}$ is $\frac{1}{4}$, and $4 \times \frac{1}{4} = \frac{4}{1} \times \frac{1}{4} = \frac{4}{4} = 1$.

c. The reciprocal of $2\frac{3}{4} = \frac{11}{4}$ is $\frac{4}{11}$, and $2\frac{3}{4} \times \frac{4}{11} = \frac{11}{4} \times \frac{4}{11} = \frac{44}{44} = 1$. ■

▶ **You Try It** Find the reciprocal. Then find the product of each fraction and its reciprocal.

9. $\frac{4}{9}$ **10.** 6 **11.** $7\frac{2}{3}$

---

2 Dividing Fractions

How do you divide the fractions below?

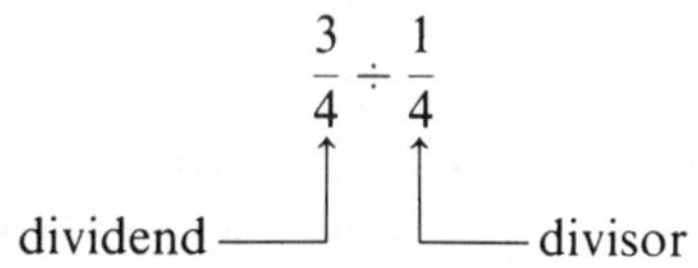

OBSERVE This is read "$\frac{3}{4}$ divided by $\frac{1}{4}$." Intuitively, the answer is 3. There are three $\frac{1}{4}$ cups in $\frac{3}{4}$ cups of milk.

To see how to divide, first write the division in fraction form.

$$\frac{3}{4} \div \frac{1}{4} = \frac{\frac{3}{4}}{\frac{1}{4}}$$

Recall, $a \div b = \frac{a}{b}$. Replace the letter a with $\frac{3}{4}$.

Replace b with $\frac{1}{4}$.

$$= \frac{\frac{3}{4}}{\frac{1}{4}} \times 1$$

Multiplication property of one.

$$= \frac{\frac{3}{4}}{\frac{1}{4}} \times \frac{\frac{4}{1}}{\frac{4}{1}}$$

Write 1 as the reciprocal of $\frac{1}{4}$ (the denominator), divided by itself.

$$= \frac{\frac{3}{4} \times \frac{4}{1}}{\frac{1}{\not{4}_1} \times \frac{\not{4}^1}{1}}$$

In the denominator, a fraction times its reciprocal equals 1.

$$= \frac{\frac{3}{4} \times \frac{4}{1}}{1}$$

$$= \frac{3}{4} \times \frac{4}{1}$$

To divide fractions, multiply the dividend by the reciprocal of the divisor.

$$= 3$$

To divide two proper fractions

1. Multiply the dividend by the reciprocal of the divisor. (You also say: invert the divisor, then multiply. To invert a fraction means write its reciprocal.)

$$\frac{a}{b} \div \frac{c}{d} = \frac{a}{b} \times \frac{d}{c}$$

2. Reduce *after* the problem has been rewritten as a multiplication.

EXAMPLE 5 a. $\frac{4}{5} \div \frac{3}{7} = \frac{4}{5} \times \frac{7}{3} = \frac{28}{15}$ or $1\frac{13}{15}$

b. $\frac{14}{15} \div \frac{7}{10} = \frac{\overset{2}{\cancel{14}}}{\underset{3}{\cancel{15}}} \times \frac{\overset{2}{\cancel{10}}}{\underset{1}{\cancel{7}}} = \frac{4}{3}$ or $1\frac{1}{3}$ ■

EXAMPLE 6 a. $4 \div \frac{2}{5} = \frac{4}{1} \div \frac{2}{5} = \frac{\overset{2}{\cancel{4}}}{1} \times \frac{5}{\underset{1}{\cancel{2}}} = \frac{10}{1}$ or 10

b. $\frac{2}{9} \div 3 = \frac{2}{9} \div \frac{3}{1} = \frac{2}{9} \times \frac{1}{3} = \frac{2}{27}$ ■

You Try It Divide. **12.** $\frac{3}{8} \div \frac{2}{5}$ **13.** $\frac{3}{4} \div \frac{5}{6}$ **14.** $\frac{5}{9} \div \frac{10}{21}$ **15.** $9 \div \frac{3}{7}$ **16.** $\frac{4}{5} \div 2$

You can divide whole numbers using fraction notation.

EXAMPLE 7 $12 \div 3 = \frac{12}{1} \div \frac{3}{1} = \frac{\overset{4}{\cancel{12}}}{1} \times \frac{1}{\underset{1}{\cancel{3}}} = \frac{4}{1} = 4$ ■

You Try It Divide using fraction notation. **17.** $18 \div 6$

CAUTION When dividing fractions, if you make the mistake of multiplying by the reciprocal of the dividend, your answer will be the reciprocal of the correct answer. Using Example 5,

$$\frac{4}{5} \div \frac{3}{7} = \frac{5}{4} \times \frac{3}{7} = \frac{15}{28}$$

is **wrong**. Observe, this answer is the reciprocal of the correct answer.

Division of fractions is not commutative. Changing the order of the fractions in the following division changes the answer.

$$\frac{3}{4} \div \frac{2}{5} \neq \frac{2}{5} \div \frac{3}{4}$$

$$\frac{3}{4} \times \frac{5}{2} = \frac{15}{8} \neq \frac{2}{5} \times \frac{4}{3} = \frac{8}{15}$$

3 Dividing Mixed Numbers

To divide with mixed numbers

1. First change each mixed number to an improper fraction.
2. Then follow the same procedure used to divide fractions.

EXAMPLE 8 a. $3\frac{2}{5} \div \frac{3}{4} = \frac{17}{5} \div \frac{3}{4} = \frac{17}{5} \times \frac{4}{3} = \frac{68}{15}$ or $4\frac{8}{15}$

b. $5\frac{5}{8} \div 2\frac{1}{4} = \frac{45}{8} \div \frac{9}{4} = \frac{\overset{5}{\cancel{45}}}{\underset{2}{\cancel{8}}} \times \frac{\overset{1}{\cancel{4}}}{\underset{1}{\cancel{9}}} = \frac{5}{2}$ or $2\frac{1}{2}$ ■

EXAMPLE 9 a. $2\frac{5}{6} \div 4 = \frac{17}{6} \div \frac{4}{1} = \frac{17}{6} \times \frac{1}{4} = \frac{17}{24}$

b. $12 \div 1\frac{1}{5} = \frac{12}{1} \div \frac{6}{5} = \frac{\overset{2}{\cancel{12}}}{1} \times \frac{5}{\underset{1}{\cancel{6}}} = \frac{10}{1}$ or 10 ■

▶ **You Try It** Divide. **18.** $1\frac{2}{7} \div \frac{5}{6}$ **19.** $3\frac{3}{10} \div 2\frac{1}{5}$ **20.** $5\frac{1}{8} \div 8$ **21.** $16 \div 1\frac{1}{3}$

OBSERVE Every division of fractions problem is solved by first rewriting it as a multiplication problem.

▶ **Answers to You Try It** 1. $\frac{7}{4}$ 2. $\frac{10}{9}$ 3. $\frac{8}{1}$ or 8 4. $\frac{1}{2}$ 5. 1 6. $\frac{1}{18}$ 7. $\frac{5}{11}$ 8. $\frac{8}{43}$ 9. $\frac{9}{4}$; 1 10. $\frac{1}{6}$; 1 11. $\frac{3}{23}$; 1 12. $\frac{15}{16}$ 13. $\frac{9}{10}$ 14. $1\frac{1}{6}$ 15. 21 16. $\frac{2}{5}$ 17. 3 18. $1\frac{19}{35}$ 19. $1\frac{1}{2}$ 20. $\frac{41}{64}$ 21. 12

SECTION 3.4 EXERCISES

1 *Find the reciprocal. Then show the product of the fraction and its reciprocal is 1.*

1. $\frac{2}{7}$ **2.** $\frac{5}{6}$ **3.** 8 **4.** 3 **5.** $\frac{1}{10}$ **6.** $\frac{1}{6}$

7. $\frac{11}{15}$ **8.** $\frac{9}{14}$ **9.** $5\frac{1}{8}$ **10.** $2\frac{8}{9}$ **11.** $\frac{7}{4}$ **12.** $\frac{6}{25}$

2 *Divide and reduce all answers to lowest terms.*

13. $\frac{3}{4} \div \frac{5}{7}$ **14.** $\frac{2}{9} \div \frac{3}{5}$ **15.** $\frac{4}{7} \div \frac{1}{3}$ **16.** $\frac{1}{2} \div \frac{4}{9}$ **17.** $\frac{2}{3} \div \frac{5}{7}$

18. $\frac{1}{8} \div \frac{1}{3}$ **19.** $\frac{7}{9} \div \frac{5}{6}$ **20.** $\frac{4}{5} \div \frac{1}{5}$ **21.** $\frac{7}{10} \div \frac{3}{5}$ **22.** $\frac{3}{4} \div \frac{6}{7}$

23. $\frac{8}{11} \div \frac{4}{5}$ **24.** $\frac{7}{12} \div \frac{3}{14}$ **25.** $\frac{5}{9} \div \frac{5}{12}$ **26.** $\frac{3}{7} \div \frac{4}{21}$ **27.** $\frac{5}{18} \div \frac{2}{9}$

28. $\frac{9}{10} \div \frac{4}{5}$ **29.** $\frac{4}{5} \div 3$ **30.** $\frac{5}{8} \div 8$ **31.** $6 \div \frac{2}{3}$ **32.** $2 \div \frac{3}{5}$

33. $\frac{15}{20} \div \frac{3}{20}$ **34.** $\frac{24}{40} \div \frac{36}{50}$ **35.** $\frac{6}{11} \div \frac{30}{55}$ **36.** $\frac{35}{49} \div \frac{10}{14}$

3 *Divide and reduce all answers to lowest terms.*

37. $3\frac{3}{4} \div 2\frac{1}{3}$ **38.** $9\frac{2}{3} \div 2\frac{1}{2}$ **39.** $2\frac{1}{10} \div 1\frac{1}{6}$ **40.** $10\frac{1}{2} \div 6\frac{3}{8}$ **41.** $4\frac{1}{2} \div \frac{1}{4}$

42. $5\frac{1}{6} \div \frac{3}{4}$ **43.** $4 \div 1\frac{1}{3}$ **44.** $9 \div 2\frac{3}{4}$ **45.** $4\frac{3}{8} \div 2$ **46.** $3\frac{4}{5} \div 5$

47. $11 \div 2\frac{1}{5}$ **48.** $5 \div 6\frac{7}{8}$ **49.** $\frac{13}{20} \div 2\frac{3}{5}$ **50.** $7\frac{1}{8} \div \frac{2}{3}$ **51.** $16\frac{1}{2} \div 4\frac{1}{8}$

52. $5\frac{2}{7} \div 1\frac{3}{10}$ **53.** $1\frac{1}{6} \div 3\frac{1}{2}$ **54.** $5\frac{1}{3} \div 16$ **55.** $4\frac{7}{11} \div 7\frac{1}{6}$ **56.** $3\frac{5}{9} \div 8\frac{4}{9}$

57. $3\frac{3}{20} \div 2\frac{4}{5}$ **58.** $17\frac{1}{2} \div 11\frac{2}{3}$ **59.** $20\frac{7}{10} \div 2\frac{1}{4}$ **60.** $13\frac{3}{8} \div 7\frac{5}{6}$

SKILLSFOCUS (Section 1.5) *Change the following improper fractions to mixed numbers.*

61. $\frac{23}{4}$ **62.** $\frac{46}{5}$ **63.** $\frac{60}{8}$ **64.** $\frac{30}{11}$

Change the following mixed numbers to improper fractions.

65. $6\frac{1}{5}$ **66.** $11\frac{2}{7}$ **67.** $1\frac{7}{9}$ **68.** $20\frac{2}{3}$

Write the following whole numbers as fractions in two different ways.

69. 7 **70.** 10 **71.** 1

EXTEND YOUR THINKING ▶▶▶

▶ TROUBLESHOOT IT

Find and correct the error.

72. $\frac{4}{5} \div \frac{3}{8} = \frac{\overset{1}{\cancel{4}}}{5} \times \frac{\overset{2}{\cancel{8}}}{3} = \frac{1 \times 2}{5 \times 3} = \frac{2}{15}$

73. $\frac{7}{9} \div \frac{1}{6} = \frac{\overset{3}{\cancel{9}}}{7} \times \frac{1}{\underset{2}{\cancel{6}}} = \frac{3}{14}$

74. $3\frac{3}{4} \div 1\frac{5}{9} = \frac{\cancel{15}}{\underset{2}{\cancel{4}}} \div \frac{\cancel{14}}{\underset{3}{\cancel{9}}} = \frac{5}{2} \div \frac{7}{3} = \frac{5}{2} \times \frac{3}{7} = \frac{15}{14}$ or $1\frac{1}{14}$

75. $\frac{1}{2} \div 2 = \frac{1}{\underset{1}{\cancel{2}}} \times \frac{\overset{1}{\cancel{2}}}{1} = \frac{1}{1} = 1$

WRITING TO LEARN ▶▶▶

76. Explain how to find the reciprocal of a mixed number. Use your own example to demonstrate your point.

77. An operation is commutative if the order of the numbers does not matter. Use an example of your own to show that division is not commutative.

78. Explain how the reciprocal of a number can be a whole number.

▶ YOU BE THE JUDGE

79. Dave claims that every number has a reciprocal. Is he correct? Explain your decision.

3.5 APPLICATIONS: MULTIPLICATION AND DIVISION OF FRACTIONS

OBJECTIVES

1. Find a fraction of an amount.
2. Solve repeated addition problems.
3. Find the area of a triangle.
4. Find how many times one amount is in another.
5. Split an amount a into b equal parts.

1 Finding a Fraction of an Amount

Taking a fraction *of* an amount means *multiply* the fraction times the amount. To see why, suppose there are 6 rooms in a house. One half of the rooms are painted. This means 3 rooms are painted because $\frac{1}{2}$ of 6 rooms is 3 rooms. Multiply $\frac{1}{2}$ times 6 to get the same result.

$$\frac{1}{2} \text{ of } 6 \text{ rooms} = \frac{1}{2} \times 6 = \frac{1}{\cancel{2}_{1}} \times \frac{\cancel{6}^{3}}{1} = 3 \text{ rooms}$$

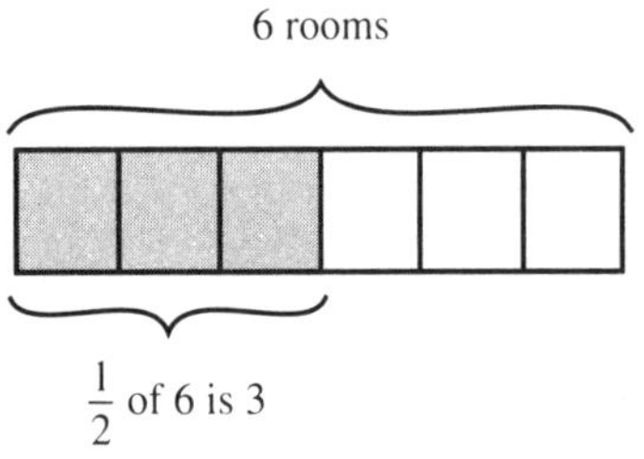

Taking a fraction of another fraction also means multiply. For example,

$$\frac{1}{2} \text{ of } \frac{3}{4} = \frac{1}{2} \times \frac{3}{4} = \frac{3}{8}.$$

To see why this is true, represent $\frac{3}{4}$ using the figure.

$\frac{3}{4} =$ 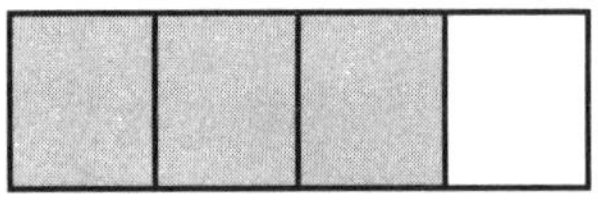

You can see that $\frac{1}{2}$ of $\frac{3}{4}$ is half of the shaded figure. This is $\frac{3}{8}$ of it, as shown in the next figure.

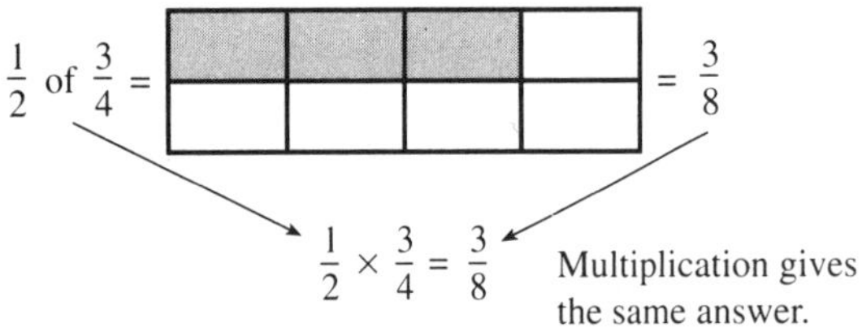

> When taking a fraction of an amount, multiply the fraction times the amount. In this context, *of* means *multiply*.

EXAMPLE 1 The gas tank on a riding mower holds $2\frac{1}{4}$ gallons of gas. If the tank is $\frac{2}{3}$ full, how many gallons of gas are in the tank?

A full tank contains $2\frac{1}{4}$ gallons of gas. When $\frac{2}{3}$ full, the tank contains $\frac{2}{3}$ of $2\frac{1}{4}$ gallons. Since this is a *fraction of an amount*, multiply the fraction times the amount.

$$\frac{2}{3} \text{ of } 2\frac{1}{4} = \frac{2}{3} \times 2\frac{1}{4} = \frac{\overset{1}{\cancel{2}}}{\underset{1}{\cancel{3}}} \times \frac{\overset{3}{\cancel{9}}}{\underset{2}{\cancel{4}}}$$

$$= \frac{3}{2} \text{ or } 1\frac{1}{2} \text{ gallons of gas left in the tank} \quad \blacksquare$$

▶ **You Try It** 1. A gas tank has a capacity of $16\frac{1}{2}$ gallons. The tank is $\frac{2}{3}$ full. How many gallons of gas are in the tank?

EXAMPLE 2 A real estate agent sold a home for \$145,500. The agent makes a commission of $\frac{7}{100}$ of the selling price of the home. What commission is made on this sale?

The agent makes $\frac{7}{100}$ of the \$145,500 selling price. Since you are taking a *fraction of an amount*, multiply the fraction times the amount.

$$\frac{7}{100} \text{ of } \$145{,}500 = \frac{7}{100} \times \$145{,}500 = \frac{7}{\underset{1}{\cancel{100}}} \times \frac{\overset{\$1{,}455}{\cancel{\$145{,}500}}}{1} = \$10{,}185$$

The agent made a \$10,185 commission on the sale of the house. ■

▶ **You Try It** 2. A real estate agent sold 5 acres of land for \$79,300. The agent made a commission of $\frac{9}{100}$ of the selling price. What commission did the agent make on this sale?

EXAMPLE 3 Kim makes \$104 for working an 8-hour day. She only worked 5 hours today. How much did she make?

Kim worked 5 hours out of 8, or $\frac{5}{8}$ of a day. She is paid \$104 a day. So she made $\frac{5}{8}$ of \$104.

$$\frac{5}{8} \text{ of } \$104 = \frac{5}{\underset{1}{\cancel{8}}} \times \frac{\overset{\$13}{\cancel{\$104}}}{1} = \$65 \quad \blacksquare$$

▶ **You Try It** 3. Irma makes \$150 for working a 12-hour shift. How much will she make if she only works 7 hours of her shift?

2 Solving Repeated Addition Problems

You buy 5 gifts costing \$8 each. Find the total cost by adding \$8 to itself 5 times.

$$\$8 + \$8 + \$8 + \$8 + \$8 = \$40$$

This is repeated addition. Recall from Section 1.4 that repeated addition can be solved quickly by multiplying.

$$5 \times \$8 = \$40$$

Now suppose Nancy uses $\frac{2}{3}$ of a tank of gas to drive to work each week. How many tanks will she use in 6 weeks? Add $\frac{2}{3}$ to itself 6 times. Since this is repeated addition, just multiply $\frac{2}{3}$ times 6.

$$\frac{2}{3} \times 6 = \frac{2}{\cancel{3}_1} \times \frac{\cancel{6}^2}{1} = 4 \text{ tanks used in 6 weeks}$$

> A problem involving repeated addition of the same number n to itself p times can be solved by multiplying $n \cdot p$.

EXAMPLE 4 One share of GulfOil stock costs $\$29\frac{3}{8}$. You purchase 20 shares. What do you pay?

One share costs $\$29\frac{3}{8}$. The cost of 20 shares is $\$29\frac{3}{8}$ added to itself 20 times. Since this is *repeated addition,* just multiply 20 times $\$29\frac{3}{8}$.

$$20 \times 29\frac{3}{8} = \frac{\cancel{20}^5}{1} \times \frac{235}{\cancel{8}_2} = \frac{5 \times 235}{1 \times 2} = \$587\frac{1}{2} \text{ for 20 shares} \quad ■$$

EXAMPLE 5 One inch on a map represents $2\frac{3}{5}$ miles. How many miles long is a road that measures $7\frac{1}{2}$ inches on the map?

One inch on the map represents $2\frac{3}{5}$ miles. $7\frac{1}{2}$ inches equals $2\frac{3}{5}$ miles added to itself $7\frac{1}{2}$ times. This is *repeated addition.* Multiply to find the length of the road.

$$7\frac{1}{2} \times 2\frac{3}{5} = \frac{\cancel{15}^3}{2} \times \frac{13}{\cancel{5}_1} = \frac{3 \times 13}{2 \times 1} = \frac{39}{2} = 19\frac{1}{2} \text{ miles long} \quad ■$$

▶ **You Try It** 4. One share of ABH stock costs $\$12\frac{3}{4}$. What do you pay for 80 shares?

5. One inch on a map represents $7\frac{1}{2}$ miles. A road measures 5 inches on the map. How long is the road?

3 Area of a Triangle

In the figures, two identical triangles are arranged to form a rectangle (or parallelogram) with the same base and height. Therefore, the area of the triangle is half the area of the rectangle (or parallelogram) with the same base and height.

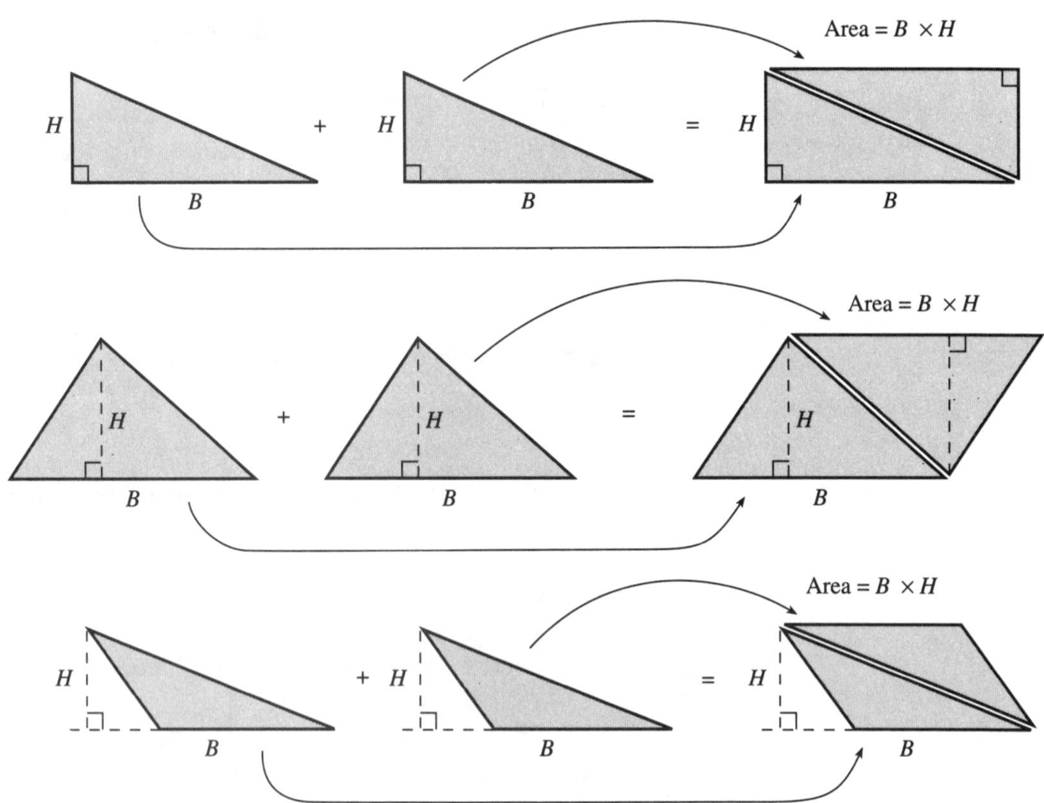

Observe that the height H of a triangle is always perpendicular to the base B, even if the base has to be extended.

The area of a triangle is the base times the height, divided by 2.

$$A = \frac{B \cdot H}{2} \quad \text{or} \quad \frac{1}{2}BH$$

H

B

The height H is always perpendicular to the base B.

EXAMPLE 6 Find the area of the triangle.

$$A = \frac{B \cdot H}{2} = \frac{6\text{ ft} \cdot 4\text{ ft}}{2} = \frac{24\text{ ft}^2}{2}$$

$$A = 12\text{ ft}^2$$

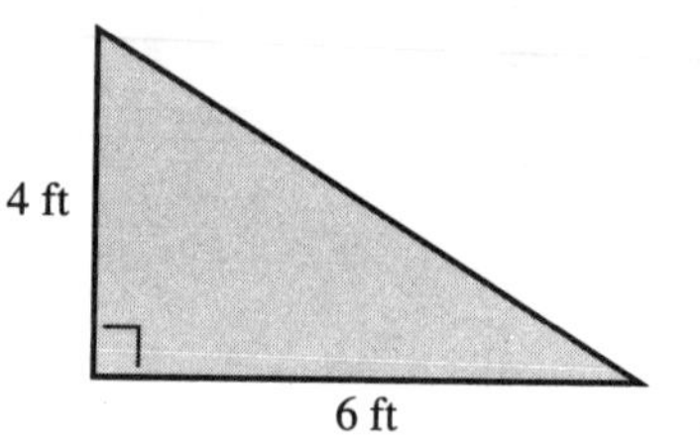

■

You Try It 6. Find the area.

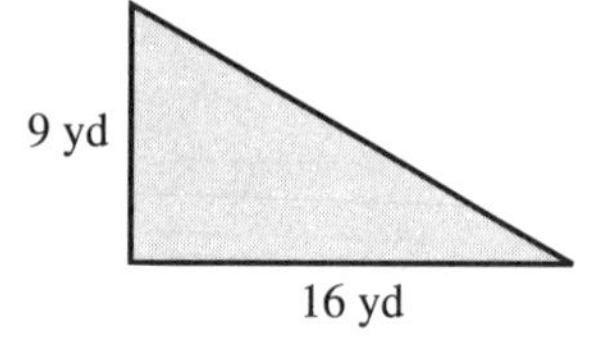

EXAMPLE 7 Find the area of the triangle.

$$A = \frac{B \cdot H}{2} = \frac{8\text{ cm} \cdot 5.2\text{ cm}}{2} = \frac{41.6\text{ cm}^2}{2}$$

$$A = 20.8\text{ cm}^2$$

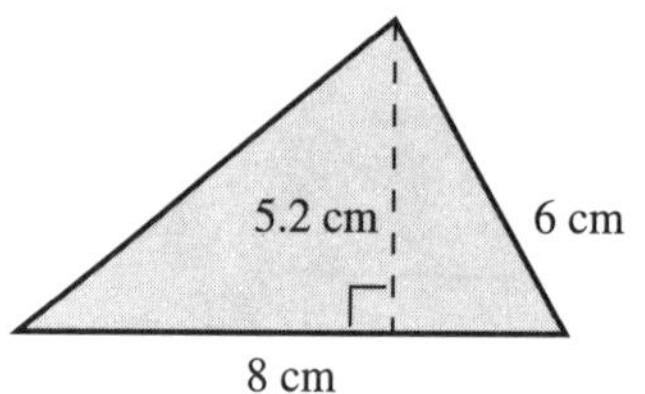

CAUTION 6 cm is not the height of the triangle, because that side is not perpendicular to the base. It is just the length of one of the sides. ■

EXAMPLE 8 Find the area and perimeter of the triangle.

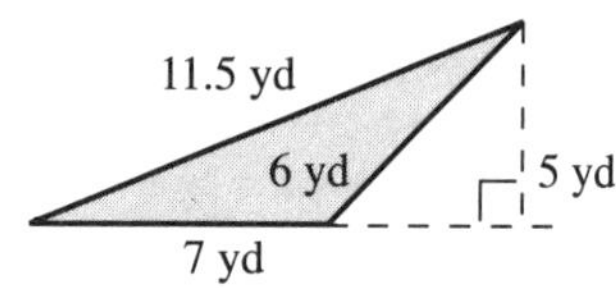

a. Area: $A = \frac{B \cdot H}{2} = \frac{7\text{ yd} \cdot 5\text{ yd}}{2} = \frac{35\text{ yd}^2}{2}$

$A = 17.5\text{ yd}^2$

CAUTION 6 yd is not the height, because that side is not perpendicular to the base. Extend the base of the triangle to draw the height.

b. Perimeter: $P = 11.5\text{ yd} + 6\text{ yd} + 7\text{ yd} = 24.5\text{ yd}$

CAUTION 5 yd is not a side of the triangle, so it is not added into the perimeter. ■

▶ **You Try It** Find the area and the perimeter.

7. 5 ft, 4 ft, 10 ft, 12 ft

8. 5 m, 9 m, 6 m, 8 m

---

4 Finding How Many Times b Is Contained in a

To find how many 3's are in 12, write $12 \div 3 = 4$. In the same way, to find how many $\frac{1}{4}$ are in $\frac{3}{4}$ write:

$$\text{How many } \frac{1}{4} \text{ are in } \frac{3}{4} = \frac{3}{4} \div \frac{1}{4} = \frac{3}{\not{4}_1} \times \frac{\not{4}^1}{1} = 3.$$

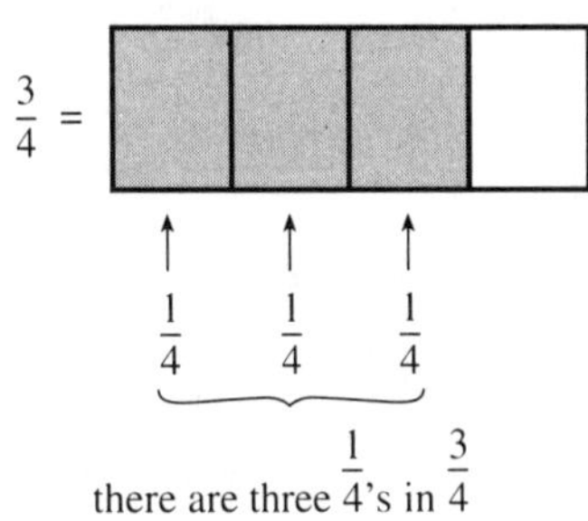

> To find how many times the number b is contained in the number a, divide
>
> $$a \div b.$$

EXAMPLE 9 How many shelves $1\frac{1}{2}$ feet long can be cut from a board 12 feet long?

To find *how many* $1\frac{1}{2}$-foot shelves *are contained in* a 12-foot board, divide 12 by $1\frac{1}{2}$.

$$\text{How many } 1\frac{1}{2} \text{ in } 12 = 12 \div 1\frac{1}{2} = \frac{12}{1} \div \frac{3}{2} = \frac{\not{12}^4}{1} \times \frac{2}{\not{3}_1} = \frac{4 \times 2}{1 \times 1} = 8 \text{ shelves.}$$

8 shelves each $1\frac{1}{2}$ feet long can be cut from a 12-foot board. ■

▶ **You Try It** **9.** How many pieces of wire $2\frac{1}{2}$ inches long can be cut from a 35-inch piece of wire?

The next example illustrates the difference between the answer to a division problem, and the answer to its associated word problem.

EXAMPLE 10 **Chapter Problem** Read the problem at the beginning of this chapter. A dump truck can deliver $6\frac{1}{4}$ tons of stone in one trip. How many trips are needed to deliver 85 tons of stone?

$6\frac{1}{4}$ tons can be delivered in one trip. To find how many trips are needed to deliver 85 tons, find *how many times* $6\frac{1}{4}$ *is contained in* 85.

$$\text{How many } 6\frac{1}{4} \text{ in } 85 = 85 \div 6\frac{1}{4} = \frac{85}{1} \div \frac{25}{4} = \frac{\overset{17}{\cancel{85}}}{1} \times \frac{4}{\underset{5}{\cancel{25}}} = \frac{17 \times 4}{1 \times 5} = 13\frac{3}{5}.$$

$$13\frac{3}{5} \text{ loads} = \underbrace{13 \text{ full loads}}_{13 \text{ trips}} + \underbrace{\frac{3}{5} \text{ of a load}}_{1 \text{ trip}} = 14 \text{ trips needed}$$

CAUTION $13\frac{3}{5}$ is the answer to the division problem. It is not the answer to the word problem. The answer to the word problem is 14 trips. ■

▶ **You Try It 10.** A truck can haul $5\frac{1}{3}$ cubic yards of topsoil. How many trips must be made to deliver 42 cubic yards?

5 Splitting an Amount *a* into *b* Equal Parts

To split \$20 into 4 equal parts, write \$20 ÷ 4 = \$5 per part.

To split an amount *a* into *b* equal parts, write

$$a \div b$$

total amount to be divided up ↙ a; b ↘ number of equal parts wanted

EXAMPLE 11 A $4\frac{1}{2}$-foot rope is cut into 3 equal pieces. How long is each piece?

$4\frac{1}{2}$ feet is the *amount to be split into 3 equal parts.*

$$4\frac{1}{2} \div 3 = \frac{9}{2} \div \frac{3}{1} = \frac{\overset{3}{\cancel{9}}}{2} \times \frac{1}{\underset{1}{\cancel{3}}} = \frac{3 \times 1}{2 \times 1} = \frac{3}{2} = 1\frac{1}{2} \text{ ft}$$

Each piece of rope will be $1\frac{1}{2}$ feet long. This is pictured in the diagram.

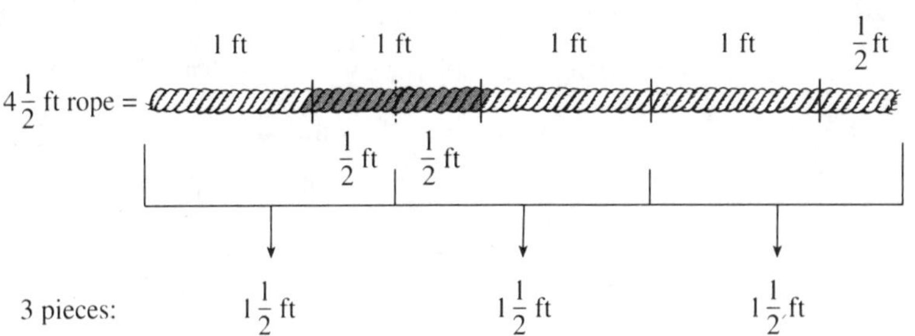

■

EXAMPLE 12 Tom purchased $7\frac{1}{2}$ pounds of ground beef to make 30 hamburgers. How much ground beef will be in each hamburger?

Tom wants to *split* $7\frac{1}{2}$ pounds of ground beef *into* 30 *equal parts.*

$$7\frac{1}{2} \div 30 = \frac{15}{2} \div \frac{30}{1} = \frac{\overset{1}{\cancel{15}}}{2} \times \frac{1}{\underset{2}{\cancel{30}}} = \frac{1 \times 1}{2 \times 2} = \frac{1}{4} \text{ lbs}$$

Each hamburger will contain $\frac{1}{4}$ pound of hamburger meat. ■

▶ **You Try It** **11.** $3\frac{1}{2}$ tons of mulch is split evenly among 5 customers. How much mulch does each customer get?

12. Susan purchased $10\frac{3}{4}$ pounds of ham to make 43 sandwiches. How much ham can she put in each sandwich?

Division Summary

Fractions stay in same order:	Fractions reverse order:
1. $\frac{2}{3}$ divided by $\frac{4}{5} = \frac{2}{3} \div \frac{4}{5}$.	**4.** How many $\frac{1}{5}$ are in $\frac{7}{8} = \frac{7}{8} \div \frac{1}{5}$.
2. Divide $\frac{2}{3}$ by $\frac{4}{5} = \frac{2}{3} \div \frac{4}{5}$.	**5.** Divide $\frac{1}{5}$ into $\frac{7}{8} = \frac{7}{8} \div \frac{1}{5}$.
3. Split $\frac{3}{4}$ into 3 parts $= \frac{3}{4} \div 3$.	

♦ **Answers to You Try It** **1.** 11 gal **2.** $7,137 **3.** $87 $\frac{1}{2}$ **4.** $1,020 **5.** $37\frac{1}{2}$ mi. **6.** 72 sq yd **7.** Area = 24 sq ft; Perimeter = 27 ft **8.** Area = 20 sq m; Perimeter = 23 m **9.** 14 pieces **10.** 8 trips **11.** $\frac{7}{10}$ ton **12.** $\frac{1}{4}$ lb

SECTION 3.5 EXERCISES

1 2 3 4 5

1. What is $\frac{1}{3}$ of 600?

2. What is $\frac{3}{4}$ of 800?

3. Find $\frac{4}{7}$ of 350 men.

4. Find $\frac{7}{10}$ of 32,000 people.

5. If $\frac{2}{9}$ of a cow's milk is butterfat, how much butterfat is contained in 12 gallons of cow's milk?

6. Mr. Malacki purchased 32 stamps and used $\frac{3}{8}$ of them to send a package to his daughter in college. How many stamps did he use?

7. How many $\frac{1}{8}$ are in 1?

8. How many $\frac{2}{5}$ are in 4?

9. How many $\frac{7}{10}$ are in $2\frac{4}{5}$?

10. How many $4\frac{3}{8}$ are in $30\frac{5}{8}$?

---

11. How many $\frac{2}{5}$-quart packages can be made from 8 quarts of berries?

12. A family orders $16\frac{1}{4}$ tons of coal. They use $\frac{5}{8}$ tons each week. For how many weeks will the coal last?

13. Justin weighs 156 pounds. On the moon he weighs $\frac{1}{6}$ of what he weighs on Earth. What would Justin weigh on the moon?

14. A real estate agent makes a commission of $\frac{7}{100}$ of the selling price of a home. Find the commission on a \$162,500 sale.

15. How many $\frac{1}{4}$-pound hamburgers can be made from $18\frac{3}{4}$ pounds of ground beef?

16. How many small bags of flour can be packed from the large sack of flour?

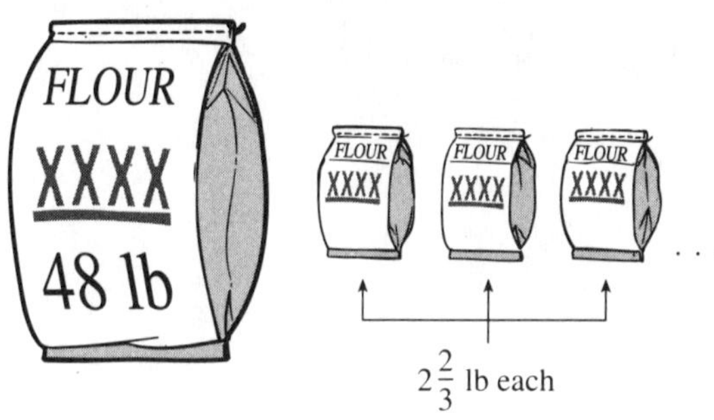

17. What is 7 divided by $\frac{4}{9}$?

18. What is $2\frac{1}{2}$ divided by $\frac{3}{10}$?

19. What is $\frac{5}{18}$ of $2\frac{4}{7}$?

20. What is $\frac{16}{21}$ of $7\frac{4}{9}$?

21. Divide $3\frac{1}{6}$ by $2\frac{2}{3}$.

22. Divide $8\frac{5}{8}$ by $3\frac{1}{4}$.

23. Lettuce is $\frac{24}{25}$ water. How much water is in 50 pounds of lettuce?

24. Peaches, when dried, lose $\frac{7}{10}$ of their fresh weight. How much weight will 45 pounds of peaches lose when dried?

25. You want to tile a hallway 24 feet long. Each tile is $\frac{3}{4}$ feet long. How many tiles must be placed end to end to extend the length of the hallway?

26. A road $17\frac{5}{8}$ miles long has signposts placed after each $\frac{3}{8}$ of a mile. How many signposts are needed?

27. A steak weighs $2\frac{1}{2}$ pounds. It costs $3\frac{1}{2}$ dollars per pound. What is the total cost of the steak?

28. A plane flies 175 miles in one hour. How far will it go in $\frac{5}{8}$ hours?

29. What is $\frac{5}{9}$ divided by $\frac{1}{9}$?

30. What is $\frac{1}{9}$ divided by $\frac{5}{9}$?

31. Chris has $3\frac{1}{2}$ pies and wants to serve 14 people. What part of a pie will each person be served?

32. How many shelves can be cut from the board shown here?

33. Al worked from noon to 8 P.M. on his history project. He spent $\frac{4}{5}$ of this time doing research. How many hours of research did he do?

34. One inch on a map represents $3\frac{3}{8}$ miles. How many miles long is a road that measures $6\frac{2}{3}$ inches on the map?

35. How many trips must the truck make to haul all of the dirt shown in the figure?

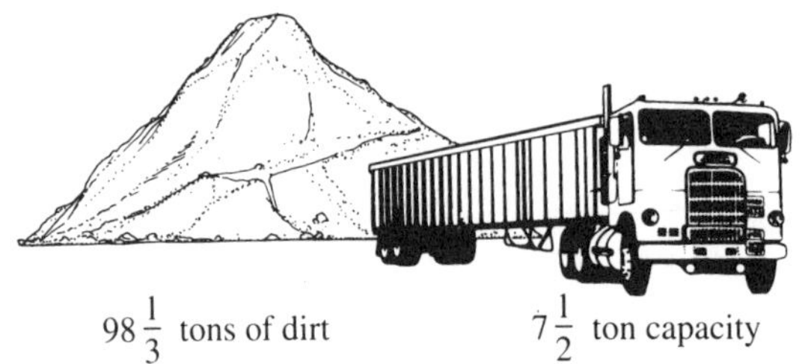

36. A total of 138 gallons of oil must be carried 40 feet. If a $4\frac{1}{5}$-gallon container is used to carry the oil, how many trips are needed?

37. Using 93 pints of cola, how many $\frac{3}{4}$-pint cups can be filled?

38. A 10-acre piece of land is subdivided into $\frac{2}{5}$-acre lots. How many lots will you get?

39. A family can save $\frac{1}{8}$ of its $1,240 monthly income.

a. How long will it take to save $1,085 for a dining room set?

b. How long will it take to save $2,635 for a used car?

40. A theater that can hold 2,730 people is $\frac{2}{3}$ full. How many people are seated in the theater?

41. What is the thickness of a stack of 24 boards if each board is $1\frac{1}{4}$ inches thick?

42. A stock sells for $\$16\frac{3}{4}$ per share. What will 45 shares cost?

43. There are 450 people who will be inoculated. If each inoculation contains $\frac{2}{9}$ of an ounce of serum, how many ounces of the serum are needed?

44. A hospital has 72 ounces of a drug. If each patient receives $\frac{3}{8}$ of an ounce, how many patients can be treated?

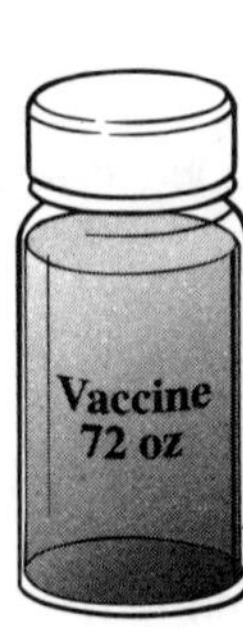

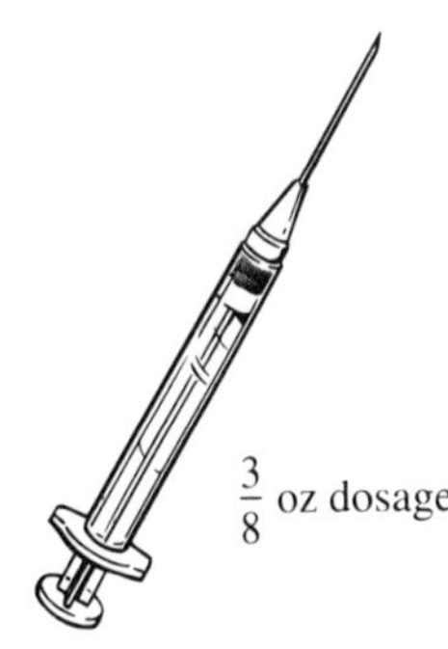

45. A couple applies for a mortgage. Their monthly payment can be no larger than $\frac{1}{4}$ of their monthly income. If their monthly income is $2,860 what maximum mortgage payment are they allowed?

46. A basketball team scored 112 points. Christie scored $\frac{2}{7}$ of the points. How many points did Christie score?

47. One tie requires $\frac{3}{16}$ of a yard of material. How many ties can be made from 24 yards of material?

48. Ellen is building a circuit board. She has 189 centimeters of wire. How many pieces of wire each $2\frac{5}{8}$ centimeters long can she cut for the board?

49. Three people out of 80 have a certain disease. In a town of 72,000 people, how many will have the disease?

50. Five out of every 8 ounces of a solution is acid. How much acid is contained in 32 ounces of the solution?

51. Tom's insurance will pay $\frac{4}{5}$ of a $2,670 medical bill.

a. How much will the insurance pay?

b. How much will Tom pay?

52. A jury awards Ms. Edwards $87,000 for injuries. Her lawyers take $\frac{1}{3}$ of this amount for their legal fees.

a. How much do the lawyers take?

b. How much does Ms. Edwards get?

Find the area.

53.

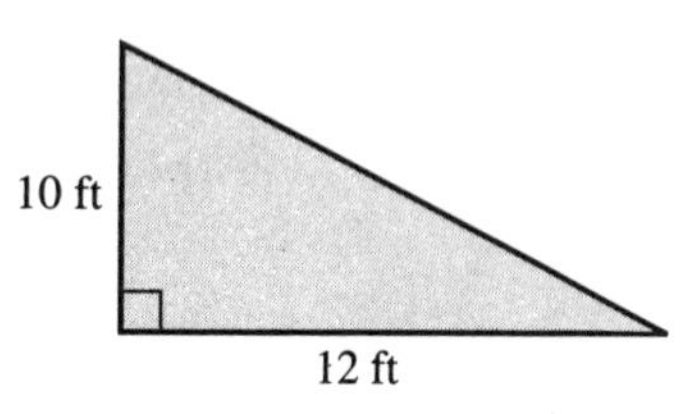

54.

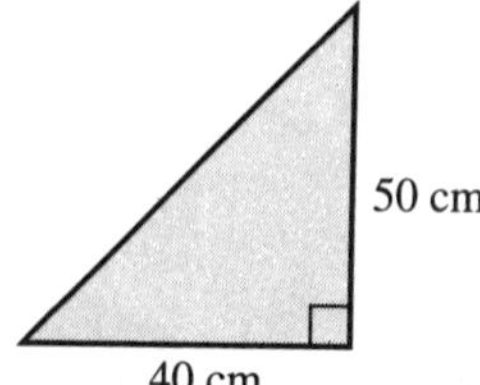

55.

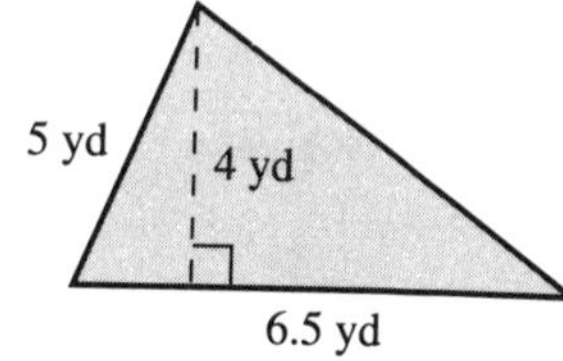

56.

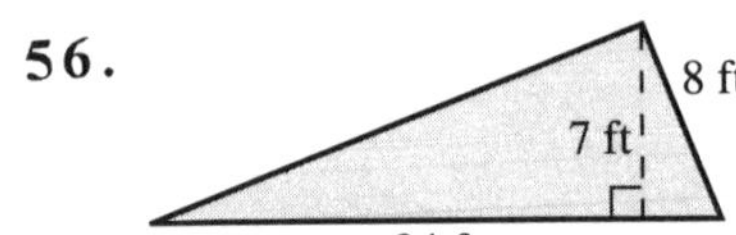

57.

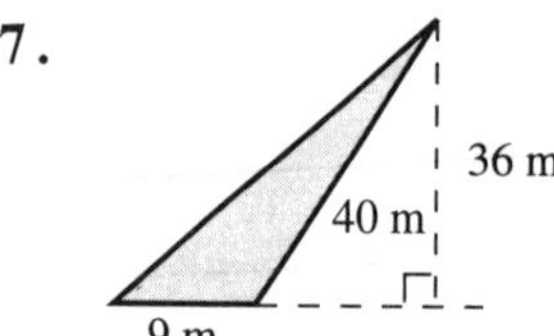

58.

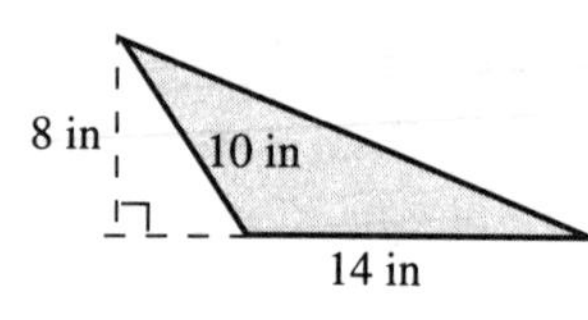

59.

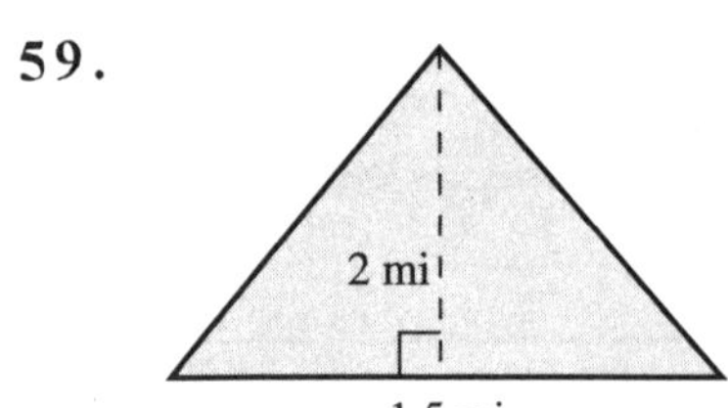

60.

61.

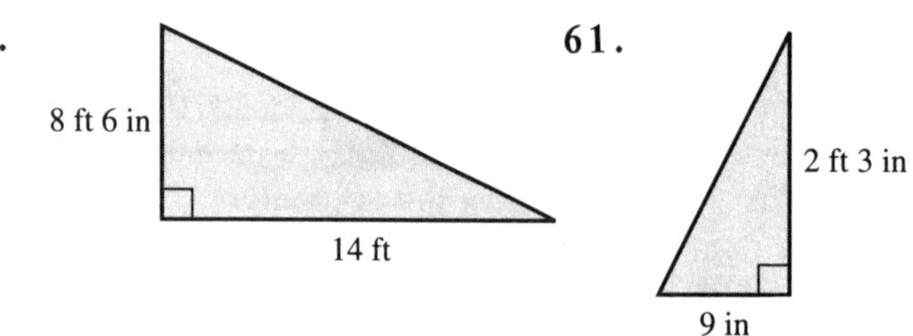

62. Find the area and perimeter.

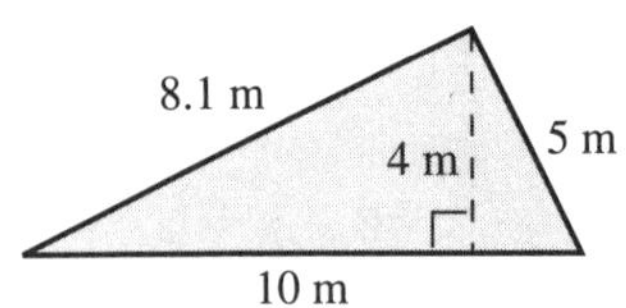

63. Find the area and perimeter.

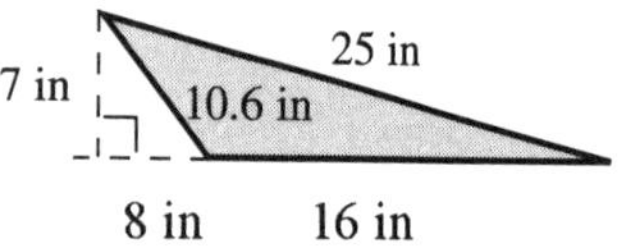

64. Find the area in sq ft and sq in.

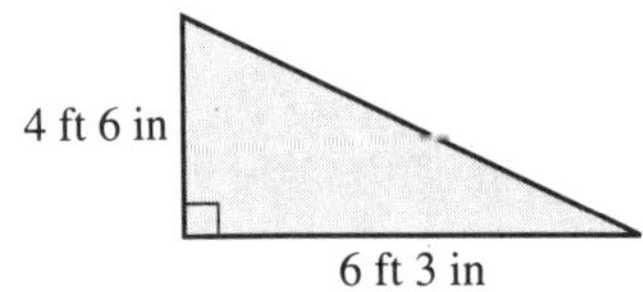

65. Find the area in sq ft and sq in.

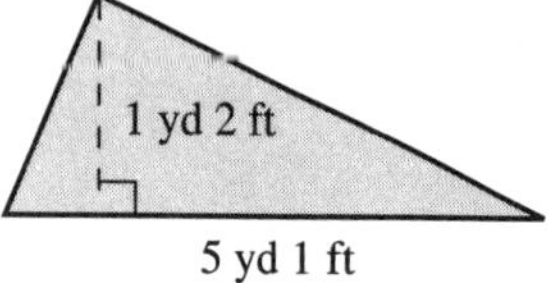

SKILLSFOCUS (Section 2.7) *Evaluate each formula.*

66. Find T when $T = 4P - 2Q$ and $P = 3$ and $Q = 5$.

67. Find M when $M = A + BC$ and $A = 2.5$, $B = 3.4$, and $C = 7$.

68. Find A when $A = 2\pi R$ and $\pi = 3.14$ and $R = 6$.

69. Find V when $V = LWH$ and $L = 5$, $W = 2$, and $H = 3.5$.

EXTEND YOUR THINKING

SOMETHING MORE

70. **Snow** A snow gauge collected $9\frac{3}{8}$ inches of snow.

a. A very dry snow gives 1 inch of water to every 30 inches of snow. How many inches of water will be in the gauge if the snow is very dry?

b. A very wet snow gives 1 inch of water to every 6 inches of snow. How many inches of water will be in the gauge if the snow is very wet?

71. One inoculation uses $\frac{3}{8}$ ounce of antibiotic and costs $2. You plan to inoculate 160,000 people.

a. How many ounces of antibiotic are needed?

b. What is the cost?

72. Two children need dessert in their lunch for 5 school days. Dad purchased a package of 12 cupcakes.

a. Is there enough to give each child $1\frac{1}{2}$ cupcakes per day for 5 days?

b. At this rate, exactly how long will the cupcakes last?

73. Ed gets 20 miles to a gallon of gas. His gas tank has a capacity of $14\frac{2}{5}$ gallons. Ed fills his tank, and uses $9\frac{3}{10}$ gallons to drive to the beach. Driving home that evening, he ran out of gas. How far was Ed from home when he ran out of gas?

74. **Recipes** A recipe that serves 6 people calls for the following.

$1\frac{1}{2}$ pints pistachio ice cream	$\frac{3}{4}$ cup whipping cream
6 maraschino cherries	$\frac{1}{2}$ cup sugar
$\frac{1}{3}$ cup pistachio nuts	1 package raspberries

a. How much of each ingredient is needed to serve 2 people?

b. How much of each ingredient is needed to serve 24 people?

75. **The Fractional Pie Chart and Budget** The Gaynor family lives on a $1,920 monthly budget. The money is spent according to the pie chart on the right. For example, $\frac{1}{6}$ of $1,920 is spent on food each month. How much of the monthly budget is spent on each of the five listed categories?

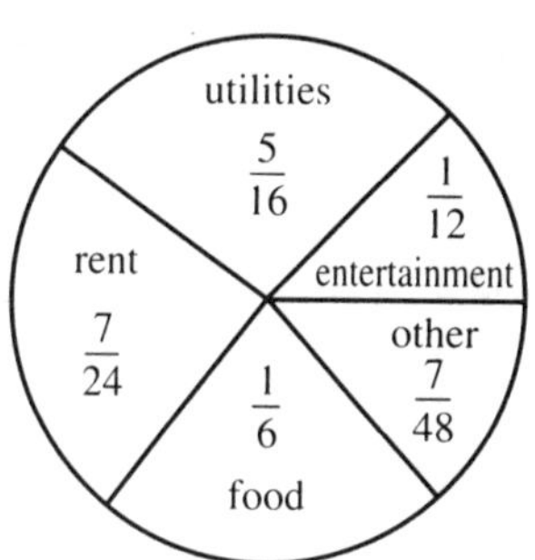

76. Mrs. Smith left \$96,000 to be divided among her 3 sons Tom, Dick, and Harry. She left $\frac{1}{4}$ to Tom, $\frac{5}{12}$ to Dick, and $\frac{1}{3}$ to Harry. Tom kept $\frac{1}{3}$ of his share and gave $\frac{1}{3}$ to each of his daughters Terry and Teresa. Dick kept $\frac{1}{2}$ of his share and split the rest evenly between his 2 sons Dan and David. Harry kept \$2,000 of his share and gave the rest to his daughters Helen, Hazel, and Harriet in an even split. How much did each person receive?

77. There are 16,000 people attending a concert. Of those attending, $\frac{4}{5}$ are from California. Of the Californians, $\frac{3}{8}$ are from L.A. How many people from L.A. are attending the concert?

78. Sue uses $1\frac{7}{10}$ gallons of gas to drive to work one way. Her gas tank has a capacity of $16\frac{1}{2}$ gallons. If Sue fills her tank, will she have enough gas to drive back and forth to work for 5 days?

▶ TROUBLESHOOT IT

Find and correct the error.

79. How many $\frac{5}{8}$ are in $\frac{2}{3} = \frac{5}{8} \div \frac{2}{3} = \frac{5}{8} \times \frac{3}{2} = \frac{15}{16}$.

80. Split $\frac{5}{6}$ pounds into 3 equal parts $= \frac{\overset{1}{\cancel{3}}}{1} \times \frac{5}{\underset{2}{\cancel{6}}} = 2\frac{1}{2}$ lb per part.

81. Divide $2\frac{1}{2}$ into $6\frac{1}{2} = 6\frac{1}{2} \div 2\frac{1}{2} = (6 \div 2) + \left(\frac{1}{2} \div \frac{1}{2}\right) = 3 + 1 = 4.$

WRITING TO LEARN ▶▶▶

82. Write a word problem about 5 sisters. The answer is $7\frac{1}{2}$. The problem must include division of fractions, and involve a supermarket. Then write the solution.

▶ YOU BE THE JUDGE

83. Yana claims if you multiply two proper fractions, the answer will be smaller than both fractions. Is she right? Explain your decision. $\left(\text{Hint: Use } \frac{1}{2} \times \frac{1}{2} = \frac{1}{4} \text{ as an example, and the fact that } \frac{1}{2} \times \frac{1}{2} = \frac{1}{2} \text{ of } \frac{1}{2}.\right)$

CHAPTER 3 REVIEW

VOCABULARY AND MATCHING

New words and phrases introduced in this chapter are shown in the left-hand column. Match each term on the left with the phrase or sentence on the right that best describes it.

A. fraction ______ the numerator and denominator

B. numerator ______ what 6 is to 30

C. fraction bar ______ write the reciprocal of the fraction

D. denominator ______ like 1/2, 5/8, or 3/5; not like 2/4, 10/16, or 6/10

E. terms of a fraction ______ the process of writing a fraction in lowest terms

F. proper fraction ______ has 3 or more divisors

G. improper fraction ______ write each term as a product of prime factors, then reduce common factors

H. mixed number ______ indicates division

I. divisor ______ whole number plus a fraction

J. successive division by primes ______ writing a counting number as a product of primes only

K. prime number ______ examples are 2/3, 4/6, 6/9, and 8/12

L. composite number ______ numerator < denominator; it is always less than 1

M. prime factorization ______ method used to prime factor a composite number

N. equivalent fractions ______ product of the numerators divided by the product of the denominators

O. lowest terms fraction ______ a portion of a whole amount; may be proper, improper, or mixed

P. reducing ______ when used to indicate a fraction of an amount, it means multiply

Q. reducing by inspection ______ number of parts you divide the whole into; bottom number of a fraction

R. reducing by prime factoring ______ numerator > denominator or numerator = denominator

S. multiplying fractions ______ multiply the dividend by the reciprocal of the divisor

T. reciprocal ______ the number of parts you take; the top number of a fraction

U. invert a fraction ______ find a common factor by looking, then divide numerator and denominator by it

V. dividing fractions ______ has exactly two divisors

W. of ______ for 2/5 is 5/2; for 7 is 1/7

REVIEW EXERCISES

The Meaning of a Fraction

1. How many sixths are in one whole?

2. How many twentieths are in one whole?

What fraction is represented by the shading in each figure?

3. 

4. 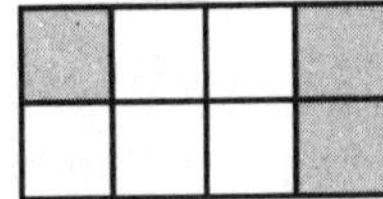

5. 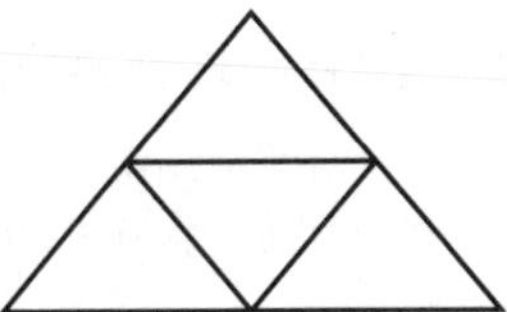

Draw and shade a figure for each fraction.

6. $\frac{5}{6}$

7. $\frac{7}{10}$

8. $\frac{9}{4}$

Simplify each fraction, or write not possible.

9. $\frac{14}{7}$

10. $\frac{11}{11}$

11. $\frac{0}{7}$

12. $\frac{2}{0}$

Proper and Improper Fractions; Mixed Numbers

Change each mixed number to an improper fraction.

13. $4\frac{1}{5}$

14. $2\frac{9}{10}$

15. $5\frac{3}{7}$

16. $15\frac{2}{3}$

Change each improper fraction to a mixed number.

17. $\frac{19}{4}$

18. $\frac{11}{9}$

19. $\frac{45}{8}$

20. $\frac{107}{12}$

3.1 Prime Numbers and Prime Factoring

Find all the divisors. Identify each as prime or composite.

21. 49

22. 26

23. 280

24. 73

25. 167

26. 68

27. 94

28. 89

Prime factor.

29. 117 **30.** 94 **31.** 56 **32.** 71

33. 365 **34.** 1,200 **35.** 273 **36.** 38

3.2 Reducing Fractions

Reduce each fraction to lowest terms.

37. $\frac{8}{12}$ **38.** $\frac{9}{15}$ **39.** $\frac{21}{24}$ **40.** $\frac{24}{30}$

41. $\frac{136}{90}$ **42.** $\frac{200}{125}$ **43.** $\frac{153}{171}$ **44.** $\frac{840}{1,200}$

3.3 Multiplying Fractions

45. $\frac{4}{5} \times \frac{3}{7}$ **46.** $\frac{7}{10} \times \frac{15}{24}$ **47.** $\frac{25}{36} \times \frac{30}{54}$ **48.** $\frac{15}{8} \times \frac{9}{12}$

49. $3\frac{2}{3} \times 2\frac{1}{4}$ **50.** $2\frac{5}{8} \times 4\frac{1}{6}$ **51.** $3\frac{5}{12} \times 9$ **52.** $6 \times 1\frac{7}{9}$

53. $\left(\frac{7}{9}\right)^3$ **54.** $\left(\frac{3}{10}\right)^3$ **55.** $\left(\frac{13}{20}\right)^2$ **56.** $\left(4\frac{1}{3}\right)^2$

3.4 Dividing Fractions

57. $\frac{5}{8} \div \frac{1}{4}$ **58.** $\frac{3}{10} \div \frac{7}{12}$ **59.** $\frac{7}{9} \div \frac{14}{15}$ **60.** $\frac{16}{25} \div \frac{1}{30}$

61. $3\frac{1}{5} \div \frac{2}{3}$ **62.** $6\frac{3}{4} \div 2\frac{5}{6}$ **63.** $6 \div \frac{2}{5}$ **64.** $1\frac{7}{8} \div 10$

65. $1 \div 4\frac{1}{4}$ **66.** $7\frac{1}{7} \div 2$

67. Twenty-four cars need an oil change. If each car requires $4\frac{2}{3}$ quarts of oil, how much oil is needed altogether?

68. What will $3\frac{1}{2}$ pounds of beef cost if one pound costs $\$2\frac{3}{4}$?

69. How many $\frac{4}{5}$ are in 25?

70. Tom uses $\frac{9}{20}$ of a gallon of gas to mow his lawn. He mows his lawn once a week. If he has $6\frac{3}{4}$ gallons of gas stored in his garage, for how many weeks can he mow his lawn before needing more gas?

71. Fred purchased $5\frac{2}{3}$ pounds of plums. If there are 6 plums per pound, how many plums did he buy?

72. Elvin's basketball team scored 135 points. If Elvin scored $\frac{2}{5}$ of his team's points, how many points did he score?

73. How many $2\frac{3}{4}$-inch wires can be cut from a 77-inch length of wire?

74. A dress can be made from $2\frac{1}{4}$ yards of fabric. How many dresses can be made from 27 yards of fabric?

Find the area.

75.

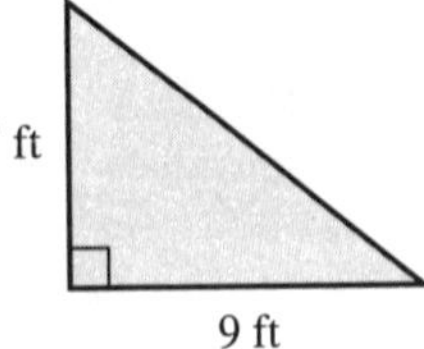

76.

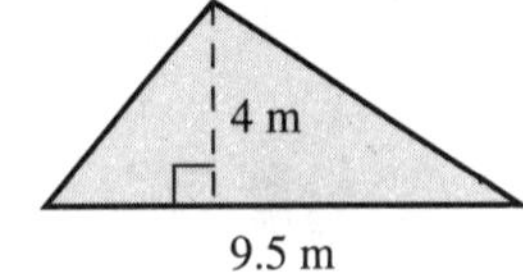

4

Addition and Subtraction of Fractions

THE NEW YORK STOCK EXCHANGE

Fractions are used extensively in finance. See this for yourself in the following segment taken from the financial page of a popular daily newspaper.

Donna purchased 400 shares of GulfOil stock at the Low price. Two hours later she sold them at the High for the day. What was her gain from the transaction?

From the NYSE summary of the 20 most active stocks, the Low price for GulfOil was $\$27\frac{3}{4}$ per share. The High was $\$30\frac{3}{8}$ per share.

Her gain per share was $\quad \$30\frac{3}{8} - \$27\frac{3}{4}$.

Her gain for 400 shares was

$$400\left(\$30\frac{3}{8} - \$27\frac{3}{4}\right).$$

In this chapter you will see how to evaluate this expression.

NYSE SUMMARY 20 MOST ACTIVE

	Sales	High	Low	Last	Chg.
Exxon s	5,338,400	$29^3/8$	$27^3/4$	29	$+\ ^7/8$
IBM	4,188,100	$59^5/8$	$57^1/8$	$57^7/8$	$-\ ^5/8$
RCA	4,012,800	$21^5/8$	$19^1/8$	$19^1/4$	$-\ ^1/8$
Tandy s	3,391,500	29	25	27	$-1^1/2$
Mobil s	2,820,600	$23^1/2$	$20^3/4$	$23^1/8$	$+2^3/8$
ATT	2,712,200	$57^1/8$	$53^1/3$	57	$+\ ^3/8$
MaOil	2,647,600	$76^1/4$	$74^1/4$	$75^5/8$	+ 8
Sears	2,466,600	$18^1/4$	$17^3/8$	$18^1/4$	$-\ ^1/2$
TexInt	2,411,500	$13^3/4$	$10^3/4$	$13^1/2$	$+1^1/4$
SFeInd s	2,347,700	$14^3/8$	13	$13^7/8$	$-\ ^1/2$
Schlimb s	2,330,200	$44^1/2$	40	$42^5/8$	$-\ ^1/4$
Digital	2,131,300	$78^3/8$	$71^3/8$	$72^1/4$	$-5^5/8$
Deltaa s	2,040,000	31	$28^1/4$	$28^1/2$	$-1^7/8$
StOInd	2,026,300	38	$34^1/8$	$37^3/4$	$+2^1/2$
K mart	1,972,400	$18^3/4$	$17^3/4$	$18^1/4$	$-\ ^3/8$
WrnCm	1,935,400	$55^1/4$	$50^1/4$	$51^5/8$	$-2^1/2$
Pennzol	1,932,800	$42^1/2$	$33^5/8$	34	$-6^1/2$
Halbtn	1,856,900	$35^3/4$	32	$35^1/2$	$+1^1/2$
GulfOil	1,812,500	$30^3/8$	$27^3/4$	$29^3/8$	$+1^5/8$
PhilPet	1,799,700	$29^5/8$	27	$28^1/4$	$-\ ^7/8$

4.1 ADDING AND SUBTRACTING LIKE FRACTIONS

OBJECTIVES

1 Add like fractions and mixed numbers.

2 Subtract like fractions and mixed numbers.

NEW VOCABULARY

like fractions

1 Adding Like Fractions

You can add quantities that have the same name.

2 cars + 3 cars = 5 cars

5 men + 4 men = 9 men

1 fourth + 2 fourths = 3 fourths

$$\frac{1}{4} + \frac{2}{4} = \frac{3}{4}$$

Like fractions have the same denominator. $\frac{1}{4}$ and $\frac{2}{4}$ are like fractions. To add like fractions, add the numerators ($1 + 2 = 3$). Place this sum over the common denominator, 4. This addition is pictured in the figure.

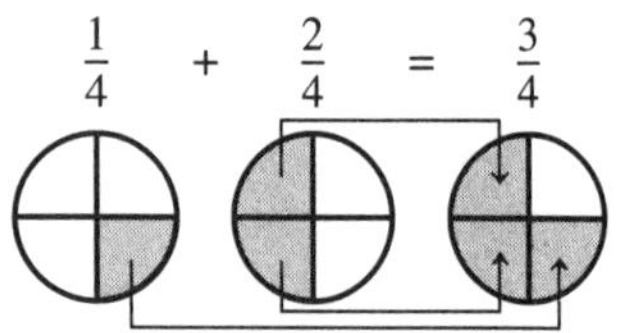

> To add like fractions
>
> **1.** Add the numerators.
> **2.** Write this sum over the common denominator.
> **3.** Reduce.

EXAMPLE 1

$$\frac{3}{5} + \frac{1}{5} = \frac{3+1}{5} = \frac{4}{5} \quad \blacksquare$$

EXAMPLE 2

$$\frac{3}{20} + \frac{7}{20} + \frac{13}{20} = \frac{3+7+13}{20} = \frac{23}{20} = 1\frac{3}{20} \quad \blacksquare$$

EXAMPLE 3

$$\frac{2}{9} + \frac{4}{9} = \frac{2+4}{9} = \frac{\overset{2}{\cancel{6}}}{\underset{3}{\cancel{9}}} = \frac{2}{3} \quad \blacksquare$$

You may add fractions vertically.

EXAMPLE 4 Add $\frac{11}{12} + \frac{5}{12}$.

$$\begin{array}{r} \frac{11}{12} \\ +\frac{5}{12} \\ \hline \frac{16}{12} \end{array} = \frac{\overset{4}{\cancel{16}}}{\underset{3}{\cancel{12}}} = \frac{4}{3} = 1\frac{1}{3} \quad \blacksquare$$

Add the numerators, 11 + 5 = 16.
Write 16 over the common denominator 12, and reduce.

CAUTION Never add the two denominators together. The addition below shows why this gives the wrong answer.

$$\frac{1}{2}+\frac{1}{2}=\frac{1+1}{2+2}=\frac{2}{4}=\frac{1}{2}$$

$\frac{1}{2}+\frac{1}{2}$ equals 1 whole, not $\frac{1}{2}$.

► **You Try It** **1.** $\frac{1}{6}+\frac{3}{6}$ **2.** $\frac{5}{8}+\frac{7}{8}$ **3.** $\frac{4}{11}+\frac{2}{11}$ **4.** $\frac{7}{10}+\frac{9}{10}$ **5.** $\frac{2}{5}+\frac{1}{5}+\frac{7}{5}+\frac{4}{5}$

Adding Like Mixed Numbers

Like mixed numbers have proper fractions with the same denominator. To add them, use the fact that addition is commutative.

$$3\frac{2}{7}+1\frac{4}{7}=\left(3+\frac{2}{7}\right)+\left(1+\frac{4}{7}\right)=(3+1)+\left(\frac{2}{7}+\frac{4}{7}\right)=4+\frac{6}{7}=4\frac{6}{7}$$

Write each mixed number as an addition. Add the whole numbers. Add the proper fractions.

To add like mixed numbers

1. Add the whole numbers. **2.** Add the like proper fractions.
3. Reduce.

EXAMPLE 5

$$4\frac{1}{5}+2\frac{3}{5}=(4+2)+\left(\frac{1}{5}+\frac{3}{5}\right)=6+\frac{4}{5}=6\frac{4}{5} \quad ■$$

EXAMPLE 6

$$7+2\frac{5}{6}=(7+2)+\frac{5}{6}=9+\frac{5}{6}=9\frac{5}{6} \quad ■$$

EXAMPLE 7

$$3\frac{1}{8}+\frac{5}{8}=3+\left(\frac{1}{8}+\frac{5}{8}\right)=3+\frac{\overset{3}{\cancel{6}}}{\underset{4}{\cancel{8}}}=3\frac{3}{4} \quad ■$$

You may add mixed numbers vertically.

EXAMPLE 8

Write the whole numbers in the same column.
Write the proper fractions in the same column.

$$\begin{array}{r} 8\frac{7}{10} \\ +5\frac{9}{10} \\ \hline 13\frac{16}{10} \end{array} = 13+1\frac{\overset{3}{\cancel{6}}}{\underset{5}{\cancel{10}}}=14\frac{3}{5} \quad ■$$

► **You Try It** **6.** $1\frac{6}{7}+4\frac{5}{7}$ **7.** $6\frac{1}{3}+4\frac{1}{3}$ **8.** $2+5\frac{2}{9}$ **9.** $7\frac{1}{4}+\frac{3}{4}$

2 Subtracting Like Fractions

You can subtract quantities that have the same name.

5 cabs − 2 cabs = 3 cabs

8 girls − 3 girls = 5 girls

7 tenths − 4 tenths = 3 tenths

$$\frac{7}{10} - \frac{4}{10} = \frac{3}{10}$$

> To subtract like fractions
>
> **1.** Subtract the two numerators.
> **2.** Write this difference over the common denominator.
> **3.** Reduce.

EXAMPLE 9

$$\frac{6}{7} - \frac{2}{7} = \frac{6-2}{7} = \frac{4}{7} \quad ■$$

EXAMPLE 10

$$\frac{7}{8} - \frac{3}{8} = \frac{7-3}{8} = \frac{\overset{1}{\cancel{4}}}{\underset{2}{\cancel{8}}} = \frac{1}{2} \quad ■$$

EXAMPLE 11

$$\frac{13}{5} - \frac{6}{5} = \frac{13-6}{5} = \frac{7}{5} = 1\frac{2}{5} \quad ■$$

EXAMPLE 12

$$\begin{array}{r} \frac{7}{12} \\ -\frac{5}{12} \\ \hline \frac{2}{12} \end{array} = \frac{\overset{1}{\cancel{2}}}{\underset{6}{\cancel{12}}} = \frac{1}{6} \quad ■$$

You Try It 10. $\frac{5}{7} - \frac{1}{7}$ 11. $\frac{9}{10} - \frac{3}{10}$ 12. $\frac{16}{3} - \frac{8}{3}$ 13. $\frac{11}{16} - \frac{7}{16}$

Subtracting Like Mixed Numbers

EXAMPLE 13

First subtract the proper fractions.

Then subtract the whole numbers.

$$\begin{array}{r} 3\frac{5}{8} \\ -1\frac{2}{8} \\ \hline 2\frac{3}{8} \end{array} \quad ■$$

> To subtract two like mixed numbers
>
> **1.** Subtract the like proper fractions. **2.** Subtract the whole numbers.
> **3.** Reduce.

EXAMPLE 14 **a.**

Drop down the whole number, 5.
Subtract the like proper fractions.

$$\begin{array}{r} 5\frac{6}{7} \\ -\ \frac{4}{7} \\ \hline 5\frac{2}{7} \end{array}$$

b.

$$\begin{array}{r} 8\frac{7}{10} \\ -2\frac{3}{10} \\ \hline 6\frac{4}{10} \end{array} = 6\frac{\overset{2}{\cancel{4}}}{\underset{5}{\cancel{10}}} = 6\frac{2}{5} \quad \blacksquare$$

▶ **You Try It** **14.** $4\frac{7}{10} - 1\frac{3}{10}$ **15.** $6\frac{8}{9} - \frac{4}{9}$ **16.** $14\frac{7}{12} - 6\frac{1}{12}$

▶ **Answers to You Try It** 1. $\frac{2}{3}$ 2. $1\frac{1}{2}$ 3. $\frac{6}{11}$ 4. $1\frac{3}{5}$ 5. $2\frac{4}{5}$ 6. $6\frac{4}{7}$ 7. $10\frac{2}{3}$ 8. $7\frac{2}{9}$ 9. 8 10. $\frac{4}{7}$ 11. $\frac{3}{5}$ 12. $2\frac{2}{3}$ 13. $\frac{1}{4}$ 14. $3\frac{2}{5}$ 15. $6\frac{4}{9}$ 16. $8\frac{1}{2}$

SECTION 4.1 EXERCISES

1 *Add the fractions, and reduce your answers.*

1. $\frac{2}{9}+\frac{5}{9}$ **2.** $\frac{3}{7}+\frac{2}{7}$ **3.** $\frac{3}{8}+\frac{1}{8}$ **4.** $\frac{2}{5}+\frac{2}{5}$

5. $\frac{7}{10}+\frac{6}{10}$ **6.** $\frac{10}{13}+\frac{8}{13}$ **7.** $\frac{1}{18}+\frac{5}{18}$ **8.** $\frac{9}{16}+\frac{23}{16}$

9. $\frac{13}{24}+\frac{7}{24}$ **10.** $\frac{13}{30}+\frac{11}{30}$ **11.** $\frac{7}{25}+\frac{10}{25}$ **12.** $\frac{27}{50}+\frac{33}{50}$

13. $\frac{1}{8}+\frac{9}{8}+\frac{4}{8}$ **14.** $\frac{5}{6}+\frac{7}{6}+\frac{3}{6}$ **15.** $\frac{3}{9}+\frac{7}{9}+\frac{8}{9}$ **16.** $\frac{1}{4}+\frac{3}{4}+\frac{2}{4}$

17. $\frac{16}{25}+\frac{21}{25}+\frac{18}{25}$ **18.** $\frac{4}{16}+\frac{9}{16}+\frac{7}{16}$ **19.** $\frac{2}{7}+\frac{6}{7}+\frac{6}{7}$ **20.** $\frac{26}{45}+\frac{14}{45}+\frac{32}{45}$

Add the mixed numbers, and reduce your answers.

21. $3\frac{2}{5}+4\frac{1}{5}$ **22.** $2\frac{1}{8}+6\frac{5}{8}$ **23.** $1\frac{1}{3}+3\frac{2}{3}$ **24.** $6\frac{1}{6}+9\frac{5}{6}$

25. $\begin{array}{r} 5\frac{4}{9} \\ +7\frac{2}{9} \\ \hline \end{array}$ **26.** $\begin{array}{r} 10\frac{4}{7} \\ +13\frac{1}{7} \\ \hline \end{array}$ **27.** $\begin{array}{r} 4\frac{3}{10} \\ +2\frac{9}{10} \\ \hline \end{array}$ **28.** $\begin{array}{r} 7\frac{11}{14} \\ +6\frac{13}{14} \\ \hline \end{array}$ **29.** $\begin{array}{r} 6\frac{7}{12} \\ +11\frac{11}{12} \\ \hline \end{array}$ **30.** $\begin{array}{r} 9\frac{3}{10} \\ +5\frac{7}{10} \\ \hline \end{array}$

31. $3\frac{8}{9}+4\frac{3}{9}+1\frac{7}{9}$ **32.** $10\frac{5}{7}+4+\frac{4}{7}+8\frac{6}{7}$

33. $23\frac{5}{12}+14\frac{11}{12}+10\frac{8}{12}$ **34.** $7\frac{5}{6}+4\frac{1}{6}+3\frac{7}{6}$

2 *Subtract the fractions, and reduce all answers.*

35. $\frac{5}{7}-\frac{2}{7}$ **36.** $\frac{3}{5}-\frac{2}{5}$ **37.** $\frac{3}{8}-\frac{1}{8}$ **38.** $\frac{4}{5}-\frac{2}{5}$

39. $\frac{9}{8}-\frac{5}{8}$ **40.** $\frac{7}{6}-\frac{3}{6}$ **41.** $\frac{7}{10}-\frac{6}{10}$ **42.** $\frac{10}{13}-\frac{8}{13}$

43. $\frac{11}{12}-\frac{7}{12}$ **44.** $\frac{9}{11}-\frac{4}{11}$ **45.** $\frac{7}{18}-\frac{1}{18}$ **46.** $\frac{23}{16}-\frac{2}{16}$

Subtract the mixed numbers, and reduce all answers.

47. $4\frac{4}{7} - 3\frac{2}{7}$

48. $6\frac{7}{8} - 3\frac{1}{8}$

49. $3\frac{2}{3} - 3\frac{1}{3}$

50. $9\frac{5}{6} - 7\frac{5}{6}$

51. $\begin{array}{r} 9\frac{8}{9} \\ -7\frac{2}{9} \\ \hline \end{array}$

52. $\begin{array}{r} 16\frac{4}{7} \\ -13\frac{1}{7} \\ \hline \end{array}$

53. $\begin{array}{r} 7\frac{11}{14} \\ -6\frac{4}{14} \\ \hline \end{array}$

54. $\begin{array}{r} 12\frac{7}{10} \\ -\ 6\frac{2}{10} \\ \hline \end{array}$

55. $\begin{array}{r} 8\frac{3}{4} \\ -\ \frac{1}{4} \\ \hline \end{array}$

56. $\begin{array}{r} 2\frac{21}{25} \\ -\ \frac{16}{25} \\ \hline \end{array}$

57. Subtract $\frac{7}{10}$ from $\frac{9}{10}$.

58. Subtract $\frac{5}{16}$ from $\frac{14}{16}$.

59. Subtract $7\frac{5}{12}$ from $16\frac{10}{12}$.

60. Subtract $2\frac{1}{6}$ from $4\frac{3}{6}$.

SKILLSFOCUS (Sections 2.2 & 2.4) *Evaluate.*

61. 7^2

62. 10^1

63. 2^4

64. 5^3

65. $\sqrt{49}$

66. $\sqrt{100}$

67. $\sqrt{81}$

68. $\sqrt{64}$

69. $3\sqrt{16}$

70. $7\sqrt{25}$

EXTEND YOUR THINKING ▶▶▶▶

▶ SOMETHING MORE

71. Four-eighteenths of a class had an A average. One-eighteenth had a D average or lower. Eight-eighteenths had a C average. What fraction of the class had a B average?

72. Two families want to split a bushel of steamed crabs in a fair way. One family has 3 adults and 2 children. The other has 6 adults and 2 children. A child will eat half as many crabs as an adult. Crabs cost $77 a bushel. How much should each family pay?

▶ TROUBLESHOOT IT

Find and correct the error.

73. $\frac{5}{8} + \frac{7}{8} = \frac{12}{16} = \frac{3}{4}$

74. $\frac{7}{9} - \frac{2}{9} = \frac{7-2}{9-9} = \frac{5}{0} = 0$

WRITING TO LEARN ▶▶▶▶

75. Explain why $\frac{1}{2} + \frac{1}{2} \neq \frac{2}{4}$. Relate your explanation to your own experience.

4.2 EQUIVALENT FRACTIONS

OBJECTIVES

1. Determine if two fractions are equivalent.
2. Expand a fraction to get an equivalent fraction.
3. Rename a fraction in terms of a larger denominator.

NEW VOCABULARY

equivalent fractions
expanding fractions
rename a fraction
lower terms
higher terms

1 Determining if Fractions Are Equivalent

Equivalent fractions are equal fractions. The fractions $\frac{2}{3}$ and $\frac{4}{6}$ are equivalent because they represent the same portion of the whole.

$\frac{2}{3} =$

$\frac{4}{6} =$

Therefore, $\frac{2}{3} = \frac{4}{6}$

$\frac{2}{3}$ is in lowest terms. $\frac{4}{6}$ reduces to $\frac{2}{3}$.

$$\frac{4}{6} = \frac{\cancel{4}^{2}}{\cancel{6}_{3}} = \frac{2}{3}$$

If two fractions reduce to the same lowest terms answer, they are equivalent.

> To determine if two fractions are equivalent
>
> 1. Reduce both to lowest terms.
> 2. They are equivalent if they reduce to the same lowest terms answer. Otherwise, they are unequal.

EXAMPLE 1 Are the fractions $\frac{6}{8}$ and $\frac{15}{20}$ equivalent?

Reduce both fractions to lowest terms.

$$\frac{6}{8} = \frac{\cancel{6}^{3}}{\cancel{8}_{4}} = \frac{3}{4} \qquad \frac{15}{20} = \frac{\cancel{15}^{3}}{\cancel{20}_{4}} = \frac{3}{4}$$

$\frac{6}{8} = \frac{15}{20}$ because both reduce to the same lowest terms answer, $\frac{3}{4}$. ■

EXAMPLE 2 Are the fractions $\frac{16}{28}$ and $\frac{24}{40}$ equivalent?

Reduce both fractions to lowest terms.

$$\frac{16}{28} = \frac{\overset{4}{\cancel{16}}}{\underset{7}{\cancel{28}}} = \frac{4}{7} \qquad \frac{24}{40} = \frac{\overset{3}{\cancel{24}}}{\underset{5}{\cancel{40}}} = \frac{3}{5}$$

Since $\frac{4}{7} \neq \frac{3}{5}$, the fractions $\frac{16}{28}$ and $\frac{24}{40}$ are not equal. ■

▶ **You Try It** Determine if the fractions are equivalent.

1. $\frac{6}{14}$ and $\frac{15}{36}$

2. $\frac{12}{27}$ and $\frac{20}{45}$

---

2 Expanding Fractions

Reducing a fraction writes the fraction with a smaller numerator and denominator. **Expanding a fraction** writes it with a larger numerator and denominator. The fraction resulting from either reducing or expanding is equivalent to the original fraction. Expanding fractions will be used later to add and subtract fractions with unlike denominators.

The number 1 can be written as any counting number divided by itself.

$$1 = \frac{1}{1} = \frac{2}{2} = \frac{3}{3} = \frac{4}{4} = \frac{5}{5} = \frac{6}{6} = \cdots$$

You expand a fraction by multiplying it by 1 written in any of these forms. The number 1 is the multiplicative identity. This means multiplying by 1 does not change the value of a fraction.

EXAMPLE 3 Expand $\frac{2}{7}$ using $\frac{5}{5}$.

$$\frac{2}{7} = \frac{2}{7} \times \frac{5}{5} = \frac{10}{35}$$

Check by reducing: $\frac{10}{35} = \frac{10 \div 5}{35 \div 5} = \frac{2}{7}$ ■

EXAMPLE 4 Expand $\frac{7}{9}$ using $\frac{6}{6}$.

$$\frac{7}{9} = \frac{7}{9} \times \frac{6}{6} = \frac{42}{54}$$

Check: $\frac{42}{54} = \frac{42 \div 6}{54 \div 6} = \frac{7}{9}$ ■

▶ **You Try It**

3. Expand $\frac{5}{8}$ using $\frac{6}{6}$.

4. Expand $\frac{1}{10}$ using $\frac{3}{3}$.

Any fraction can be written in an unlimited number of equivalent ways. For example, $\frac{1}{2}$ can be written in the following equivalent ways by multiplying it by 1 written as $\frac{2}{2}, \frac{3}{3}, \frac{4}{4}$, and so on.

$$\frac{1}{2} \times \frac{1}{1} = \frac{1}{2}$$

$$\frac{1}{2} \times \frac{2}{2} = \frac{2}{4}$$

$$\frac{1}{2} \times \frac{3}{3} = \frac{3}{6}$$

$$\frac{1}{2} \times \frac{4}{4} = \frac{4}{8}$$

$$\frac{1}{2} \times \frac{5}{5} = \frac{5}{10}$$

. . . and so on.

Each fraction in this column equals 1. Multiplying by 1 does not change the value of a fraction.

Each fraction in this column is equivalent to $\frac{1}{2}$. As a check, reduce each of these fractions to lowest terms. You get $\frac{1}{2}$.

EXAMPLE 5 Write four fractions equivalent to $\frac{3}{5}$.

$$\frac{3}{5}, \frac{3}{5} \times \frac{2}{2} = \frac{6}{10}, \quad \frac{3}{5} \times \frac{3}{3} = \frac{9}{15}, \quad \frac{3}{5} \times \frac{4}{4} = \frac{12}{20}, \quad \frac{3}{5} \times \frac{5}{5} = \frac{15}{25}$$

fractions equivalent to $\frac{3}{5}$: $\frac{3}{5}, \frac{6}{10}, \frac{9}{15}, \frac{12}{20}, \frac{15}{25}$ ■

EXAMPLE 6 Write five fractions equivalent to $\frac{3}{8}$.

$$\frac{3}{8}, \frac{6}{16}, \frac{9}{24}, \frac{12}{32}, \frac{15}{40}, \frac{18}{48}$$ ■

▶ **You Try It**

5. Write four fractions equivalent to $\frac{1}{6}$.
6. Write five fractions equivalent to $\frac{5}{7}$.

To **rename a fraction** means to write an equivalent fraction. You can rename a fraction in two ways.

1. **Reducing:** Divide numerator and denominator by the same counting number, n. This gives an equivalent fraction in **lower terms.**

$$\frac{a}{b} = \frac{a \div n}{b \div n}$$

2. **Expanding:** Multiply numerator and denominator by the same counting number, n. This gives an equivalent fraction in **higher terms** (with a larger numerator and denominator).

$$\frac{a}{b} = \frac{a \cdot n}{b \cdot n}$$

3 Renaming Fractions in Terms of a Larger Denominator

Multiplying a fraction by 1 gives an equivalent fraction. To write a fraction in higher terms, multiply the fraction by 1 written in one of the forms

$$1 = \frac{1}{1} = \frac{2}{2} = \frac{3}{3} = \frac{4}{4} = \frac{5}{5} = \frac{6}{6} = \cdots.$$

> To write a fraction in terms of a larger denominator
>
> 1. Divide the larger denominator by the smaller.
> 2. Multiply the original fraction by 1 written as the quotient in Step 1 divided by itself.

EXAMPLE 7 Find the missing numerator. $\frac{2}{3} = \frac{?}{12}$

Divide the denominators. $12 \div 3 = 4$. Multiply $\frac{2}{3}$ by $1 = \frac{4}{4}$.

$$\frac{2}{3} = \frac{2}{3} \times \frac{4}{4} = \frac{8}{12}.$$

The missing numerator is 8. ■

▶ **You Try It** 7. Find the missing numerator. $\frac{4}{7} = \frac{?}{28}$

EXAMPLE 8 Find the missing numerator. $\frac{5}{6} = \frac{?}{42}$

Divide the denominators. $42 \div 6 = 7$. Multiply $\frac{5}{6}$ by $1 = \frac{7}{7}$.

$$\frac{5}{6} = \frac{5}{6} \times \frac{7}{7} = \frac{35}{42}.$$

The missing numerator is 35. ■

EXAMPLE 9 Find the missing numerator. $7\frac{3}{8} = 7\frac{?}{40}$

Find the missing numerator so that $\frac{3}{8} = \frac{?}{40}$. Divide the denominators.

$40 \div 8 = 5$. Multiply $\frac{3}{8}$ by $1 = \frac{5}{5}$.

$$7\frac{3}{8} = 7 + \frac{3}{8} \times \frac{5}{5} = 7 + \frac{15}{40} = 7\frac{15}{40}.$$

The missing numerator is 15. ■

▶ **You Try It** Find the missing numerator. **8.** $\frac{3}{5} = \frac{?}{45}$ **9.** $5\frac{3}{10} = 5\frac{?}{60}$

▶ **Answers to You Try It** 1. no: $\frac{3}{7} \neq \frac{5}{12}$ 2. yes: $\frac{4}{9} = \frac{4}{9}$ 3. $\frac{30}{48}$
4. $\frac{3}{30}$ 5. $\frac{2}{12}, \frac{3}{18}, \frac{4}{24}, \frac{5}{30}$ 6. $\frac{10}{14}, \frac{15}{21}, \frac{20}{28}, \frac{25}{35}, \frac{30}{42}$ 7. 16 8. 27 9. 18

SECTION 4.2 EXERCISES

1 *Determine if the fractions are equivalent or not by reducing each fraction to lowest terms.*

1. $\frac{1}{2}, \frac{7}{14}$ **2.** $\frac{2}{3}, \frac{8}{12}$ **3.** $\frac{3}{4}, \frac{16}{24}$ **4.** $\frac{1}{3}, \frac{6}{18}$

5. $\frac{2}{8}, \frac{12}{40}$ **6.** $\frac{12}{36}, \frac{18}{27}$ **7.** $\frac{6}{10}, \frac{21}{35}$ **8.** $\frac{12}{15}, \frac{21}{24}$

9. $\frac{4}{12}, \frac{8}{18}$ **10.** $\frac{20}{28}, \frac{24}{42}$ **11.** $\frac{12}{14}, \frac{18}{21}$ **12.** $\frac{20}{32}, \frac{24}{40}$

13. $\frac{36}{60}, \frac{32}{48}$ **14.** $\frac{44}{55}, \frac{12}{15}$ **15.** $\frac{35}{50}, \frac{60}{100}$ **16.** $\frac{64}{80}, \frac{15}{20}$

17. $\frac{150}{200}, \frac{80}{120}$ **18.** $\frac{14}{24}, \frac{21}{36}$ **19.** $\frac{8}{20}, \frac{10}{25}$ **20.** $\frac{6}{16}, \frac{16}{40}$

2 *Expand each fraction using the form given for* 1.

21. $\frac{3}{4}$ using $\frac{6}{6}$ **22.** $\frac{1}{3}$ using $\frac{7}{7}$ **23.** $\frac{5}{6}$ using $\frac{4}{4}$ **24.** $\frac{2}{5}$ using $\frac{12}{12}$

25. $\frac{7}{8}$ using $\frac{8}{8}$ **26.** $\frac{5}{9}$ using $\frac{3}{3}$ **27.** $\frac{3}{8}$ using $\frac{6}{6}$ **28.** $\frac{7}{10}$ using $\frac{7}{7}$

29. $\frac{8}{15}$ using $\frac{2}{2}$ **30.** $\frac{7}{12}$ using $\frac{5}{5}$ **31.** $\frac{6}{7}$ using $\frac{3}{3}$ **32.** $\frac{16}{25}$ using $\frac{4}{4}$

Write five fractions equivalent to the given fraction.

33. $\frac{1}{3}$ **34.** $\frac{4}{5}$ **35.** $\frac{7}{10}$

36. $\frac{5}{8}$ **37.** $\frac{3}{4}$ **38.** $\frac{5}{6}$

3 *Find the missing numerator. Check each answer by reducing.*

39. $\frac{2}{3} = \frac{?}{6}$ **40.** $\frac{3}{4} = \frac{?}{8}$ **41.** $\frac{4}{5} = \frac{?}{20}$ **42.** $\frac{1}{6} = \frac{?}{18}$

43. $\frac{4}{9} = \frac{?}{36}$ **44.** $\frac{7}{8} = \frac{?}{24}$ **45.** $\frac{2}{7} = \frac{?}{35}$ **46.** $\frac{3}{10} = \frac{?}{50}$

47. $\frac{1}{3} = \frac{?}{18}$

48. $\frac{5}{7} = \frac{?}{28}$

49. $\frac{2}{9} = \frac{?}{45}$

50. $\frac{5}{8} = \frac{?}{56}$

51. $\frac{12}{11} = \frac{?}{33}$

52. $\frac{17}{10} = \frac{?}{50}$

53. $\frac{8}{5} = \frac{?}{35}$

54. $\frac{25}{12} = \frac{?}{60}$

55. $\frac{5}{2} = \frac{?}{70}$

56. $\frac{7}{4} = \frac{?}{28}$

57. $\frac{9}{6} = \frac{?}{54}$

58. $\frac{5}{4} = \frac{?}{52}$

59. $\frac{11}{8} = \frac{?}{48}$

60. $\frac{7}{5} = \frac{?}{30}$

61. $\frac{17}{6} = \frac{?}{54}$

62. $\frac{41}{10} = \frac{?}{90}$

63. $3\frac{7}{9} = 3\frac{?}{27}$

64. $2\frac{5}{8} = 2\frac{?}{64}$

65. $6\frac{2}{3} = 6\frac{?}{24}$

66. $5\frac{7}{12} = 5\frac{?}{84}$

SKILLSFOCUS (Section 3.1) *Prime factor.*

67. 40 **68.** 72 **69.** 18 **70.** 21 **71.** 33

EXTEND YOUR THINKING ▶▶▶▶

▶ TROUBLESHOOT IT

Find and correct the error.

72. The following fractions are equivalent to $\frac{3}{7}$: $\frac{6}{14}, \frac{18}{35}, \frac{9}{21}, \frac{36}{63}, \frac{21}{49}$.

WRITING TO LEARN ▶▶▶▶

73. Explain the difference between expanding a fraction and reducing a fraction.

4.3 FINDING THE LEAST COMMON DENOMINATOR

OBJECTIVES

1. Define common denominator and why you need it.
2. Find the LCD by inspection.
3. Find the LCD by prime factoring.
4. Find the LCD by successive division by primes.

NEW VOCABULARY

common denominator
least common denominator (LCD)
common multiple
least common multiple (LCM)
inspection

1 The Common Denominator and Why You Need It

You cannot add 3 dimes to 2 quarters and get 5 dimes, or 5 quarters, because the quantities are unlike. To add dimes to quarters you must first write both in terms of some common name, such as cents.

$$\begin{aligned} \text{3 dimes} &= \text{30 cents} \\ +\ \text{2 quarters} &= \text{50 cents} \\ \hline \text{3 dimes} + \text{2 quarters} &= \text{80 cents} \end{aligned}$$

In the same way, you cannot add unlike fractions. For example, add $\frac{1}{2} + \frac{1}{3}$. The problem is pictured in the figure.

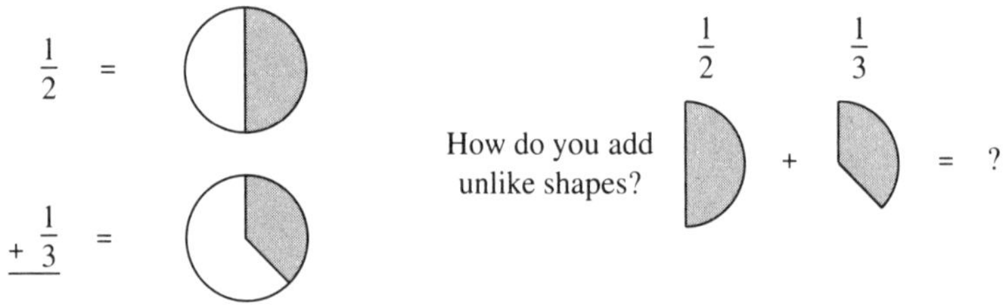

$\frac{1}{2} =$ (circle) $\quad + \frac{1}{3} =$ (circle)

How do you add unlike shapes? $\frac{1}{2} + \frac{1}{3} = ?$

Halves and thirds are as different as dimes and quarters. You cannot add them until you first rewrite each in terms of a common name, or common denominator. Begin by writing fractions equivalent to $\frac{1}{2}$ and $\frac{1}{3}$.

$$\text{one half} = \frac{1}{2} = \frac{2}{4} = \frac{3}{6} = \frac{4}{8} = \frac{5}{10} = \frac{6}{12} = \frac{7}{14} = \frac{8}{16} = \frac{9}{18} = \cdots$$

$$\text{one third} = \frac{1}{3} = \frac{2}{6} = \frac{3}{9} = \frac{4}{12} = \frac{5}{15} = \frac{6}{18} = \frac{7}{21} = \frac{8}{24} = \cdots$$

To add $\frac{1}{2}$ to $\frac{1}{3}$, add a fraction equivalent to $\frac{1}{2}$ to a fraction equivalent to $\frac{1}{3}$ that have the same denominator. Three possible choices are shown using the denominators 6, 12, and 18.

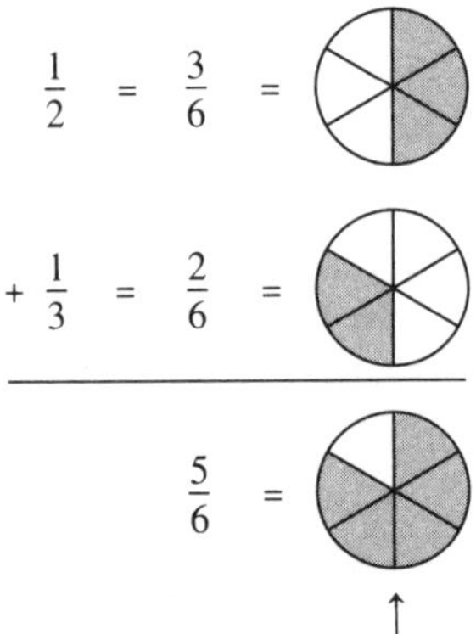

$$\frac{1}{2} = \frac{3}{6} = \text{(circle)} \qquad + \frac{1}{3} = \frac{2}{6} = \text{(circle)} \qquad \frac{5}{6} = \text{(circle)}$$

$$\frac{1}{2} = \frac{6}{12} \qquad +\frac{1}{3} = \frac{4}{12} \qquad \frac{10}{12} = \frac{\overset{5}{\cancel{10}}}{\underset{6}{\cancel{12}}} = \frac{5}{6}$$

$$\frac{1}{2} = \frac{9}{18} \qquad +\frac{1}{3} = \frac{6}{18} \qquad \frac{15}{18} = \frac{\overset{5}{\cancel{15}}}{\underset{6}{\cancel{18}}} = \frac{5}{6}$$

To add irregularly shaped figures, divide them into pieces of the same size and shape. Divide each circle into sixths and add the sixths.

Same answer as on the left. But the terms are larger, and each answer had to be reduced.

In each addition, the answer is the same. The numbers 6, 12, and 18 are called common denominators. A **common denominator** for two fractions is any number divisible by both denominators.

The number 6 is called the least common denominator because it is the smallest. The **least common denominator (LCD)** for two fractions is the smallest number that both denominators divide into exactly.

Use the LCD to add and subtract fractions. It is easier to use because it is the smallest possible common denominator. Smaller numbers mean there is less chance of an addition or reducing error. Furthermore, when you use the LCD, your answer will sometimes be in lowest terms. If you use a denominator larger than the LCD, you will *always* have to reduce your answer.

A common denominator is also called a common multiple. 6, 12, and 18 are multiples of 2. They are also multiples of 3. As such, 6, 12, and 18 are called **common multiples** for 2 and 3. Because 6 is the smallest, it is called the **least common multiple (LCM)**. Finding the LCD for two fractions is the same as finding the LCM for the two denominators. Three methods are discussed in this section.

2 Finding the LCD by Inspection

If by reading a problem you can determine the LCD for two fractions, you are using the method of **inspection**.

EXAMPLE 1 The LCD for $\frac{1}{2}$ and $\frac{3}{5}$ is 10 because 10 is the smallest number that both 2 and 5 divide into exactly. ■

EXAMPLE 2 The LCD for $\frac{3}{4}$ and $\frac{2}{7}$ is 28 because 28 is the smallest number that both 4 and 7 divide into exactly. ■

▶ **You Try It** Find the LCD by inspection.

1. $\frac{2}{3}$ and $\frac{5}{6}$

2. $\frac{1}{4}$ and $\frac{2}{5}$

3 Finding the LCD by Prime Factoring

Find the LCD for $\frac{7}{12}$ and $\frac{11}{15}$. It is the smallest number that both 12 and 15 divide into exactly. To find it, prime factor both denominators.

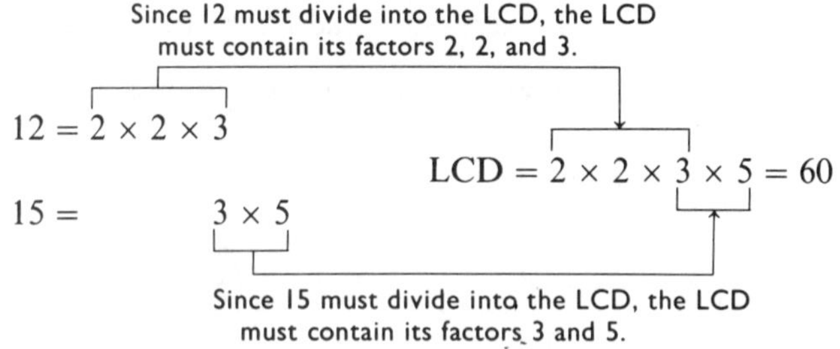

Notice that 12 and 15 each contain one factor of 3. But you did not need to include the factor 3 twice in the LCD. The LCD contains the least number of factors so that both prime factorizations are contained in it.

To find the LCD for two fractions using prime factoring

1. Prime factor both denominators.
2. Write like prime factors in the same column.
3. With each like prime, circle the greatest number of times it occurs in its column.
4. The LCD is the product of the circled factors.

EXAMPLE 3 Find the LCD for $\frac{7}{12}$ and $\frac{11}{15}$.

$$\begin{aligned} 12 &= (2 \times 2) \times (3) \\ 15 &= \quad\quad\quad 3 \times (5) \\ \text{LCD} &= 2 \times 2 \times 3 \times 5 = 60 \end{aligned}$$

Prime factor both denominators. Since 3 is the only factor common to both, only circle one factor of 3.
Drop the circled factors down into the LCD, and multiply. ■

▶ **You Try It** 3. Use prime factoring to find the LCD for $\frac{7}{9}$ and $\frac{5}{12}$.

EXAMPLE 4 Find the LCD for $\frac{17}{24}$ and $\frac{9}{40}$.

$$\begin{aligned} 24 &= (2 \times 2 \times 2) \times (3) \\ 40 &= 2 \times 2 \times 2 \quad \times (5) \\ \text{LCD} &= 2 \times 2 \times 2 \times 3 \times 5 = 120 \end{aligned}$$ ■

Prime factor both denominators. 2 is a factor 3 times in both prime factorizations. Circle one occurrence.

EXAMPLE 5 Find the LCD for $\frac{11}{30}$ and $\frac{4}{45}$.

$$\begin{aligned} 30 &= (2) \times 3 \quad \times (5) \\ 45 &= \quad (3 \times 3) \times 5 \\ \text{LCD} &= 2 \times 3 \times 3 \times 5 = 90 \end{aligned}$$ ■

Since 3 occurs twice in 45, circle the 3×3 in 45. Do not circle the single occurence of 3 in 30.

▶ **You Try It** Find the LCD by prime factoring.

4. $\frac{1}{36}$ and $\frac{7}{45}$

5. $\frac{3}{20}$ and $\frac{9}{28}$

EXAMPLE 6 Find the LCD for $\frac{2}{45}$, $\frac{5}{36}$, and $\frac{13}{75}$.

$$\begin{aligned} 36 &= (2 \times 2) \times (3 \times 3) \\ 45 &= \quad\quad 3 \times 3 \times 5 \\ 75 &= \quad\quad 3 \quad \times (5 \times 5) \\ \text{LCD} &= 2 \times 2 \times 3 \times 3 \times 5 \times 5 = 900 \end{aligned}$$ ■

Prime factor the three denominators. Since 3 occurs twice in two denominators, only circle one 3×3.

▶ **You Try It** 6. Find the LCD by prime factoring.

$\frac{7}{21}$, $\frac{4}{35}$, and $\frac{9}{30}$

4 Finding the LCD by Successive Division by Primes

This procedure is a graphic approach to prime factorization.

> To find the LCD by using successive division by primes
>
> **1.** Write the denominators on the same horizontal line.
> **2.** Find a prime number that exactly divides two (or more) of the denominators. Write the quotients on the line below. (If a denominator is not divisible by the prime number, move it down unchanged.)
> **3.** Keep repeating division by primes until no prime can be found that divides two (or more) of the quotients.
> **4.** Multiply "around" for the LCD (that is, multiply all the divisors times the final quotients for the LCD).

EXAMPLE 7 Solve Example 3 using successive division by primes. Find the LCD for $\frac{7}{12}$ and $\frac{11}{15}$.

$$\begin{array}{r|rr} 3 & 12 & 15 \\ \hline & 4 & 5 \end{array}$$

Write the denominators on the same line.
The prime 3 divides into both 12 and 15.
Write the quotients 4 and 5 directly below.
Since no prime divides 4 and 5, you are done.
Multiply "around" for the LCD.

$\text{LCD} = 3 \times 4 \times 5 = 60$ ■

OBSERVE The factors 3, 4 (as 2×2), and 5 used in Example 7 to get the LCD are the same factors found in Example 3 using prime factoring.

▶ **You Try It** 7. Use successive division by primes to find the LCD.

$$\frac{7}{15} \text{ and } \frac{4}{27}$$

EXAMPLE 8 Find the LCD for $\frac{5}{18}$ and $\frac{11}{24}$.

$$\begin{array}{r|rr} 2 & 18 & 24 \\ \hline 3 & 9 & 12 \\ \hline & 3 & 4 \end{array}$$

The prime 2 divides 18 and 24 exactly. Write the quotients 9 and 12 below.
The prime 3 divides 9 and 12 exactly. Write the quotients 3 and 4 below.
Since no prime divides both 3 and 4, you are done. Multiply around for the LCD.

$\text{LCD} = 2 \times 3 \times 3 \times 4 = 72$ ■

EXAMPLE 9 Find the LCD for $\frac{2}{7}$ and $\frac{8}{9}$.

$$\begin{array}{|rr} 7 & 9 \\ \hline \end{array}$$

Since no prime divides both 7 and 9, the LCD is just the product of the two denominators.

$\text{LCD} = 7 \times 9 = 63$ ■

▶ You Try It Use successive division by primes to find the LCD.

8. $\frac{2}{9}$ and $\frac{5}{18}$ **9.** $\frac{9}{32}$ and $\frac{11}{24}$ **10.** $\frac{13}{36}$ and $\frac{1}{60}$

The next two examples demonstrate how to find the LCD for 3 or more fractions.

EXAMPLE 10 Find the LCD for $\frac{5}{48}$, $\frac{3}{20}$, and $\frac{19}{30}$.

2	48	20	30	The prime 2 divides all three exactly.
2	24	10	15	2 divides 24 and 10 exactly. 2 does not divide 15. Drop down 15 as is.
3	12	5	15	3 divides 12 and 15. 3 does not divide the 5. Drop 5 down as is.
5	4	5	5	5 divides the two fives. 5 does not divide the 4. Drop down the 4.
	4	1	1	Since 4, 1, and 1 have no common factors, multiply around for the LCD.

LCD = 2 × 2 × 3 × 5 × 4 × 1 × 1 = 240 ■

OBSERVE In Example 10, you clearly see the value of finding the *least* common denominator. Multiplying the original denominators always gives a common denominator. But 48 × 20 × 30 = 28,800! Compared to 240, the denominator 28,800 is much too large to be practical.

EXAMPLE 11 Find the LCD for $\frac{3}{14}$, $\frac{10}{21}$, and $\frac{4}{5}$.

7	14	21	5	No prime divides all three denominators. The prime 7 divides 14 and 21.
	2	3	5	Write the quotients 2 and 3 below. Drop down the 5.

LCD = 7 × 2 × 3 × 5 = 210 ■

▶ You Try It Use successive division by primes to find the LCD.

11. $\frac{3}{40}$, $\frac{7}{25}$, and $\frac{11}{90}$ **12.** $\frac{5}{22}$, $\frac{7}{33}$, and $\frac{1}{66}$

▶ Answers to You Try It 1. 6 2. 20 3. 36 4. 180 5. 140 6. 210 7. 135 8. 18 9. 96 10. 180 11. 1,800 12. 66

SECTION 4.3 EXERCISES

1 2 3 4 *Find the LCD for the given denominators using any method.*

1. 8 and 16

2. 21 and 15

3. 12 and 15

4. 10 and 18

5. 6 and 10

6. 9 and 16

7. 8 and 24

8. 7 and 14

9. 10 and 12

10. 14 and 22

11. 16 and 20

12. 15 and 25

13. 12 and 36

14. 20 and 40

15. 10 and 21

16. 6 and 25

17. 15 and 20

18. 10 and 36

19. 6 and 9

20. 14 and 16

21. 8 and 40

22. 4 and 24

23. 4 and 7

24. 5 and 11

25. 25 and 50

26. 33 and 11

27. 18 and 30

28. 24 and 36

29. 36 and 60

30. 30 and 45

31. 42 and 63

32. 39 and 26

33. 80 and 64

34. 90 and 120

35. 240 and 400

36. 77 and 121

37. 3, 5, and 10

38. 4, 8, and 12

39. 4, 5, and 7

40. 2, 3, and 5

41. 3, 4, and 6

42. 6, 8, and 18

43. 10, 15, and 20

44. 8, 16, and 40

45. 36, 24, and 30

46. 72, 40, and 54

47. 10, 17, and 12

48. 160, 144, and 96

49. 4, 8, 12, and 16

50. 12, 15, 90, and 45

SKILLSFOCUS (Section 1.5) *Change to an improper fraction.*

51. $4\frac{1}{6}$

52. $2\frac{7}{8}$

53. $6\frac{3}{4}$

54. $12\frac{2}{5}$

EXTEND YOUR THINKING

TROUBLESHOOT IT

Find and correct the error.

55. By inspection, the LCD for $\frac{3}{4}$ and $\frac{5}{8}$ is 32.

56. Find the LCD for 6, 9, and 12.

$$\begin{array}{r|rrr} 3 & 6 & 9 & 12 \\ \hline & 2 & 3 & 4 \end{array} \quad \text{LCD} = 3 \times 2 \times 3 \times 4 = 72$$

WRITING TO LEARN

57. Explain how successive division by primes is similar to the method of prime factoring when finding the LCD.

58. Explain the difference between a common denominator and a least common denominator.

59. Make up your own problem to find the LCD for two fractions using successive division by primes. Then solve it. Explain each step.

4.4 ADDING FRACTIONS WITH UNLIKE DENOMINATORS

OBJECTIVES

1. Add unlike proper fractions.
2. Add unlike mixed numbers.

1 Adding Unlike Proper Fractions

To add unlike fractions, rewrite them as like fractions. To do this, find the LCD, then rewrite each fraction as an equivalent fraction with the LCD as its denominator.

EXAMPLE 1 Add $\frac{1}{3} + \frac{2}{5}$.

By inspection, the LCD for 3 and 5 is 15. Rewrite each fraction as an equivalent fraction with a denominator of 15.

$$\frac{1}{3} = \frac{1}{3} \times \frac{5}{5} = \frac{5}{15}$$ Since $15 \div 3 = 5$, multiply $\frac{1}{3}$ by $1 = \frac{5}{5}$.

$$+\frac{2}{5} = \frac{2}{5} \times \frac{3}{3} = \frac{6}{15}$$ Since $15 \div 5 = 3$, multiply $\frac{2}{5}$ by $1 = \frac{3}{3}$.

$$\frac{11}{15}$$ Add the like fractions. ■

To add unlike fractions

1. Find the LCD for the denominators.
2. Rewrite each fraction as an equivalent fraction with the LCD as its denominator.
3. Add the like fractions. Reduce if needed.

EXAMPLE 2 What is $\frac{3}{4} + \frac{5}{7}$?

By inspection, the LCD for 4 and 7 is 28.

$$\frac{3}{4} = \frac{3}{4} \times \frac{7}{7} = \frac{21}{28}$$

$$+\frac{5}{7} = \frac{5}{7} \times \frac{4}{4} = \frac{20}{28}$$

$$\frac{41}{28} = 1\frac{13}{28}$$ ■

▶ **You Try It**

1. Add $\frac{5}{6} + \frac{3}{8}$.

2. Add $\frac{2}{9} + \frac{3}{5}$.

EXAMPLE 3 Add $\frac{7}{30} + \frac{5}{18}$.

$$30 = (2) \times 3 \times (5)$$
$$18 = 2 \times (3 \times 3)$$
$$\text{LCD} = 2 \times 3 \times 3 \times 5 = 90$$

Find the LCD for 30 and 18 by prime factoring.

Rewrite each fraction as an equivalent fraction with a denominator of 90.

$$\begin{array}{r} \dfrac{7}{30} = \dfrac{7}{30} \times \dfrac{3}{3} = \dfrac{21}{90} \\ + \dfrac{5}{18} = \dfrac{5}{18} \times \dfrac{5}{5} = \dfrac{25}{90} \\ \hline \end{array}$$

$$\frac{46}{90} = \frac{\overset{23}{\cancel{46}}}{\underset{45}{\cancel{90}}} = \frac{23}{45} \quad \blacksquare$$

▶ **You Try It** 3. Add $\frac{7}{12} + \frac{11}{30}$.

EXAMPLE 4 Add $\frac{2}{3} + \frac{1}{4} + \frac{5}{6}$.

By inspection, the LCD for 3, 4, and 6 is 12.

$$\begin{array}{r} \dfrac{2}{3} = \dfrac{2}{3} \times \dfrac{4}{4} = \dfrac{8}{12} \\ \dfrac{1}{4} = \dfrac{1}{4} \times \dfrac{3}{3} = \dfrac{3}{12} \\ + \dfrac{5}{6} = \dfrac{5}{6} \times \dfrac{2}{2} = \dfrac{10}{12} \\ \hline \end{array}$$

$$\frac{21}{12} = \frac{\overset{7}{\cancel{21}}}{\underset{4}{\cancel{12}}} = \frac{7}{4} \text{ or } 1\frac{3}{4} \quad \blacksquare$$

EXAMPLE 5 Add $\frac{7}{24} + \frac{1}{18} + \frac{4}{15}$.

Find the LCD using successive division by primes.

$$\begin{array}{r|rrr} 2 & 24 & 18 & 15 \\ \hline 3 & 12 & 9 & 15 \\ \hline & 4 & 3 & 5 \end{array}$$

$$\text{LCD} = 2 \times 3 \times 4 \times 3 \times 5 = 360$$

$$\begin{array}{r} \dfrac{7}{24} = \dfrac{7}{24} \times \dfrac{15}{15} = \dfrac{105}{360} \\ \dfrac{1}{18} = \dfrac{1}{18} \times \dfrac{20}{20} = \dfrac{20}{360} \\ + \dfrac{4}{15} = \dfrac{4}{15} \times \dfrac{24}{24} = \dfrac{96}{360} \\ \hline \dfrac{221}{360} \end{array} \quad \blacksquare$$

▶ **You Try It** 4. Add $\frac{1}{6} + \frac{4}{5} + \frac{7}{15}$. 5. Add $\frac{4}{45} + \frac{1}{30} + \frac{3}{50}$.

You may reduce a fraction to lowest terms before adding.

EXAMPLE 6 Add $\frac{9}{16} + \frac{21}{24}$.

$$\begin{array}{r} \dfrac{9}{16} = \dfrac{9}{16} = \dfrac{9}{16} \\ + \dfrac{21}{24} = \dfrac{\cancel{21}^{7}}{\cancel{24}_{8}} = \dfrac{7}{8} \times \dfrac{2}{2} = \dfrac{14}{16} \\ \hline \dfrac{23}{16} = 1\dfrac{7}{16} \end{array} \quad \blacksquare$$

You Try It 6. Add $\frac{20}{36} + \frac{4}{27}$.

2 Adding Unlike Mixed Numbers

To add unlike mixed numbers, add the proper fractions. Then add the whole numbers.

EXAMPLE 7 Add $2\frac{1}{4} + 6\frac{3}{5}$.

Rewrite each proper fraction with LCD = 20.

$$2\frac{1}{4} = 2 + \frac{1}{4} = 2 + \frac{1}{4} \times \frac{5}{5} = 2 + \frac{5}{20}$$

$$+6\frac{3}{5} = 6 + \frac{3}{5} = 6 + \frac{3}{5} \times \frac{4}{4} = 6 + \frac{12}{20}$$

$$8 + \frac{17}{20} = 8\frac{17}{20} \quad ■$$

> To add mixed numbers
>
> 1. Find the LCD for the proper fractions.
> 2. Rewrite each proper fraction as an equivalent fraction with the LCD as denominator.
> 3. Add the equivalent fractions.
> 4. Add the whole numbers.
> 5. Reduce if possible.

EXAMPLE 8 Add $4\frac{2}{3} + 1\frac{5}{6}$.

The LCD for the proper fractions 3 and 6 is 6.

$$4\frac{2}{3} = 4 + \frac{2}{3} \times \frac{2}{2} = 4 + \frac{4}{6}$$

$$+1\frac{5}{6} = 1 + \frac{5}{6} \quad = 1 + \frac{5}{6}$$

$$5 + \frac{9}{6} = 5 + 1\frac{\cancel{3}^{1}}{\cancel{6}_{2}} = 5 + 1\frac{1}{2} = 6\frac{1}{2} \quad ■$$

You Try It 7. Add $4\frac{5}{7} + 6\frac{2}{3}$.

8. Add $5\frac{4}{9} + 3\frac{7}{15}$.

EXAMPLE 9 Add $5\frac{3}{16} + 3\frac{7}{12}$.

2	16	12
2	8	6
	4	3

LCD = 2 × 2 × 4 × 3 = 48

$$5\frac{3}{16} = 5 + \frac{3}{16} \times \frac{3}{3} = 5 + \frac{9}{48}$$

$$+3\frac{7}{12} = 3 + \frac{7}{12} \times \frac{4}{4} = 3 + \frac{28}{48}$$

$$8 + \frac{37}{48} = 8\frac{37}{48} \quad ■$$

EXAMPLE 10 Add $2\frac{1}{6} + 4\frac{5}{9} + 5\frac{11}{24}$.

$$\begin{array}{r|rrr} 3 & 6 & 9 & 24 \\ 2 & 2 & 3 & 8 \\ \hline & 1 & 3 & 4 \end{array}$$

$$\text{LCD} = 3 \times 2 \times 1 \times 3 \times 4 = 72$$

$$\begin{aligned} 2\frac{1}{6} &= 2 + \frac{1}{6} \times \frac{12}{12} = 2 + \frac{12}{72} \\ 4\frac{5}{9} &= 4 + \frac{5}{9} \times \frac{8}{8} = 4 + \frac{40}{72} \\ +5\frac{11}{24} &= 5 + \frac{11}{24} \times \frac{3}{3} = 5 + \frac{33}{72} \\ \hline & \qquad 11 + \frac{85}{72} = 12\frac{13}{72} \end{aligned}$$

■

▶ **You Try It** 9. Add $14\frac{5}{18} + 16\frac{1}{12}$.

10. Add $1\frac{3}{4} + 2\frac{5}{8} + 5\frac{3}{10}$.

Adding Unlike Mixed Numbers by First Writing Them as Improper Fractions

Another way to add unlike mixed numbers is to first change them to improper fractions. Then add the improper fractions just as you added proper fractions. Solve Example 7 again.

EXAMPLE 11 Add $2\frac{1}{4} + 6\frac{3}{5}$ by first changing each mixed number to an improper fraction.

The LCD for 4 and 5 is 20.

$$\begin{aligned} 2\frac{1}{4} &= \frac{9}{4} = \frac{9}{4} \times \frac{5}{5} = \frac{45}{20} \\ +6\frac{3}{5} &= \frac{33}{5} = \frac{33}{5} \times \frac{4}{4} = \frac{132}{20} \\ \hline & \qquad\qquad \frac{177}{20} = 8\frac{17}{20} \end{aligned}$$

This is the same answer as in Example 7. ■

Compare Example 11 to Example 7. Adding by first changing mixed numbers to improper fractions will always give the correct answer. The disadvantage is having to work with larger numerators.

▶ **You Try It** 11. Add by first changing each mixed number to an improper fraction.

$$3\frac{1}{2} + 2\frac{4}{5}$$

▶ **Answers to You Try It** 1. $1\frac{5}{24}$ 2. $\frac{37}{45}$ 3. $\frac{19}{20}$ 4. $1\frac{13}{30}$ 5. $\frac{41}{225}$ 6. $\frac{19}{27}$ 7. $11\frac{8}{21}$ 8. $8\frac{41}{45}$ 9. $30\frac{13}{36}$ 10. $9\frac{27}{40}$ 11. $6\frac{3}{10}$

SECTION 4.4 EXERCISES

1 *Add the fractions. Reduce all answers to lowest terms.*

1. $\frac{2}{3}+\frac{1}{2}$ **2.** $\frac{3}{4}+\frac{1}{2}$ **3.** $\frac{3}{5}+\frac{1}{4}$ **4.** $\frac{2}{3}+\frac{5}{6}$ **5.** $\frac{1}{3}+\frac{2}{9}$

6. $\frac{1}{6}+\frac{5}{12}$ **7.** $\frac{3}{4}+\frac{1}{3}$ **8.** $\frac{3}{8}+\frac{5}{16}$ **9.** $\frac{4}{5}+\frac{7}{10}$ **10.** $\frac{11}{12}+\frac{5}{6}$

11. $\frac{3}{7}+\frac{5}{14}$ **12.** $\frac{5}{9}+\frac{5}{12}$ **13.** $\frac{2}{3}+\frac{7}{12}$ **14.** $\frac{1}{6}+\frac{4}{15}$ **15.** $\frac{3}{10}+\frac{8}{15}$

16. $\frac{3}{16}+\frac{5}{24}$ **17.** $\frac{1}{12}+\frac{1}{18}$ **18.** $\frac{14}{15}+\frac{17}{30}$ **19.** $\frac{13}{20}+\frac{3}{32}$ **20.** $\frac{27}{40}+\frac{13}{24}$

21. $\frac{1}{36}+\frac{1}{48}$ **22.** $\frac{8}{21}+\frac{4}{35}$ **23.** $\frac{5}{36}+\frac{10}{27}$ **24.** $\frac{17}{20}+\frac{3}{50}$ **25.** $\frac{7}{10}+\frac{13}{30}$

26. $\frac{4}{25}+\frac{8}{75}$ **27.** $\frac{9}{16}+\frac{3}{20}$ **28.** $\frac{1}{100}+\frac{3}{250}$ **29.** $\frac{5}{16}+\frac{3}{32}$ **30.** $\frac{31}{48}+\frac{17}{80}$

31. $\frac{1}{4}+\frac{1}{6}+\frac{1}{2}$ **32.** $\frac{2}{3}+\frac{5}{6}+\frac{1}{9}$ **33.** $\frac{1}{4}+\frac{5}{8}+\frac{11}{16}$ **34.** $\frac{1}{6}+\frac{7}{9}+\frac{5}{18}$

35. $\frac{3}{25}+\frac{13}{50}+\frac{7}{15}$ **36.** $\frac{4}{9}+\frac{1}{18}+\frac{25}{36}$ **37.** $\frac{11}{12}+\frac{17}{18}+\frac{1}{24}$

38. $\frac{9}{10}+\frac{7}{20}+\frac{1}{30}$ **39.** $\frac{5}{8}+\frac{7}{12}+\frac{9}{20}$ **40.** $\frac{13}{24}+\frac{11}{36}+\frac{31}{60}$

2 *Add the mixed numbers. Reduce all answers to lowest terms.*

41. $2\frac{1}{3}+3\frac{1}{2}$ **42.** $4\frac{3}{4}+1\frac{2}{5}$ **43.** $3\frac{4}{7}+5\frac{2}{3}$ **44.** $1\frac{4}{5}+2\frac{1}{4}$

45. $5\frac{7}{8}+3\frac{3}{4}$ **46.** $4\frac{2}{9}+1\frac{1}{3}$ **47.** $2\frac{3}{8}+4\frac{7}{12}$ **48.** $5\frac{1}{6}+2\frac{7}{10}$

49. $5\frac{1}{2}+\frac{3}{4}$ **50.** $4\frac{5}{12}+\frac{7}{8}$ **51.** $16\frac{3}{8}+1\frac{3}{4}$ **52.** $2\frac{4}{9}+5\frac{7}{27}$

53. $4 + 3\frac{5}{7}$

54. $3\frac{5}{31} + 7$

55. $7\frac{2}{5} + 4\frac{3}{8}$

56. $10\frac{5}{8} + 12\frac{11}{18}$

57. $25\frac{1}{2} + 13\frac{3}{5}$

58. $50\frac{5}{7} + 40\frac{1}{8}$

59. $4\frac{7}{9} + 8$

60. $12 + 16\frac{13}{20}$

61. $6\frac{16}{25} + 8\frac{3}{10}$

62. $5\frac{4}{15} + 7\frac{7}{9}$

63. $3\frac{5}{6} + 9\frac{8}{21}$

64. $60\frac{2}{5} + 4\frac{1}{8}$

65. $3\frac{1}{2} + 5\frac{3}{4} + 2\frac{2}{3}$

66. $5\frac{3}{4} + 6\frac{1}{3} + 4\frac{4}{5}$

67. $4\frac{5}{25} + 1\frac{7}{60} + 3\frac{9}{40}$

68. $7\frac{7}{20} + 4\frac{9}{10} + 3\frac{1}{5}$

SKILLSFOCUS (Section 2.3) *Perform the indicated operation.*

69. 4.561×100

70. 0.0876×10^3

71. $56 \cdot 1{,}000$

72. $3.4(1000)$

EXTEND YOUR THINKING ▶▶▶▶

▶ TROUBLESHOOT IT

Find and correct the error.

73.

$$\frac{2}{5} = \frac{2}{5} \times \frac{3}{3} = \frac{6}{15}$$
$$+\frac{2}{3} = \frac{2}{3} \times \frac{5}{5} = \frac{10}{15}$$
$$\frac{16}{30} = \frac{8}{15}$$

74.

$$5\frac{1}{2} = 5 + \frac{1}{2} \times \frac{5}{5} = 5 + \frac{6}{10}$$
$$+2\frac{4}{5} = 2 + \frac{4}{5} \times \frac{2}{2} = 2 + \frac{6}{10}$$
$$7 + \frac{12}{10} = 8\frac{1}{5}$$

WRITING TO LEARN ▶▶▶▶

75. Create a word problem requiring you to add two mixed numbers. It must involve gallons of gasoline and driving to work. Then solve it, explaining each step you use in the solution.

76. You carry when you add whole numbers. Explain how you carry when you add fractions. Use your own example to show your point.

▶ YOU BE THE JUDGE

77. A perfect number (see Exercise 67, Section 3.1), equals the sum of all its divisors less than itself. Six is perfect because the divisors of 6 are 1, 2, 3, and 6, and 6 = 1 + 2 + 3.

a. Len claims that the sum of the reciprocals of all the divisors of the perfect number 6 equals 2. That is, $\frac{1}{1} + \frac{1}{2} + \frac{1}{3} + \frac{1}{6} = 2$. Is he correct? Support your decision with calculations.

b. Another perfect number is 28. Can the same assertion be made for 28? Support your decision with calculations.

4.5 SUBTRACTING FRACTIONS WITH UNLIKE DENOMINATORS

OBJECTIVES

1. Subtract unlike proper fractions.
2. Subtract unlike mixed numbers.
3. Subtract mixed numbers with borrowing.
4. Subtract by changing mixed numbers to improper fractions.

NEW VOCABULARY

borrowing

1 Subtracting Unlike Proper Fractions

EXAMPLE 1 Subtract $\frac{5}{8} - \frac{1}{3}$.

The LCD for 8 and 3 is 24.

Rewrite each fraction with an LCD of 24.
Then subtract the like fractions.

$$\begin{array}{r} \frac{5}{8} = \frac{5}{8} \times \frac{3}{3} = \frac{15}{24} \\ -\frac{1}{3} = \frac{1}{3} \times \frac{8}{8} = \frac{8}{24} \\ \hline \frac{7}{24} \end{array}$$ ■

To subtract fractions with unlike denominators

1. Find the LCD for the denominators.
2. Rewrite each fraction as an equivalent fraction with the LCD as denominator.
3. Subtract the like fractions. Reduce if needed.

EXAMPLE 2 Subtract $\frac{5}{6} - \frac{7}{10}$.

$$\begin{array}{rl} 6 = & (2) \times (3) \\ 10 = & 2 \qquad \times (5) \\ \hline \text{LCD} = & 2 \times 3 \times 5 \\ = & 30 \end{array}$$

$$\begin{array}{r} \frac{5}{6} = \frac{5}{6} \times \frac{5}{5} = \frac{25}{30} \\ -\frac{7}{10} = \frac{7}{10} \times \frac{3}{3} = \frac{21}{30} \\ \hline \frac{4}{30} = \frac{\overset{2}{\cancel{4}}}{\underset{15}{\cancel{30}}} = \frac{2}{15} \end{array}$$ ■

▶ **You Try It** 1. Subtract $\frac{3}{7} - \frac{1}{4}$. 2. Subtract $\frac{7}{8} - \frac{5}{6}$.

EXAMPLE 3 Subtract $\frac{7}{16} - \frac{3}{8}$.

The LCD for 16 and 8 is 16.

$$\begin{aligned} \frac{7}{16} &= &= \frac{7}{16} \\ -\frac{3}{8} &= \frac{3}{8} \times \frac{2}{2} &= \frac{6}{16} \\ \hline & & \frac{1}{16} \end{aligned}$$

■

EXAMPLE 4 Subtract $\frac{11}{24} - \frac{7}{30}$.

$$\begin{array}{r|rr} 2 & 24 & 30 \\ 3 & 12 & 15 \\ \hline & 4 & 5 \end{array}$$

$$\begin{aligned} \text{LCD} &= 2 \times 3 \times 4 \times 5 \\ &= 120 \end{aligned}$$

$$\begin{aligned} \frac{11}{24} &= \frac{11}{24} \times \frac{5}{5} = \frac{55}{120} \\ -\frac{7}{30} &= \frac{7}{30} \times \frac{4}{4} = \frac{28}{120} \\ \hline & \frac{27}{120} = \frac{\overset{9}{\cancel{27}}}{\underset{40}{\cancel{120}}} = \frac{9}{40} \end{aligned}$$

■

You Try It 3. Subtract $\frac{11}{15} - \frac{4}{9}$. 4. Subtract $\frac{13}{27} - \frac{17}{45}$.

---

2 Subtracting Unlike Mixed Numbers

EXAMPLE 5 Subtract $3\frac{2}{5} - 2\frac{1}{4}$.

The LCD for 5 and 4 is 20.

$$\begin{aligned} 3\frac{2}{5} &= 3\frac{2}{5} \times \frac{4}{4} = 3\frac{8}{20} \\ -2\frac{1}{4} &= -2\frac{1}{4} \times \frac{5}{5} = -2\frac{5}{20} \\ \hline & 1\frac{3}{20} \end{aligned}$$

■

Rewrite each proper fraction with an LCD of 20.

Subtract the proper fractions. Then subtract the whole numbers.

> To subtract mixed numbers
>
> 1. Find the LCD for the proper fractions.
> 2. Rewrite each proper fraction as an equivalent fraction with the LCD as the denominator.
> 3. Subtract the equivalent fractions.
> 4. Subtract the whole numbers.
> 5. Reduce if possible.

EXAMPLE 6 Subtract $24\frac{5}{6} - 17\frac{3}{8}$.

The LCD for 6 and 8 is 24.

$$\begin{aligned} 24\frac{5}{6} &= 24\frac{5}{6} \times \frac{4}{4} = 24\frac{20}{24} \\ -17\frac{3}{8} &= -17\frac{3}{8} \times \frac{3}{3} = -17\frac{9}{24} \\ \hline & \qquad\qquad\qquad 7\frac{11}{24} \end{aligned}$$
■

▶ **You Try It** **5.** Subtract $4\frac{2}{3} - 2\frac{1}{8}$. **6.** $7\frac{9}{10} - 2\frac{4}{15}$.

---

3 Borrowing

The next three examples demonstrate **borrowing** a 1 from the whole number in order to subtract.

EXAMPLE 7 Subtract $5\frac{1}{8} - 2\frac{5}{8}$.

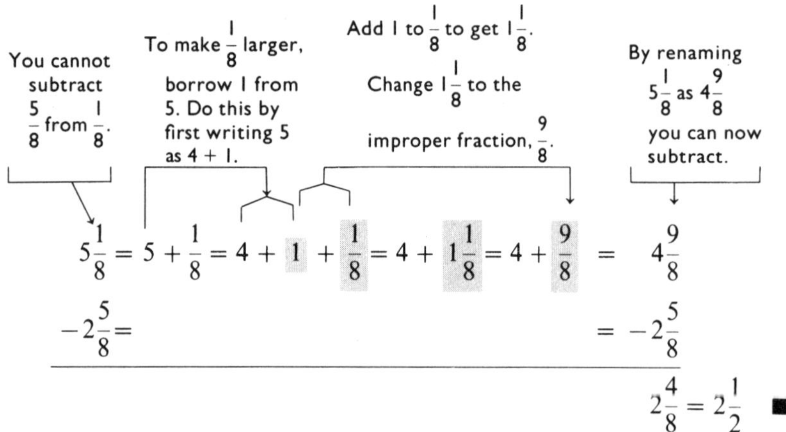

$$\begin{aligned} 5\frac{1}{8} &= 5 + \frac{1}{8} = 4 + 1 + \frac{1}{8} = 4 + 1\frac{1}{8} = 4 + \frac{9}{8} = 4\frac{9}{8} \\ -2\frac{5}{8} &= \qquad\qquad\qquad\qquad\qquad\qquad\quad = -2\frac{5}{8} \\ \hline & \qquad\qquad\qquad\qquad\qquad\qquad\quad 2\frac{4}{8} = 2\frac{1}{2} \end{aligned}$$
■

EXAMPLE 8 Subtract $8\frac{1}{6} - 5\frac{7}{9}$.

The LCD is 18.

$$\begin{aligned} 8\frac{1}{6} &= 8\frac{1}{6} \times \frac{3}{3} = 8\frac{3}{18} = 7\frac{21}{18} \\ -5\frac{7}{9} &= 5\frac{7}{9} \times \frac{2}{2} = 5\frac{14}{18} = 5\frac{14}{18} \\ \hline & \qquad\qquad\qquad\qquad\quad 2\frac{7}{18} \end{aligned}$$
■

You can't subtract $\frac{14}{18}$ from $\frac{3}{18}$. To make $\frac{3}{18}$ larger, borrow 1 from the 8, leaving 7. Add the 1 to $\frac{3}{18}$ to get $1\frac{3}{18} = \frac{21}{18}$. Now you can subtract.

▶ **You Try It** **7.** Subtract $8\frac{4}{9} - 2\frac{7}{9}$. **8.** Subtract $2\frac{3}{10} - 1\frac{3}{4}$.

EXAMPLE 9 Find $7 - 2\frac{3}{5}$.

$$\begin{array}{rcl} 7 & = & 6\frac{5}{5} \\ -2\frac{3}{5} & = & -2\frac{3}{5} \\ \hline & & 4\frac{2}{5} \end{array}$$

To subtract, first write a fraction above $\frac{3}{5}$. To do this, borrow 1 from 7. This leaves 6. Write the 1 as $\frac{5}{5}$, using the same denominator as $\frac{3}{5}$. Now you can subtract the like fractions. ■

▶ **You Try It** 9. Subtract $15 - 9\frac{5}{12}$.

---

4 Subtracting Mixed Numbers by First Changing Them to Improper Fractions

You can subtract two mixed numbers by first writing both as improper fractions. You do not have to borrow 1 from the whole number if you use this method.

EXAMPLE 10 Subtract by changing to improper fractions. $4\frac{1}{8} - 1\frac{2}{3}$

Write each mixed number as an improper fraction.

Write each fraction in terms of the LCD = 24. Then subtract.

$$\begin{array}{rcccl} 4\frac{1}{8} = & \frac{33}{8} = & \frac{33}{8} \times \frac{3}{3} = & \frac{99}{24} \\ -1\frac{2}{3} = & \frac{5}{3} = & \frac{5}{3} \times \frac{8}{8} = & \frac{40}{24} \\ \hline & & & \frac{59}{24} = 2\frac{11}{24} \end{array}$$ ■

This method will always give the correct answer. The disadvantage is having to work with larger numerators.

EXAMPLE 11 Subtract $8\frac{11}{12} - 5\frac{7}{9}$.

$$\begin{array}{rcccl} 8\frac{11}{12} = & \frac{107}{12} = & \frac{107}{12} \times \frac{3}{3} = & \frac{321}{36} \\ -5\frac{7}{9} = & \frac{52}{9} = & \frac{52}{9} \times \frac{4}{4} = & \frac{208}{36} \\ \hline & & & \frac{113}{36} = 3\frac{5}{36} \end{array}$$ ■

▶ **You Try It** Subtract by changing to improper fractions.

10. $3\frac{1}{2} - 1\frac{3}{5}$ **11.** $20\frac{1}{6} - 12\frac{5}{8}$

▶ **Answers to You Try It** 1. $\frac{5}{28}$ 2. $\frac{1}{24}$ 3. $\frac{13}{45}$ 4. $\frac{14}{135}$ 5. $2\frac{13}{24}$ 6. $5\frac{19}{30}$ 7. $5\frac{2}{3}$ 8. $\frac{11}{20}$ 9. $5\frac{7}{12}$ 10. $1\frac{9}{10}$ 11. $7\frac{13}{24}$

SECTION 4.5 EXERCISES

1 *Subtract the fractions. Reduce all answers to lowest terms.*

1. $\frac{3}{4} - \frac{1}{2}$
2. $\frac{5}{6} - \frac{2}{3}$
3. $\frac{7}{8} - \frac{1}{4}$
4. $\frac{4}{5} - \frac{3}{10}$
5. $\frac{3}{4} - \frac{3}{8}$
6. $\frac{9}{10} - \frac{2}{5}$
7. $\frac{1}{3} - \frac{1}{6}$
8. $\frac{1}{2} - \frac{1}{4}$
9. $\frac{1}{6} - \frac{1}{8}$
10. $\frac{8}{9} - \frac{5}{6}$
11. $\frac{5}{8} - \frac{1}{10}$
12. $\frac{7}{8} - \frac{1}{2}$
13. $\frac{11}{12} - \frac{3}{8}$
14. $\frac{9}{14} - \frac{2}{7}$
15. $\frac{5}{9} - \frac{5}{12}$
16. $\frac{7}{12} - \frac{3}{20}$
17. $\frac{23}{36} - \frac{5}{9}$
18. $\frac{5}{16} - \frac{9}{32}$
19. $\frac{17}{18} - \frac{5}{12}$
20. $\frac{1}{2} - \frac{7}{30}$
21. $\frac{5}{7} - \frac{2}{5}$
22. $\frac{8}{13} - \frac{2}{9}$
23. $\frac{20}{27} - \frac{5}{18}$
24. $\frac{49}{60} - \frac{24}{36}$
25. $\frac{24}{35} - \frac{8}{21}$
26. $\frac{7}{24} - \frac{2}{15}$
27. $\frac{18}{24} - \frac{9}{18}$
28. $\frac{45}{80} - \frac{16}{40}$
29. $\frac{28}{35} - \frac{15}{25}$
30. $\frac{16}{21} - \frac{15}{28}$
31. $\frac{11}{30} - \frac{5}{36}$
32. $\frac{27}{50} - \frac{9}{40}$

2 **3** **4** *Subtract the mixed numbers. Reduce answers to lowest terms.*

33. $3\frac{2}{3} - 1\frac{1}{6}$
34. $4\frac{3}{4} - 1\frac{5}{8}$
35. $8\frac{4}{7} - 5\frac{1}{3}$
36. $11\frac{4}{5} - 2\frac{1}{4}$
37. $6\frac{1}{7} - 2\frac{5}{7}$
38. $8\frac{1}{9} - 3\frac{7}{9}$
39. $5\frac{7}{8} - 3\frac{3}{4}$
40. $4\frac{5}{9} - 1\frac{1}{3}$
41. $12\frac{3}{4} - 4\frac{2}{3}$
42. $5\frac{2}{3} - 4\frac{3}{5}$
43. $6\frac{3}{8} - 4\frac{7}{12}$
44. $5\frac{1}{6} - 2\frac{7}{10}$
45. $5\frac{1}{2} - \frac{3}{4}$
46. $4\frac{5}{12} - \frac{7}{8}$
47. $6\frac{5}{9} - \frac{1}{3}$
48. $2\frac{4}{5} - \frac{4}{9}$
49. $5\frac{1}{6} - 3\frac{5}{6}$
50. $8\frac{15}{21} - 5\frac{20}{21}$
51. $16\frac{3}{8} - 1\frac{3}{4}$
52. $20\frac{2}{9} - 5\frac{7}{27}$

53. $4 - 3\frac{5}{7}$

54. $13 - 8\frac{5}{31}$

55. $7\frac{2}{5} - 4\frac{3}{8}$

56. $16\frac{5}{8} - 12\frac{5}{18}$

57. $25\frac{1}{2} - 13\frac{3}{5}$

58. $50\frac{5}{7} - 40\frac{13}{14}$

59. $15\frac{7}{9} - 8$

60. $16\frac{13}{20} - 11$

61. $8\frac{16}{25} - 8\frac{3}{10}$

62. $7\frac{4}{15} - 7\frac{2}{9}$

63. $9\frac{5}{6} - 2\frac{8}{21}$

64. $60\frac{2}{5} - 4\frac{1}{8}$

65. $24 - 5\frac{7}{12}$

66. $50 - 37\frac{1}{2}$

67. $7\frac{5}{12} - 4\frac{4}{7}$

68. $15\frac{6}{13} - 3\frac{3}{4}$

69. $8\frac{3}{7} - 4\frac{6}{14}$

70. $4\frac{15}{18} - 4\frac{10}{12}$

SKILLSFOCUS (Section 1.2) *Solve each problem.*

71. What is the sum of 35 and 26?

72. Find 10 increased by 3.

73. Subtract 37 from 52.

74. What is 9 less than 9?

EXTEND YOUR THINKING ▶▶▶▶

▶ TROUBLESHOOT IT

Find and correct the error.

75.
$$\begin{array}{rl} 3\frac{1}{5} = 3\frac{1}{5} \times \frac{8}{8} = 3\frac{8}{40} = & 2\frac{18}{40} \\ -1\frac{3}{8} = 1\frac{3}{8} \times \frac{5}{5} = 1\frac{15}{40} = & 1\frac{15}{40} \\ \hline & 1\frac{3}{40} \end{array}$$

76. $\frac{5}{8} - \frac{2}{5} = \frac{5-2}{8-5} = \frac{3}{3} = 1$

WRITING TO LEARN ▶▶▶▶

77. Create a subtraction problem where you must borrow. Solve the problem. Explain why you borrow, and show how it is done.

78. How is borrowing in the whole number subtraction 25 – 19 like borrowing in the fraction subtraction $3\frac{1}{4} - 1\frac{2}{3}$? How is it different?

4.6 APPLICATIONS: ADDITION AND SUBTRACTION OF FRACTIONS

OBJECTIVE

Solve word problems.

Solving Word Problems

EXAMPLE 1 Yesterday one share of XYZ stock sold for $\$27\frac{1}{2}$. Today the price for one share increased by $\$2\frac{5}{8}$. What is the new price for one share?

The phrase *increased by* indicates addition.

yesterday's price:	$27\frac{1}{2} = 27 + \frac{1}{2} \times \frac{4}{4} =$	$27\frac{4}{8}$
increase in price:	$+2\frac{5}{8} =$	$= +2\frac{5}{8}$
today's price:		$\$29\frac{9}{8} = \$30\frac{1}{8}$ per share ■

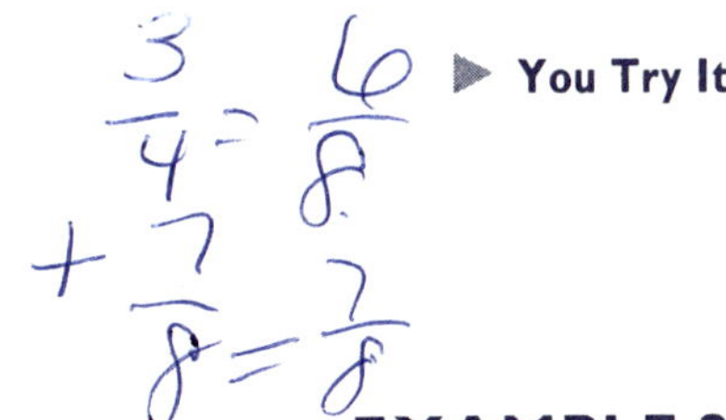

▶ **You Try It** 1. Yesterday 1 share of TRW stock sold for $\$64\frac{3}{4}$. Today the price for one share increased by $\$1\frac{7}{8}$. What is the current price per share?

EXAMPLE 2 Three pieces of clothing require 3, $1\frac{3}{4}$, and $\frac{5}{6}$ yards of material, respectively.

a. How many yards of material are needed to make the clothing?

To find the total amount of material needed, add the three amounts together. The LCD for 4 and 6 is 12.

Write 3 yards in the whole number column.

$$
\begin{aligned}
3 &= 3 && = 3 \\
1\frac{3}{4} &= 1 + \frac{3}{4} \times \frac{3}{3} && = 1\frac{9}{12} \\
+\frac{5}{6} &= \frac{5}{6} \times \frac{2}{2} && = +\frac{10}{12} \\
\hline
& && 4\frac{19}{12} = 5\frac{7}{12} \text{ yards of material needed}
\end{aligned}
$$

b. What do you pay for the material if it costs $15 per yard?

You want to buy $5\frac{7}{12}$ yards of material *at* $15 per yard. The word *at* indicates multiplication.

$$5\frac{7}{12} \text{ yds } at \text{ \$15 per yard} = 5\frac{7}{12} \times 15 = \frac{67}{\cancel{12}_{4}} \times \frac{\cancel{15}^{5}}{1} = \$83\frac{3}{4} \quad ■$$

You Try It 2. Three pieces of clothing require 8, $3\frac{5}{8}$, and $\frac{7}{10}$ yards of material, respectively.

a. How many total yards of material are needed?
b. What is the cost of the material at $40 per yard?

EXAMPLE 3 A newly designed cylinder has a volume of $63\frac{4}{9}$ cubic inches. The old model had a volume of $55\frac{7}{10}$ cubic inches. How much more volume does the new cylinder have?

To find the difference in the two volumes, subtract them. The LCD for 9 and 10 is 90.

$$\begin{array}{lrl} \text{new model} & 63\frac{4}{9} = 63 + \frac{4}{9} \times \frac{10}{10} = 63\frac{40}{90} = & 62\frac{130}{90} \\ \text{old model} & -55\frac{7}{10} = 55 + \frac{7}{10} \times \frac{9}{9} = 55\frac{63}{90} = & -55\frac{63}{90} \\ \hline & & 7\frac{67}{90} \text{ cubic inches difference} \end{array}$$

← Borrow 1 from 63. $1\frac{40}{90} = \frac{130}{90}$

The new model cylinder has $7\frac{67}{90}$ cubic inches more volume. ■

You Try It 3. A newly designed freezer has a volume of $10\frac{1}{6}$ cubic feet. The old model had a volume of $9\frac{2}{5}$ cubic feet. How much more volume does the new model have?

EXAMPLE 4 **Chapter Problem** Read the problem at the beginning of this chapter.

Donna purchased 400 shares at $\$27\frac{3}{4}$ each. She then sold them for $\$30\frac{3}{8}$ each. Finding her gain is a two-part problem.

Step #1: Find the gain for 1 share. This is $\$30\frac{3}{8} - \$27\frac{3}{4}$.

$$\begin{array}{rl} 30\frac{3}{8} = \qquad\qquad\qquad 30\frac{3}{8} = & 29\frac{11}{8} \\ -27\frac{3}{4} = 27 + \frac{3}{4} \times \frac{2}{2} = 27\frac{6}{8} = & -27\frac{6}{8} \\ \hline & \$2\frac{5}{8} \text{ gain per share} \end{array}$$

← Borrow 1 from 30. $1\frac{3}{8} = \frac{11}{8}$

Step #2: The gain on one share is $\$2\frac{5}{8}$. The gain on 400 shares is 400 times this amount.

$$400 \times \$2\frac{5}{8} = \frac{\overset{50}{\cancel{400}}}{1} \times \frac{21}{\underset{1}{\cancel{8}}} = \frac{50 \times 21}{1 \times 1} = \$1{,}050$$

Donna gained $1,050 on this transaction. ■

▶ **You Try It** 4. Read the problem at the beginning of this chapter. Danielle bought 400 shares of Sears stock at the High. She sold them at the Low. How much did she lose on this transaction?

EXAMPLE 5 It takes Susan 4 hours to paint a room. It takes Beth 6 hours to paint the same room. If they work together, how fast can they paint the room?

Step #1: If Susan can paint the room in 4 hours, in one hour she can paint $\frac{1}{4}$ of it.

If Beth can paint the room in 6 hours, in one hour she can paint $\frac{1}{6}$ of it.

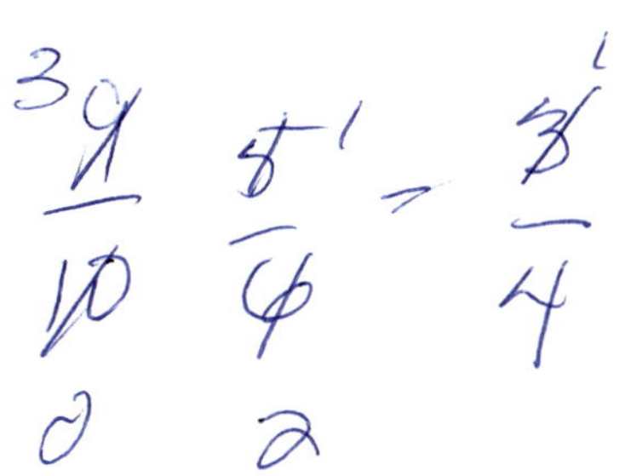

If they work together, in one hour they can paint $\frac{1}{4}+\frac{1}{6}$ of the room.

$$\frac{1}{4}=\frac{1}{4}\times\frac{3}{3}=\frac{3}{12}$$
$$+\frac{1}{6}=\frac{1}{6}\times\frac{2}{2}=\frac{2}{12}$$
$$\frac{5}{12}$$

In one hour they can paint $\frac{5}{12}$ of the room.

Step #2: To find how many hours it takes to paint 1 whole room, find how many $\frac{5}{12}$ are in 1 whole.

How many $\frac{5}{12}$ are in $1 = 1 \div \frac{5}{12} = \frac{1}{1}\times\frac{12}{5}=\frac{12}{5}=2\frac{2}{5}$ hrs

Working together, Susan and Beth can paint the room in $2\frac{2}{5}$ hours.

OBSERVE The time it takes them to paint the room $\left(\frac{12}{5}\right)$ is the reciprocal of the fraction of the room they can paint in one hour $\left(\frac{5}{12}\right)$. ■

▶ **You Try It** 5. It takes Tom 5 hours to paint a room. It takes Steve 8 hours to paint the same room. If they work together, how fast can they paint the room?

▶ Answers to You Try It 1. $\$66\frac{5}{8}$ per share 2. a. $12\frac{13}{40}$ yards b. \$493 3. $\frac{23}{30}$ cubic feet 4. \$350 5. $3\frac{1}{13}$ hours

SECTION 4.6 EXERCISES

Solve each word problem. Reduce answers to lowest terms.

1. Ron purchased $16\frac{9}{10}$ gallons of gas this morning and another $13\frac{1}{5}$ gallons this afternoon. How much gasoline did Ron buy today?

2. In one week a family drank $3\frac{4}{5}$ gallons of milk. They drank $4\frac{2}{3}$ gallons the following week. How much milk did they drink for the two weeks?

3. A nail $3\frac{1}{6}$ inches long is driven into a board $2\frac{5}{8}$ inches thick. How much of the nail will protrude from the other side of the board?

4. A share of PG stock sold for $\$62\frac{1}{8}$ one month ago. Today it sells for $\$15\frac{3}{4}$ less. What is today's price for one share of PG?

5. A carpenter needs $22\frac{5}{8}$ feet of wood to build a table and $16\frac{7}{12}$ feet to build a bench. How much wood is needed altogether?

6. A sofa requires $16\frac{5}{6}$ yards of material to be reupholstered. A love seat requires $10\frac{3}{4}$ yards. How much material is needed for the two pieces?

7. A tank contains $7\frac{1}{2}$ gallons of gas.

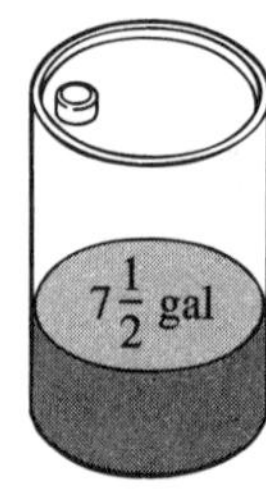

   a. How much more gas is needed to fill the tank?

   b. What will this additional gas cost if one gallon costs $\$1\frac{1}{5}$?

8. A tractor requires $8\frac{2}{3}$ quarts of oil. It now holds $5\frac{1}{2}$ quarts.

   a. How many more quarts of oil does the tractor need?

   b. What will this additional oil cost if one quart runs for $\$1\frac{1}{2}$?

9. One share of BMI sold for $\$46\frac{3}{4}$ at 10 A.M. One share sold for $\$1\frac{3}{8}$ more at 3 P.M. the same day. Sandra owns 160 shares.

   a. How much did one share sell for at 3 P.M.?

   b. How much more were Sandra's shares worth at 3 P.M. compared to 10 A.M.?

10. A share of XRT stock sold for $\$24\frac{5}{8}$ yesterday. Today one share is selling for $\$5\frac{1}{4}$ less. James owns 1,200 shares.

a. How much is one share selling for today?

b. How much less are James's shares worth today compared to yesterday?

11. A recipe calls for $\frac{1}{3}$ cup of walnuts, $\frac{3}{4}$ cups of almonds, and $\frac{1}{2}$ cup of peanuts. How many cups of nuts are used in the recipe?

12. A new business purchased $\frac{5}{8}$ of a page of advertising in Monday's paper, $\frac{5}{12}$ of a page in Tuesday's paper, and $\frac{1}{4}$ of a page in Wednesday's paper. How many pages of advertising did the business buy?

13. What must be added to $3\frac{2}{5}$ to get $5\frac{1}{6}$?

14. What must $\frac{2}{3}$ be increased by to get $\frac{9}{10}$?

15. Subtract $\frac{5}{12}$ from $\frac{1}{2}$.

16. Subtract $\frac{3}{8}$ from $\frac{11}{15}$.

17. Hector purchased 12 gallons of punch for a party. By midnight, $3\frac{2}{5}$ gallons were left. How much punch was consumed by midnight?

18. A $5\frac{7}{12}$-foot piece of wood is cut from a 16-foot board. What length of wood is left?

19. Two boards each $\frac{7}{8}$ inches thick are stacked on a third board $2\frac{1}{4}$ inches thick. How thick is the stack?

20. A bedroom required $2\frac{1}{3}$ gallons of paint. An adjoining sitting room needed $1\frac{3}{4}$ gallons. How much paint did the two rooms need?

21. Al owns $\frac{1}{3}$ of a business. Bob owns $\frac{2}{5}$. Cy owns the rest. What fraction of the business does Cy own?

22. In a student election, $\frac{1}{2}$ of the people voted for Cheryl. $\frac{3}{7}$ voted for Jayne. The rest abstained. What fraction of the voters abstained?

23. The City Fair record for eating ice cream is $9\frac{1}{8}$ quarts in one hour. In a practice session George ate $7\frac{2}{3}$ quarts. How many quarts was George short of the record?

24. A customer purchased $42\frac{7}{10}$ yards of wire from a roll that was 210 yards long. How much wire was left on the roll?

25. Tom can paint a room in 8 hours. Ed can paint it in 10 hours. How long will it take them to paint the room if they work together?

26. Pat can erect a jungle gym in 12 hours. Nancy can do it in 9 hours. It takes Rowan 10 hours. How long will it take them to erect the jungle gym if they work together?

27. Find the missing dimensions for the figure below.

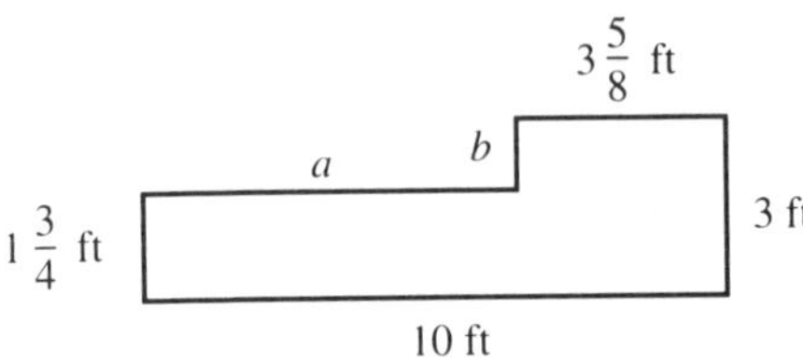

28. What is the inside diameter of the pipe in the figure?

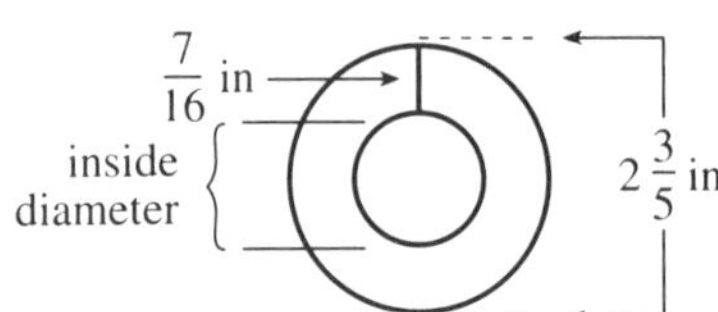

29. Anna inflated 60 balloons for a party. Of these, $\frac{2}{5}$ were popped, $\frac{4}{15}$ were used as party favors, and $\frac{1}{6}$ were given to the children next door. The rest are hanging from the chandelier.

a. What fraction of the balloons are hanging from the chandelier?

b. How many balloons are hanging?

30. Jayne spent $\frac{1}{3}$ of her day working, $\frac{1}{4}$ sleeping, and $\frac{1}{6}$ with her family. The rest was devoted to attending classes and doing homework.

a. What fraction of Jayne's day was devoted to her education?

b. How many hours is this out of a 24-hour day?

SKILLSFOCUS (Section 2.6) *Simplify each expression.*

31. $4 + 3 \times 5$

32. $25 \div 5 \times 5$

33. $4(9 - 2)$

34. $5 \cdot 2^2$

EXTEND YOUR THINKING ▶▶▶

35. A cook has four cup measures of $1, \frac{1}{2}, \frac{1}{3}$, and $\frac{1}{4}$ cups, respectively. Describe the steps the cook uses to measure

a. exactly $\frac{1}{6}$ of a cup of oil.

b. exactly $\frac{1}{12}$ of a cup of lime juice.

36. **Foresight in the Stock Market** Sheila owns 480 shares of TLQ stock. Anticipating a downturn in the market soon, she sold them for $\$45\frac{3}{8}$ per share. The next week the stock market did turn down, and one share of TLQ dropped to $\$30\frac{1}{4}$. Sheila then spent all the money from the previous week's sale (assume no commissions or taxes were deducted from her sale) and bought all TLQ stock at the lower price.

a. How many shares of TLQ does Sheila now own?

b. In six months, the market recovered to the point that TLQ was again selling for $\$45\frac{3}{8}$ per share. By how much did the value of Sheila's investment in TLQ stock increase as a result of her anticipating the market downturn?

37. A father owns 5 acres of land. The land is in the shape of a square. He keeps $\frac{1}{4}$ of it for himself. He splits the rest among his four sons.

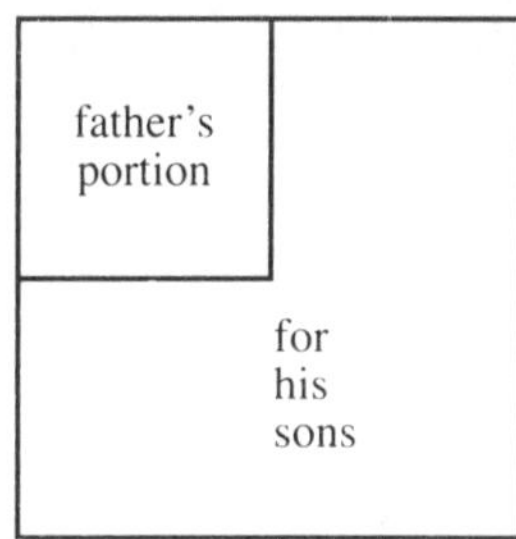

a. How many acres does each son get?

b. How does the father subdivide the land so that each son gets a piece with the exact same size and shape? (Hint: Divide the entire square into sixteenths.)

WRITING TO LEARN ▶▶▶▶

38. Write a word problem whose answer is $\$12\frac{3}{4}$. The problem must be about food, and must include a subtraction of fractions.

▶ YOU BE THE JUDGE

39. George flew his single-engine plane from Baltimore to Richmond and used $\frac{3}{8}$ of a tank of gas. He used another $\frac{1}{6}$ of a tank to get from Richmond to Norfolk. At the halfway point in his flight from Norfolk directly back to Baltimore, he used another $\frac{11}{24}$ of a tank of gas. George started with a full tank, and has not refueled since he left Baltimore. Does he have reason to be alarmed? Explain your decision.

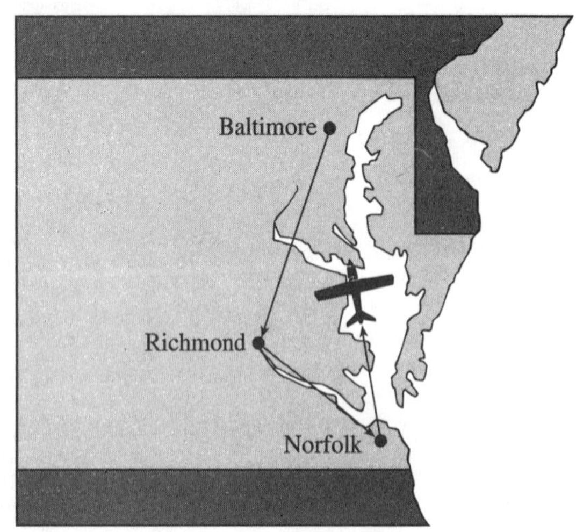

4.7 COMBINED OPERATIONS; ORDER

OBJECTIVES

1. Apply the order of operations to fractions.
2. Order fractions.
3. Solve problems involving fractions and decimals.

NEW VOCABULARY

simple fraction

complex fraction

1 Applying the Order of Operations to Fractions

The order of operations applied to whole numbers in Section 2.6 also applies to fractions. You may insert parentheses to highlight the operation to be performed first.

EXAMPLE 1

$$\frac{2}{3}+\frac{1}{8}\times\frac{4}{5}$$

$$=\frac{2}{3}+\left(\frac{1}{\cancel{8}_{2}}\times\frac{\cancel{4}^{1}}{5}\right)$$

Multiply before you add. You may insert parentheses around the multiplication to indicate it is done first.

$$=\frac{2}{3}+\frac{1}{10}$$

Add. The LCD for 3 and 10 is 30.

$$=\frac{20}{30}+\frac{3}{30}=\frac{23}{30}$$ ■

EXAMPLE 2

$$\frac{4}{27}\div\frac{5}{9}\times\frac{3}{5}$$

$$=\left(\frac{4}{\cancel{27}_{3}}\times\frac{\cancel{9}^{1}}{5}\right)\times\frac{3}{5}$$

Multiply and divide in order from left to right. Insert parentheses to indicate you divide first.

$$=\frac{4}{\cancel{15}_{5}}\times\frac{\cancel{3}^{1}}{5}=\frac{4}{25}$$ ■

EXAMPLE 3

$$3\frac{1}{2}\left(\frac{3}{4}+\frac{7}{8}\right)$$

$$=\frac{7}{2}\cdot\left(\frac{6}{8}+\frac{7}{8}\right)$$

Write the dot for implied multiplication. Simplify within parentheses first. The LCD = 8.

$$=\frac{7}{2}\cdot\frac{13}{8}=\frac{91}{16}\text{ or }5\frac{11}{16}$$ ■

▶ **You Try It**

1. $\frac{5}{6}+\frac{3}{4}\cdot\frac{2}{9}$
2. $\frac{9}{10}\div\frac{6}{5}\cdot\frac{5}{6}$
3. $6\frac{3}{5}\left(\frac{4}{5}+\frac{7}{10}\right)$

EXAMPLE 4

$$\left(2\frac{1}{3}\right)^2 - 3 \cdot 1\frac{3}{5}$$

$$= \left(\frac{7}{3}\right)^2 - \frac{3}{1} \cdot \frac{8}{5}$$ Write the mixed numbers as improper fractions. Evaluate the power.

$$= \frac{49}{9} - \frac{3}{1} \cdot \frac{8}{5}$$ Multiply.

$$= \frac{49}{9} - \frac{24}{5}$$ Subtract. The LCD for 9 and 5 is 45.

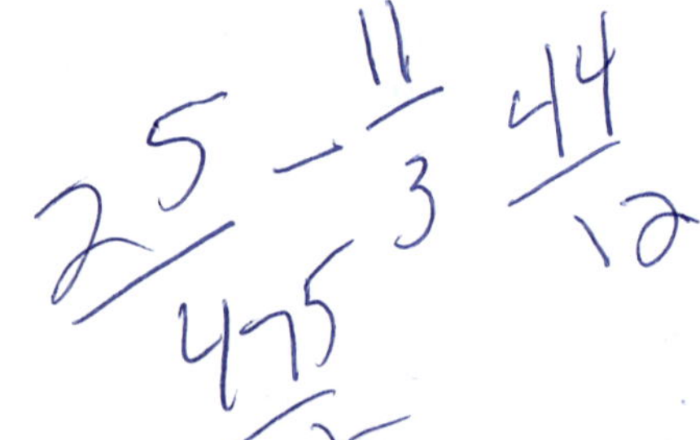

$$= \frac{49}{9} \cdot \frac{5}{5} - \frac{24}{5} \cdot \frac{9}{9}$$

$$= \frac{245}{45} - \frac{216}{45} = \frac{29}{45}$$ ■

EXAMPLE 5

$$2 \cdot 3 + 5 \div 10$$

$$= 2 \cdot 3 + \frac{5}{10} = 6 + \frac{\overset{1}{\cancel{5}}}{\underset{2}{\cancel{10}}} = 6\frac{1}{2}$$ ■

▶ **You Try It** 4. $\left(2\frac{1}{2}\right)^2 - 3 \cdot 1\frac{2}{9}$

5. $(4 \cdot 5) + (10 \div 6)$

EXAMPLE 6 Evaluate $16t + 3$ when $t = \frac{3}{4}$.

$$16t + 3 = 16 \cdot t + 3$$ Write the dot for the implied multiplication.

$$= 16 \cdot \frac{3}{4} + 3$$ Substitute $\frac{3}{4}$ for t.

$$= \left(\frac{\overset{4}{\cancel{16}}}{1} \cdot \frac{3}{\underset{1}{\cancel{4}}}\right) + 3$$

$$= 12 + 3 = 15$$ ■

▶ **You Try It** 6. Evaluate $5m - 2$ when $m = 1\frac{3}{8}$.

A **simple fraction** has whole numbers for its numerator and denominator. A **complex fraction** has a fraction in its numerator, or denominator, or both. Therefore, a complex fraction has more than one fraction bar.

Recall that $\frac{12}{6}$ means $12 \div 6$. If you replace the numerator 12 with $\frac{3}{4}$ and the denominator 6 with $\frac{2}{5}$, you can write the following complex fraction.

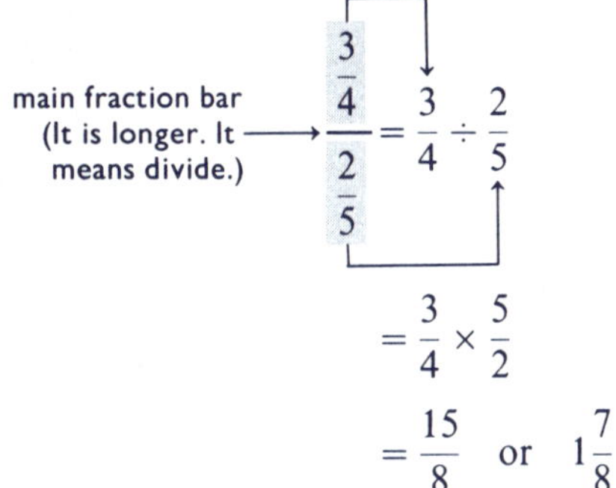

$$\frac{\frac{3}{4}}{\frac{2}{5}} = \frac{3}{4} \div \frac{2}{5} = \frac{3}{4} \times \frac{5}{2} = \frac{15}{8} \text{ or } 1\frac{7}{8}$$

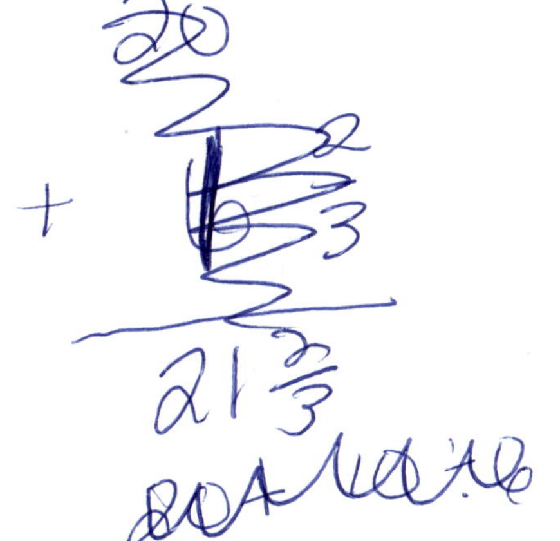

OBSERVE In short, when a fraction is in the denominator, multiply the numerator by the reciprocal of the denominator.

$$\frac{\frac{3}{4}}{\frac{2}{5}} = \frac{3}{4} \times \frac{5}{2}$$

Multiply by reciprocal.

EXAMPLE 7

$$\frac{8}{\frac{4}{5}} = 8 \div \frac{4}{5} = \frac{\overset{2}{\cancel{8}}}{1} \cdot \frac{5}{\underset{1}{\cancel{4}}} = 10 \quad \blacksquare$$

EXAMPLE 8

$$\frac{6\frac{1}{4}}{\frac{3}{10}} = 6\frac{1}{4} \div \frac{3}{10} = \frac{25}{\underset{2}{\cancel{4}}} \cdot \frac{\overset{5}{\cancel{10}}}{3} = \frac{125}{6} \text{ or } 20\frac{5}{6} \quad \blacksquare$$

▶ **You Try It** Simplify.

7. $\dfrac{\frac{7}{10}}{\frac{3}{5}}$

8. $\dfrac{6}{\frac{7}{9}}$

9. $\dfrac{2\frac{5}{12}}{\frac{1}{3}}$

EXAMPLE 9 Three bottles contain $\frac{5}{8}$ ounce, $1\frac{1}{2}$ ounces, and $\frac{3}{4}$ ounce of the same drug. If a pharmacist combines the contents of the three bottles, how many $\frac{1}{4}$-ounce doses of the drug can she make?

Step #1: Find the total number of ounces of the drug.

$$\text{total ounces of drug} = \frac{5}{8} + 1\frac{1}{2} + \frac{3}{4} = \frac{5}{8} + \frac{3}{2} + \frac{3}{4} = \frac{5}{8} + \frac{12}{8} + \frac{6}{8} = \frac{23}{8} \text{ oz}$$

Step #2: To find how many $\frac{1}{4}$-ounce doses can be made, divide this sum by $\frac{1}{4}$.

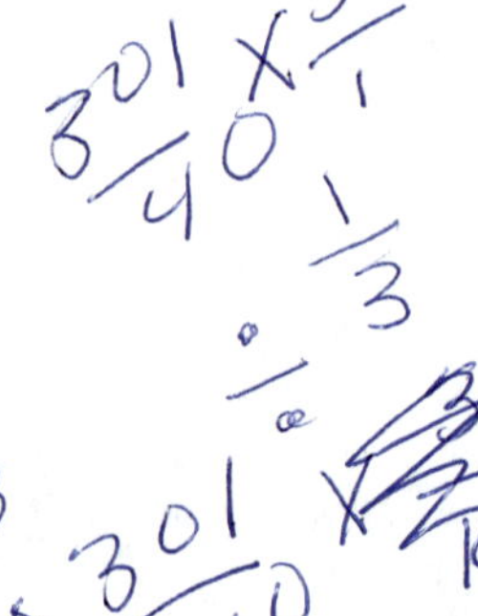

$$\text{number of } \frac{1}{4}\text{-ounce doses} = \frac{23}{8} \div \frac{1}{4} = \frac{23}{\cancel{8}_2} \cdot \frac{\cancel{4}^1}{1} = \frac{23}{2} \quad \text{or} \quad 11\frac{1}{2}$$

Note, $11\frac{1}{2}$ doses = 11 doses + $\frac{1}{2}$ of a dose. 11 doses of $\frac{1}{4}$ ounces each can be made. There is only enough drug left over to make $\frac{1}{2}$ of a twelfth dose. The answer is 11 doses. ■

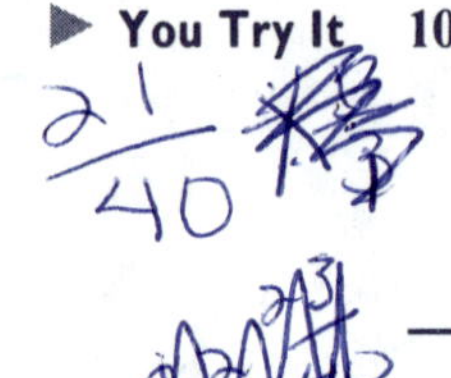

▶ **You Try It** **10.** Three bottles contain $\frac{9}{10}$ ounce, $4\frac{3}{4}$ ounces, and $1\frac{7}{8}$ ounces of the same drug. How many $\frac{1}{3}$-ounce doses of the drug can be made if the contents of the three bottles are combined?

2 Ordering Fractions

To order fractions, first write each fraction with the same common denominator. Then order the numerators.

EXAMPLE 10 Which fraction is larger, $\frac{4}{7}$ or $\frac{3}{5}$?

Rewrite each fraction in terms of the LCD 35.

$$\frac{4}{7} = \frac{4}{7} \cdot \frac{5}{5} = \frac{20}{35}$$

$$\frac{3}{5} = \frac{3}{5} \cdot \frac{7}{7} = \frac{21}{35}$$

Compare the numerators. Since $\frac{20}{35} < \frac{21}{35}$, then $\frac{4}{7} < \frac{3}{5}$. $\frac{3}{5}$ is larger. ■

> Given two or more fractions, to put them in order
>
> 1. Rewrite each fraction as an equivalent fraction with the same common denominator.
> 2. Compare the numerators. The largest fraction has the largest numerator. The smallest fraction has the smallest numerator.

EXAMPLE 11 Arrange the fractions $\frac{11}{14}, \frac{5}{7}$, and $\frac{3}{4}$ in order from smallest to largest.

Find the LCD.

$$\begin{array}{r|rrr} 7 & 14 & 7 & 4 \\ \hline 2 & 2 & 1 & 4 \\ \hline & 1 & 1 & 2 \end{array}$$

$$\text{LCD} = 7 \times 2 \times 1 \times 1 \times 2$$
$$= 28$$

Rewrite each fraction in terms of the LCD 28.

$$\frac{11}{14} = \frac{11}{14} \cdot \frac{2}{2} = \frac{22}{28}$$

$$\frac{5}{7} = \frac{5}{7} \cdot \frac{4}{4} = \frac{20}{28}$$

$$\frac{3}{4} = \frac{3}{4} \cdot \frac{7}{7} = \frac{21}{28}$$

Compare the numerators. Since $\frac{20}{28} < \frac{21}{28} < \frac{22}{28}$, then $\frac{5}{7} < \frac{3}{4} < \frac{11}{14}$. ■

▶ **You Try It** Order the fractions from smallest to largest.

11. $\frac{3}{4}$ and $\frac{7}{10}$

12. $\frac{2}{5}, \frac{3}{10}$, and $\frac{7}{25}$

---

3 Solving Problems Involving Fractions and Decimals

To add or subtract a fraction and a decimal, first change the fraction to a decimal, or change the decimal to a fraction.

EXAMPLE 12 What is $0.4 + \frac{1}{2}$?

Solution #1: Change 0.4 to the fraction $\frac{4}{10}$. Then add.

$$0.4 + \frac{1}{2} = \frac{4}{10} + \frac{1}{2} = \frac{4}{10} + \frac{1}{2} \cdot \frac{5}{5} = \frac{4}{10} + \frac{5}{10} = \frac{9}{10} \quad \text{or} \quad 0.9$$

Solution #2: Change $\frac{1}{2}$ to the decimal 0.5. Then add.

$$0.4 + \frac{1}{2} = 0.4 + 0.5 = 0.9 \quad \text{or} \quad \frac{9}{10}$$ ■

OBSERVE The answers to both solutions are the same.

▶ **You Try It** **13.** $0.6 + \frac{3}{4}$ **14.** $3.91 + \frac{7}{10}$

To multiply a fraction and a decimal, first write the decimal over 1. Then multiply as you would multiply two fractions.

EXAMPLE 13 A mechanic charges \$28.75 per hour for labor. It takes $4\frac{1}{3}$ hours to overhaul an engine. What is the labor charge?

Multiply \$28.75 per hour times $4\frac{1}{3}$ hours of labor.

$$\text{Labor charge} = \$28.75 \cdot 4\frac{1}{3} = \frac{\$28.75}{1} \cdot \frac{13}{3} = \frac{\$373.75}{3} \doteq \$124.58 \quad \blacksquare$$

EXAMPLE 14 How many $\frac{5}{6}$-ounce mini-bags of chips can be filled from a large 12.5-ounce bag?

Find how many $\frac{5}{6}$-ounce bags are in 12.5 ounces. Divide 12.5 by $\frac{5}{6}$.

$$\begin{array}{l}\text{Number of } \frac{5}{6}\text{-ounce} \\ \text{bags in 12.5 ounces}\end{array} = 12.5 \div \frac{5}{6} = \frac{12.5}{1} \div \frac{5}{6} = \frac{\overset{2.5}{\cancel{12.5}}}{1} \cdot \frac{6}{\underset{1}{\cancel{5}}} = 15 \text{ bags} \quad \blacksquare$$

▶ **You Try It** **15.** An electrician charges \$34.80 per hour for labor. It takes $12\frac{5}{6}$ hours for him to wire a clubroom. Find the labor charge.

16. One share of stock costs $\$9\frac{1}{8}$. How many shares can you buy for \$2,591.50?

♦**Answers to You Try It** **1.** 1 **2.** $\frac{5}{8}$ **3.** $9\frac{9}{10}$ **4.** $2\frac{7}{12}$ **5.** $21\frac{2}{3}$ **6.** $4\frac{7}{8}$ **7.** $1\frac{1}{6}$ **8.** $7\frac{5}{7}$ **9.** $7\frac{1}{4}$ **10.** 22 doses **11.** $\frac{7}{10} < \frac{3}{4}$ **12.** $\frac{7}{25} < \frac{3}{10} < \frac{2}{5}$ **13.** $\frac{27}{20}$ or 1.35 **14.** $4\frac{61}{100}$ or 4.61 **15.** \$446.60 **16.** 284 shares

SECTION 4.7 EXERCISES

1 *Simplify each expression using the order of operations.*

1. $\frac{3}{5} + \frac{2}{3} - \frac{1}{6}$

2. $\frac{3}{8} + \frac{1}{2} - \frac{1}{4}$

3. $\frac{5}{8} + \frac{1}{2} \cdot \frac{3}{4}$

4. $\frac{5}{9} + \frac{2}{3} \cdot \frac{5}{6}$

5. $\frac{1}{6} \cdot 3\frac{1}{3} + 1\frac{1}{5}$

6. $\frac{5}{8} \cdot \frac{4}{15} + 2\frac{2}{9}$

7. $\frac{7}{10} \cdot \frac{2}{5} \div \frac{3}{8}$

8. $\frac{5}{6} \cdot 2\frac{1}{4} \div 5\frac{1}{3}$

9. $8 \div 1\frac{3}{4} \cdot 2$

10. $10 \div \frac{5}{8} \cdot 2\frac{1}{4}$

11. $7\frac{1}{2} \div 2 \div \frac{3}{4}$

12. $\frac{7}{9} \div \frac{2}{3} \div \frac{1}{6}$

13. $6 \cdot \frac{3}{4} + 1\frac{5}{6} \cdot 2$

14. $\frac{5}{4} \cdot 3\frac{3}{7} + \frac{3}{8} \cdot 4$

15. $4\frac{1}{2} \cdot 4^2$

16. $\frac{5}{8} \cdot 5^2$

17. $\frac{1}{2} + \frac{1}{2} \cdot \frac{1}{2} + \frac{1}{2}$

18. $\frac{2}{5} + \frac{3}{5} \cdot \frac{1}{5} + \frac{4}{5}$

19. $\left(2\frac{3}{5}\right)^2 + \frac{4}{5} \cdot \frac{3}{10}$

20. $\left(7\frac{1}{2}\right)^2 + 3\frac{1}{4} \cdot \frac{5}{2}$

21. $\frac{1}{8}\left(\frac{3}{8} + \frac{5}{8}\right)$

22. $\frac{3}{10}\left(\frac{7}{10} + \frac{8}{10}\right)$

23. $6\left(\frac{4}{5} - \frac{2}{15}\right)$

24. $14\left(\frac{3}{7} - \frac{5}{21}\right)$

25. Evaluate $3n + 4$ when $n = 2\frac{1}{3}$.

26. Evaluate $7p - 4$ when $p = \frac{5}{6}$.

27. Find E if $E = 4h + 5$ and $h = \frac{7}{8}$.

28. Find F if $F = 4A$ and $A = 5\frac{1}{3}$.

29. What is the product of $\frac{3}{7}$ and 280, increased by 50?

30. The income tax is $\frac{2}{7}$ of \$28,000 plus \$1,250. Find the tax.

Simplify.

31. $\dfrac{\frac{2}{3}}{\frac{1}{2}}$

32. $\dfrac{\frac{5}{8}}{\frac{15}{16}}$

33. $\dfrac{12}{\frac{4}{5}}$

34. $\dfrac{7}{\frac{3}{14}}$

35. $\dfrac{\frac{3}{8}}{30}$

36. $\dfrac{\frac{7}{9}}{21}$

37. $\dfrac{2\frac{5}{6}}{1\frac{1}{6}}$

38. $\dfrac{5\frac{8}{21}}{3\frac{2}{21}}$

39. $\dfrac{5\frac{5}{8}}{\frac{5}{6}}$

40. $\dfrac{\frac{4}{9}}{2\frac{6}{7}}$

Solve.

41. What is the average of $\frac{3}{4}$, $1\frac{5}{8}$, $\frac{1}{6}$, and $\frac{2}{3}$?

42. What is the average of $3\frac{1}{2}$, $2\frac{3}{4}$, 3, and $\frac{7}{8}$?

43. Three bottles contain $3\frac{1}{4}$ ounces, $1\frac{5}{8}$ ounces, and 2 ounces of the same drug. If a pharmacist combines the contents of the three bottles, how many $\frac{2}{3}$-ounce doses of the drug can she make?

44. Ed has $4\frac{3}{8}$ pounds of hamburger in his refrigerator, and $2\frac{1}{3}$ pounds in his freezer. If he combines the two amounts, how many hamburgers weighing $\frac{1}{4}$ pound each can he make?

45. Tom owns 40 shares of EZ stock. One share sells for $\$16\frac{3}{8}$. Tom tells his broker to sell all the shares. His broker deducts a \$25 fee for making the sale. Tom splits the remainder of the money with his brother, Dick. How much does each man get?

46. 250 shares of GF stock are sold for $\$62\frac{1}{4}$ each. A broker deducts a \$300 fee for making the sale. Three sisters split the remainder of the money. How much does each sister receive?

2 *Order the fractions from smallest to largest.*

47. $\frac{3}{5}, \frac{5}{8}$

48. $\frac{4}{7}, \frac{1}{2}$

49. $\frac{11}{12}, \frac{9}{10}$

50. $\frac{2}{9}, \frac{1}{6}$

51. $\frac{4}{5}, \frac{5}{6}, \frac{3}{4}$

52. $\frac{1}{4}, \frac{2}{5}, \frac{1}{3}$

53. $\frac{2}{4}, \frac{7}{15}, \frac{1}{5}$

54. $\frac{10}{15}, \frac{15}{25}, \frac{14}{20}$

55. $\frac{5}{16}, \frac{8}{20}, \frac{1}{8}$

56. $\frac{7}{12}, \frac{13}{24}, \frac{1}{3}$

57. $\frac{5}{8}, \frac{1}{4}, \frac{3}{5}, \frac{2}{3}$

58. $\frac{1}{3}, \frac{1}{6}, \frac{3}{10}, \frac{1}{4}$

59. $\frac{3}{4}, \frac{7}{12}, \frac{5}{6}, \frac{5}{8}$

60. $\frac{2}{5}, \frac{1}{4}, \frac{3}{8}, \frac{3}{10}$

61. Three drill bits have sizes $\frac{5}{8}$, $\frac{11}{16}$, and $\frac{3}{4}$ inches, respectively. Which is the smallest?

62. Screws are purchased with sizes $\frac{1}{4}$, $\frac{5}{16}$, and $\frac{3}{8}$ inches, respectively. Which is the largest?

63. Company A pays a $\frac{7}{60}$ commission on sales. Company B pays a $\frac{9}{80}$ commission. Which company pays the higher commission?

64. Company A gave $\frac{3}{8}$ of its employees a promotion. Company B gave $\frac{4}{9}$ of its employees a promotion. Which company gave a larger portion of its employees a promotion?

3 *Perform the indicated operation.*

65. $0.5 + \frac{3}{4}$

66. $4\frac{1}{2} + 3.75$

67. $5\frac{1}{4} - 3.85$

68. $1.2 - \frac{4}{5}$

69. $\frac{1}{2} \cdot 0.62$

70. $3\frac{5}{8} \cdot 45.2$

71. $25.8 \div \frac{6}{7}$

72. $8.05 \div 3\frac{1}{2}$

73. $4.6 \cdot \frac{1}{3}$ round to tenths

74. $4\frac{3}{7} \cdot 56.1$ round to hundredths

75. $7.2 \div 2\frac{1}{3}$ round to hundredths

76. $0.451 \div 2\frac{5}{6}$ round to thousandths

77. What will $2\frac{1}{3}$ pounds of Swiss cheese cost at $4.59 a pound?

78. Alice worked $14\frac{5}{6}$ hours at $8.64 per hour. What did she earn?

79. A refrigerator normally selling for $659 is on sale for $\frac{1}{4}$ off.

a. How much is being taken off the price of the refrigerator?

b. What is the sale price?

80. A dress normally selling for $39.50 is on sale for $\frac{1}{3}$ off.

a. How much is being taken off the price? Round to cents.

b. What is the sale price?

SKILLSFOCUS (Section 2.5) *Divide by the power of ten.*

81. $3{,}400 \div 10$ **82.** $52{,}000 \div 100$ **83.** $640 \div 10$ **84.** $327{,}000 \div 1{,}000$

EXTEND YOUR THINKING ♦♦♦

♦SOMETHING MORE

85. Al made a commission of $\frac{1}{7}$ of the price, plus \$10, on the sale of a \$350 bike. Bob made a commission of $\frac{1}{10}$ of the price, plus \$25, on the sale of a \$360 bike. Who made the greater commission?

♦TROUBLESHOOT IT

Find and correct the error.

86. $\frac{1}{2}+\frac{1}{2}\cdot 6 = 1\cdot 6 = 6$

87. $\frac{(5+3)}{(11+3)} = \frac{(5+\overset{1}{\cancel{3}})}{(11+\underset{1}{\cancel{3}})} = \frac{6}{12} = \frac{1}{2}$

WRITING TO LEARN ♦♦♦

88. Explain the difference between a simple fraction and a complex fraction.

89. Explain what happens to the size of a fraction when the denominator gets smaller, but the numerator stays the same. Use your own examples.

90. Calculate the answer to each problem.

a. $\frac{10}{(2\cdot 5)}$ **b.** $10 \div 2\cdot 5$ **c.** $10\div 2\cdot 5$

Explain why the answer in **c** is different from the answer in **a** and **b**.

CHAPTER 4 REVIEW

VOCABULARY AND MATCHING

New words and phrases introduced in this chapter are shown in the left-hand column. Match each term on the left with the phrase or sentence on the right that best describes it.

A. like fractions	______	involves dividing the denominators by prime factors
B. equivalent fractions	______	rewrite a fraction by either expanding or reducing
C. expanding a fraction	______	any number that 2 or more denominators divide into exactly
D. rename a fraction	______	numerator and denominator are whole numbers
E. lower terms	______	done with a mixed number, to make its proper fraction improper
F. higher terms	______	write each fraction in terms of the LCD, then find the difference between the numerators
G. common denominator	______	the smallest number that 2 or more denominators divide into exactly
H. least common denominator	______	look different, but have the same value
I. find LCD by inspection	______	has more than one fraction bar
J. find LCD by prime factoring	______	when a fraction is reduced, its numerator and denominator are in this
K. find LCD by successive division by primes	______	involves multiplying the most frequently occurring prime factors in each prime factorization
L. unlike fractions	______	have the same name, or denominator
M. adding fractions	______	requires looking at the denominators, and *seeing* what it is
N. subtracting fractions	______	writing an equivalent fraction with a larger numerator and denominator
O. borrowing	______	arrange them by first writing each in terms of the same denominator
P. simple fraction	______	fractions with different denominators
Q. complex fraction	______	write each fraction in terms of the LCD, then sum the numerators
R. ordering fractions	______	when a fraction is expanded, its numerator and denominator are in this

REVIEW EXERCISES

4.1 Adding and Subtracting Like Fractions

1. $\frac{3}{8}+\frac{7}{8}$ **2.** $\frac{7}{10}+\frac{5}{10}$ **3.** $\frac{2}{7}+\frac{5}{7}$ **4.** $\frac{3}{20}+\frac{7}{20}$

5. $4\frac{2}{5}+1\frac{1}{5}$ **6.** $6\frac{4}{9}+5\frac{8}{9}$ **7.** $4+1\frac{7}{12}$ **8.** $3\frac{5}{6}+4$

9. $\frac{7}{9}-\frac{5}{9}$ **10.** $\frac{13}{15}-\frac{4}{15}$ **11.** $\frac{8}{25}-\frac{3}{25}$ **12.** $\frac{5}{16}-\frac{1}{16}$

13. $5\frac{7}{8}-2\frac{3}{8}$ **14.** $4\frac{2}{3}-1\frac{1}{3}$ **15.** $3\frac{11}{18}-\frac{5}{18}$ **16.** $1\frac{37}{60}-\frac{23}{60}$

4.2 Equivalent Fractions

Determine which fractions are equivalent by reducing each to lowest terms.

17. $\frac{12}{18}$ and $\frac{15}{24}$ **18.** $\frac{12}{15}$ and $\frac{20}{25}$ **19.** $\frac{15}{20}$ and $\frac{27}{36}$ **20.** $\frac{25}{60}$ and $\frac{20}{45}$

Expand each fraction using the form for 1 given.

21. $\frac{2}{5}$ using $\frac{6}{6}$ **22.** $\frac{2}{3}$ using $\frac{8}{8}$ **23.** $\frac{7}{10}$ using $\frac{3}{3}$ **24.** $\frac{11}{12}$ using $\frac{15}{15}$

Write four fractions equivalent to the given fraction.

25. $\frac{1}{8}$ **26.** $\frac{6}{7}$ **27.** $\frac{3}{5}$ **28.** $\frac{7}{9}$

Find the new numerator for the given denominator.

29. $\frac{5}{6} = \frac{}{30}$ **30.** $\frac{5}{9} = \frac{}{27}$ **31.** $\frac{7}{8} = \frac{}{72}$ **32.** $\frac{7}{10} = \frac{}{70}$

4.3 Finding the Least Common Denominator

The numbers shown are denominators for fractions. Find the LCD for each group of denominators using any method.

33. 12 and 18 **34.** 15 and 25 **35.** 20 and 32 **36.** 60 and 84

37. 12, 18, and 24 **38.** 15, 30, and 75 **39.** 7, 8, and 21

4.4 Adding Fractions with Unlike Denominators

40. $\frac{5}{6} + \frac{7}{8}$ **41.** $\frac{3}{10} + \frac{11}{12}$ **42.** $\frac{5}{9} + \frac{5}{6}$ **43.** $\frac{11}{18} + \frac{4}{45}$

44. $3\frac{1}{5} + 1\frac{3}{4}$ **45.** $5\frac{7}{10} + 6\frac{1}{6}$ **46.** $4\frac{8}{9} + 6\frac{7}{8}$ **47.** $2\frac{5}{18} + 7\frac{11}{24}$

4.5 Subtracting Fractions with Unlike Denominators

48. $\frac{5}{6} - \frac{3}{8}$ **49.** $\frac{4}{9} - \frac{1}{4}$ **50.** $\frac{7}{12} - \frac{3}{8}$ **51.** $\frac{15}{28} - \frac{4}{21}$

52. $3\frac{4}{7} - 1\frac{1}{4}$ **53.** $5\frac{2}{9} - 4\frac{8}{15}$ **54.** $6\frac{3}{8} - 4\frac{5}{6}$ **55.** $8 - 5\frac{3}{10}$

4.6 Applications: Addition and Subtraction of Fractions

56. A share of stock sold for $\$23\frac{7}{8}$ a month ago. Today it sells for $\$26\frac{3}{4}$. How much has the price for one share increased?

57. One board is 12 feet long. The other is $8\frac{7}{12}$ feet long. What is the difference in length between the two boards?

58. On vacation, Beth filled her gas tank twice, buying $10\frac{4}{5}$ and $12\frac{7}{10}$ gallons of gas respectively. How much gas did Beth buy on her vacation?

59. On four successive days, a jogger ran $6\frac{3}{4}$, $2\frac{7}{10}$, $5\frac{3}{5}$, and 8 miles. What total mileage did the jogger run?

60. One-quarter of a fortune was invested in real estate, and $\frac{4}{9}$ was invested in the stock market. The rest was given to charity. What fraction of the fortune was given to charity?

61. Marty purchased 400 shares of a stock at $\$56\frac{1}{2}$ a share, and sold them at $\$48\frac{7}{8}$ a share. How much money did Marty lose?

4.7 Combined Operations; Order

62. $\frac{5}{6} + \frac{2}{3} \cdot \frac{4}{5}$

63. $\frac{6}{8} \div \frac{3}{4} \div \frac{1}{8}$

64. $\left(\frac{2}{5}\right)^2 + \frac{1}{6} \cdot \frac{3}{4}$

65. $4 \cdot 3\frac{1}{2} - 5$

66. $\dfrac{\frac{3}{4}}{\frac{5}{9}}$

67. $\dfrac{6}{\frac{2}{7}}$

68. $4.73 + \frac{5}{8}$

69. $2\frac{3}{5} - 1.972$

70. $\frac{6}{7} \cdot 4.732$

71. $5.86 \div \frac{2}{9}$

72. What will $2\frac{3}{4}$ pounds of roast beef cost at $5.29 a pound? Round to the nearest cent.

Arrange each set of fractions in order from smallest to largest.

73. $\frac{1}{4}, \frac{3}{8}, \frac{1}{6}$

74. $\frac{2}{3}, \frac{7}{9}, \frac{11}{15}$

75. $\frac{3}{10}, \frac{8}{25}, \frac{7}{20}$

76. $\frac{5}{18}, \frac{9}{32}, \frac{7}{24}$

The Integers

Have you heard of pinochle or rummy? They are examples of games where you may at times have a negative score. When you keep score in such games, you are adding positive and negative numbers, called integers.

What is each team's score in the game below?
Who has the higher score?

WE	THEY
10	−15
+15	+10
25	
−30	
−5	
−15	
−20	
+10	

In this chapter, you will study operations on a new set of numbers called integers. The integers are positive and negative whole numbers, including 0. You already have seen and used integers. A negative temperature, such as −10°, is an example of an integer. A checking account that is overdrawn by fifteen dollars can have its balance represented by the negative integer −$15. Have you ever kept score in a game and found yourself with a negative score? If so, then you have already performed addition of integers. In this chapter you will learn to add, subtract, multiply, and divide integers, with applications.

5.1 INTEGERS AND THE NUMBER LINE

OBJECTIVES

1. Define the integers.
2. Compare integers.
3. Find the absolute value of an integer.
4. Find the opposite of an integer.

NEW VOCABULARY

negative integers, negative sign, integers, absolute value, positive sign, opposite

1 The Integers

The number line was introduced in Chapter 1.

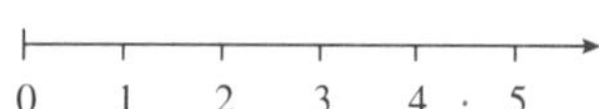

However, there are numbers you cannot represent on this line. For example, a winter temperature of 5 degrees below zero. Or a checkbook balance that is $40 overdrawn.

These numbers are less than 0. To represent them on the number line, extend the number line to the left of 0. These new numbers are called **negative integers**.

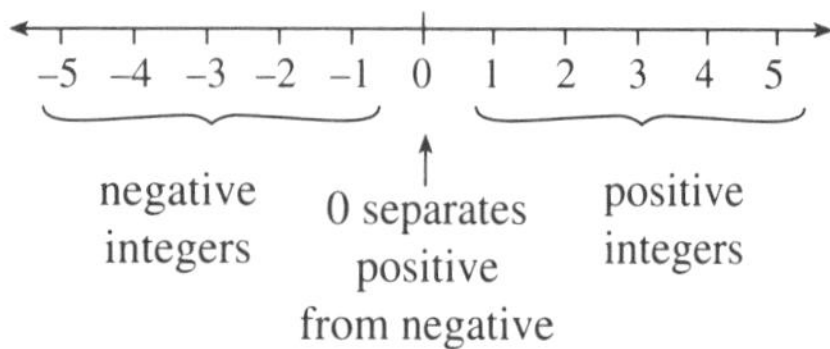

A **positive sign** (+) means a number is to the right of 0. A number without a sign is understood to be positive.

$$4 = +4$$
$$7 = +7$$

A **negative sign** (−) means a number is to the left of 0.

The numbers . . . , −5, −4, −3, −2, −1, 0, 1, 2, 3, 4, 5, . . . are called the **integers**. Integers are also called signed numbers. The integer 0 is neither positive nor negative.

EXAMPLE 1

3 is read three, or positive three.
+9 is read positive nine, or just nine.
−7 is read negative 7. ■

▶ You Try It Read each integer, and write it in words.

1. 8 **2.** +4 **3.** −12

EXAMPLE 2 Negative integers are used in everyday life.

a. The lowest point in Death Valley, California, is 282 ft below sea level, written −282 ft.

b. A checkbook balance overdrawn $40 is written −$40.

$20	A positive $20 balance means you are *in the black*.
−$60	You write a check for $60.
−$40	A negative balance means you are *in the red*.

c. A temperature 5 degrees below zero is written $-5°$.

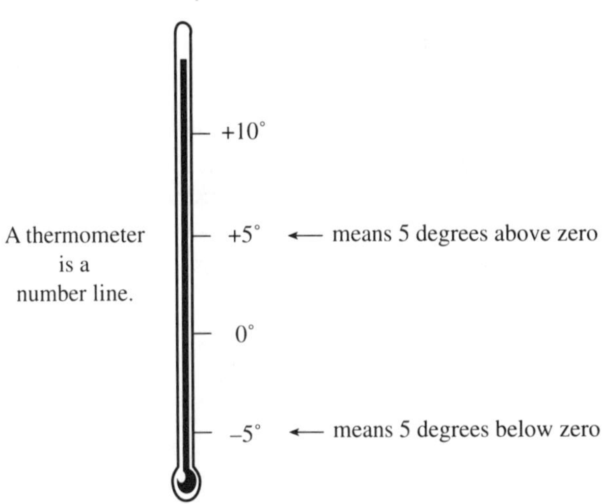

■

▶ **You Try It** Write an integer for each statement.

4. A cave is 410 feet below sea level.

5. Your business is \$3,522 in the red.

6. The windchill is 28° below zero.

2 Comparing Integers

Given two integers, the smaller one is farther to the left on the number line. The larger one is farther to the right.

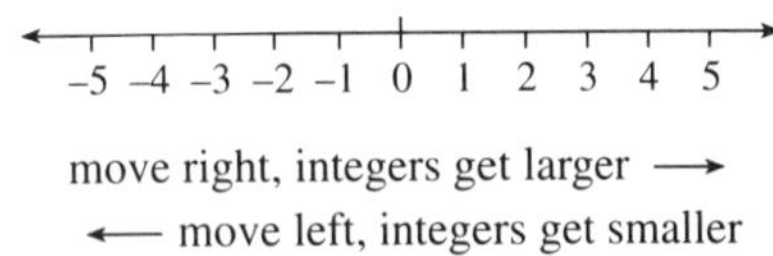

EXAMPLE 3 Comparing integers.

a. $-4 < -2$: -4 is smaller because it is farther to the left on the number line. (Which temperature is lower, $-4°$ or $-2°$? Since $-4°$ is lower, write $-4° < -2°$.)

b. $-1 > -5$: -1 is larger because it is farther to the right on the number line. (Which is the higher checkbook balance, $-\$1$ or $-\$5$? Since $-\$1$ is higher, write $-\$1 > -\$5$.)

c. $0 > -3$: 0 is larger because it is farther to the right on the number line. (A score of 0 in a game is higher than a score of -3.) ■

▶ **You Try It** Write $<$ or $>$ between the integers.

7. $-6 \quad -8$ **8.** $-4 \quad 0$ **9.** $-9 \quad -3$ **10.** $1 \quad -12$

3 Absolute Value

The **absolute value** of an integer n is written $|n|$. It is the distance n is from 0 on the number line. Since distances are positive, the absolute value of any number is always positive.

For example, $|-5|$ is read "the absolute value of -5."

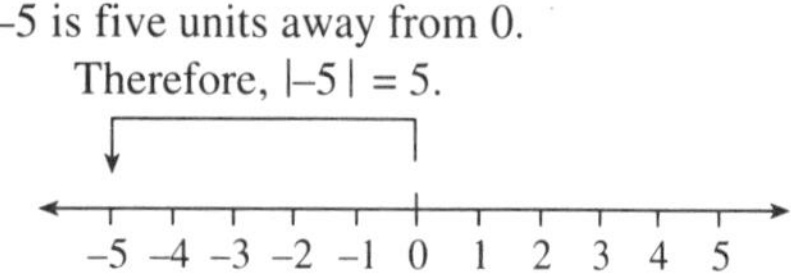

EXAMPLE 4

a. $|-3| = 3$ because -3 is 3 units from 0 on the number line
b. $|+7| = 7$ because $+7$ is 7 units from 0 on the number line
c. $4 + |-6| = 4 + 6 = 10$ ■

▶ **You Try It** Evaluate each absolute value.

11. $|-8|$ **12.** $|-2|$ **13.** $|9|$ **14.** $|+3|$

---

4 The Opposite of a Number

The **opposite** of a number is a second number the same distance from 0 on the number line, but on the opposite side. The opposite of $+4$ is -4. -4 and $+4$ are both 4 units from 0 on the number line, but on opposite sides.

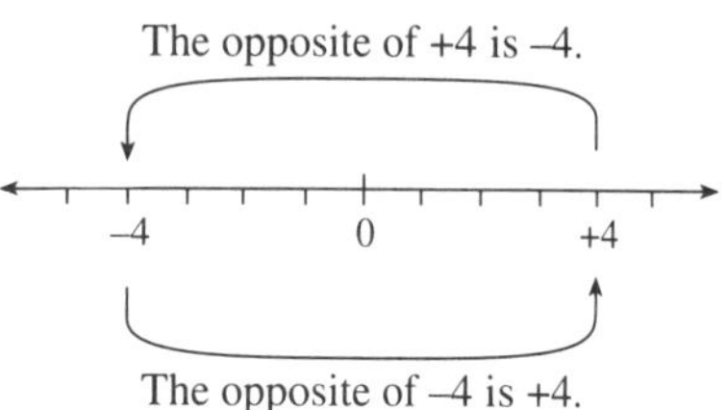

Opposite numbers have the same absolute value. For example, $|-4| = 4$ and $|+4| = 4$.

EXAMPLE 5

The opposite of $+7$ is -7.
The opposite of 5 is -5.
} The opposite of a positive number is negative.

The opposite of -2 is $+2$, or just 2.
The opposite of -100 is $+100$, or 100.
} The opposite of a negative number is positive.

The opposite of 0 is 0.} 0 is its own opposite. Note, $0 = -0 = +0$. ■

▶ **You Try It** Find the opposite of each integer.

15. $+20$ **16.** -14 **17.** -1 **18.** 300

A negative sign, $-$, can be read "the opposite of." For example, -5 is read "the opposite of 5." Then,

$$-(-5) = \text{the opposite of } -5 = +5$$

or,

$$-(-5) = +5.$$

EXAMPLE 6 **a.** $-(+9) =$ the opposite of $+9 = -9$ So, $-(+9) = -9$.

b. $-(+15) = -15$

c. $-(-7) =$ the opposite of $-7 = +7$ So, $-(-7) = 7$.

d. $-(-36) = 36$ ■

▶ **You Try It** Simplify.

19. $-(+4)$ **20.** $-(-8)$ **21.** $-(-48)$ **22.** $-(+2)$

▶ **Answers to You Try It** 1. eight, or positive eight 2. positive four, or four 3. negative 12 4. -410 ft 5. $-\$3{,}522$ 6. $-28°$ 7. > 8. < 9. < 10. > 11. 8 12. 2 13. 9 14. 3 15. -20 16. 14 17. 1 18. -300 19. -4 20. 8 21. 48 22. -2

SECTION 5.1 EXERCISES

1 *Write a positive or negative integer for each statement.*

1. The temperature is 40 degrees above zero.

2. Today's low temperature in Nome, Alaska, was 54 degrees below zero.

3. The submarine is 35 ft below sea level.

4. The plane is 33,000 ft above sea level.

5. Tom's checkbook balance is $110 in the black.

6. Alice scored 28 points in the last hand.

7. George lost 23 points this hand.

8. Deduct a check for $29.50 from your checkbook balance.

9. As of today, his business is operating $4,000 in the red.

10. The elephant's weight dropped 36 pounds in one week.

2 **3** *Find the absolute value, and simplify.*

11. $|-13|$ **12.** $|7|$ **13.** $|+15|$ **14.** $|-25|$

15. $5 + |-3|$ **16.** $|+7| + 7$ **17.** $8 + |-8|$ **18.** $|-50| + 50$

19. $|-1| + |-3| + |-6|$ **20.** $|-10| + 10 + |10|$

Insert the correct symbol ($<$, $>$, $=$) *between the integers.*

21. $-3 \quad -5$ **22.** $-6 \quad -2$ **23.** $-4 \quad +3$ **24.** $-7 \quad +10$

25. $-30 \quad -25$ **26.** $-50 \quad +1$ **27.** $|-3| \quad -3$ **28.** $|0| \quad 0$

29. $|-9| \quad 9$ **30.** $2 \quad |-5|$ **31.** $0 \quad -1$ **32.** $17 \quad -62$

33. $-3 \quad -4$ **34.** $-23 \quad -19$ **35.** $|6| \quad 8$ **36.** $|-2| \quad -8$

4

37. The opposite of 8 is ________.

38. The opposite of -8 is ________.

39. The opposite of -45 is ________.

40. The opposite of $+19$ is ________.

41. $-(+6)$

42. $-(+10)$

43. $-(-6)$

44. $-(-38)$

45. $-(-(-7))$

46. $-(+(-(+8)))$

SKILLSFOCUS (Section 2.8) *Change each fraction or mixed number to a decimal number.*

47. $\frac{3}{4}$

48. $\frac{1}{8}$

49. $\frac{5}{6}$

50. $3\frac{1}{2}$

EXTEND YOUR THINKING ▶▶▶▶

▶ TROUBLESHOOT IT

Find and correct the error.

51. $|-9| + |-6| = 9 - 6 = 3$

52. $-(-9) = -(+9) = -9$

WRITING TO LEARN ▶▶▶▶

53. If two numbers are opposites, explain why their absolute values are equal.

54. Explain how the number line is used in the launch of the space shuttle. Let 0 be the moment of lift-off. Then what will negative numbers, such as -5 and -3, represent? What will positive numbers, such as $+1$ and $+60$, represent?

5.2 ADDITION OF INTEGERS

OBJECTIVES

1 Add integers with like signs.

2 Add integers with unlike signs.

NEW VOCABULARY

additive inverse

Have you ever had a negative score in a game? Have you ever found a negative balance in your checkbook? If so, then you have added integers.

For example, check the scores shown here. Based on this, can you state some rules for adding integers?

like signs

The sum of two positive integers is positive.

$$\begin{array}{r} 10 \\ +\ 5 \\ \hline 15 \end{array}$$

The sum of two negative integers is negative.

$$\begin{array}{r} -5 \\ -15 \\ \hline -20 \end{array}$$

unlike signs

$$\begin{array}{r} 15 \\ -20 \\ \hline -5 \end{array} \qquad \begin{array}{r} -20 \\ +25 \\ \hline 5 \end{array}$$

When adding two integers with unlike signs, the sign of the answer is sometimes positive and sometimes negative. Can you tell when it will be positive? Or when it will be negative?

1 Adding Integers with Like Signs

EXAMPLE 1 Different ways to write addition of signed numbers.

a. $(+7) + (+5)$ is read positive 7 added to positive 5

(+7) is read positive 7; + means addition; (+5) is read positive 5

OBSERVE $(+7) + (+5)$ is another way to write $7 + 5$.

b. $(-2) + (-9)$ is read negative 2 added to negative 9

(−2) negative 2; + addition; (−9) negative 9

c. $6 + (-8)$ is read 6 added to negative 8

6 six, or positive 6; + addition; (−8) negative 8

d. $-4 + 9 = (-4) + (+9)$ ⟵ You may rewrite the problem this way. This is read negative 4 added to 9.

−4 negative 4; + addition; 9 positive 9 ■

▶ You Try It Write each addition in words.

1. $(+2) + (+6)$ **2.** $(-3) + (-4)$ **3.** $1 + (-9)$ **4.** $-12 + 4$

EXAMPLE 2 Add $(+7) + (+5)$ using the number line.

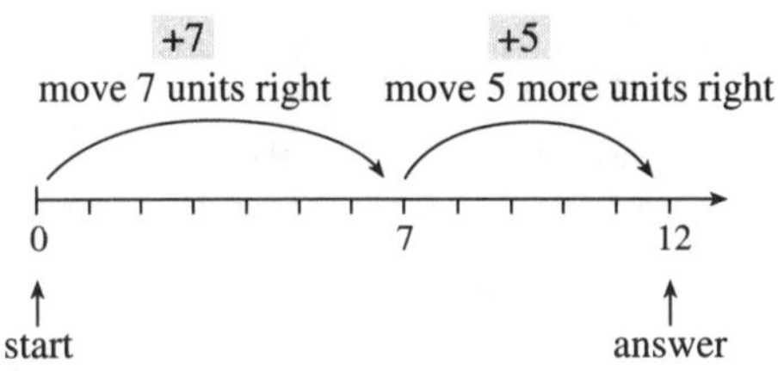

Therefore, $(+7) + (+5) = +12$. ■

> To add two integers using the number line
>
> **1.** Start at 0.
> **2.** Count to the right if the first addend is positive. Count to the left if it is negative.
> **3.** Starting where you left off with the first addend, count right if the second addend is positive. Count left if it is negative.
> **4.** Your final stopping point is the answer.

EXAMPLE 3 Add $-4 + (-2)$ using the number line.

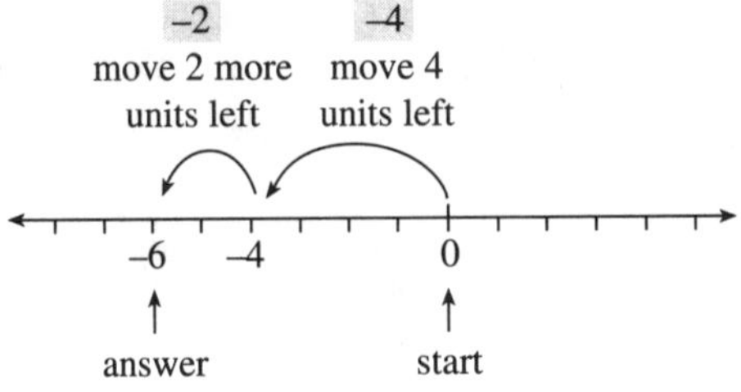

Therefore, $-4 + (-2) = -6$. ■

▶ **You Try It** Add using the number line.

5. $(+3) + (+5)$ **6.** $+1 + (+5)$ **7.** $(-8) + (-3)$ **8.** $-6 + (-7)$

In Examples 2 and 3, the sum has the same sign as the addends.

> To add two integers with like signs
>
> **1.** Add the absolute values of the two integers.
> **2.** If both integers are positive, the sum is positive.
> If both integers are negative, the sum is negative.

EXAMPLE 4 Add $(+7)+(+9)$.

Signs are the same.
Add the absolute values.

$$|+7|+|+9|=7+9=16$$

$$(+7)+(+9)=+16$$

Both integers are positive, so the sum is positive. ■

EXAMPLE 5 Add $-6+(-8)$.

The signs are the same.
Add the absolute values.

$$|-6|+|-8|=6+8=14$$

$$(-6)+(-8)=-14$$

Both integers are negative, so the sum is negative. ■

▶ **You Try It** Add the integers.

9. $(+4)+(+11)$ **10.** $(+5)+(+5)$ **11.** $-7+(-9)$ **12.** $(-8)+(-5)$

---

2 Adding Integers with Unlike Signs

EXAMPLE 6 Add $(-5)+(+8)$ using the number line.

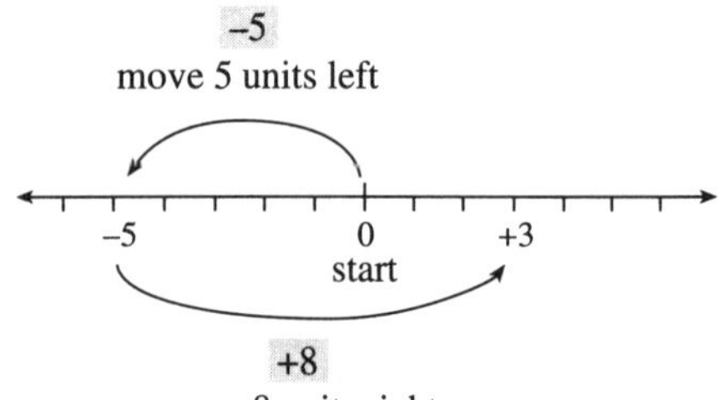

Therefore, $(-5)+(+8)=+3$. ■

EXAMPLE 7 Add $6+(-10)$ using the number line.

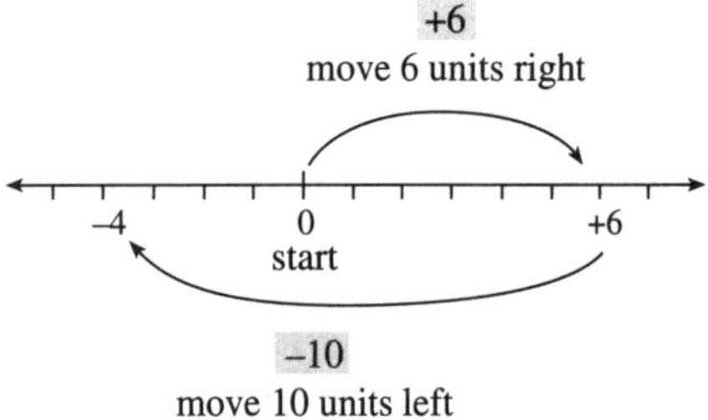

Therefore, $6+(-10)=-4$. ■

You Try It Add using the number line.

13. $(+2) + (-6)$ **14.** $(-3) + (+4)$ **15.** $-1 + (+7)$ **16.** $-12 + 4$

In Examples 6 and 7, the sum has the sign of the integer with the larger absolute value.

> To add two integers with unlike signs
>
> **1.** Subtract the smaller absolute value from the larger.
> **2.** Give this difference the sign of the integer with the larger absolute value.

EXAMPLE 8 Add $(-6) + (+5)$.

The signs are unlike. Subtract the absolute values.

$$|-6| = 6, |+5| = 5, \text{ and } 6 - 5 = 1$$

$$(-6) + (+5) = -1$$

Since $6 > 5$, -6 has the larger absolute value. Use the negative sign for the answer.

EXAMPLE 9 Add $8 + (-3)$.

The signs are unlike. Subtract the absolute values.

$$|8| = 8, |-3| = 3, \text{ and } 8 - 3 = 5$$

$$8 + (-3) = +5$$

Since $8 > 3$, 8 has the greater absolute value. Use the positive sign for the answer. ■

You Try It Add.

17. $(+2) + (-9)$ **18.** $(+13) + (-4)$ **19.** $14 + (-6)$ **20.** $(-18) + (+15)$

EXAMPLE 10 Rita's score in a game was -6. She just made 10 points. What is her new score?

Rita's new score is her old score, -6, plus 10.

$$\text{new score} = -6 + 10 = (-6) + (+10)$$
$$= 4 \text{ points}$$ ■

CAUTION In Example 10, it is an error to write $-6 + 10 = -16$. Why? You may insert your own parentheses around integers to help you add them.

You Try It **21.** At 6 A.M., the temperature was $-15°$. By noon the temperature rose $9°$. What was the noon temperature?

Add an integer to its opposite and you get 0. The opposite of -8 is $+8$, and

$$(-8) + (+8) = 0.$$

> The sum of an integer and its opposite is 0.
> The opposite of an integer is called its **additive inverse**.

EXAMPLE 11

a. The opposite of 6 is (-6) and $6 + (-6) = 0$.
-6 is also called the additive inverse of 6.

b. The opposite of -1 is $+1$ and $-1 + (+1) = 0$.
$+1$, or 1, is the additive inverse of -1. ■

▶ **You Try It** Find the opposite of each integer. Then show that the sum of each integer and its opposite is 0.

22. 5 **23.** -8 **24.** -2 **25.** $+7$

Adding Three or More Integers

To add three or more integers together, follow the order of operations. Add the two integers on the left. Then add the next integer to this sum. Continue this way until all integers have been added.

EXAMPLE 12 Add $(+6) + (-11) + (+8)$.

$$(+6) + (-11) + (+8) = +3$$

Add $+6$ to -11. You get -5.

Add -5 to $+8$. You get $+3$.

-5

$+3$ (answer) ■

EXAMPLE 13 Add $-12 + 3 + (-6) + 1$.

$$-12 + 3 + (-6) + 1 = -14$$

-9

-15

-14 ■

▶ **You Try It** Add.

26. $-4 + 7 + (-8)$ **27.** $8 + (-10) + (-5) + 9$

EXAMPLE 14 **Chapter Problem** Read the problem at the beginning of this chapter. The score is:

WE	THEY
-20	-15
$+10$	$+10$
-10	-5

Since $-10 < -5$, "THEY" are leading. ■

Summary

To add two integers

1. If the integers have the same sign, add the absolute values, keep the sign.
2. If the integers have different signs, subtract the smaller absolute value from the larger. Use the sign of the integer with the larger absolute value.

► **Answers to You Try It** 1. positive 2 added to positive 6 2. negative 3 added to negative 4 3. 1 added to negative 9 4. negative 12 added to 4 5. $+8$ 6. $+6$ 7. -11 8. -13 9. $+15$ 10. $+10$ 11. -16 12. -13 13. -4 14. $+1$ 15. $+6$ 16. -8 17. -7 18. $+9$ 19. $+8$ 20. -3 21. $-6°$ 22. -5; $5 + (-5) = 0$ 23. $+8$; $-8 + (+8) = 0$ 24. $+2$; $-2 + (+2) = 0$ 25. -7; $+7 + (-7) = 0$ 26. -5 27. $+2$

SECTION 5.2 EXERCISES

1 **2** *Write the addition in words, then add.*

1. $8 + 5$

2. $7 + 12$

3. $-6 + 4$

4. $-10 + 2$

5. $3 + (-4)$

6. $8 + (-11)$

7. $-5 + 16$

8. $-2 + 3$

9. $6 + (-4)$

10. $5 + (-1)$

11. $8 + (-8)$

12. $1 + (-1)$

13. $-7 + (-13)$

14. $-8 + (-2)$

15. $(-9) + (+5)$

16. $(4) + (-14)$

Add.

17. $-5 + (+5)$

18. $(-5) + (5)$

19. $-6 + 0$

20. $-2 + 0$

21. $0 + (-12)$

22. $0 + (+6)$

23. $(7) + (-12)$

24. $(-3) + (5)$

25. $-16 + 16$

26. $-4 + 4$

27. $12 + (-5)$

28. $13 + (-14)$

29. $-3 + 2$

30. $-12 + 3$

31. $(+2) + (+9)$

32. $(-8) + (-8)$

33. $-10 + (-12)$

34. $-1 + (-1)$

35. $15 + (-9)$

36. $-9 + (15)$

37. $7 + (-4)$

38. $-4 + (7)$

39. $-7 + 2$

40. $-8 + 12$

41. $-6 + (-12)$

42. $-2 + (-2)$

43. $+9 + 7$

44. $+4 + 9$

45. $-121 + 84$

46. $-82 + 106$

47. $-80 + 70$

48. $-80 + 90$

49. $-9 + (-4)$

50. $-2 + (-5)$

51. $20 + (-14)$

52. $10 + (-14)$

53. $40 + (-40)$

54. $80 + (-20)$

55. $(-65) + 52$

56. $(-33) + 50$

57. $-6 + (-3) + (-6)$

58. $7 + (-4) + 5$

59. $7 + (-10) + (4) + (-2)$

60. $-8 + (-5) + 6 + (-6)$

61. $-8 + 8 + 9 + (-9)$

62. $7 + 8 + (-7) + (-8)$

63. $-1 + (-1) + 1$

64. $6 + 9 + (-12)$

65. $-10 + (-12) + (-21) + (-15)$

66. $-23 + (-40) + 34 + (-18)$

67. The temperature in St. Helena at 6 A.M. was $-19°$. It is expected to rise $31°$ by noon. What is the expected noon temperature?

68. The midnight temperature in Nome, Alaska, was $-41°$. By noon the next day, it had risen $27°$. What was the noon temperature?

69. Tom's score in a game of cards is -16. He just won 25 points. What is his new score?

70. Maria's score in a game is -29. She just made 13 points. What is her new score?

71. Edna's checking account balance is $-\$42$. She deposits a \$27 check in her account. What is her new balance?

72. The county payroll office has a balance of $-\$250{,}680$. Tax revenues recorded this morning added \$1,078,000 to this account. What is the new balance?

SKILLSFOCUS (Section 1.7) *Use the correct symbol, < or >, to compare the following numbers.*

73. 0.3 0.033

74. $\frac{1}{2}$ 0

75. 50.7 51.07

76. $12\frac{1}{5}$ 13

EXTEND YOUR THINKING ▶▶▶▶

▶ TROUBLESHOOT IT

Find and correct the error.

77. $-5 + 10 = -15$

78. $-4 + (-2) + (-8) = (-6) + (-8) = -2$

WRITING TO LEARN ▶▶▶▶

79. **a.** Explain how to add two integers with like signs.

b. Explain how to add two integers with unlike signs.

80. Use the number line, and the addition $(+5) + (-8)$, to show that addition of integers is commutative.

81. Explain why you start at 0 when adding integers using the number line.

82. In the CAUTION following Example 10, explain why $-6 + 10 = -16$ is an error.

5.3 SUBTRACTION OF INTEGERS

OBJECTIVES

1. Subtract integers.
2. Applications.

1 Subtracting Integers

Addition of integer problems are written:	Subtraction of integer problems are written:
$4 + 7$	$4 - 7$
$(-3) + (-5)$	$(-3) - (-5)$
$-6 + (+2)$	$-6 - (+2)$
↑ addition sign	↑ subtraction sign

EXAMPLE 1 Write each subtraction problem in words.

a. $4 - 7$ is read 4 minus 7. You can also write

(− means subtraction)

$4 - 7 = (+4) - (+7)$

or $4 - 7 = (4) - (7)$

b. $(-3) - (-5)$ is read negative 3 minus negative 5

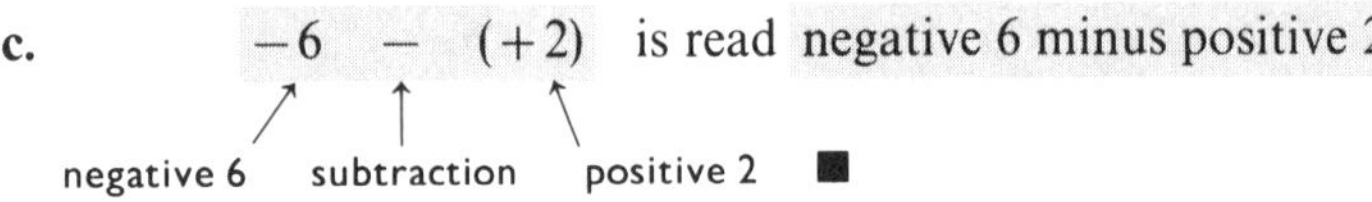

c. $-6 - (+2)$ is read negative 6 minus positive 2

(negative 6, subtraction, positive 2) ■

▶ **You Try It** Write each subtraction problem in words.

1. $4 - 9$ **2.** $(-10) - (-7)$ **3.** $-2 - (+5)$

You subtract integers by adding the opposite of the subtrahend.

For example, show that the subtraction $(+5) - (-3)$ gives the same answer as the addition $(+5) + (+3)$.

Subtraction	*Addition*
The distance from -3 to $+5$ on the number line is written $(+5) - (-3)$. This distance is 8.	
	$(+5) + (+3) = 8$

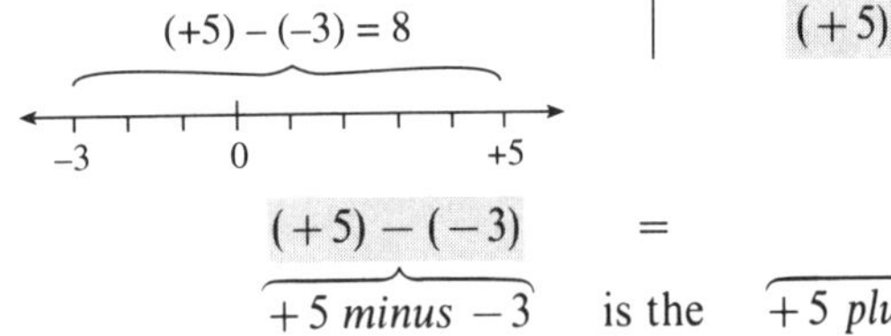

$(+5) - (-3)$ = $(+5) + (+3)$

$+5$ *minus* -3 is the same as $+5$ *plus* the opposite of -3

> To subtract two integers
>
> **1.** Change the subtraction sign to addition.
> **2.** Change the sign of the subtrahend to its opposite.
> **3.** Add.

To subtract two integers, *add the opposite* of the subtrahend.

EXAMPLE 2 Find $-14-(-9)$.

Change subtraction to addition. — Change the sign of the subtrahend to its opposite.

$$\begin{aligned} &-14-(-9)\\ =&-14+(+9)\\ =&\quad -5 \quad ■ \end{aligned}$$

EXAMPLE 3 Find $-20-(+70)$.

Change subtraction to addition. — Change the sign of the subtrahend to its opposite.

$$\begin{aligned} &-20-(+70)\\ =&-20+(-70)\\ =&\quad -90 \quad ■ \end{aligned}$$

EXAMPLE 4 What is $10-18$?

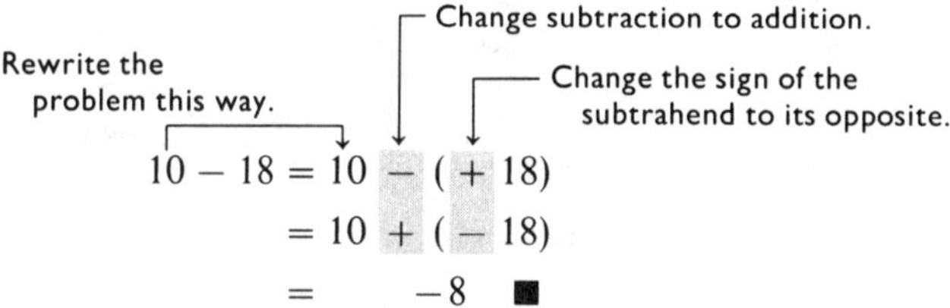

▶ **You Try It**

4. $-18-(-12)$	**5.** $-4-(-7)$	**6.** $-15-(+20)$
7. $-25-(+16)$	**8.** $6-11$	**9.** $8-9$
10. $3-(-10)$	**11.** $50-(-20)$	

EXAMPLE 5 Subtract -16 from 12.

$$\begin{aligned} \text{subtract } -16 \text{ from } 12 &= 12-(-16)\\ &= 12+(+16)\\ &= 28 \quad ■ \end{aligned}$$

▶ **You Try It** **12.** Subtract -8 from -5. **13.** Subtract -4 from 6.

> When two or more additions and subtractions occur together
>
> **1.** Change each subtraction to addition by adding the opposite.
> **2.** Add the integers two at a time from left to right.

EXAMPLE 6 Simplify $6-(-5)+2$.

Write the subtraction as addition of the opposite.

$$\begin{aligned} &6-(-5)+2\\ =&\underbrace{\underbrace{6+(+5)}_{+11}+2}_{+13} = 13 \quad ■ \end{aligned}$$

EXAMPLE 7 Simplify $-5-(-2)+8-(4)-3$.

Write each of the three subtractions as addition of the opposite.

$$-5-(-2)+8-(4)-3$$
$$=-5+(+2)+8+(-4)+(-3)=-2$$

$-5+(+2) = -3$; $-3+8 = +5$; $+5+(-4) = +1$; $+1+(-3) = -2$ ■

▶ **You Try It**

14. $8-(-3)+4$

15. $10-(-7)-5$

16. $-1-(-2)+(-6)+3$

17. $-8-(-5)+6-6-(+5)$

2 Applications

What practical meaning can $5-(-3)$ have? Where is such a subtraction used in real life? Suppose the temperature this morning was $-3°$. This afternoon the temperature was $5°$. How much did the temperature rise?

The rise in temperature is the higher temperature minus the lower.

$$\text{rise} = 5°-(-3°)$$
$$= 5°+(+3°)$$
$$= 8°$$

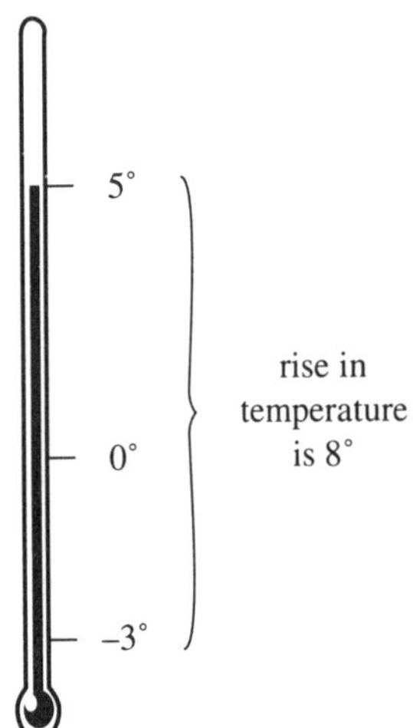

> To find the difference between two numbers on a ruler, thermometer, or any other form of number line, subtract the lower number from the higher.

Draw and label a number line to help you solve problems containing negative integers.

EXAMPLE 8 Ed's bank balance was $-\$27$ yesterday. After a deposit, the balance was \$49. How large was the deposit?

Draw a number line for the bank balances. The amount of the deposit is the difference between \$49 and $-\$27$.

original balance — new balance

-\$27 — \$0 — \$49

$$\begin{aligned}\text{deposit} &= \$49 - (-\$27)\\ &= \$49 + (+\$27)\\ &= \$76\end{aligned}$$

Check: $-\$27 + \$76 = \$49$ ■

▶ **You Try It** **18.** Tom's score in a game was -34. An hour later his score was 27. By how many points did his score rise?

▶ **Answers to You Try It** 1. 4 minus 9 2. negative 10 minus negative 7 3. negative 2 minus positive 5 4. -6 5. 3 6. -35 7. -41 8. -5 9. -1 10. 13 11. 70 12. 3 13. 10 14. 15 15. 12 16. -2 17. -8 18. 61 points

SECTION 5.3 EXERCISES

1 *Subtract by adding the opposite of the subtrahend.*

1. $8 - 5$
2. $12 - 9$
3. $3 - 7$
4. $1 - 8$
5. $-5 - 5$
6. $0 - 3$
7. $0 - 10$
8. $-12 - 12$
9. $7 - (-4)$
10. $5 - (-8)$
11. $-9 - (-4)$
12. $-1 - (-4)$
13. $-10 - (10)$
14. $-2 - (2)$
15. $-8 - (-8)$
16. $-3 - (-3)$
17. $16 - 21$
18. $26 - 28$
19. $0 - (-1)$
20. $0 - (-15)$
21. $-12 - (-4)$
22. $-25 - (-20)$
23. $16 - (-2)$
24. $2 - (-7)$
25. $-80 - (-80)$
26. $-1 - (-1)$
27. $-100 - (-3)$
28. $-9 - (-4)$
29. $-40 - (-140)$
30. $-9 - (-79)$
31. $16 - (-14)$
32. $4 - (-24)$
33. $8 - 0$
34. $-5 - 0$
35. $3 - (-3)$
36. $9 - (-9)$
37. $0 - (-50)$
38. $0 - (-12)$
39. $-8 + 9$
40. $-12 + 8$
41. $-10 + 10$
42. $-23 + (23)$
43. $-30 - (-45)$
44. $6 - (-40)$
45. $-9 + 4 - (-5)$
46. $-2 + 6 - (-4)$
47. $8 - (-3) - 8$
48. $20 - (-5) - 20$
49. $4 - (-7) - 1$
50. $3 - (-5) - 9$
51. $-4 + 7 + 2 - 8 - 3$
52. $+5 - 1 - 12 + 5 - 9$
53. $8 - (-12) - 8 + 3 - 7$
54. $16 + (-12) - (-14) - 2$

2 *Draw and label a number line for each problem.*

55. Al's checkbook balance yesterday was −\$57. After a deposit his balance was \$140. How large was the deposit?

56. George had a score of −38 in a game of cards. His score is now 14. By how many points did his score rise?

57. Zelda had a checkbook balance of \$8. She wrote checks for \$26 and \$41. She then made a \$60 deposit. What is her new balance?

58. A business had a bank balance of −\$400. The owner wrote checks for \$120, \$67, and \$240 and made a \$900 deposit. What was the new balance?

59. At 2 A.M. the temperature was −21°. By noon, it was 17°. By how many degrees did the temperature rise?

60. The temperature at an Arctic station at midnight was −56°. By noon the next day the temperature was −18°. By how many degrees did the temperature rise?

61. The ocean surface is at 0 ft elevation. A diver is underwater at the base of a rock formation. Her elevation is −178 ft. The top of the formation has an elevation of −113 ft. What is the height of the rock formation?

62. An underwater missile was launched from a −251 ft elevation. It reached a height of 1,375 ft above the surface of the water. What vertical distance did the missile travel?

SKILLSFOCUS (Section 2.5) *Perform the division the easy way.*

63. $245.367 \div 100$ 64. $\frac{345}{1000}$ 65. $10\overline{)0.0546}$ 66. $89.01 \div 1000$ 67. $\frac{67.1}{100}$

WRITING TO LEARN ▶▶▶

68. Write a short essay to explain how to subtract (−8) − (−3).

69. Write a word problem whose answer is given by the expression 21 − (−6). Then solve it.

70. What is (−9) − (−14)? Reverse the integers. What is (−14) − (−9)?
   a. What is similar about the two answers?
   b. How are the two answers different?
   c. What happens when you reverse the order of a subtraction?
   d. Is subtraction of integers commutative?

5.4 MULTIPLYING AND DIVIDING INTEGERS

OBJECTIVES

1. Multiply two integers with unlike signs.
2. Multiply two integers with like signs.
3. Multiply by 1, −1, and 0.
4. Divide two integers.

NEW VOCABULARY

multiplicative identity

1 Multiplying Two Integers with Unlike Signs

The product of two positive integers is positive. What sign do you give the product of a negative integer and a positive integer? Find the answer by looking for the pattern in the following products.

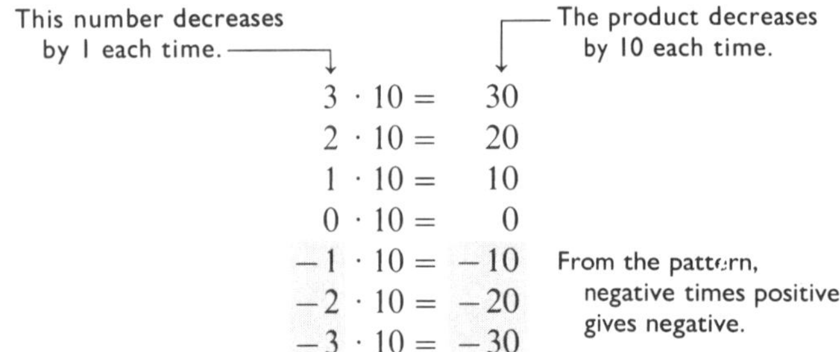

> To multiply a negative integer times a positive integer
> 1. Multiply the absolute values.
> 2. Attach a negative sign to the product.

EXAMPLE 1 Multiply $(-5) \cdot (4)$.

$(-5) \cdot (4) = -20$ — Multiply the absolute values: $5 \cdot 4 = 20$. Since one integer is negative and the other positive, attach a negative sign to the product. ■

EXAMPLE 2 $7 \cdot (-3) = -21$ — Multiply the absolute values: $7 \cdot 3 = 21$. Attach a negative sign. ■

EXAMPLE 3 $-6 \cdot 8 = -48$ ■

EXAMPLE 4 $(+10)(-10) = -100$ — Multiplication is implied. ■

EXAMPLE 5 $5(-2) = -10$ — Multiplication is implied. ■

OBSERVE In Example 3, the multiplication $-6 \cdot 8$ can be rewritten in any of the following equivalent ways.

$-6 \cdot 8$	$(-6) \cdot (+8)$	$-6(8)$	$-6(+8)$
$-6 \cdot (8)$	$(-6) \cdot (8)$	$(-6)(8)$	$(-6)(+8)$
$-6 \cdot (+8)$	$(-6) \cdot 8$	$(-6)8$	

▶ **You Try It**

1. $(-3) \cdot (7)$	**2.** $(-6) \cdot (+2)$	**3.** $5 \cdot (-9)$
4. $8 \cdot (-8)$	**5.** $(+1)(-1)$	**6.** $(-10)(+5)$
7. $(9)(-6)$	**8.** $(-4) \cdot 0$	**9.** $(-7)5$

EXAMPLE 6 What is -4 added to itself 42 times?

This is repeated addition. Multiply for the answer. (See Section 1.4.)

$$-4 \text{ added to itself 42 times} = -4 \cdot 42 = -168 \quad \blacksquare$$

▶ **You Try It** **10.** What is -7 added to itself 36 times?

2 Multiplying Two Integers with Like Signs

What sign do you give the product of a negative integer and a negative integer? Find the answer by looking for the pattern in the following products.

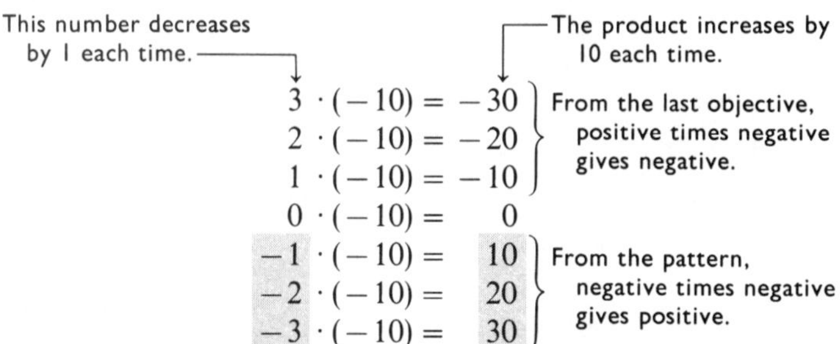

> To multiply a negative integer times a negative integer
>
> **1.** Multiply the absolute values.
> **2.** Attach a positive sign to the product.

EXAMPLE 7 Multiply $-6 \cdot (-5)$.

$-6 \cdot (-5) = +30$ or 30 — Multiply the absolute values: $6 \cdot 5 = 30$. Since both integers are negative, attach a positive sign to the product. ■

EXAMPLE 8 $(-4) \cdot (-7) = +28$ or 28 — Multiply the absolute values: $4 \cdot 7 = 28$. Attach a positive sign to the product. ■

EXAMPLE 9 $-3(-4) = 12$ — The multiplication is implied. ■

EXAMPLE 10 $(-8)(-10) = 80$ ■

▶ **You Try It**

11. $-8 \cdot (-3)$ **12.** $-6 \cdot (-2)$ **13.** $(-4) \cdot (-1)$
14. $(-10)(-10)$ **15.** $(-9)(-7)$ **16.** $-20(-2)$

To multiply three or more integers, multiply the two on the left. Multiply this product by the next integer on the right, and so on.

EXAMPLE 11 Multiply $-4(-2)(3)(-1)$.

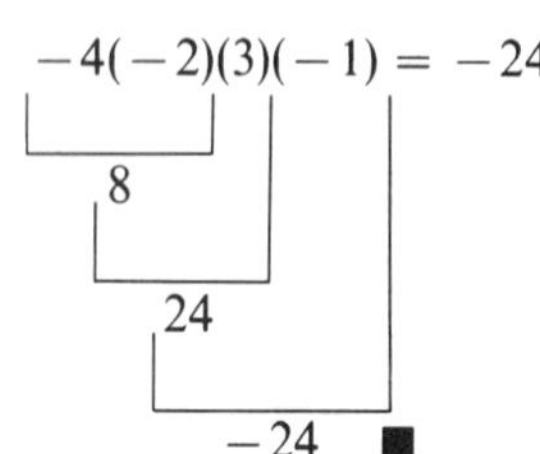

▶ **You Try It** 17. $-6(-4)(-2)$ 18. $-8(5)(-1)\cdot 3$

3 Multiplying by 1, −1, and 0

The **multiplicative identity** for integers is the number 1. Multiplying 1 times any integer gives the same integer.

EXAMPLE 12

a. $1\cdot(-4)=-4$ **b.** $(-4)\cdot 1=-4$
c. $(-5)1=-5$ **d.** $1(-5)=-5$
e. $1\cdot x=x$ **f.** $x\cdot 1=x$
g. $1y=y$ **h.** $y1=y$ ■

Multiplying an integer by −1 gives the opposite of the integer.

EXAMPLE 13

a. $-1\cdot(+6)=-6$ **b.** $-1\cdot(-6)=+6$ or 6
c. $-1(-15)=15$ **d.** $4(-1)=-4$
e. $(-7)(-1)=7$ **f.** $-1\cdot p=-p$ ■

0 times any integer equals 0.

EXAMPLE 14

a. $0\cdot(-6)=0$ **b.** $(0)(+5)=0$
c. $(-8)0=0$ ■

▶ **You Try It**

19. $1\cdot(-8)$ 20. $(-6)1$ 21. $1\cdot m$
22. $-1\cdot(3)$ 23. $-1\cdot(-5)$ 24. $(-9)(-1)$
25. $0\cdot(+2)$ 26. $(-25)0$

4 Dividing Two Integers

Recall from Chapter 2, every division has a related multiplication. This fact can be used to show that the rules for dividing integers are the same as the rules for multiplying integers.

Division		*Related Multiplication*	
$\frac{10}{2}=5$	because	$2\cdot 5=10$	Normal check for division.
$\frac{-10}{2}=-5$	because	$2\cdot(-5)=-10$	Dividing a positive and a negative gives a negative.
$\frac{10}{-2}=-5$	because	$-2\cdot(-5)=10$	
$\frac{-10}{-2}=+5$	because	$-2\cdot(+5)=-10$	Dividing a negative by a negative gives a positive.

> To divide (or multiply) two integers
>
> **1.** Divide (or multiply) the absolute values.
> The sign of the answer is
> **a.** positive if both integers are positive or both are negative.
> **b.** negative if one integer is negative and the other is positive.

EXAMPLE 15 Divide $\frac{-40}{5}$.

$$\frac{-40}{5} = -8$$ Divide the absolute values: $\frac{40}{5} = 8$. The signs are unlike, so the answer is negative.

Check: $5 \times (-8) = -40$ ■

▶ **You Try It** **27.** $\frac{-64}{8}$ **28.** $\frac{+30}{-6}$

EXAMPLE 16 $\frac{-24}{-4} = +6$ or 6 Divide the absolute values: $\frac{24}{4} = 6$.

The signs are both negative, so the answer is positive.

Check: $(-4)(6) = -24$ ■

EXAMPLE 17 $-36 \div (-9) = 4$

Check: $(-9) \cdot 4 = -36$ ■

EXAMPLE 18 $\frac{63}{-7} = -9$

Check: $(-7) \cdot (-9) = 63$ ■

▶ **You Try It** **29.** $\frac{-27}{-9}$ **30.** $\frac{30}{-2}$ **31.** $-28 \div (-7)$ **32.** $\frac{-10}{10}$

EXAMPLE 19 How many -2 are in -38?

To find how many a are in b, write $b \div a$. (See Section 2.5.)

$$\begin{aligned}\text{The number of } -2 \text{ in } -38 &= -38 \div -2 \\ &= 19\end{aligned}$$ ■

▶ **You Try It** **33.** How many -3 are in -24?

Summary

To multiply or divide two integers

1. Multiply or divide the two absolute values.
2. The sign of this answer is
   a. positive if both integers are positive or both are negative.
   b. negative if one integer is negative and the other is positive.

▶ **Answers to You Try It** 1. −21 2. −12 3. −45 4. −64 5. −1 6. −50 7. −54 8. 0 9. −35 10. −252 11. 24 12. 12 13. 4 14. 100 15. 63 16. 40 17. −48 18. 120 19. −8 20. −6 21. m 22. −3 23. 5 24. 9 25. 0 26. 0 27. −8 28. −5 29. 3 30. −15 31. 4 32. −1 33. 8

SECTION 5.4 EXERCISES

1 2 3 *Perform each multiplication.*

1. $5 \cdot (+7)$ **2.** $6 \cdot (+4)$ **3.** $(3)(2)$ **4.** $(+5)(+5)$

5. $-6 \cdot (+5)$ **6.** $-4 \cdot (+7)$ **7.** $-8 \cdot 2$ **8.** $-5 \cdot 3$

9. $4(-5)$ **10.** $7(-2)$ **11.** $-1 \cdot (9)$ **12.** $-1 \cdot (1)$

13. $(-10) \cdot 0$ **14.** $(-3) \cdot 0$ **15.** $-8(-7)$ **16.** $-10(-4)$

17. $-15(-1)$ **18.** $-3(-1)$ **19.** $(-6) \cdot (-6)$ **20.** $(-4)(-3)$

21. $(8) \cdot (-3)$ **22.** $(5) \cdot (-6)$ **23.** $-9(3)$ **24.** $-2(7)$

25. $0(-1)$ **26.** $0(-120)$ **27.** $(-3) \cdot (-9)$ **28.** $(-10) \cdot (-10)$

29. $6(-12)$ **30.** $8(-6)$ **31.** $(-12)(-2)$ **32.** $(-3)(-10)$

33. $-2 \cdot (-4)(-2)$ **34.** $-3 \cdot (-1)(-5)$ **35.** $-1 \cdot (-6)(2) \cdot (4)$ **36.** $-9 \cdot (-8)(7)(0)$

37. $5(-2)(-3)(-1)$ **38.** $2(-3)(-3)(-2)$ **39.** $-15(-12)(0)$ **40.** $-8(-2)(2)$

41. What is -7 added to itself 60 times?

42. What is -3 added to itself 45 times?

43. What is $6 \cdot t$ if $t = -8$?

44. What is $-3 \cdot v$ if $v = -4$?

45. What is $-5 \cdot n$ if $n = -7$?

46. What is $-10 \cdot y$ if $y = +6$?

4 *Divide. Check using the related multiplication.*

47. $\frac{-30}{5}$ **48.** $\frac{-18}{6}$ **49.** $\frac{24}{-3}$ **50.** $\frac{40}{-8}$

51. $\frac{-45}{-9}$ **52.** $\frac{-30}{-6}$ **53.** $\frac{8}{-1}$ **54.** $\frac{6}{-1}$

55. $\frac{60}{-5}$ **56.** $\frac{100}{-10}$ **57.** $\frac{-90}{0}$ **58.** $\frac{0}{-25}$

59. $\frac{-80}{10}$

60. $\frac{-65}{5}$

61. $\frac{200}{-40}$

62. $\frac{300}{-3}$

63. $\frac{-72}{-8}$

64. $\frac{-84}{-7}$

65. $\frac{0}{7}$

66. $\frac{+64}{0}$

67. $\frac{-35}{-1}$

68. $\frac{-7}{-1}$

69. $\frac{-480}{60}$

70. $\frac{-600}{-50}$

71. $12 \div (-2)$

72. $30 \div (0)$

73. $-36 \div 0$

74. $-70 \div (-7)$

75. $(-8) \div (-1)$

76. $-90 \div (-1)$

77. $0 \div (-2)$

78. $0 \div (-5)$

79. Find $\frac{k}{-4}$ when $k = 28$.

80. Find $\frac{p}{-12}$ when $p = -36$.

81. What is $\frac{-35}{t}$ when $t = -5$?

82. What is $\frac{-88}{y}$ when $y = 11$?

83. How many -3 are in 60?

84. How many 8 are in -72?

85. How many -1 are in -15?

86. How many 10 are in -300?

SKILLSFOCUS (Section 3.2) *Reduce.*

87. $\frac{12}{18}$

88. $\frac{21}{36}$

89. $\frac{72}{27}$

90. $\frac{84}{240}$

EXTEND YOUR THINKING ▶▶▶▶

▶ TROUBLESHOOT IT

Find and correct the error.

91. $(-2)(-3)(-7) = (-5)(-7) = +35$

92. $\frac{-100}{+50} = -50$

WRITING TO LEARN ▶▶▶▶

93. In your own words, explain how to multiply -4 by -5.

94. Explain the difference between the two divisions: $\frac{0}{-4}$ and $\frac{-4}{0}$. Write your answer for each with a reason why.

▶ YOU BE THE JUDGE

95. Is division of integers commutative? Explain your decision using your own examples.

96. If $3 \cdot \$10 = \$30$ can be interpreted as selling 3 tickets at \$10 each, for a total of \$30, how can you interpret $3 \cdot (-\$10) = -\$30$?

5.5 ORDER OF OPERATIONS

OBJECTIVES

1 Evaluate powers of integers.

2 Apply the order of operations to integers.

3 Evaluate algebraic expressions involving integers.

1 Powers of Integers

Recall from Section 2.2 that a power is a short way to write a repeated multiplication.

EXAMPLE 1 Simplify the powers.

a. $5^2 = 5 \cdot 5 = 25$

b. $(+4)^3 = (+4)(+4)(+4) = +64$

$(+4)(+4) = +16$; $(+16)(+4) = +64$

c. $(-6)^2 = (-6)(-6) = 36$

d. $(-2)^5 = (-2)(-2)(-2)(-2)(-2) = -32$

$(-2)(-2) = +4$; $(+4)(-2) = -8$; $(-8)(-2) = +16$; $(+16)(-2) = -32$

e. $(-3)^4 = (-3)(-3)(-3)(-3) = 81$

$(-3)(-3) = +9$; $(+9)(-3) = -27$; $(-27)(-3) = +81$

f. $(-10)^3 = (-10)(-10)(-10) = -1{,}000$

$(-10)(-10) = +100$; $(+100)(-10) = -1{,}000$ ■

> A negative integer raised to an even power is positive (see Examples 1c and 1e).
> A negative integer raised to an odd power is negative (see Examples 1d and 1f).

▶ You Try It Evaluate.

1. 8^2	**2.** $(+5)^3$	**3.** $(-9)^2$
4. $(-1)^9$	**5.** $(-3)^3$	**6.** $(-1)^8$
7. $(-10)^5$	**8.** $(7)^3$	**9.** $(-2)^7$

EXAMPLE 2 Show that $(-3)^2 \neq -3^2$.

$$(-3)^2 = (-3)\cdot(-3) = 9$$

The negative sign is in the parentheses with the 3. So -3 is multiplied twice.

$$-3^2 = -(3\cdot 3) = -(9) = -9$$

The negative sign is not written in parentheses with the 3. So only the 3 is multiplied twice.

Therefore, $(-3)^2 \neq -3^2$. ■

OBSERVE You may also write $-3^2 = -1\cdot 3^2 = -1\cdot 9 = -9$.

EXAMPLE 3 **a.** $(-6)^2 = (-6)\cdot(-6) = 36$

b. $-6^2 = -(6\cdot 6) = -36$ ■

▶ **You Try It** Evaluate.

10. $(-8)^2$ **11.** -8^2 **12.** $(-1)^2$ **13.** -1^2

14. -7^2 **15.** $(-7)^2$ **16.** -2^2 **17.** $(-2)^2$

2 Order of Operations

The order of operations for integers is the same as for whole numbers. Review the procedure in Section 2.6.

EXAMPLE 4 Simplify $-40 \div 8(-4)$.

$$\begin{aligned} -40 \div 8(-4) &= \underbrace{-40 \div 8}\cdot(-4) && \text{Insert the multiplication symbol.}\\ &= -5\cdot(-4) && \text{Multiply and divide in order from left to right. Divide first.}\\ &= 20 \end{aligned}$$ ■

EXAMPLE 5 Evaluate $-6 + 6(-2)$.

$$\begin{aligned} -6 + 6(-2) &= -6 + \underbrace{6\cdot(-2)}\\ &= -6 + (-12) && \text{Do multiplication first. Then add.}\\ &= -18 \end{aligned}$$ ■

▶ **You Try It** **18.** $-60 \div 10(-2)$ **19.** $-8 + 8(-3)$

EXAMPLE 6 Find $7 - 5 - 3(-4)$.

$$\begin{aligned} 7 - 5 - 3(-4) &= 7 - 5 - \underbrace{3\cdot(-4)}\\ &= 7 - 5 - (-12) && \text{Multiply first: } 3\cdot(-4) = -12.\\ &= 2 - (-12) && \text{Subtract from left to right.}\\ &= 2 + (+12)\\ &= 14 \end{aligned}$$ ■

▶ **You Try It** 20. $10 - 6 - 4(-7)$

EXAMPLE 7 Evaluate $(-6)^2(2) + \sqrt{49}(-10)$.

$$\begin{aligned}(-6)^2(2) + \sqrt{49}(-10) &= \underbrace{(-6)^2} \cdot (2) + \underbrace{\sqrt{49}} \cdot (-10) && \text{Powers and roots}\\ &= \underbrace{(+36) \cdot (2)} + \underbrace{7 \cdot (-10)} && \text{Multiply.}\\ &= 72 + (-70) && \text{Add.}\\ &= 2 \quad \blacksquare\end{aligned}$$

▶ **You Try It** 21. $(-4)^2(3) + \sqrt{81}(-6)$

3 Evaluating Algebraic Expressions

Algebraic expressions can also be evaluated using negative numbers. To avoid confusion with signs, use parentheses when replacing a letter with a negative number.

EXAMPLE 8 Evaluate $5 - y$ when $y = -8$.

subtraction sign — negative sign

$$\begin{aligned}5 - y &= 5 - (-8) && \text{Replace } y \text{ with } -8. \text{ Enclose } -8 \text{ in parentheses to keep the subtraction sign and the negative sign separate.}\\ &= 5 + (+8)\\ &= 13 \quad \blacksquare\end{aligned}$$

▶ **You Try It** 22. Evaluate $7 - x$ when $x = -9$.

EXAMPLE 9 Find $5k + 2m$ when $k = -2$ and $m = 4$.

$$\begin{aligned}5k + 2m &= 5 \cdot k + 2 \cdot m\\ &= \underbrace{5 \cdot (-2)} + \underbrace{2 \cdot (4)}\\ &= -10 + (8)\\ &= -2 \quad \blacksquare\end{aligned}$$

▶ **You Try It** 23. Evaluate $3p + 7w$ when $p = -4$ and $w = 3$.

EXAMPLE 10 Evaluate $-7x^2$ when $x = -3$.

$$\begin{aligned}-7x^2 &= -7 \cdot x^2\\ &= -7 \cdot (-3)^2 && \text{Replace } x \text{ with } -3.\\ &= -7 \cdot (+9) && \text{Evaluate the exponent before multiplying.}\\ &= -63 \quad \blacksquare\end{aligned}$$

▶ **You Try It** 24. Evaluate $-5w^2$ when $w = -4$.

EXAMPLE 11 Evaluate $b^2 - 4ac$ when $b = 3$, $a = 1$, and $c = -4$.

$$\begin{aligned} b^2 - 4ac &= b^2 - 4 \cdot a \cdot c \\ &= \underbrace{3^2} - 4 \cdot 1 \cdot (-4) \quad \text{Evaluate the exponent first.} \\ &= 9 - \underbrace{(4 \cdot 1)} \cdot (-4) \\ &= 9 - \underbrace{(4) \cdot (-4)} \\ &= 9 - (-16) \\ &= 9 + (+16) \\ &= 25 \quad \blacksquare \end{aligned}$$

▶ **You Try It** **25.** Evaluate $b^2 - 4ac$ when $b = 6$, $a = 2$, and $c = -5$.

▶ **Answers to You Try It** 1. 64 2. 125 3. 81 4. -1 5. -27 6. 1 7. $-100{,}000$ 8. 343 9. -128 10. 64 11. -64 12. 1 13. -1 14. -49 15. 49 16. -4 17. 4 18. 12 19. -32 20. 32 21. -6 22. 16 23. 9 24. -80 25. 76

SECTION 5.5 EXERCISES

1 *Evaluate each expression.*

1. $(+5)^2$

2. $(+8)^2$

3. $(-6)^2$

4. $(-7)^2$

5. $(-10)^2$

6. $(-1)^2$

7. -6^2

8. -7^2

9. $(-3)^3$

10. $(-2)^4$

11. $(-4)^2$

12. $(-1)^7$

13. -3^3

14. -2^4

15. -4^2

16. -1^7

2

17. $6(3-9)$

18. $4(6-8)$

19. $-3(1-4+2)$

20. $-7(8-5-3)$

21. $6\cdot(-3)^2$

22. $4\cdot(-5)^2$

23. $10\div(-2)\cdot 5$

24. $-60\div(6)\cdot(-10)$

25. $4+6(-2)$

26. $6+3(-7)$

27. $-8+4(2)$

28. $-1+1(1)$

29. $-9+3\cdot(-4)$

30. $-4+4\cdot(-5)$

31. $9-9(-4)$

32. $6-8(-1)$

33. $-3-3(-3)$

34. $-8-5(-2)$

35. $-4\cdot(6)\div(-3)$

36. $8\cdot(-9)\div(-6)$

37. $-100\div(50)(2)$

38. $-80\div(20)\cdot(-4)$

39. $-6+5^2$

40. $-4+(-2)^2$

41. $(-2)(-4)+(-3)(-8)$

42. $(-1)(-2)+(-4)(5)$

43. $60\div(-10)+55\div(11)$

44. $(-36)\div(-9)+(-2)(9)$

45. $(-6)(-6)-(-4)(-4)$

46. $(-1)(-1)-(-2)(-3)$

47. $\dfrac{-40}{(-5)(8)}$

48. $\dfrac{(-9)(-8)}{6(-6)}$

49. $\dfrac{(-6)-(-2)}{(-8)-(-4)}$

50. $\dfrac{(10)-(-10)}{(-2)-(3)}$

51. $\dfrac{(-8)(-6)}{(-4)(-3)}$

52. $\dfrac{(-40)(60)}{(20)(-30)}$

3 *Evaluate each expression using* $x=-2$, $y=3$, $a=-4$, *and* $b=-1$.

53. $8-x$

54. $3-b$

55. $2y-7b$

56. $-4a+1$

57. $4-a-y$

58. $12-y-1$

59. $6a^2$

60. $-4x^2$

61. $b^2 - 4ay$

62. $y^2 - 4xb$

63. $1 - 3x$

64. $-8 - 4b$

65. axy

66. abx

67. $b - 9$

68. $x + a$

69. $-7bx$

70. $3ab$

71. $b - 1$

72. $x - 2y$

73. $(x + y)^2$

74. $(b - a)^3$

75. $x^2 - a^2$

76. $(ax)/(2b)$

SKILLSFOCUS (Section 2.7) *Use the correct geometric formula to find the quantity indicated.*

77. The perimeter of a rectangle with length 4.5 feet and width 3.1 feet.

78. The volume of a box with length 2 yards, width 4.2 yards and height 1.6 yards.

79. The area of a triangle with base 4 in. and height $2\frac{1}{2}$ in.

80. The hypotenuse of a right triangle with sides 15 meters and 20 meters.

EXTEND YOUR THINKING

SOMETHING MORE

81. Simplify $(20 - (19 - (18 - (\ldots(3 - (2 - 1))\ldots))))$.

82. Given $a = 4$, $b = -5$, $x = 3$, and $y = -2$, do the following.
   a. Write three expressions which, when evaluated, give 13. (Hint: $5x + y$ is an example because $5x + y = 5 \cdot 3 + (-2) = 15 + (-2) = 13$.)
   b. Write an expression that evaluates to 0.
   c. Write an expression that evaluates to -20.
   d. Write an expression that evaluates to -37.

TROUBLESHOOT IT

Find and correct the error.

83. $(-4)^3 = (-4)(-4)(-4) = -12$

84. $-5^2 = (-5)(-5) = 25$

85. $(-100) \div (10) \cdot (-10) = (-100) \div (-100) = 1$

WRITING TO LEARN

86. Evaluate $-w$ when $w = -6$. Explain why the sign of the answer is positive.

YOU BE THE JUDGE

87. Maria claims the product of an even number of negative integers always gives a positive answer. Is she correct? Explain your decision using your own examples.

88. Kim claims the product of an odd number of negative integers always gives a negative answer. Is she correct? Explain your decision using your own examples.

5.6 SCIENTIFIC NOTATION

OBJECTIVES

1. Define scientific notation.
2. Write numbers greater than 1 in scientific notation.
3. Write numbers between 0 and 1 in scientific notation.

NEW VOCABULARY

scientific notation

1 Scientific Notation

Scientific notation is a convenient way to express very large or very small numbers. Computers and some calculators display answers in scientific notation. It is used extensively in science. For example, in chemistry, Avogadro's number (the number of atoms in 16 grams of oxygen) is

602,000,000,000,000,000,000,000.

It is much more convenient to express this number in scientific notation as

$$6.02 \times 10^{23}.$$

must be a number 1 or larger, but less than 10 (6.02) — times (×) — a power of 10 (10^{23})

> A number is written in scientific notation if it is written in the form
>
> $$a \times 10^n.$$
>
> a is greater than (or equal to) 1 and less than 10. — The power n is an integer.

The number 5.2×10^3 is written in scientific notation. To change it to decimal notation, apply the order of operations.

$5.2 \times 10^3 = 5.2 \times 1{,}000$ — Evaluate the power.

$= 5{,}200$ — Multiply.

$10^3 = 1{,}000$ has 3 zeros. To multiply 5.2 by 1,000, just move the decimal point 3 places right (see Section 2.1).

$5.2 \times 1{,}000 = 5\ 200.$ or 5,200

3 zeros — 3 places right

> To multiply by a power of 10 when the exponent on 10 is positive, move the decimal point to the right the number of places equal to the power on 10.

EXAMPLE 1 Change from scientific notation to decimal notation.

a. $6 \times 10^2 = 6\ 00. = 600$

2 zeros — 2 places right

b. $1.076 \times 10^5 = 1\ 07600. = 107{,}600$

5 zeros — 5 places right ■

▶ You Try It Change to decimal notation.

1. 8.93×10^4

2. 9×10^7

2 Writing Numbers Greater Than 1 in Scientific Notation

Write 6,800 in scientific notation.

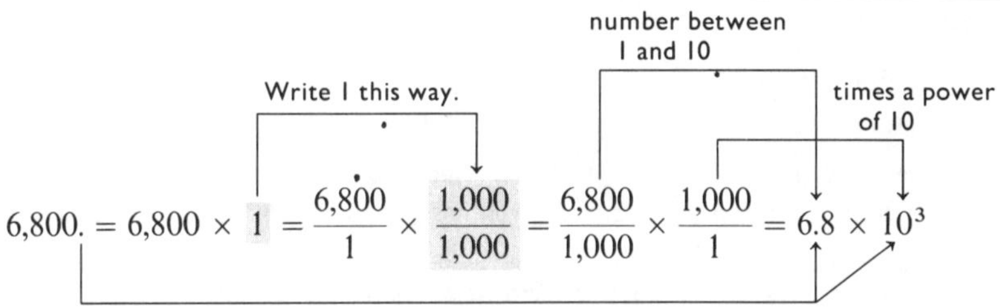

In short, move the decimal point 3 places left. Multiply by 10 raised to the power of 3.

In summary,

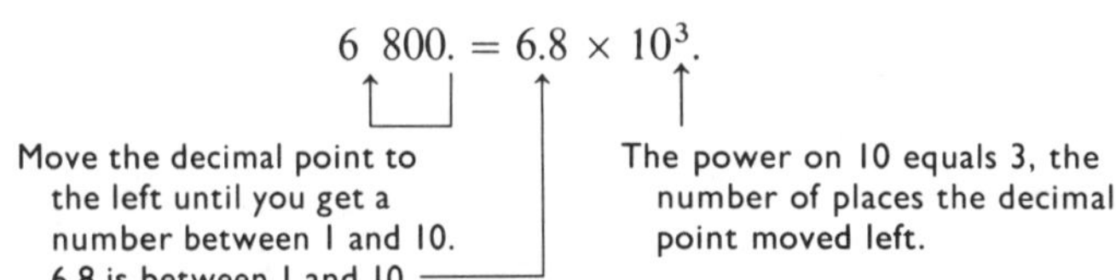

Check: $6.8 \times 10^3 = 6.8 \times 1{,}000 = 6{,}800$

> To write a decimal number greater than 1 in scientific notation
>
> **1.** Move the decimal point to the left until you get a number 1 or larger and less than 10. (At this point, there will be one digit to the left of the decimal point.)
>
> **2.** Count the number of places the decimal point moved left. Write this number as the power on 10.
>
> **3.** Multiply the number in Step 1 by the power of 10 in Step 2.

EXAMPLE 2 Write each number in scientific notation.

a. $45.3 = 4\ 5.3 = 4.53 \times 10^1$

Move the decimal point 1 place left to get 4.53, a number between 1 and 10.

The power on 10 is 1, because the decimal point moved 1 place left.

b. $300 = 3\ 00. = 3 \times 10^2$

number between 1 and 10

2 places left

c. $62{,}490{,}000 = 6\ 2\ 490\ 000. = 6.249 \times 10^7$

number between 1 and 10

7 places left ■

You Try It Write each number in scientific notation.

3. 73.2 **4.** 506 **5.** 810,000 **6.** 6,073.4

---

3 Writing Numbers between 0 and 1 in Scientific Notation

10^{-1}, 10^{-2}, 10^{-3}, and so on, are needed to write numbers between 0 and 1 in scientific notation. To understand what these numbers mean, try to find the pattern with the following powers of 10.

Pattern: Using your calculator, divide by 10 to get the power of 10 on the next line.

$$1{,}000 = 10 \times 10 \times 10 = 10^3$$
$$100 = 10 \times 10 = 10^2$$
$$10 = 10 = 10^1$$
$$1 = 1 = 10^0$$
$$0.1 = \frac{1}{10} = \frac{1}{10^1} = 10^{-1}$$
$$0.01 = \frac{1}{100} = \frac{1}{10^2} = 10^{-2}$$
$$0.001 = \frac{1}{1{,}000} = \frac{1}{10^3} = 10^{-3}$$

Pattern: The powers on 10 decrease by 1 each time, following the number line from right to left.

Therefore, $10^{-1} = \frac{1}{10^1}$, $10^{-2} = \frac{1}{10^2}$, $10^{-3} = \frac{1}{10^3}$, and so on. As a result, you can write

$$\frac{4.3}{10^2} = \frac{4.3}{1} \times \frac{1}{10^2} = 4.3 \times 10^{-2}.$$

Dividing by 10^2 is the same as multiplying by 10^{-2}.

Write 0.043 in scientific notation. To do this, write 0.043 as a number between 1 and 10, times a power of 10.

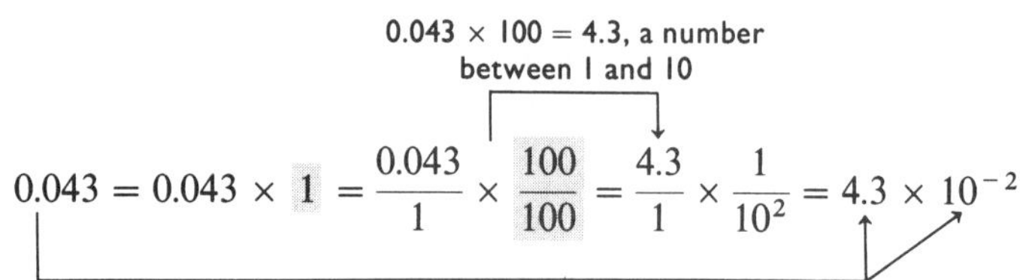

This is summarized as follows.

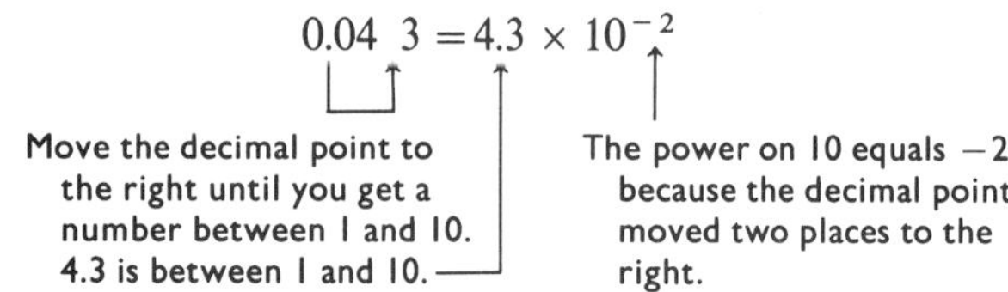

To write a decimal number between 0 and 1 in scientific notation

1. Move the decimal point to the right until you get a number between 1 and 10. (At this point, there will be one nonzero digit to the left of the decimal point.)
2. Count the number of places the decimal point moved right. Write this number as a negative power on 10.
3. Multiply the number in Step 1 by the power of 10 in Step 2.

EXAMPLE 3 Write each number in scientific notation.

a. $0.00952 = 0.009\ 52 = 9.52 \times 10^{-3}$

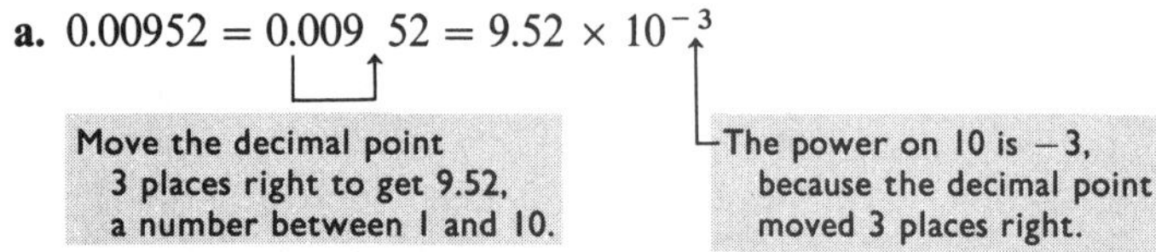

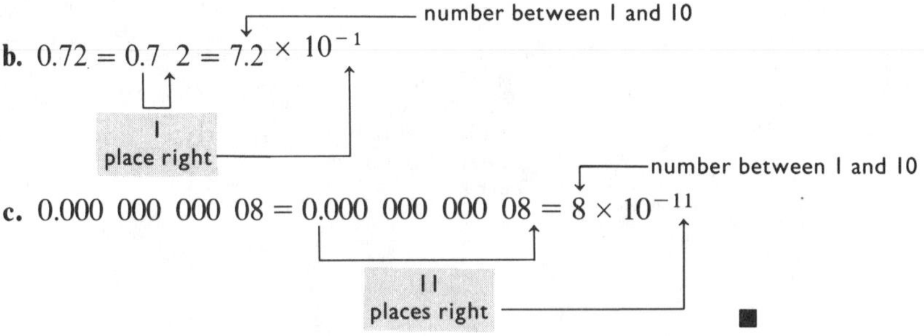

c. $0.000\ 000\ 000\ 08 = 0.000\ 000\ 000\ 08 = 8 \times 10^{-11}$

(number between 1 and 10; 11 places right) ■

OBSERVE From Example 2, a decimal number between 0 and 1 has a negative power on 10 when written in scientific notation.

▶ **You Try It** Write each number in scientific notation.

7. 0.00006 **8.** 0.904 **9.** 0.000 000 027 **10.** 0.013

> To change a number from scientific notation to decimal notation when the power on 10 is negative, move the decimal point to the left the number of places indicated by the power on 10.

EXAMPLE 4 Change from scientific to decimal notation.

a. $4.9 \times 10^{-2} = 0.04\ 9 = 0.049$

−2 power means move the decimal point 2 places left

b. $3.706 \times 10^{-5} = 0.00003\ 706 = 0.00003706$

−5 power means move the decimal point 5 places left

c. $8 \times 10^{-3} = 0.008 = 0.008$

−3 power means move the decimal point 3 places left

▶ **You Try It** Write each number in decimal notation.

11. 6.02×10^{-4} **12.** 1.73×10^{-1} **13.** 3×10^{-7} **14.** 9.043×10^{-9}

OBSERVE From Example 3, a negative power on 10 means the number in scientific notation is between 0 and 1 when changed to decimal notation.

▶ **Answers to You Try It** 1. 89,300 2. 90,000,000 3. 7.32×10^1 4. 5.06×10^2 5. 8.1×10^5 6. 6.0734×10^3 7. 6×10^{-5} 8. 9.04×10^{-1} 9. 2.7×10^{-8} 10. 1.3×10^{-2} 11. 0.000602 12. 0.173 13. 0.0000003 14. 0.000000009043

CALCULATOR TIPS

The Opposite Key, [+/−], and Integers Some calculators have an opposite key, [+/−]. Each time you press [+/−], you change the sign of the number on the display to its opposite. Such a key is called a *toggle*.

4 [+/−] Displays −4 on the calculator.
−4 [+/−] Displays 4 on the calculator.

EXAMPLE 1 Simplify −(−7) using the [+/−] key.

−(−7): Enter 7 Display shows 7.
Press [+/−] Display shows −7.
Press [+/−] Display shows 7.
Therefore, −(−7) = 7. ■

EXAMPLE 2 Add −6 + (+2).

6 [+/−] [+] 2 [=] −4 ■

EXAMPLE 3 Add −432 + (−68) + 140 + (−275) + 600.

432 [+/−] [+] 68 [+/−] [+] 140 [+] 275 [+/−] [+] 600 [=] −35 ■

EXAMPLE 4 Use your calculator to subtract −8 − (−3).

−8 − (−3) = 8 [+/−] [−] 3 [+/−] [=] −5 ■

EXAMPLE 5 Multiply and divide.

a. −8(−4) = 8 [+/−] [×] 4 [+/−] [=] 32
b. (4)(−5)(−2) = 4 [×] 5 [+/−] [×] 2 [+/−] [=] 40
c. $\frac{-120}{8}$ = 120 [+/−] [÷] 8 [=] 15 ■

Use your calculator to simplify each problem.

1. −(−9)
2. −(+4)
3. (−6) + (−5)
4. (+8) + (−17)
5. (−260) + (−427) + 216 + (−73) + (−92)
6. 12 − (−3)
7. (−7) − (−7)
8. (+10)(−8)
9. (−2)(+6)(−3)(−5)
10. $\frac{-60}{-4}$
11. $\frac{-267}{89}$
12. Explain why 0 [−] 9 [=] gives the same result as 9 [+/−].

Square Roots of Negative Integers You know $\sqrt{9} = 3$ because $3^2 = 3 \cdot 3 = 9$ (see Section 2.4). But what is $\sqrt{-9}$? Is the answer $+3$ or -3? Now $(+3)^2 = (+3)(+3) = 9$, not -9. And $(-3)^2 = (-3)(-3) = 9$, not -9. So, $\sqrt{-9}$ has no integer answer. You can demonstrate this on your calculator. To find $\sqrt{-9}$, enter

9 [+/−] [√].

You see E on your calculator display. E means error. Your calculator is telling you $\sqrt{-9}$ has no integer answer. The square root of any negative integer is not an integer.

Calculators and Scientific Notation Use your calculator to multiply $5{,}000{,}000 \times 7{,}000{,}000$. If your calculator displays the letter E (for error), then your calculator cannot express answers in scientific notation. The E means the answer is too big to fit on the calculator's display.

If instead your answer was displayed as

[3.5 13]

then your calculator can express answers in scientific notation.

[3.5 13] means 3.5×10^{13}

number between 1 and 10 (3.5)

The power on 10 (13). *If this number is positive*, move the decimal point this many places right. *If this number is negative*, move the decimal point this many places left.

So, [3.5 13] = 3.5000000000000 = 35,000,000,000,000 (13 places)

EXAMPLE Write the calculator display in decimal notation.

a. [3.426 8] = 3.42600000 = 342,600,000 (8 places)

b. [7.03 −11] = .00000000007 03 = 0.0000000000703 ■ (11 places)

Write each calculator display in decimal notation.

1. [4.6 10]

2. [2.07 −12]

3. [8 −9]

SECTION 5.6 EXERCISES

1 *Write each number in decimal notation.*

1. 4.6×10^2 **2.** 3.7×10^1 **3.** 5.92×10^2 **4.** 6.30×10^3

5. 8×10^4 **6.** 1×10^6 **7.** 9.932×10^5 **8.** 5.768×10^4

9. 6.002×10^7 **10.** 5.514×10^1 **11.** 7.3×10^9 **12.** 4.021×10^0

13. The circumference of the Earth is 2.5×10^4 miles.

14. The diameter of the Earth is 1.3×10^7 meters.

2 *Write each number in scientific notation.*

15. 56 **16.** 33 **17.** 902 **18.** 526

19. 5,800 **20.** 6,040 **21.** 38.6 **22.** 923.7

23. 60,000 **24.** 4,000,000 **25.** 7.8 **26.** 1.2

27. 465,000 **28.** 14,700 **29.** 255,000,000 **30.** 71,380,000

31. The speed of light is 186,000 miles per second.

32. There are 31,600,000 seconds in one year.

33. The Earth is 4,500,000,000 years old.

34. Light travels 5,878,000,000,000 miles in one year.

35. The average distance between the Earth and the Sun is 92,600,000 miles.

36. The Earth weighs 6,000,000,000,000,000,000,000 metric tons.

3 *Write each number in scientific notation.*

37. 0.43 **38.** 0.88 **39.** 0.063 **40.** 0.051

41. 0.00953 **42.** 0.00812 **43.** 0.000052 **44.** 0.0000803

45. 0.1 **46.** 0.9 **47.** 0.000007 **48.** 0.004

49. 0.006045 **50.** 0.000006824 **51.** 0.00056 **52.** 0.000011

53. 0.0000061 **54.** 0.00000702 **55.** 0.0925 **56.** 0.06251

57. One inch equals 0.0000157 miles.

58. One hour equals 0.0001141 years.

59. A 150-pound man weighs 0.075 tons.

60. The fraction of radon gas present is 0.00000045.

61. Planck's constant is 0.000000000000000000000000006624.

62. The mass of a hydrogen atom is 0.0000000000000000000000016617 g.

Write each number in decimal notation.

63. 3.6×10^{-2}

64. 2.7×10^{-1}

65. 7.03×10^{-3}

66. 9.405×10^{-2}

67. 1.026×10^{-1}

68. 6.6×10^{-3}

69. 9.04×10^{-5}

70. 3.85×10^{-4}

71. 4.8×10^{-2}

72. 9.93×10^{-5}

73. 8.507×10^{-4}

74. 6.204×10^{-3}

75. 5.4×10^{-12}

76. 2.2×10^{-7}

77. 1.077×10^{-6}

78. 6×10^{-9}

79. The mass of the oxygen atom is 2.6372×10^{-23}

80. The Leaning Tower of Pisa leans another 2.2×10^{-5} inches each day.

SKILLSFOCUS (Section 2.6) *Simplify.*

81. $7 + 4 \cdot 3$

82. $5 \cdot 2 + 6 \cdot 4$

83. $90 \div 30 \cdot 3$

84. $4 + 3(9 - 4)$

WRITING TO LEARN ▶▶▶▶

85. Explain each step you would use to change 45,300,000 into scientific notation.

86. Explain each step you would use to change 0.00048 into scientific notation.

▶ YOU BE THE JUDGE

87. Nicole claims when you change a number from scientific to decimal notation, the number of zeros you attach is the absolute value of the power, minus 1. Is she right? Explain your decision with your own examples.

CHAPTER 5 REVIEW

VOCABULARY AND MATCHING

New words and phrases introduced in this chapter are shown in the left-hand column. Match each term on the left with the phrase or sentence on the right that best describes it.

A. integers	______	add the opposite of the subtrahend
B. positive integers	______	attached to each number on the right side of 0 on the number line
C. negative integers	______	-25
D. 0	______	have positive integers as powers on 10 when written in scientific notation
E. positive sign	______	$+1$
F. negative sign	______	positive and negative counting numbers and 0
G. absolute value	______	product of two like signs is positive; product of two unlike signs is negative
H. opposite of a number	______	lie to the left of 0 on the number line
I. adding integers	______	have negative integers as powers on 10 when written in scientific notation
J. subtracting integers	______	distance between a number and 0 on the number line
K. multiplicative identity	______	numbers in the form $a \times 10^n$; a is 1 or larger and less than 10; n is an integer
L. multiply integers	______	$+25$
M. divide integers	______	attached to each number on the left side of 0 on the number line
N. $(-5)^2$	______	this number is the same distance from 0 on the number line, but on the other side
O. -5^2	______	the only integer that is neither positive nor negative
P. scientific notation	______	same sign rules as for multiplying two integers
Q. numbers greater than 1	______	two like signs, keep the sign; two unlike signs, use the sign of the larger absolute value
R. numbers between 0 and 1	______	same as the counting numbers

REVIEW EXERCISES

5.1 Integers and the Number Line

1. Write an integer for 37° below zero.

2. Write an integer for a checkbook balance \$25 below zero.

3. $|-3|$ **4.** $|+32|$ **5.** $|-16|$

6. Insert the proper symbol $<$, $=$, or $>$.
a. $-7 \quad -10$ **b.** $-20 \quad -15$ **c.** $|-6| \quad -6$ **d.** $12 \quad |-12|$

7. The opposite of -6 is ___. **8.** The opposite of $+15$ is ____.

9. $-(-22)$ **10.** $-(+30)$

5.2 Addition of Integers

11. $5 + (-3)$

12. $6 + (-13)$

13. $-3 + (-7)$

14. $-10 + (-5)$

15. $-7 + (-7)$

16. $-15 + (+9)$

17. $-20 + (-25)$

18. $-8 + (8)$

19. $-10 + (-12)$

20. $(30) + (-25)$

21. $-4 + (-7)$

22. $-11 + (20)$

23. $-3 + (-18) + (32) + (+11)$

24. $-40 + (-36) + (-100) + (50)$

25. The temperature in Nome, Alaska, was -47° this morning. It rose 31° by noon. What was the noon temperature?

26. Tom has a score of -25. He made 42 points. What is his new score?

5.3 Subtraction of Integers

27. $7 - 6$

28. $9 - 12$

29. $-16 - (-16)$

30. $-9 - (+4)$

31. $+13 - (-17)$

32. $60 - (-45)$

33. $-5 - (+12)$

34. $-10 - (-10)$

35. $-100 - (-90)$

36. $-17 - (-27)$

37. $-21 - (30)$

38. $-1 - (+3)$

39. $-5 - (-10) - (+9) + (-3)$

40. $-10 - (-36) + (-5) - (+12)$

41. The temperature at midnight was -56°. At noon it was -13°. By how many degrees did the temperature rise?

42. Sally had a checkbook balance of $-\$136$. After a deposit her new balance was \$48. How much did she deposit?

5.4 Multiplying and Dividing Integers

43. $(+5)(+5)$

44. $(+8)(-8)$

45. $(-7)(-3)$

46. $(-2)(-9)$

47. $-8(+10)$

48. $-12(+7)$

49. $-6(-6)$

50. $-15(-4)$

51. $+20(-3)$

52. $+30(-6)$

53. $-1(+1)$

54. $-7(0)$

55. $-2(-5)(-2)$

56. $-4(2)(-2)$

57. $(-1)(-2)(-3)(-4)$

58. What is -5 added to itself 60 times?

59. What is -25 added to itself 18 times?

60. $-6 \div (+2)$

61. $-80 \div (-10)$

62. $(20) \div (-4)$

63. $30 \div (+6)$

64. $\frac{-45}{-9}$

65. $\frac{+50}{-25}$

66. $\frac{-24}{+24}$

67. $\frac{-8}{-1}$

68. $\frac{-90}{-3}$

69. $\frac{-40}{5}$

70. $\frac{150}{-30}$

71. $\frac{-54}{-6}$

72. What is -350 divided by 7?

73. How many -8 are in -96?

5.5 Order of Operations

74. $(-4)^3$ **75.** $(-9)^2$ **76.** -6^2 **77.** -10^3

78. $-4 + 7 - 9 - 3$ **79.** $-7 + 4(-3)$ **80.** $-5 - 5(-5)$ **81.** $-3(-5) + (-2)(6)$

82. $(-40) \div 8 \cdot (-5)$ **83.** $-3(-4)^2$

84. Evaluate using $x = -4$, $y = -2$, $a = 3$, and $b = 10$.

a. $5xy$ **b.** $-4y$ **c.** $4y - 3b$

d. $2b - x + (-3)$ **e.** $-x$ **f.** $\frac{b^2}{x}$

5.6 Scientific Notation

Change each number to decimal notation.

85. 7.05×10^4 **86.** 4×10^{-2} **87.** 1.8×10^{-5}

88. 9.88×10^7 **89.** 6.025×10^1 **90.** 7.6003×10^0

91. 9.4×10^3 **92.** 3.24×10^{-6}

Write each number in scientific notation.

93. 15,000 **94.** 477 **95.** 0.9

96. 0.0023 **97.** 507,000 **98.** 0.00308

99. 0.0000048 **100.** 1,060,000,000 **101.** 6,400

102. 40.7 **103.** 0.00011 **104.** 0.000000000041

6

Ratio and Proportion

HOUSING PRICES AND FAMILY INCOME

Ratios are used in the housing table shown here. The table shows the average price of a new house and the average family income for each year listed. As you can see, average income has risen faster than new home prices from 1900 to 1970. But this trend reversed from 1970 to 1990. Comparisons of these two amounts are listed as ratios in the column on the far right. What does a ratio of 9.9 mean? This question will be answered in this chapter.

Year	*Average Price for a New House*	*Average Family Income*	*Price to Income Ratio (to 1)*
1900	$ 4,881	$ 490	9.9
1910	5,377	630	8.5
1920	6,296	1,489	4.2
1930	7,146	1,360	5.2
1940	6,558	1,300	5.0
1950	9,446	3,319	2.8
1960	16,652	5,620	2.9
1970	23,400	9,867	2.3
1980	64,600	21,023	3.1
1990	122,900	35,191	3.5

SOURCE: National Association of Home Builders and the U.S. Census Bureau.

Ratios are used to compare quantities. In the table, new home prices are compared to family income. Ratios are also used in unit pricing, the stock market, construction, business, and finance. Proportions are used to help you solve a wide variety of problems using ratios. In this chapter these problems deal with blueprints, recipes, and estimation, such as estimating the number of people in a city infected with a virus, or estimating the height of a building using the length of its shadow. Ratios also provide the foundation for Chapter 7 on percents.

6.1 SOLVING EQUATIONS

OBJECTIVES

1. Determine whether a value is a solution to an equation.
2. Solve equations using the addition/subtraction property of equality.
3. Solve equations using the multiplication/division property of equality.
4. Solve equations using both properties of equality.

NEW VOCABULARY

solution
addition/subtraction property
multiplication/division property

1 EQUATIONS

We first discussed equations in Chapter 1, so you may remember the definition of an equation:

> An **equation** is a statement that two mathematical quantities are equal.

An equation can be true or false. For example, the equation $2 + 3 = 5$ is a true equation because the quantity $2 + 3$ (the left-hand side of the equation) has the same value as 5 (the right-hand side of the equation). However, the equation $2 + 3 = 10$ is a false equation because the quantity $2 + 3$ (the left-hand side of the equation) has a different value from 10 (the right-hand side of the equation). On the other hand, the equation $x + 2 = 7$ is neither true nor false. Since the left-hand side of the equation contains a variable, x, we cannot determine the value of that side of the equation. If we are given a value for x we can determine whether the equation is true or false. Naturally, values that make the equation true are of more interest to us than values that make the equation false; the values of the variable that make the equation true are called *solutions* to the equation.

> A **solution** to an equation is a value of the variable that makes the equation true.

EXAMPLE 1 Determine whether -2 is a solution to the equation $x+4=-2$.

Replace x by -2 and evaluate:

$$x+4=-2$$
$$-2+4\stackrel{?}{=}-2$$
$$2\neq-2$$

Since the left and right sides of the equation do not have the same value, -2 **is not** a solution to the equation.

EXAMPLE 2 Determine whether 4 is a solution to the equation $-6x=-24$.

Replace x by 4 and evaluate:

$$-6x=-24$$
$$-6(4)\stackrel{?}{=}-24$$
$$-24=-24$$

Since both sides have the same value, 4 **is** a solution to the equation.

♦You Try It

1. Determine whether 8 is a solution to the equation $5-x=-3$.

2. Determine whether -5 is a solution to the equation $4x=20$.

2 THE ADDITION/SUBTRACTION PROPERTY

In this section, we will develop some techniques for solving simple equations. The ability to solve equations is essential for solving proportions later in this chapter and for solving percent problems in Chapter 7. Some equations have solutions that you might consider "obvious". For example, if $x+2=7$, then x is clearly 5 because $5 + 2 = 7$. Of course, most equations do not have such obvious solutions so we need to develop techniques for determining what the solution is.

To begin with, let's agree that the simplest equation of all is one in which one side of the equation is a variable by itself and the other side is a number, like $x = 4$. At this point everyone would agree that the solution is 4. Our goal will be to take an equation whose solution is not obvious and transform it into an equivalent equation of this simple type. An *equivalent equation* is an equation with the same solution. For example, the equations $4+x=11$ and $x=7$ are *equivalent* because they both have the same solution, 7. In order to see how to transform an equation into a simpler equivalent equation, it is helpful to compare an equation to a balance scale. If the left and right sides of an equation are equal, they "balance" each other. On a balance scale, if you add 2 pounds to one side of the scale, you can only keep the scale in balance if you add 2 pounds to the other side as well. Similarly, if you remove one pound from one side of the scale, you can preserve the balance only if you remove one pound from the other side. The same principal works for equations. You can add the same number to both sides of an equation and you will have an equivalent equation (one with the same solution). Also, you can subtract the same number from both sides of an equation and you will have an equivalent equation. This is an important property of equality.

The Addition/Subtraction Property of Equality

1. Adding the same number to both sides of an equation produces an equivalent equation.

2. Subtracting the same number from both sides of an equation produces an equivalent equation.

Now we need to see how to make use of this property to solve equations. Consider the equation

$$x-9=-4$$

Remember that our goal is to produce an equivalent equation with x alone on one side. That is, we want to *isolate* x on one side of the equation. To do that we need to look at the equation and see what is connected to x and how it is connected. In this equation, 9 is connected to x and it is connected by *subtraction*. If we want to isolate x, we need to "get rid of" the 9 by using the *opposite* operation, addition. We will use the addition property and **add 9** to **both** sides of the equation.

$$x-9+\mathbf{9}=-4+\mathbf{9}$$

Now simplify each side: $x=5$

Thus, the solution is **5**.

Check: We can check this solution by substituting 5 for x in the original equation and then evaluating:

$$x-9=-4$$

Replace x by 5: $5-9\stackrel{?}{=}-4$

Simplify: $-4=-4$ ✓

EXAMPLE 3 Solve the equation $-4 = 6 + t$.

Here we want to isolate t and we see that 6 is *added* to t. In order to isolate t, we need to *subtract* 6 from both sides of the equation.

$$-4 = 6 + t$$

Subtract 6 from both sides: $-4 - \mathbf{6} = 6 - \mathbf{6} + t$

Simplify $-10 = t$

Thus, the solution is **−10.**

Check: We can check this solution by substituting −10 for t in the original equation and then evaluating:

$$-4 = 6 + t$$

Replace t by −10: $-4 \overset{?}{=} 6 + (-10)$

Simplify: $-4 = -4$ ✓

♦**You Try It**

3. Solve the equation $y + 10 = 7$.

4. Solve the equation $20 = -11 + m$.

3 THE MULTIPLICATION/DIVISION PROPERTY

The Addition/Subtraction Property of Equality is useful only when we want to remove a number that is added to or subtracted from a variable. If the variable is being multiplied or divided by the number we want to remove, then a different property is needed. Here is the second property of equality.

The Multiplication/Division Property of Equality

1. Multiplying both sides of an equation by the same nonzero number produces an equivalent equation

2. Dividing both sides of an equation by the same nonzero number produces an equivalent equation.

Notice that we specify that the number we multiply or divide by must not be zero. Can you see why?

EXAMPLE 4 Solve the equation $6k = -54$.

We want to isolate k, and it is connected to 6 by multiplication. In order to isolate k, we will remove the 6 by using the *opposite* operation, division. We will divide both sides by 6.

$$6k = -54$$

Divide both sides by 6: $\frac{6k}{6} = \frac{-54}{6}$

Simplify: $k = -9$

Thus, the solution is **−9.**

Check: We can check this solution by substituting –9 for k in the original equation and then evaluating:

$$6k = -54$$

Replace k by –9: $6(-9) \stackrel{?}{=} -54$

Simplify: $-54 = -54$ ✓

♦**You Try It**

5. Solve the equation $-2x = 15$.

6. Solve the equation $-40 = 8t$.

EXAMPLE 5 Solve the equation $\frac{1}{3}w = \frac{1}{6}$.

Here we want to isolate w and $\frac{1}{3}$ is connected to w by multiplication, so it looks like we should divide both sides of the equation by $\frac{1}{3}$. However, dividing by $\frac{1}{3}$ is the same as multiplying by 3, so in this case we will *multiply* both sides by 3.

$$\frac{1}{3}w = \frac{1}{6}$$

Multiply both sides by 3: $3 \cdot \frac{1}{3}w = 3 \cdot \frac{1}{6}$

Simplify: $\cancel{3} \cdot \frac{1}{\cancel{3}}w = \cancel{3} \cdot \frac{1}{\cancel{6}_2}$

$$w = \frac{1}{2}$$

Thus, the solution is $\frac{1}{2}$.

Check: We can check this solution by substituting $\frac{1}{2}$ for w in the original equation and then evaluating:

$$\frac{1}{3}w = \frac{1}{6}$$

Replace w by $\frac{1}{2}$: $\frac{1}{3} \cdot \frac{1}{2} \stackrel{?}{=} \frac{1}{6}$

Simplify: $\frac{1}{6} = \frac{1}{6}$ ✓

♦You Try It 7. Solve the equation $\frac{1}{4}y = 5$.

8. Solve the equation $\frac{1}{8} = \frac{3}{4}n$.

4 USING BOTH PROPERTIES

So far the equations we have solved have required only one operation. Naturally, there are many equations that are more complicated and require more than one step to solve. We are not going to consider any terribly complicated equations at this time but we will look at a few that require the use of both properties of equality in order to find the solution.

EXAMPLE 6 Solve the equation $2x - 3 = 5$.

We still have the same goal -- to isolate the variable. Examining the left side of the equation, we see a multiplication and a subtraction surrounding x. Which one should we "undo" first? Recall the rule for order of operations. This rule guided us through the steps needed to *evaluate* an expression and said that all multiplications and divisions should be completed before any additions and subtractions. Here, however, we are not trying to evaluate an expression, but to take it apart. In a sense, we are reversing what order of operations is all about. So we will apply the order of operations rule in reverse: we will undo any additions or subtractions first, and then undo any remaining multiplications or divisions.

$$2x - 3 = 5$$

To remove the 3, add 3 to both sides: $2x - 3 + 3 = 5 + 3$

Simplify: $2x = 8$

Now, to remove the 2, divide both sides by 2: $\frac{2x}{2} = \frac{8}{2}$

Simplify $x = 4$

Thus, the solution is 4.

Check: We can check this solution by substituting 4 for x in the original equation and then evaluating:

$$2x - 3 = 5$$

Replace x by 4: $2(4) - 3 \stackrel{?}{=} 5$

Simplify: $5 = 5$ ✓

EXAMPLE 7 Solve the equation $-4 = 7 - 2p$.

$$-4 = 7 - 2p$$

Remove the 7 first by subtracting 7 from both sides:

$$-4 - 7 = 7 - 7 - 2p$$

Simplify:

$$-11 = -2p$$

Now divide both sides by -2:

$$\frac{-11}{-2} = \frac{-2p}{-2}$$

Simplify:

$$\frac{11}{2} = p$$

Thus, the solution is $\frac{11}{2}$.

Check: We can check this solution by substituting $\frac{11}{2}$ for p in the original equation and then evaluating:

$$-4 = 7 - 2p$$

Replace p by $\frac{11}{2}$:

$$-4 \stackrel{?}{=} 7 - 2\left(\frac{11}{2}\right)$$

Simplify:

$$-4 \stackrel{?}{=} 7 - 11$$

$$-4 = -4 \checkmark$$

♦ **You Try It**

9. Solve the equation $3t + 8 = 2$.

10. Solve the equation $-7 = 8n - 3$.

Answers to You Try It **1.** 8 is a solution. **2.** -5 is not a solution. **3.** $y = -3$ **4.** $m = 31$ **5.** $x = -\frac{15}{2}$ **6.** $t = -5$ **7.** $y = 20$ **8.** $n = \frac{1}{6}$ **9.** $t = -2$ **10.** $n = -\frac{1}{2}$

SECTION 6.1 EXERCISES

1 *Determine whether the number is a solution for the given equation.*

1. $7 - x = 2$ **a.** $x = 9$ **b.** $x = 5$

2. $4y + 2 = -10$ **a.** $y = -5$ **b.** $y = -3$

3. $6 = 2w - 4$ **a.** $w = 5$ **b.** $w = 1$

4. $2t + 5 = 6$ **a.** $t = \frac{1}{2}$ **b.** $t = \frac{11}{2}$

5. $3 - 4x = x + 3$ **a.** $x = 1$ **b.** $x = 0$

2 *Solve each equation and check the solution.*

6. $t - 4 = 6$

7. $x - 7 = 2$

8. $y + 6 = 2$

9. $w + 10 = 4$

10. $12 = a - 2$

11. $9 = b - 7$

12. $-4 = x + 1$

13. $t + 8 = -2$

14. $y - 5 = -1$

15. $s - 8 = 8$

16. $k + 9 = -9$

17. $s - 8 = -8$

18. $k + 9 = 9$

19. $0 = x + 2$

20. $t - 10 = 0$

21. $a - 7 = -17$

22. $-6 = 4 + x$

23. $a - 17 = -7$

24. $4 = -6 + x$

25. $n + 4 = 11$

26. $n + 12 = 3$

27. $-40 = y - 3$

28. $0 - t - 16$

29. $k + 24 = 0$

3 *Solve each equation. Check each answer.*

30. $3x = 24$

31. $5x = 50$

32. $2y = -4$

33. $7y = -28$

34. $-6t = 30$

35. $-9t = 54$

36. $-48 = -3t$

37. $-14 = -7k$

38. $6x = 0$

39. $10x = 0$

40. $-4t = 0$

41. $-8t = 0$

42. $0 = 11y$

43. $0 = -6y$

44. $\frac{1}{4}x = 3$

45. $\frac{1}{6}x = 2$

46. $-\frac{3}{4}n = 12$

47. $-\frac{2}{3}n = 16$

48. $4 = \frac{2}{5}y$

49. $14 = \frac{7}{8}y$

50. $\frac{5}{6}t = 0$

51. $\frac{4}{5}t = 0$

52. $6x = 3$

53. $20x = 5$

54. $-24y = 3$

55. $-16y = -8$

56. $-18n = 2$

4 *Solve each equation. Check each answer.*

57. $2x + 1 = 5$

58. $3x + 2 = 11$

59. $4y - 1 = 23$

60. $5y - 4 = 16$

61. $-2t + 3 = -1$

62. $-6t + 8 = -10$

63. $4 = -8 + 6k$

64. $16 = -9 + 5k$

65. $3 - 2x = 11$

66. $7 - 10x = 47$

67. $20y - 9 = 1$

68. $12y - 2 = 4$

69. $-3n + 7 = -1$

70. $-10n + 8 = 14$

71. $9 - 5p = -4$

SKILLSFOCUS (Section 3.3) *Multiply and reduce your answers.*

72. $\frac{4}{9} \times \frac{15}{8}$

73. $\frac{1}{4} \cdot \frac{3}{7} \cdot \frac{8}{15}$

74. $2\frac{2}{3} \times 1\frac{1}{8}$

75. $\left(\frac{4}{5}\right)^2$

EXTEND YOUR THINKING ♦ ♦ ♦

♦ TROUBLESHOOT IT

Find and correct the error.

76. Solve for x: $x + 3 = 0$

$$x = 3$$

77. Solve for t: $6 - 4t = -10$

$$2t = -10$$
$$t = -5$$

78. Solve for y: $\frac{1}{4}n = 12$

$$n = \frac{1}{4} \cdot 12$$
$$n = 3$$

WRITING TO LEARN ♦ ♦ ♦

79. In your own words, explain what the addition rule says. Give an example to illustrate your explanation.

80. In your own words, explain what the multiplication rule says. Give an example to illustrate your explanation.

81. Write three equations each with a solution of –2.

6.2 THE MEANING OF RATIO

OBJECTIVES	NEW VOCABULARY
1 Define ratio.	ratio
2 Reduce ratios.	order
3 Eliminate like names in a ratio.	terms of a ratio

1 Defining Ratios

A compact car is 12 feet long and 5 feet wide. The ratio of length to width is

length to width

12 feet to 5 feet.

> A **ratio** is a comparison of two quantities using division.

EXAMPLE 1 There are 10 men and 3 women on a committee. The ratio of men to women is

men to women

10 to 3.

The reverse ratio of women to men is

women to men

3 to 10. ■

> **Order** is important when stating a ratio. Write the numbers in a ratio in the same order as the word names.

EXAMPLE 2 In a small school there are 55 male students, 70 female students, and 8 instructors.

a. The ratio of female students to male students is 70 to 55.
b. The ratio of instructors to male students is 8 to 55.
c. The ratio of students to instructors is 125 to 8 (55 + 70 = 125 students). ■

▶ **You Try It**

1. There are 40 cars and 9 trucks on a lot.
  a. What is the ratio of cars to trucks?
  b. What is the ratio of trucks to cars?

2. A park has 33 acres of grass, 45 acres of trees, and a 12-acre pond.
  a. What is the ratio of acres of grass to acres of trees?
  b. What is the ratio of acres of pond to acres of grass?
  c. What is the ratio of land acres to pond acres?

Suppose for every $8 in income there are $3 in taxes. The ratio of income to taxes can be written three ways.

Words	*Colon*	*Fraction*
8 to 3	8:3	$\frac{8}{3}$ or 8/3
eight to three	":" is read "to"	the fraction bar is read "to"

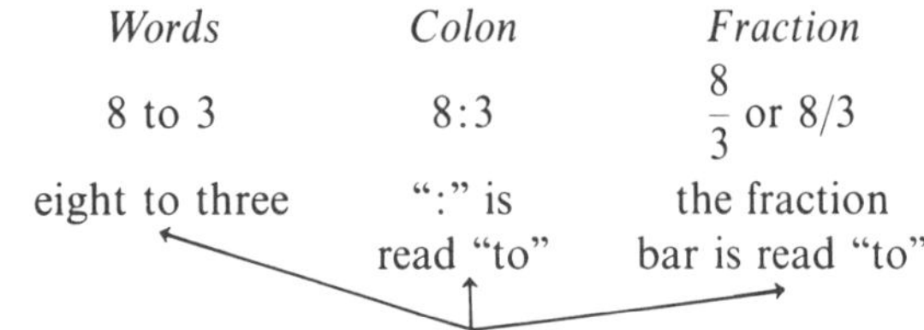

8 and 3 are called the **terms of the ratio**. 8 is called the *first term*. 3 is called the *second term*.

Any ratio can be written in the following three ways.

1. with the word "to"
2. by use of a colon, ":"
3. as a fraction

EXAMPLE 3 Ken made 30 hits in his last 50 times at bat. Write the ratio of *hits* to *at bats* in three ways.

using words: 30 to 50, or thirty to fifty
using a colon: 30:50
as a fraction: 30/50 ■

▶ **You Try It**

3. There are 4 CD players to every 3 tape players in a music store. Write the ratio of CD players to tape players in three ways.
4. Terri threw the horseshoe 21 times and made 8 ringers. Write the ratio of ringers to throws in three ways.

Any fraction can be written as a ratio. In fact, fractions are called *ratio*nal numbers.

The fraction $\frac{a}{b}$ is read as "the ratio of the numerator a to the denominator b."

$$\frac{a}{b} = a \text{ to } b$$

EXAMPLE 4

a. $\frac{5}{8}$ is the ratio 5 to 8 or 5:8.

b. $5\frac{1}{4} = \frac{21}{4}$ is the ratio 21 to 4 or 21:4. ■

▶ **You Try It** Write each fraction as a ratio in two other ways.

5. $\frac{7}{10}$ **6.** $4\frac{3}{7}$

2 Reducing Ratios

Any ratio can be written as a fraction. Therefore, it can be reduced to lowest terms in the same way fractions are reduced. For example,

$$6 \text{ to } 4 = \frac{6}{4} = \frac{6 \div 2}{4 \div 2} = \frac{3}{2} = 3 \text{ to } 2.$$

The ratio 6 to 4 is equivalent to the lowest terms ratio 3 to 2.

To reduce a ratio to lowest terms

1. Divide both terms of the ratio by the same factor.
2. The ratio is in lowest terms when the only number that divides evenly into both terms is 1.

EXAMPLE 5 Reduce each ratio to lowest terms.

a. 20 to 15 = $\overset{4}{\cancel{20}}$ to $\overset{3}{\cancel{15}}$ = 4 to 3 reduce using 5

b. $40:100 = \overset{2}{\cancel{40}}:\overset{5}{\cancel{100}} = 2:5$ reduce using 20

c. $\frac{9}{54} = \frac{9 \div 9}{54 \div 9} = \frac{1}{6}$ ■

You Try It Reduce each ratio to lowest terms.

7. 35 to 25

8. 24:18

9. $\frac{12}{30}$

Reducing ratios gives smaller numbers that are easier to understand.

EXAMPLE 6 Reduce the ratio of 6,000 students to 240 instructors to lowest terms. Interpret the answer.

students	to	instructors	
6,000	to	240	reduce using 10
600	to	24	reduce using 6
100	to	4	reduce using 4
25	to	1	

A ratio of 6,000 students to 240 instructors means there are 25 students per instructor. ■

You Try It 10. Write the ratio of 435 representatives to 100 senators in lowest terms.

3 Eliminating Like units in a Ratio

If both terms in a ratio have the same unit, eliminate the units by reducing. You reduce units in a ratio in the same way you reduce common factors when you reduce a fraction.

$$8 \text{ feet to } 5 \text{ feet} = \frac{8 \cancel{\text{feet}}}{5 \cancel{\text{feet}}} = \frac{8}{5} = 8 \text{ to } 5$$

EXAMPLE 7 $25 \text{ yards}:10 \text{ yards} = \frac{\overset{5}{\cancel{25}} \cancel{\text{yards}}}{\underset{2}{\cancel{10}} \cancel{\text{yards}}} = 5:2$ ■

You Try It Express in simplest form.

11. 40 miles to 13 miles

12. 32 hours:20 hours

If the units in a ratio are of the same type, write one unit in terms of the other. Then eliminate the units.

EXAMPLE 8 Eliminate the units in the ratio 7 inches to 1 foot.

7 inches	to	1 foot	Inches and feet are units of the same type. Both are lengths. Note, 1 foot = 12 inches.
7 ~~inches~~	to	12 ~~inches~~	Replace 1 foot with 12 inches and eliminate like units.
7	to	12	

The ratio 7 inches to 1 foot is a 7 to 12 ratio. ■

EXAMPLE 9 Eliminate the units in the ratio 2 hours to 45 minutes.

2 hours	to	45 minutes	Since 1 hour = 60 minutes, 2 hours = 2 × 60 minutes = 120 minutes.
120 ~~minutes~~	to	45 ~~minutes~~	Replace 2 hours with 120 minutes and eliminate the units.
120	to	45	Reduce using 5.
24	to	9	Reduce using 3.
8	to	3	

The ratio 2 hours to 45 minutes is an 8 to 3 ratio. ■

▶ **You Try It** Eliminate the units in each ratio.

13. 2 yards to 4 feet (Note: 1 yard = 3 feet)

14. 4 minutes to 30 seconds (Note: 1 minute = 60 seconds)

EXAMPLE 10 **Blueprint Scaling** A wall is 9 inches long on a blueprint. The actual wall is 12 feet long. What is the scale for the blueprint?

The blueprint scale is the ratio of blueprint measurement to actual measurement.

blueprint	to	actual	
9 inches	to	12 feet	Reduce using 3.
3 inches	to	4 feet	Since 1 foot = 12 inches, 4 feet = 4 × 12 inches = 48 inches.
3 ~~inches~~	to	48 ~~inches~~	Eliminate units. Reduce using 3.
1	to	16	Scale for blueprint. ■

OBSERVE 1 inch on the blueprint corresponds to 16 inches on the actual wall. The final ratio 1 to 16, or 1:16, is called the *scale* of the blueprint. If the blueprint could be blown up to 16 times its size, it would then have the exact same dimensions as the building it represents.

In addition, the ratio of blueprint to actual is 1 to 16. This means 1 inch on the blueprint corresponds to 16 inches on the actual wall. It also means 1 foot on the blueprint corresponds to 16 feet on the actual wall, or 1 blueprint meter corresponds to 16 meters on the wall. When the terms in a ratio have no unit, you may insert your own units, as long as you give each term the same unit.

▶ **You Try It** **15.** An entrance 2 inches wide on a blueprint will actually be 6 feet wide. Calculate the scale for the blueprint.

▶ **Answers to You Try It** 1. a. 40 to 9 b. 9 to 40 2. a. 33 to 45 b. 12 to 33 c. 78 to 12 3. 4 to 3, 4:3, $\frac{4}{3}$ 4. 8 to 21, 8:21, $\frac{8}{21}$ 5. 7 to 10, 7:10 6. 31 to 7, 31:7 7. 7 to 5 8. 4:3 9. $\frac{2}{5}$ 10. 87 to 20 11. 40 to 13 12. 8:5 13. 3 to 2 14. 8 to 1 15. 1 to 36 or 1:36

SECTION 6.2 EXERCISES

I

1. There were 23 mothers, 21 fathers, and 51 children at a picnic. Write the ratio of
   a. fathers to mothers.
   b. mothers to fathers.
   c. mothers to children.
   d. children to adults.

2. A debutante's wardrobe consists of 33 gowns, 32 hats, 26 pairs of evening shoes, and 11 pairs of lounging shoes. Write the ratio of
   a. gowns to hats.
   b. hats to gowns.
   c. pairs of lounging shoes to pairs of evening shoes.
   d. gowns to total pairs of shoes.

3. During visiting hours, a maternity ward consisted of 24 mothers, 23 fathers, 12 male babies, and 17 female babies. Write the ratio of
   a. male babies to female babies.
   b. fathers to mothers.
   c. babies to parents.
   d. mothers to female babies.

4. An art exposition consists of 9 artists, 7 critics, and 250 guests. Write the ratio of
   a. critics to artists.
   b. guests to critics.
   c. artists to guests.
   d. guests to artists.

---

5. Forty boys and 27 girls went on a canoe trip. Write the ratio of boys to girls in three ways.

6. A builder ordered 4,000 bricks and 633 blocks. Write the ratio of bricks to blocks in three ways.

7. Tom has 15 suits and 49 shirts. Write the ratio of shirts to suits in three ways.

8. Amanda has 10 dolls and 17 doll dresses. Write the ratio of dresses to dolls in three ways.

9. A grandmother is 80 years old. Her granddaughter is 21. Write the ratio of their ages (older to younger) in three ways.

10. Justin has 3 A's and 5 B's on his report card. Write the ratio of A's to B's in three ways.

11. A baseball team consists of 24 players and 5 coaches.

a. Write the ratio of players to coaches as a fraction.

b. Write the ratio of coaches to players using a colon.

c. Write the ratio of players to players and coaches, in words.

12. A military base has 11 ships, 103 planes, and 240 tanks stationed nearby.

a. Write the ratio of planes to ships in words.

b. Write the ratio of ships to tanks as a fraction.

c. Write the ratio of ships to planes and tanks using a colon.

Write each ratio in two other forms.

13. 5 to 3 **14.** 14 to 11 **15.** 6/5 **16.** 4/21

17. 40:1 **18.** 5:18 **19.** $\frac{1}{64}$ **20.** $\frac{35}{2}$

21. 1 to 1 **22.** 7 to 3 **23.** 8:9 **24.** 27:10

2 *Reduce each ratio to lowest terms.*

25. 22 to 16 **26.** 90 to 20 **27.** 38 to 20 **28.** 35 to 42

29. 48:12 **30.** 40:5 **31.** 15:45 **32.** 24:28

33. $\frac{30}{9}$ **34.** $\frac{64}{36}$ **35.** $\frac{52}{65}$ **36.** $\frac{14}{16}$

37. 7 to 4 **38.** 61 to 30 **39.** 100 to 28 **40.** 150 to 90

41. 900 to 250 **42.** 450 to 180 **43.** 128:640 **44.** 37:111

45. Tom weighs 180 pounds on Earth. NASA says he will weigh 30 pounds on the moon. Write the ratio of Earth weight to moon weight in lowest terms.

46. A 540-page textbook has 80 pages of problems. Write the lowest terms ratio of pages in the book to pages of problems.

47. $6,000 in profit is made on a $2,500 investment. Write the ratio of investment to profit in lowest terms.

48. A building 600 feet tall casts an 840-foot shadow. Write the ratio of building height to shadow length in lowest terms.

3 *Eliminate the units in each ratio and reduce to lowest terms.*

49. 9 feet to 10 feet

50. 4 inches to 3 inches

51. 4 feet to 9 inches

52. 15 inches to 2 feet

53. 5 yards:6 feet

54. 1 foot:3 yards

55. 12 seconds to 5 minutes

56. 1 minute to 40 seconds

57. 25 minutes/1 hour

58. 5 hours/20 minutes

59. 7 weeks to 14 days

60. 52 weeks to 364 days

61. 3 years:9 months

62. 2 months:2 years

63. $4 to 60 cents

64. 35 cents to $3.15

65. 4 hours 20 minutes to 30 minutes

66. 1 hour 15 minutes to 2 hours

67. 2 feet 4 inches to 7 inches

68. 5 feet 3 inches to 1 yard

69. 3 yards 2 feet 10 inches to 2 feet

70. 2 hours 15 minutes 30 seconds to 1 hour

71. John's waist measures 30 inches. His height is 6 feet. What is the ratio of John's waist measure to his height?

72. Sally can build a model in 2 hours, and paint it in 25 minutes. What is the ratio of build time to paint time?

73. A room 3 inches wide on a blueprint is actually 12 feet long. What is the scale for the blueprint?

74. A building is 160 feet long. On its blueprint, it is 8 inches long. What is the scale of the blueprint?

SKILLSFOCUS (Section 1.5) *Find three fractions equivalent to the given fraction.*

75. $\frac{2}{3}$ **76.** $\frac{7}{5}$ **77.** $\frac{1}{10}$ **78.** $\frac{3}{16}$

EXTEND YOUR THINKING ▶▶▶▶

▶ SOMETHING MORE

79. To prepare the fuel for a gas saw, 4 quarts and 1 pint of gasoline are mixed with 8 ounces of oil. If 1 quart = 32 ounces, and 1 pint = 16 ounces, write the ratio of gasoline to oil in the mixture.

80. D-day The biggest wartime amphibious landing ever was D-day, June 6, 1944, when about 150,000 U.S. and allied troops landed in France, supported by 2 battleships, 23 cruisers, 105 destroyers, 1,076 additional warships, 2,700 transports, 5,200 bombers, 5,500 fighters, and 2,400 transport planes. Write the lowest terms ratio of troops to total support equipment.

81. It takes 8 hours to put together a gym set. What fraction of the job will be completed in 2 hours 40 minutes?

▶ TROUBLESHOOT IT

Find and correct the error.

82. There are 20 cats, 5 ponies, and 8 dogs on a farm. The ratio of ponies to cats and dogs is 28 to 5.

83. 15 to 40 $= \frac{40 \div 5}{15 \div 5} = \frac{8}{3} =$ 8 to 3 in lowest terms.

WRITING TO LEARN ▶▶▶▶

84. In your own words, explain each step you would use to simplify the ratio \$750 to \$450.

▶ YOU BE THE JUDGE

85. A baseball player came to bat 12 times without getting a hit.

a. Write the ratio of hits to times at bat in three ways. Are these ratios valid? Explain your decision.

b. Write the ratio of at bats to hits in three ways. Are these ratios valid? Explain your decision.

6.3 RATIO AS A RATE

OBJECTIVES

1 Define a rate.

2 Change a unit name in a rate.

NEW VOCABULARY

rate

1 Defining Rates

When the units in a ratio are of different types, you cannot write one unit in terms of the other. So you cannot eliminate the units. Such a ratio is called a **rate**.

For example, the ratio

150 miles to 3 hours

is a rate. Miles cannot be written in terms of hours. Therefore, you cannot eliminate the units. This rate can be written using the word *per*.

150 miles *to* 3 hours = 150 miles *per* 3 hours

It is also common to write a rate in fraction form.

$$150 \text{ miles } per \text{ } 3 \text{ hours} = 150 \text{ miles}/3 \text{ hours} = \frac{150 \text{ miles}}{3 \text{ hours}}$$

fraction bar is read *"per"*

Other common rates are miles per gallon, dollars per pound, and dollars per hour.

> A ratio is called a rate when the units of the terms cannot be eliminated. You can write any rate using the word *per*.

EXAMPLE 1 Examples of rates.

a. 55 miles per hour = 55 miles per 1 hour = $\frac{55 \text{ miles}}{1 \text{ hour}}$

b. \$3.29 per pound = \$3.29 per 1 pound = $\frac{\$3.29}{1 \text{ pound}}$

c. Meg drove 210 miles on 13 gallons of gas. This is the rate

210 miles per 13 gallons = $\frac{210 \text{ miles}}{13 \text{ gallons}}$. ■

▶ You Try It Write each rate using the fraction bar. **1.** 25 students per course **2.** \$1.19 per gallon **3.** 43 seconds per commercial

---

When the second term of a rate is 1, the rate can be written without the 1 by using *per* or a slash, /.

EXAMPLE 2 Examples of rates with a second term of 1.

a. \$5 to 1 hr = \$5 per 1 hr
= \$5 per hr or \$5/hr

b. 30 mi to 1 hr = 30 mi per 1 hr
= 30 mi per hr or 30 mi/hr or 30 mph

c. \$7.35 to 1 lb = \$7.35 per 1 lb
= \$7.35 per lb or \$7.35/lb ■

▶ You Try It Write each rate using *per*, and using a slash.

4. 24.2 miles to 1 gallon **5.** \$32 to 1 ounce **6.** 50 people to 1 room

---

A rate is reduced in the same way a ratio is reduced.

EXAMPLE 3 Write \$450 for 20 radios as a rate in lowest terms.

\$450 for 20 radios = \$450 per 20 radios — Reduce using 10.
\$45 per 2 radios — Lowest terms. ■

EXAMPLE 4 Write \$72 for 8 hours work as dollars per hour.

\$72 for 8 hours = \$72 per 8 hours — Reduce using 8.
\$9 per 1 hour — Lowest terms.
\$9 per hour ■

EXAMPLE 5 Jayne traveled 240 miles on 16 gallons. Write this as a rate in fraction form.

$$240 \text{ mi on } 16 \text{ gal} = \frac{240 \text{ mi}}{16 \text{ gal}} = \frac{15 \text{ mi}}{1 \text{ gal}} = 15 \text{ mi/gal} \quad \text{or} \quad 15 \text{ mpg}$$ ■

▶ **You Try It** Write the following rates in lowest terms using "per."

7. Write the rate in lowest terms using *per*: \$60 to 18 pounds.

8. \$120 for 4 hours **9.** The satellite travels 24,000 miles in 90 minutes.

---

2 Changing the units in a Rate

Though you cannot reduce the units in a rate, you can change the unit of one or both terms in a rate.

EXAMPLE 6 Change feet to inches in the rate 5 feet to 6 seconds.

$$5 \text{ ft to } 6 \text{ sec} = \frac{5 \text{ ft}}{6 \text{ sec}} = \frac{60 \text{ in}}{6 \text{ sec}} = \frac{10 \text{ in}}{1 \text{ sec}} = 10 \text{ in per sec}$$

Since 1 ft = 12 in,
5 ft = 5 × 12 in = 60 in. ■

EXAMPLE 7 Liz has 6 hours to pack 20 boxes. Write the rate of hours per boxes, and change hours to minutes.

6 hrs per 20 boxes — Reduce using 2.
3 hrs per 10 boxes — Change hours to minutes. Since 1 hr = 60 min, 3 hr = 3 × 60 min = 180 min.
180 min per 10 boxes — Reduce using 10.
18 min per 1 box
18 min /box

If Liz has 6 hours to pack 20 boxes, she must pack at a rate of 18 minutes per box. ■

▶ **You Try It** **10.** Change yards to feet, and reduce: 60 yards to 9 seconds.

11. A supermarket paid \$360 for 240 pounds of butter. Write the rate of dollars to pounds, change dollars to cents, and reduce.

▶ **Answers to You Try It** 1. $\frac{25 \text{ students}}{1 \text{ course}}$ 2. $\frac{\$1.19}{1 \text{ gal}}$ 3. $\frac{43 \text{ sec}}{1 \text{ commercial}}$ 4. 24.2 mi per gal, 24.2 mi/gal 5. \$32 per oz, \$32/oz 6. 50 people per room, 50 people/room 7. \$10 per 3 lb 8. \$30 per hr, \$30/hr 9. 800 mi per 3 min 10. 20 ft per 1 sec, 20 ft/sec 11. 150¢ per 1 lb, 150¢/lb

SECTION 6.3 EXERCISES

1 *Reduce each rate to lowest terms.*

1. 40 engines to 12 cars

2. 15 cars to 6 trucks

3. \$200 to 15 hours

4. \$280 to 40 hours

5. 400 miles to 24 gallons

6. 252 miles to 14 gallons

7. 550 miles in 11 hours

8. 3,000 miles to 160 hours

9. 640 customers to 280 toys

10. \$45,000 to 800 hours

11. 4,500 books to 1,050 readers

12. 5,100 pounds of rations to 300 soldiers

13. 9,000 miles to 450 gallons

14. 250 million people to 3,500,000 square miles

15. In 8 hours a machine can bind 1,880 books. Write the rate of books bound per hour.

16. In 21 seasons Babe Ruth hit 714 home runs. Write the rate of home runs per season.

17. The Fahrenheit scale has 180 degrees. The Celsius scale has 100 degrees. Write the ratio of Fahrenheit degrees to Celsius degrees.

18. 500 liters of a solution contains 20 liters of acid. Write the ratio of liters of acid to liters of solution.

2 *Change the unit in the rate as indicated. Then reduce.*

19. 30 miles to 1 hour: change hours to minutes

20. 1 foot to 2 minutes: change feet to inches

21. \$90 to 150 pounds: change dollars to cents

22. \$320 to 7 pounds: change pounds to ounces

23. 8 hours to 24 boxes: change hours to minutes

24. 6 spark plugs to 4 minutes: change minutes to seconds

25. \$5 to 12 ounces: change dollars to cents

26. \$80 to 5 dozen: change dozen to units

27. 40 dozen eggs cost a food store \$24.
   a. Write the rate of dollars to dozens of eggs.
   b. Change dollars to cents and reduce.
   c. Rewrite the rate in part b using 1 dozen = 12 eggs.

28. 12 boxes of cigars cost \$200.
   a. Write the rate of boxes to dollars.
   b. Rewrite this rate as cigars to dollars if there are 24 cigars in each box.
   c. Rewrite the rate in part b as cigars to cents.

29. A \$17,500 car weighs 3,500 pounds.
   a. Write the rate of dollars to pounds.
   b. Rewrite this rate as cents to pounds.
   c. Rewrite the rate in part b as cents to ounces.

30. Mike lost 20 pounds in 48 days.
   a. Write the rate of pounds to days.
   b. Rewrite this rate as ounces to days.
   c. Rewrite the rate in part b as ounces to hours.

31. A man paid \$720 in taxes on a 2-acre tract of land.
   a. Write the rate of tax dollars to acres.
   b. 1 acre = 43,560 square feet. Rewrite the rate of tax dollars to square feet.
   c. Rewrite the rate in part b by changing dollars to cents.

32. Greta can run 10 miles in 1 hour and 12 minutes.
   a. Write the rate of miles run to minutes.
   b. Rewrite the rate by changing miles to feet.
   c. Rewrite the rate in part b by changing minutes to seconds.

SKILLSFOCUS (Section 1.4) *Find the average.*

33. 14, 10, 20, 18, 8, 24, 11

34. 2.3, 4, 1.06, 2.09, 0.612

EXTEND YOUR THINKING ▶▶▶▶

▶ TROUBLESHOOT IT

Find and correct the error.

35. 15 miles to 3 hours $= \frac{15 \text{ mi}}{3 \text{ hr}} = \frac{15 \text{ mi}}{180 \text{ min}} = \frac{1 \text{ mi}}{12 \text{ min}} = 12 \text{ min per } 1 \text{ mi}$

WRITING TO LEARN ▶▶▶▶

36. Explain the difference between a ratio and a rate.

37. Write a word problem whose answer in lowest terms is \$10.50/hr. The problem asks you to make and reduce a rate.

6.4 REWRITING RATIOS INVOLVING FRACTIONS AND DECIMALS AS RATIOS OF WHOLE NUMBERS

OBJECTIVES

1. Write a ratio as a division.
2. Write a ratio whose terms are fractions as a ratio of whole numbers.
3. Write a ratio of decimals as a ratio of whole numbers.

1 Writing a Ratio as a Division

Any ratio can be written in fraction form. A fraction is a division. Therefore, a ratio is a division. For example,

$$8 \text{ to } 5 = \frac{8}{5} = 8 \div 5.$$

> A ratio is a division.
>
> $a \text{ to } b = a \div b \qquad \frac{a}{b} = a \div b$
>
> $a:b = a \div b \qquad a \text{ per } b = a \div b$

EXAMPLE 1 Write each ratio as a division.

a. $50 \text{ to } 15 = 50 \div 15$
b. $6:11 = 6 \div 11$
c. $3/10 = 3 \div 10$
d. \$20 per 8 hours = \$20 ÷ 8 hours ■

▶ **You Try It** Write each ratio as a division.

1. 40 to 7 **2.** 25:10 **3.** 3/16 **4.** 40 miles per 3 gallons

---

2 Writing a Ratio Whose Terms Are Fractions as a Ratio of Whole Numbers

Write the ratio $\frac{1}{4}$ to $\frac{3}{8}$ as a ratio of two whole numbers.

Write the ratio as a division.

$$\frac{1}{4} \text{ to } \frac{3}{8} = \frac{1}{4} \div \frac{3}{8} = \frac{1}{\cancel{4}_1} \times \frac{\cancel{8}^2}{3} = \frac{2}{3} = 2 \text{ to } 3$$

The ratio $\frac{1}{4}$ to $\frac{3}{8}$ is equivalent to 2 to 3. A ratio is easier to understand when written in terms of whole numbers. For instance, a profit to cost ratio of $\$\frac{1}{4}$ to $\$\frac{3}{8}$ means there are \$2 profit to every \$3 cost.

> If one or both terms in a ratio are fractions, rewrite the ratio in terms of whole numbers by dividing the first term by the second.

EXAMPLE 2 Write 2 to $\frac{1}{6}$ as a ratio of whole numbers.

$$2 \text{ to } \frac{1}{6} = \frac{2}{1} \div \frac{1}{6} = \frac{2}{1} \times \frac{6}{1} = \frac{12}{1} = 12 \text{ to } 1$$

The ratio 2 to $\frac{1}{6}$ is a 12 to 1 ratio. ■

▶ **You Try It** 5. Write the ratio 8 to $\frac{3}{4}$ as a ratio of whole numbers.

EXAMPLE 3 Write $2\frac{1}{4}$ to $3\frac{3}{5}$ as a ratio of whole numbers.

$$2\frac{1}{4} \text{ to } 3\frac{3}{5} = 2\frac{1}{4} \div 3\frac{3}{5} = \frac{9}{4} \div \frac{18}{5} = \frac{\overset{1}{\cancel{9}}}{4} \times \frac{5}{\underset{2}{\cancel{18}}} = \frac{5}{8} = 5 \text{ to } 8$$ ■

▶ **You Try It** 6. Write $3\frac{5}{8}$ to $4\frac{5}{6}$ as a ratio of whole numbers.

EXAMPLE 4 The old computer can process a payroll in $2\frac{3}{4}$ hours. The new computer can process the same payroll in $\frac{1}{4}$ hour. Compare the speed with which the two computers process a payroll.

The ratio of old computer processing time to new computer processing time is $2\frac{3}{4}$ hours to $\frac{1}{4}$ hour.

old to *new*

$$2\frac{3}{4}\cancel{\text{hr}} \text{ to } \frac{1}{4}\cancel{\text{hr}} = \frac{11}{4} \div \frac{1}{4} = \frac{11}{\underset{1}{\cancel{4}}} \times \frac{\overset{1}{\cancel{4}}}{1} = \frac{11}{1} = 11 \text{ to } 1$$

The ratio 11 to 1 means the work the old computer could do in 11 minutes can be done by the new computer in 1 minute. The old computer took 11 times longer to process the same payroll. ■

▶ **You Try It** 7. Before training, Chris ran a sprint in $16\frac{1}{2}$ seconds. After training, she ran it in 11 seconds. What is the ratio of her pretraining to posttraining times?

---

3 Writing a Ratio Involving Decimals as a Ratio of Whole Numbers

Write 9.6 to 0.48 as a ratio of whole numbers.

$$9.6 \text{ to } 0.48 = \frac{9.6}{0.48} = \frac{9.6 \times 100}{0.48 \times 100} = \frac{960}{48} = \frac{\overset{20}{\cancel{960}}}{\underset{1}{\cancel{48}}} = \frac{20}{1} = 20 \text{ to } 1$$

Multiply numerator and denominator by the power of 10 that changes both decimals into whole numbers.

The ratio 9.6 to 0.48 is a 20 to 1 ratio.

To write a ratio of decimal numbers as a ratio of two whole numbers

1. Multiply both terms of the ratio by the power of 10 having as many zeros as the decimal with the greater number of decimal places.
2. Reduce.

EXAMPLE 5 Write 2.4 to 0.036 as a ratio of whole numbers.

$$2.4 \text{ to } 0.036 = \frac{2.4}{0.036} = \frac{2.4 \times 1{,}000}{0.036 \times 1{,}000} = \frac{\overset{200}{\cancel{2{,}400}}}{\underset{3}{\cancel{36}}} = 200 \text{ to } 3 \quad ■$$

► **You Try It**

8. Write 5.4 to 0.042 as a ratio of whole numbers.
9. Write 18 to 3.6 as a ratio of whole numbers.

► **Answers to You Try It** 1. 40 ÷ 7 2. 25 ÷ 10 3. 3 ÷ 16 4. 40 mi ÷ 3 gal 5. 32 to 3 6. 3 to 4 7. 3 to 2 8. 900 to 7 9. 5 to 1

SECTION 6.4 EXERCISES

1 **2** *Rewrite each ratio as a ratio of whole numbers.*

1. $\frac{1}{2}$ to $\frac{1}{4}$

2. $\frac{9}{24}$ to $\frac{6}{8}$

3. $\frac{7}{5}$ to $\frac{21}{10}$

4. $\frac{9}{4}$ to $\frac{7}{6}$

5. $\frac{5}{11}$ to 6

6. $3\frac{3}{4}$ to 9

7. 5 to $7\frac{1}{5}$

8. 10 to $\frac{5}{8}$

9. $\frac{4}{7}:\frac{3}{5}$

10. $\frac{3}{10}:\frac{5}{6}$

11. $6\frac{2}{3}$ to $1\frac{1}{9}$

12. $4\frac{5}{7}$ to $1\frac{6}{13}$

13. 12 to $2\frac{2}{5}$

14. 7 to $10\frac{1}{2}$

15. $\frac{5}{12}$ to $\frac{1}{12}$

16. $\frac{2}{6}$ to $\frac{5}{6}$

17. $\frac{1}{9}$ to $\frac{1}{10}$

18. $\frac{1}{20}$ to $\frac{1}{200}$

19. $2\frac{1}{4}$ to $1\frac{1}{8}$

20. $4\frac{1}{3}$ to $8\frac{2}{3}$

21. $4\frac{1}{2}$ to 1

22. 1 to $6\frac{3}{4}$

23. $\frac{1}{8}$ to $\frac{1}{8}$

24. $7\frac{4}{9}$ to $7\frac{4}{9}$

---

25. A share of XYZ stock gained $\$\frac{5}{8}$. A share of ABC gained $\$1\frac{1}{4}$. Find the whole number ratio of ABC gain to XYZ gain.

26. In an engine, wheel V turns 14 times when wheel W turns $2\frac{1}{3}$ times. Find the whole number ratio of turns in wheel V to turns in wheel W.

27. Tom walks $5\frac{3}{4}$ miles in 2 hours. Write the rate of miles per hour in terms of whole numbers.

28. Write the rate of \$3 to $3\frac{3}{10}$ gallons of gas in terms of whole numbers.

29. Write 3 yards to $4\frac{1}{2}$ inches as a whole number ratio.

145 seconds

30. Write $2\frac{1}{4}$ minutes to $4\frac{1}{2}$ seconds as a whole number ratio.

$\frac{9}{4}$

3 *Rewrite each ratio as a ratio of whole numbers.*

31. 0.4 to 0.06

32. 0.08 to 0.36

33. 0.006 to 18

34. 0.1 to 1

35. 5 to 0.09

36. 20 to 0.004

37. 0.09 to 0.03

38. 0.4 to 0.6

39. 0.1 to 0.09

40. 6.05 to 2.5

41. 7.7 to 1.1

42. 0.8 to 0.008

43. 4.8 to 0.048

44. 5.5 to 0.44

45. 0.001 to 0.01

46. 0.0014 to 0.00007

47. 0.00009 to 0.027

48. 0.064 to 0.00032

49. 3.6 to 5

50. 12.5 to 20

51. 1 to 0.06

52. 4 to 8.4

53. The fuel for a gas saw consists of 1.5 parts oil to 31.5 parts gas. Write the whole number ratio of oil to gas.

54. A widget takes 22.5 hours to construct and 4.5 hours to finish. Find the whole number ratio of construction time to finish time.

55. Write the ratio of 7.25 gallons to \$$8\frac{3}{4}$ as a ratio of two whole numbers.

56. Write the ratio of $2\frac{3}{4}$ hours to 5.5 miles as a ratio of two whole numbers.

SKILLSFOCUS (Section 6.1) *Solve the equation.*

57. $4.5x = 27$

58. $7y = 30$

59. $\frac{t}{8} = 1.25$

60. $\frac{k}{5.48} = 2.35$

EXTEND YOUR THINKING ▶▶▶

▶ TROUBLESHOOT IT

Find and correct the error.

61. $3\frac{1}{2}$ to $7 = 3\frac{1}{2} \div 7 = \frac{7}{2} \div \frac{7}{1} = \frac{2}{\cancel{7}_1} \times \frac{\cancel{7}^1}{1} = \frac{2 \times 1}{1 \times 1} = 2$ to 1

62. 0.06 to $1.8 = \frac{0.06 \times 100}{1.8 \times 100} = \frac{\cancel{6}^1}{\cancel{180}_{30}} = 30$ to 1

WRITING TO LEARN ▶▶▶

63. Explain the difference between the ratios 2 to 1 and 1 to 2.

64. Explain each step you use to write 12.5 to 0.375 as a ratio of whole numbers.

6.5 APPLICATIONS OF RATIOS

OBJECTIVES

1. Change a ratio or rate to a unit ratio or rate.
2. Calculate unit price.
3. Compare rates to solve problems.

NEW VOCABULARY

unit ratio
unit rate
unit price

1 Unit Ratio and Unit Rate

A **unit ratio** or **unit rate** is a ratio or rate with a second term of 1.

EXAMPLE 1 Examples of unit ratios.

a. 20 to 1 **b.** 7:1 **c.** 8/1 **d.** 100 feet to 1 foot ■

EXAMPLE 2 Examples of unit rates.

a. 5 men to 1 car
b. 30 mpg = 30 miles per 1 gallon
c. $2 per 1 pound
d. 50 mph = 50 miles/1 hour ■

▶ **You Try It** Label each as a unit ratio, unit rate, or neither.

1. 5 feet to 1 second **2.** $9:$4 **3.** 8 inches/1 inch

4. 40 miles:3 gallons **5.** $12 to 1 hour **6.** $\frac{6}{1}$

Write the ratio 5 to 2 as a unit ratio.

Write the ratio in fraction form. Divide both terms of the fraction by the denominator 2.

$$5 \text{ to } 2 = \frac{5}{2} = \frac{5 \div 2}{2 \div 2} = \frac{2.5}{1} = 2.5 \text{ to } 1$$

> To write any ratio or rate as a unit ratio or unit rate, divide both terms by the second term.

EXAMPLE 3 Write 7 feet to 4 feet as a unit ratio.

Divide each term by the second term, 4.

$$7 \text{ feet to } 4 \text{ feet} = \frac{7 \cancel{\text{feet}}}{4 \cancel{\text{feet}}} = \frac{7 \div 4}{4 \div 4} = \frac{1.75}{1} = 1.75 \text{ to } 1 \quad ■$$

EXAMPLE 4 Pat lost 15 pounds in 12 days. Write the unit rate.

$$15 \text{ pounds to } 12 \text{ days} = \frac{15 \text{ lb} \div 12}{12 \text{ days} \div 12} = \frac{1.25 \text{ lb}}{1 \text{ day}} = 1.25 \text{ lb/day}$$

Pat lost weight at a rate of 1.25 pounds per day. ■

▶ **You Try It**

7. Write 8 hours to 5 hours as a unit ratio.
8. Ricardo drove 800 miles in 2 hours. Write the unit rate.

EXAMPLE 5 A small town has 871 families with 1,643 children. Find children per family (round to tenths).

1,643 children per 871 families

$$\frac{1{,}643 \text{ children}}{871} \text{ per } \frac{871 \text{ families}}{871}$$

1.9 children per family ■

▶ **You Try It** **9.** A $60,000 prize was split among 7 winners. Find dollars per winner. Round to dollars.

2 Unit Pricing

Unit price is the price for one unit of an item. Unit price is a unit rate.

EXAMPLE 6 Examples of unit prices.

a. 50 cents per foot $= \dfrac{50 \text{ cents}}{1 \text{ foot}}$

b. $3.49 per pound $= \dfrac{\$3.49}{1 \text{ lb}}$

c. $1.19 per gallon $= \dfrac{\$1.19}{1 \text{ gal}}$ ■

▶ **You Try It** Write each unit price in fraction form.

10. $3.89 per yard **11.** $17 per hour **12.** 5¢ per pretzel

> To find unit price, divide the price paid by the quantity purchased.
>
> $$\text{unit price} = \frac{\text{price}}{\text{quantity}}$$

EXAMPLE 7 Peaches cost $7.25 for 5 pounds. Find the unit price.

$$\text{unit price} = \frac{\text{price}}{\text{quantity}} = \frac{\$7.25}{5 \text{ lb}} = \$1.45 \text{ per lb} \quad ■$$

▶ **You Try It** **13.** Alicia paid $42.80 for 8 pounds of seed. Find the unit price.

EXAMPLE 8 A 28-ounce jar of honey costs $2.29. Find the unit price rounded to the nearest tenth of a cent.

$$\text{unit price} = \frac{\text{price}}{\text{quantity}} = \frac{\$2.29}{28 \text{ oz}} = \frac{229¢}{28 \text{ oz}} \doteq 8.2¢/\text{oz} \quad ■$$

EXAMPLE 9 7.62 acres of land sold for $165,900. Find the unit price. Round to the nearest hundred dollars.

$$\text{unit price} = \frac{\text{price}}{\text{quantity}} = \frac{\$165{,}900}{7.62 \text{ acres}} \doteq \$21{,}800 \text{ per acre} \quad ■$$

♦ **You Try It**

14. An 18-ounce box of cereal costs \$3.19. Find the unit price in cents per ounce. Round to the nearest tenth of a cent.

15. 450 acres of land were sold for \$750,000. Find the unit price. Round to the nearest ten dollars.

3 COMPARING RATES TO SOLVE PROBLEMS

Unit price gives the consumer a tool for comparing prices.

EXAMPLE 10 **Regular vs Jumbo** Donna bought an 18-ounce jar of jam for \$1.59. Her friend Chris spent \$2.79 on the 30-ounce jumbo size. Who made the better buy?

The person who paid less per ounce made the better buy. To decide who this is, compute the unit prices and compare them.

$$\text{Donna's unit price} = \frac{\text{price}}{\text{quantity}} = \frac{\$1.59}{18 \text{ oz}} = \frac{159¢}{18 \text{ oz}} \doteq 8.8¢ \text{ per oz}$$

$$\text{Chris's unit price} = \frac{\text{price}}{\text{quantity}} = \frac{\$2.79}{30 \text{ oz}} = \frac{279¢}{30 \text{ oz}} \doteq 9.3¢ \text{ per oz}$$

Donna paid 8.8¢ per ounce. Chris paid 9.3¢ . Donna made the better buy. She saved

$9.3¢ - 8.8¢ = 0.5¢$ or $\frac{1}{2}$ cent per ounce by purchasing the smaller 18-ounce jar.

♦ **You Try It**

16. Benjamin bought 144 baseball cards for \$29.50. Sebastian bought 240 for \$53.75. Assuming each card has the same value, who made the better buy? Round to cents.

When comparing two rates, such as the unit price of two items, be sure that each one is measured in the same units.

EXAMPLE 11 **Regular vs Family Pack** A 14.2-ounce bag of chips costs \$1.89. An eight-pack of chips costs \$1.29 and consists of 8 bags of chips, each bag weighing $\frac{3}{4}$ of an ounce. Compare the prices.

$$\text{Unit price for the 14.2-ounce bag} = \frac{\text{price}}{\text{quantity}} = \frac{\$1.89}{14.2 \text{ oz}} = \frac{189¢}{14.2 \text{ oz}} \doteq 13.3¢ \text{ per oz}$$

The eight-pack consists of 8 bags of chips. Each bag weighs $\frac{3}{4}$ of an ounce.

Quantity in eight pack $= 8 \times \frac{3}{4}$ oz $= \frac{\overset{2}{\cancel{8}}}{1} \times \frac{3}{\underset{1}{\cancel{4}}}$ oz $= 6$ oz of chips in the eight-pack

Unit price for the eight-pack $= \frac{\text{price}}{\text{quantity}} = \frac{\$1.29}{6\text{ oz}} = \frac{129¢}{6\text{ oz}} \doteq 21.5¢$ per oz

You are paying $21.5¢ - 13.3¢ = 8.2¢$ more per ounce for the eight-pack because of the convenience and the extra packaging.

♦You Try It **17.** A 22.8-ounce box of pretzels costs \$2.79. A ten-pack of pretzels costs \$1.87 and consists of 10 bags, each weighing $1\frac{4}{5}$ ounces. Compare the prices. Round to the nearest tenth of a cent.

EXAMPLE 12 **Tune-ups and Miles per Gallon** Ella drove 361 miles on 15.7 gallons of gas. After a tune-up, she drove 312 miles on 11.2 gallons of gas under the same driving conditions. Use ratios to decide if the tune-up gave Ella better gas mileage.

Before tune-up: 361 mi per 15.7 gal $= \frac{361\text{ mi}}{15.7\text{ gal}} = 23.0$ miles per gallon

After tune-up: 312 mi per 11.2 gal $= \frac{312\text{ mi}}{11.2\text{ gal}} = 27.9$ miles per gallon

Ella got $27.9 - 23.0 = 4.9$ more miles per gallon after her tune-up.

♦You Try It **18.** Vince drove 186 miles in the city and used 9.6 gallons of gas. He drove 397 miles on the highway and used 17.8 gallons. Use ratios to compare Vince's city mileage to his highway mileage. Round to tenths.

♦Answers to You Try It **1.** unit rate **2.** neither **3.** unit ratio **4.** neither **5.** unit rate **6.** unit ratio **7.** 1.6 to 1 **8.** 400 mi/hr **9.** \$8,571 per winner

10. $\frac{\$3.89}{1\text{ yd}}$ **11.** $\frac{\$17}{1\text{ hr}}$ **12.** $\frac{5¢}{1\text{ pretzel}}$ **13.** \$5.35/lb **14.** 17.7¢ /lb

15. \$1,670/acre **16.** Benjamin paid 20¢ per card; Sebastian paid 22¢ per card. Sebastian paide 2¢ more per card. **17.** Unit price for the box is 12.2¢ /oz; for the 10-pack, it is 10.4¢ /oz. You pay 1.8¢ more per ounce for the box. **18.** Vince got 19.4 mpg in the city and 22.3 mpg on the highway. He got 2.9 more miles per gallon on the highway.

SECTION 6.5 EXERCISES

1 *Find the unit ratio or unit rate.*

1. 60 to 16

2. 20 to 25

3. 300 to 20

4. 72 to 24

5. 84 children to 40 families

6. $180 to 50 hours

7. 80 dozen eggs to 12 chickens

8. $420 to 12 women

---

9. According to a Census Bureau report, in 1980 there were 600,000 millionaires out of a population of 226,000,000 Americans. Find the ratio of population to 1 millionaire.

10. The ratio of Chinese to Americans is 1,000,000,000 to 250,000,000. Find the ratio of Chinese to 1 American.

11. The ratio of nitrogen to oxygen is 0.075 grams to 0.0003 grams. What is the ratio of nitrogen to 1 unit of oxygen?

12. The ratio of cost to profit is 0.72 to 0.012. What is the ratio of dollars of cost to 1 dollar of profit?

---

2 *Calculate the unit price.*

13. $48 is paid for 16 pounds of butter.

14. $440 is charged for 50 hours of bookkeeping.

15. You pay $7.28 for 5.2 gallons of gas.

16. 75 cents is paid for 5 marbles.

17. $90 was paid for 18 plastic models.

18. $70 is the cost for 28 pens.

19. The cost for $\frac{1}{4}$ pound of lunch meat was $1.27.

20. The cost for $16\frac{9}{10}$ gallons of gas is $20.28.

21. 2.945 acres of land sold for $84,000. Round to hundreds.

22. 0.28 acres of land costs $34,000. Round to thousands.

23. 20 ounces of chips cost $2.19. Round to tenth of a cent.

24. 6.37 pounds of chicken cost $10.13. Round to nearest cent.

---

25. A dozen eggs cost 89 cents. Find the price per egg to the nearest tenth of a cent.

26. A pack of 20 cigarettes cost $1.75. What is the price per cigarette to the nearest tenth of a cent?

27. 2 pounds 8 ounces of bacon cost $4.79. What is the price per ounce to the nearest tenth of a cent (1 lb = 16 oz)?

28. A board 5 feet 6 inches long costs $7.31. What is the price per 1 foot of board to the nearest cent $\left(6 \text{ in} = \frac{1}{2} \text{ ft}\right)$?

29. Meg planted 176 bulbs in 8 hours. Susan planted 144 bulbs in 6 hours. Who planted more bulbs per hour?

30. Zack painted 75 figurines in 15 hours. Zeke painted 48 in 12 hours. Who painted more figurines per hour?

31. Last year Gail scored 437 points in 23 basketball games. This year she scored 378 points in 18 games. In which year was her point per game average better?

32. Last year Johnny passed for 2,688 yards in 16 football games. This year he passed for 1,794 yards in 13 games. In which season did he pass for more yards per game?

33. Jayne purchased 12.6 gallons of unleaded gas for $14.52. Her friend paid $20.95 for 16.8 gallons of unleaded.

a. Determine who paid the better price per gallon. Round to the nearest cent.

b. How much less was paid per gallon at the better price?

34. The jumbo 64-ounce box of detergent costs $4.59. The regular 40-ounce box sells for $3.19.

a. Find the better buy. Round to the nearest tenth of a cent.

b. How much less per ounce was paid at the better price?

35. Mike bought a 12-ounce can of tomato sauce for 79¢ . Debbie bought the 28-ounce size for $1.69.

a. Who made the better buy? Round to tenths of a cent.

b. How much less per ounce was paid at the better price?

36. Paul bought a 32-ounce can of fruit for $1.19. Kim bought the 14-ounce size for 53¢ .

a. Find the better buy. Round to tenths of a cent.

b. How much less per ounce was paid at the better price?

37. Six 12-ounce cans of soda sell for $1.79. The 2-liter bottle (67.6 ounces) sells for $0.99.

a. Find the better buy by computing the unit price per ounce of soda. Round to the nearest tenth of a cent.

b. How much less per ounce was paid at the better price?

38. The 12-pack of bar soap costs $2.79. The 5-pack costs $1.09.

a. Find the better buy by computing the unit price per bar of soap. Round to the nearest tenth of a cent.

b. How much less per bar was paid at the better price?

39. Recently the population of the former Soviet Union was estimated to be 290,178,000. Its land area was 8,650,000 square miles. At the same time the population of China was estimated to be 1,117,700,000. China's land area is 3,678,000 square miles.

a. Find the number of people per square mile in each country. Round to the nearest person.

b. About how many times more populated is China per square mile than the former Soviet Union?

40. Recently the population of New York state was estimated at 18,142,000. Its land area is 49,108 square miles. At the same time the population of California was estimated to be 25,481,000. Its land area is 158,706 square miles.

a. Find the number of people per square mile in each state. Round to the nearest person.

b. About how many times more populated is New York per square mile than California?

SKILLSFOCUS (Section 4.5) *Subtract.*

41. $\frac{5}{6}-\frac{2}{5}$

42. $\frac{7}{10}-\frac{3}{8}$

43. $4\frac{1}{3}-\frac{8}{9}$

44. $7\frac{1}{12}-4\frac{7}{8}$

EXTEND YOUR THINKING ♦♦♦

WRITING TO LEARN ♦♦♦

45. Explain how to calculate unit price.

6.6 CIRCLES

OBJECTIVES

1. Calculate the circumference of a circle.
2. Applications of circumference.
3. Calculate the area of a circle.
4. Applications of area.

NEW VOCABULARY

circle	diameter
center	circumference
radius	pi

1 Circumference of a Circle

Place each person in your math class exactly 5 feet away from you. What geometric figure will they form? They will form a circle. 5 feet is called the radius of the circle.

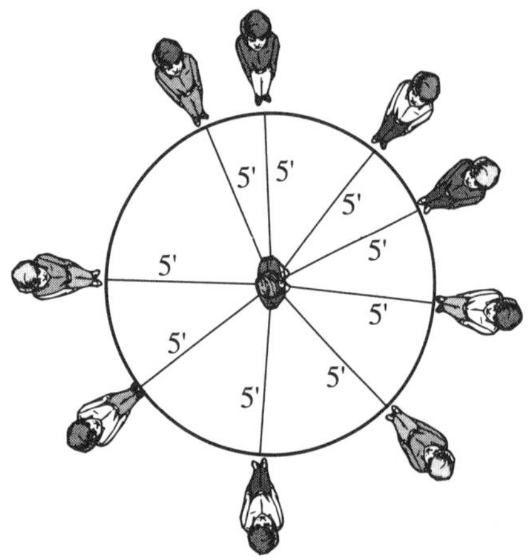

A **circle** is the set of points the same distance from a fixed point. The fixed point is the **center** of the circle. The **radius** R is the distance from the center to the circle. The **diameter** D is the distance from one point on the circle, through the center, to another point. The diameter is two times the radius, $D = 2R$.

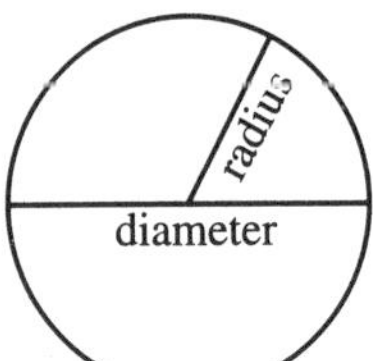

R = radius

D = diameter

$D = 2 \cdot R$

The **circumference** (*circum* means around) of a circle is the distance around the circle. It is similar to perimeter. If you broke the circle at one point and stretched it out, that length is the circumference.

Divide the circumference of any circle by its diameter. In every case, your answer is about 3.14. This number is so common in mathematics, it is given its own name. It is called π, or **pi**.

$$\pi = \frac{\text{circumference}}{\text{diameter}} \doteq 3.14$$

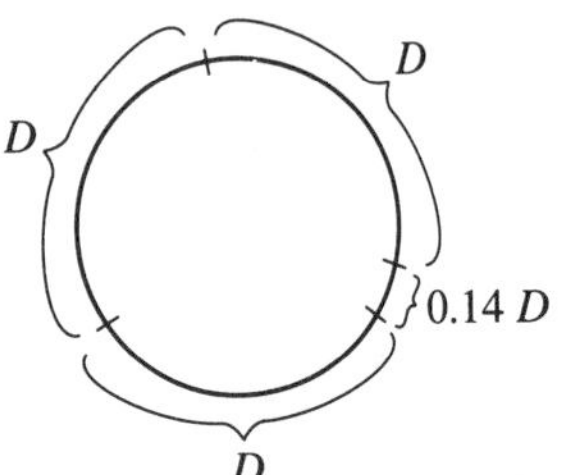

This means there are about 3.14 diameters in one circumference. That is, $C = 3.14 \cdot D$ or $C = \pi D$. Since $D = 2R$,

$$
\begin{aligned}
C &= \pi \cdot D \\
&= \pi \cdot 2R \quad \longleftarrow \text{Substitute } 2R \text{ for } D. \\
&= 2\pi R.
\end{aligned}
$$

> The circumference of a circle is
>
> $$C = \pi D \qquad \text{or} \qquad C = 2\pi R$$
>
> where D = diameter
> R = radius.

EXAMPLE 1 Find the circumference of the circle with a 7-in diameter.

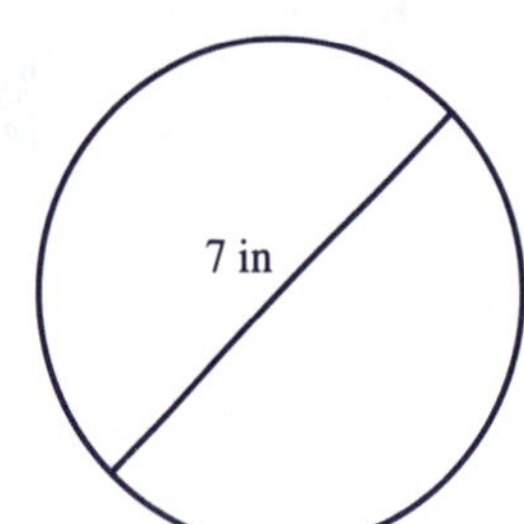

$$
\begin{aligned}
C &= \pi D \\
&= 3.14 \cdot 7 \text{ in} \\
&= 21.98 \text{ in} \quad ■
\end{aligned}
$$

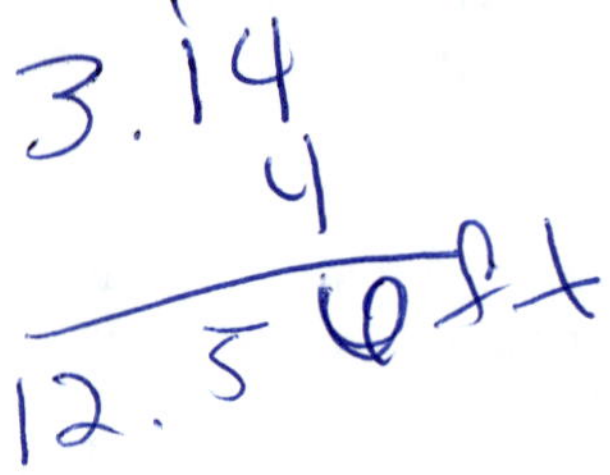

▶ **You Try It** **1.** Find the circumference.

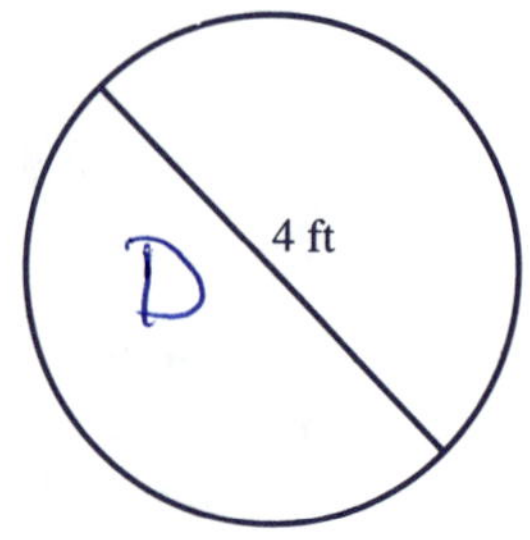

EXAMPLE 2 Find the circumference of a circle with radius 5 ft.

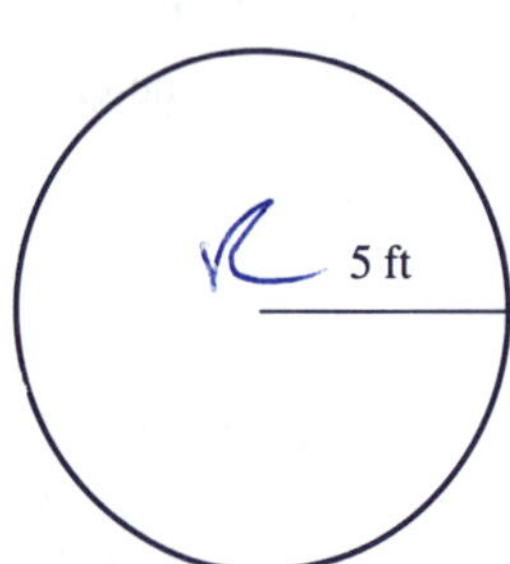

$$
\begin{aligned}
C &= 2\pi R \qquad \text{Replace } R \text{ with 5 ft.} \\
&= 2 \cdot 3.14 \cdot 5 \text{ ft} \\
&= 31.4 \text{ ft} \quad ■
\end{aligned}
$$

▶ **You Try It** **2.** Find the circumference.

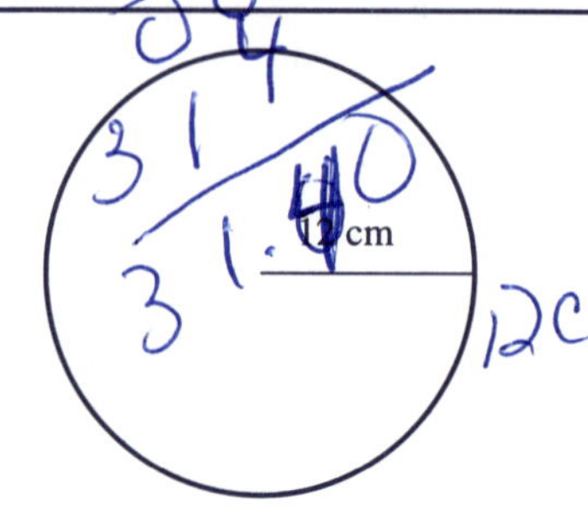

2 Applications of Circumference

EXAMPLE 3 A decorative metal guardrail will be placed around a circular fountain. The rail costs \$17.58 per foot installed. The radius of the fountain is 18 ft. What is the total cost?

Step #1: To find how much rail is needed, find the circumference of the fountain.

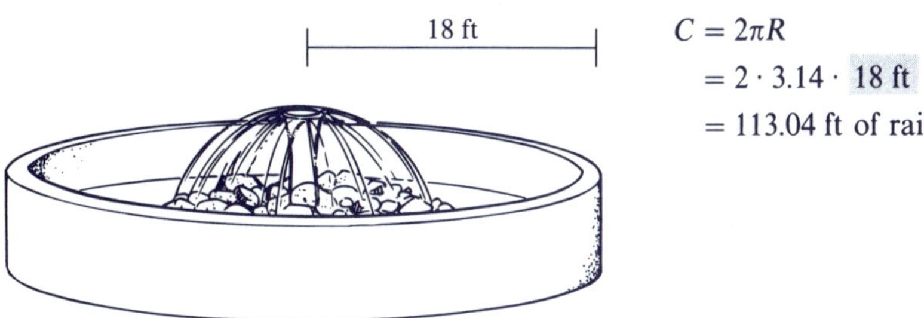

$$\begin{aligned} C &= 2\pi R \\ &= 2 \cdot 3.14 \cdot 18 \text{ ft} \\ &= 113.04 \text{ ft of rail} \end{aligned}$$

Step #2: Cost = 113.04 ft of rail · \$17.58 per ft

$\doteq$ \$1,987.24 installed ■

▶ **You Try It** 3. A curtain will be installed around a circular stage. The radius of the stage is 14 ft. What is the total cost of the curtain at \$8.29 per foot installed?

EXAMPLE 4 An automobile tire has a 15-in radius.

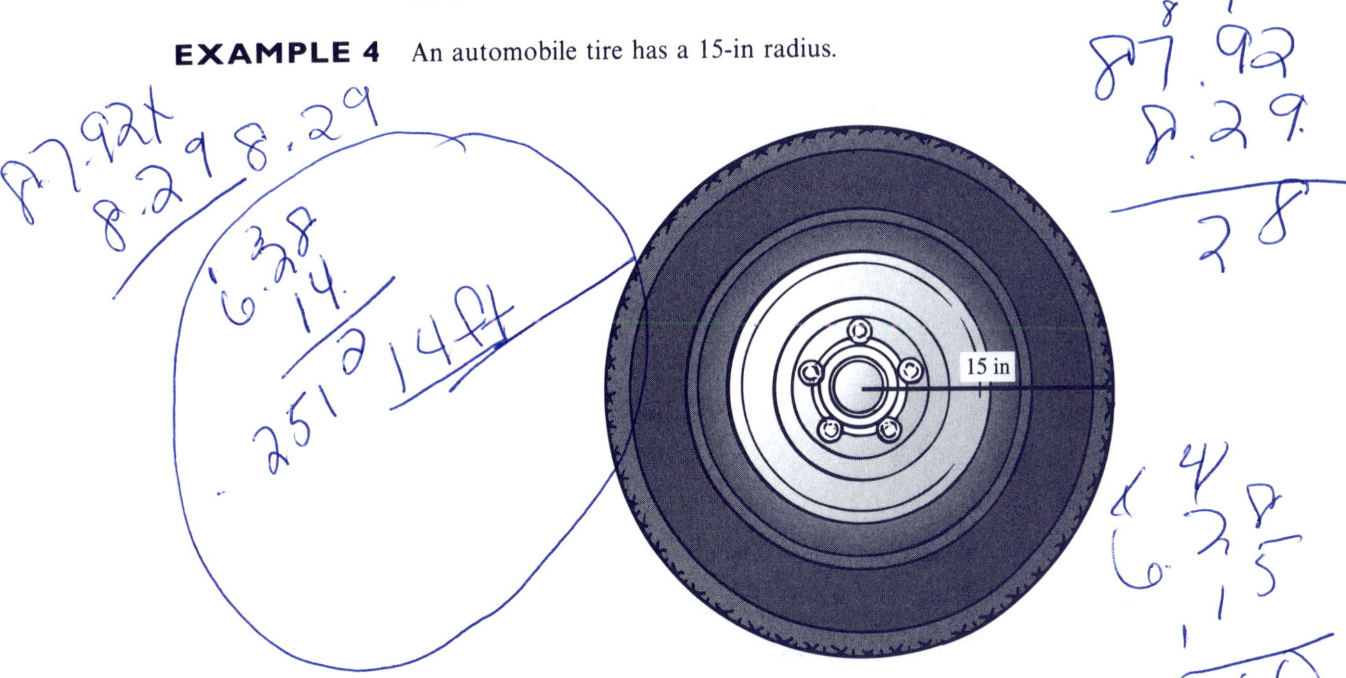

a. What is the circumference of the tire?

$$\begin{aligned} C &= 2\pi R \\ &= 2 \cdot 3.14 \times 15 \text{ in} \\ &= 94.2 \text{ in (or 7 ft 10.2 in)} \end{aligned}$$

b. How far will the car go when the wheel turns one revolution?

94.2 inches. The distance traveled in one revolution is the circumference.

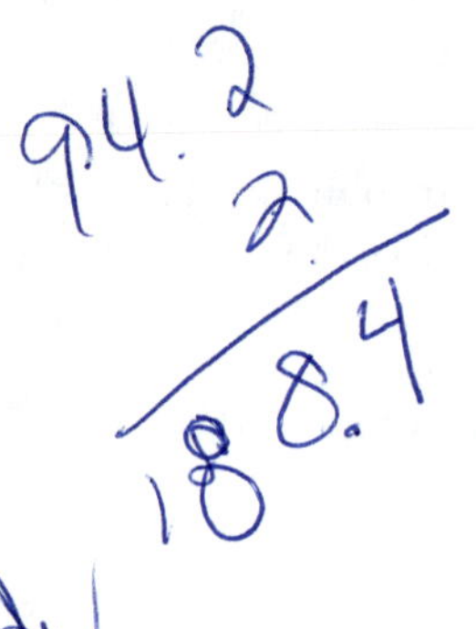

c. How many revolutions will the tire make when the car travels 1 mile?

$$1 \text{ mile} = 5{,}280 \text{ ft} = \frac{5{,}280 \cancel{\text{ft}}}{1} \cdot \frac{12 \text{ in}}{1 \cancel{\text{ft}}} = 63{,}360 \text{ in}$$

The car travels 94.2 inches in one tire revolution. To find the number of tire revolutions per mile, find how many times 94.2 in divides into 63,360 in.

$$\text{number of revolutions} = \frac{63{,}360 \cancel{\text{in}}}{94.2 \cancel{\text{in}}}$$

$$\doteq 673 \text{ revolutions of the tire in 1 mile} \quad \blacksquare$$

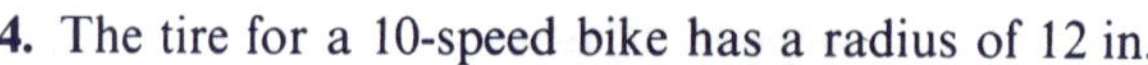

▶ You Try It

4. The tire for a 10-speed bike has a radius of 12 in.
a. What is the circumference of the tire?
b. How far will the bike go when the tire turns one revolution?
c. How many revolutions will the tire make when the bike travels 1 mile?

3 Area of a Circle

Cut a circle into equal slices as shown below left. Rearrange the slices into the approximate figure of a parallelogram. The base of the parallelogram is approximately half the circumference, $\left(\frac{1}{2}\right) \cdot 2\pi R$. The height is the radius R. The area is base times height.

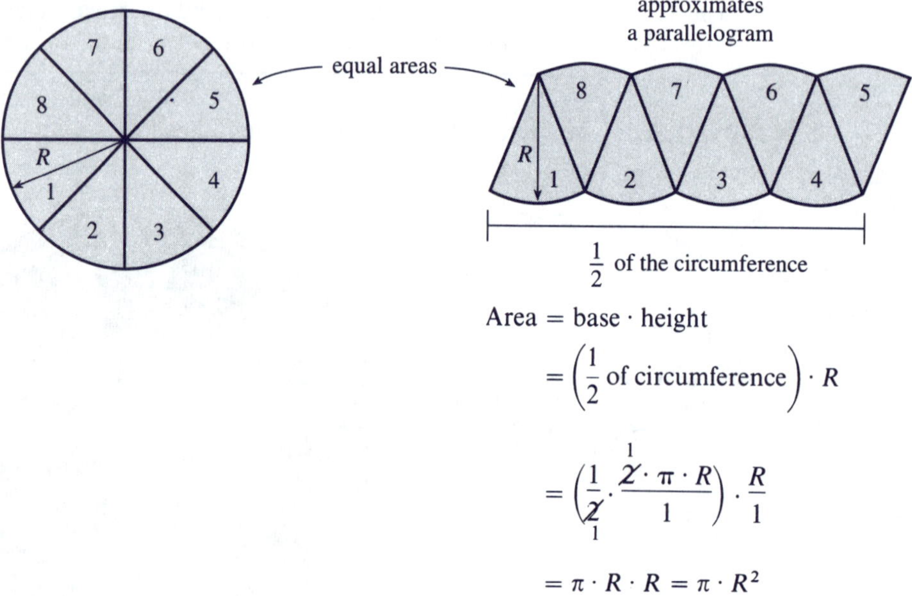

$$\text{Area} = \text{base} \cdot \text{height}$$

$$= \left(\frac{1}{2} \text{ of circumference}\right) \cdot R$$

$$= \left(\frac{1}{\cancel{2}_1} \cdot \frac{\cancel{2}^1 \cdot \pi \cdot R}{1}\right) \cdot \frac{R}{1}$$

$$= \pi \cdot R \cdot R = \pi \cdot R^2$$

> The area of a circle is π times the square of the radius R.
>
> $$A = \pi \cdot R^2$$
>
> or, $$A = \pi \cdot R \cdot R$$

EXAMPLE 5 Find the area of the circle with radius 6 yd.

$$A = \pi \cdot R^2$$
$$= 3.14 \cdot (6 \text{ yd})^2$$
$$= 3.14 \cdot 36 \text{ yd}^2$$
$$= 113.04 \text{ yd}^2$$

6 yd

■

▶ **You Try It** 5. Find the area.

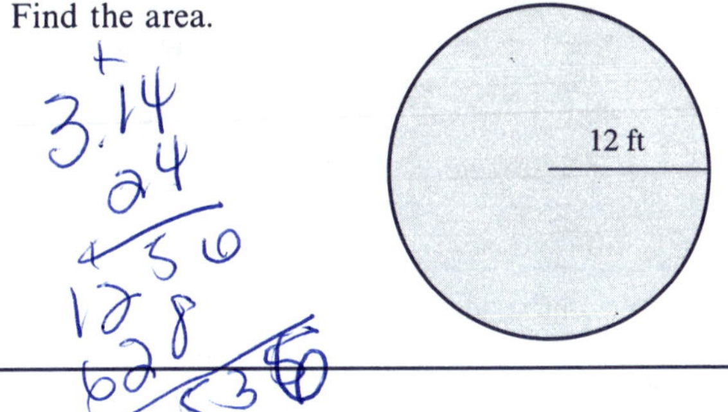

EXAMPLE 6 Find the area of a 16-inch circle.

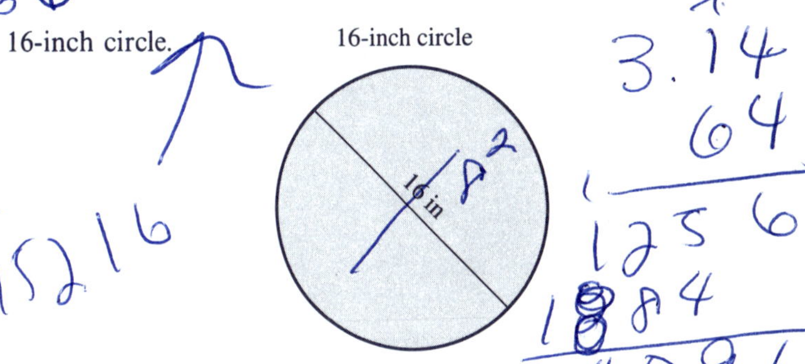

A 16-inch circle is a circle with diameter 16 in. The radius is half the diameter, or $R = 16 \text{ in} \div 2 = 8 \text{ in}$.

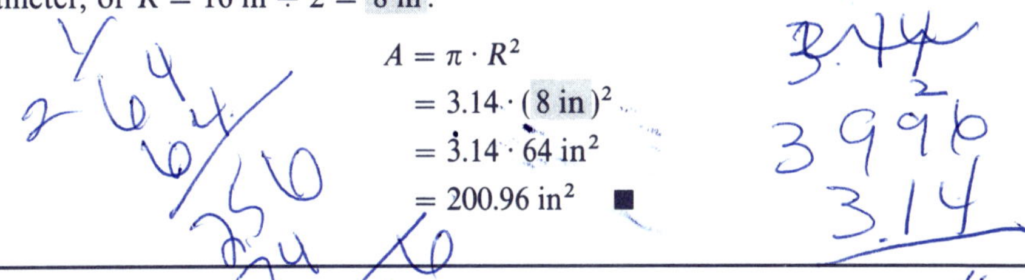

$$\begin{aligned} A &= \pi \cdot R^2 \\ &= 3.14 \cdot (8 \text{ in})^2 \\ &= 3.14 \cdot 64 \text{ in}^2 \\ &= 200.96 \text{ in}^2 \quad \blacksquare \end{aligned}$$

▶ **You Try It** 6. Find the area of an 18-yard circle. Draw and label a figure.

4 Applications of Area

EXAMPLE 7 All residents within 5 miles of a toxic spill will be evacuated. How many square miles is this?

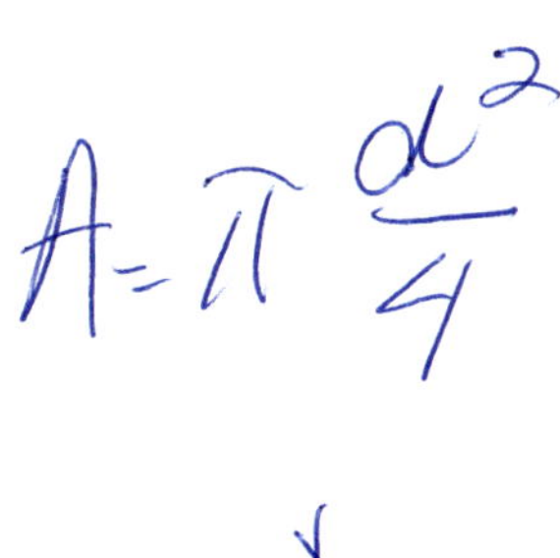

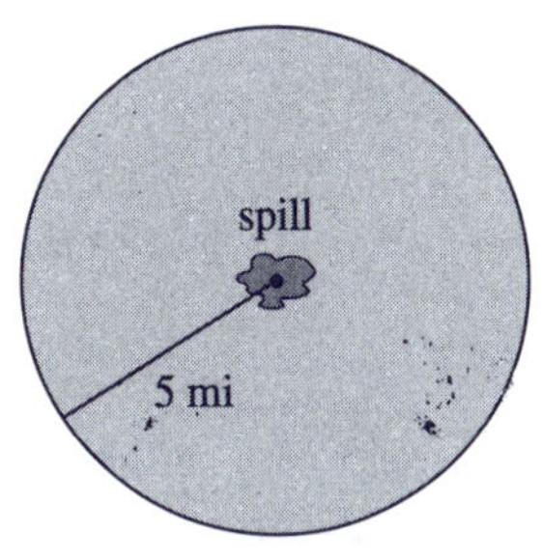

The region within 5 miles of the spill forms a circle. The spill is the center. 5 miles is the radius.

$$\begin{aligned} A &= \pi \cdot R^2 \\ &= 3.14 \cdot (5 \text{ mi})^2 \\ &= 3.14 \cdot 25 \text{ mi}^2 \\ &= 78.5 \text{ mi}^2 \text{ of land to be evacuated} \quad \blacksquare \end{aligned}$$

OBSERVE The answer in Example 7 may help in determining how much equipment and personnel will be needed to assist the evacuation. Suppose 2 people and 1 vehicle are needed to evacuate 3 sq mi in 4 hr. Then approximately $78.5 \div 3 = 26$ vehicles and 52 people will be needed for the evacuation.

▶ **You Try It** 7. A rotating sprinkler system will water the lawn within a radius of 30 feet. How many square feet of lawn is this?

EXAMPLE 8 In the figure, a circle is inscribed within a square. Find the area of the shaded region.

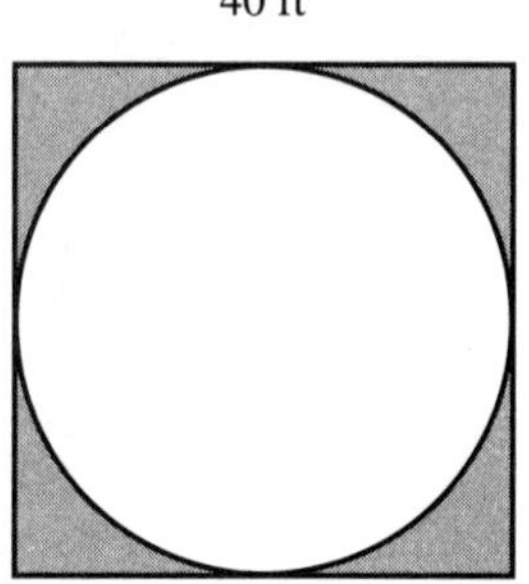

From the figure, the length of the side of the square is 40 ft. The diameter of the circle is 40 ft. So the radius of the circle is 20 ft.

$$\text{area of shaded region} = \text{area of square} - \text{area of circle}$$

$$\begin{aligned} A &= S^2 - \pi R^2 \\ &= (40 \text{ ft})^2 - 3.14 \cdot (20 \text{ ft})^2 \\ &= 1{,}600 \text{ ft}^2 - 3.14 \cdot 400 \text{ ft}^2 \\ &= 344 \text{ ft}^2 \quad \blacksquare \end{aligned}$$

▶ **You Try It** 8. Given the circle and square arranged as shown, find the area of the shaded region.

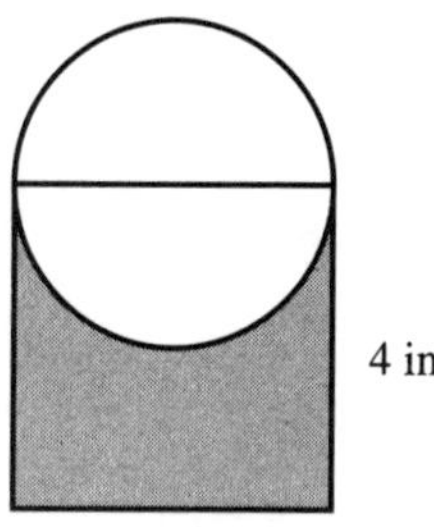

▶ **Answers to You Try It** 1. 12.56 ft 2. 75.36 cm 3. \$728.86 4. a. 75.36 in b. 75.36 in c. about 841 revolutions 5. 452.16 ft² 6. 254.34 yd² 7. 2,826 ft² 8. 9.72 in²

SECTION 6.6 EXERCISES

1 *Find the circumference. If not provided, draw and label a diagram.*

1.

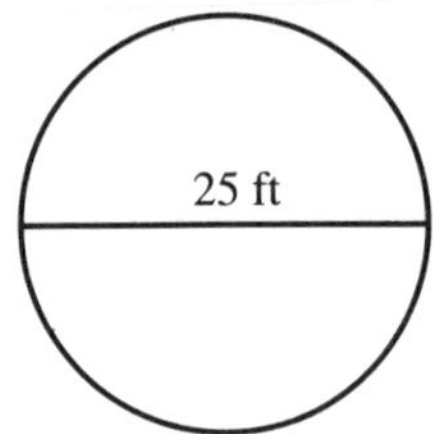

2.

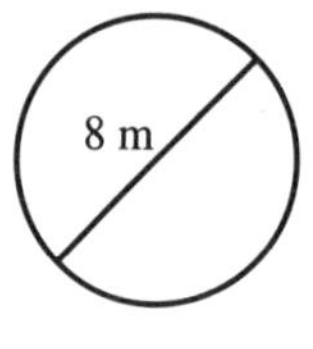

3.

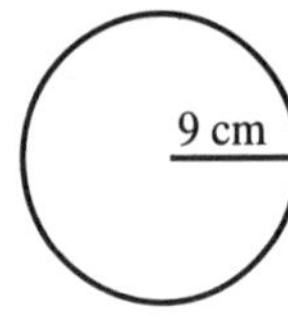

4.

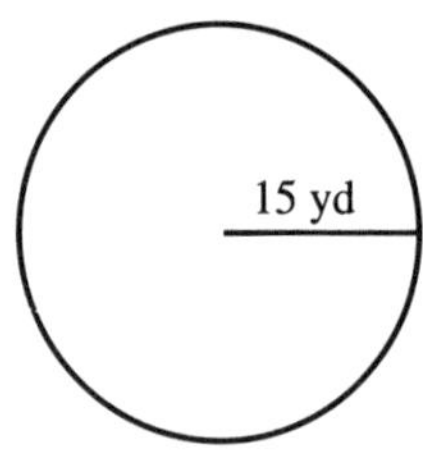

5. Find the circumference of a circle with diameter 4 ft.

6. Find the circumference of a circle with diameter 1 mile.

7. What is the circumference of a circle with radius 32 ft?

8. What is the circumference of a circle with radius 80 cm?

9. A circle has a 4 ft 6 in diameter. Find its circumference in

a. inches.

b. feet.

10. A circle has a 6 yd 1 ft radius. What is its circumference in

a. feet?

b. inches?

2

11. A figure skater must make the figure eight shown. It consists of two identical circles. Each has radius 10 ft. How far will the skater go one time around the eight?

12. How many linear feet of wood are needed to build the window in the figure? Each line in the figure is made of wood. The figure at the top of the window is a semicircle (half a circle).

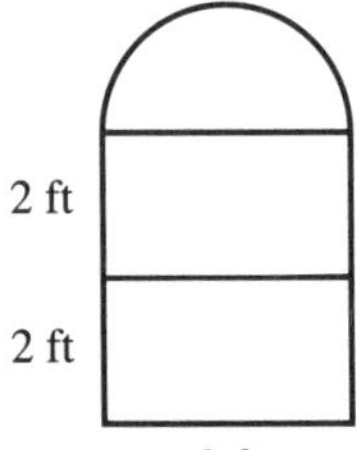

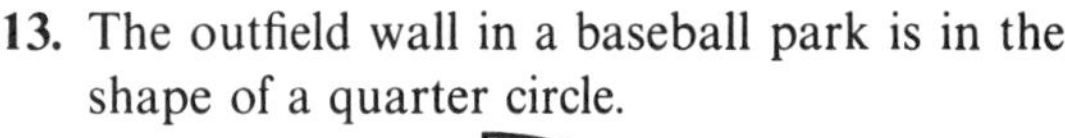

13. The outfield wall in a baseball park is in the shape of a quarter circle.

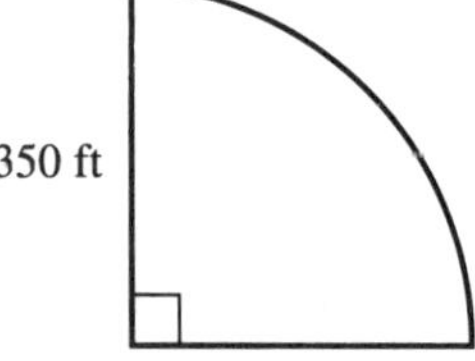

a. How long is the wall?

b. What will it cost to paint the wall at $24.80 per foot?

14. Refer to the indoor, circular racetrack shown in the figure.

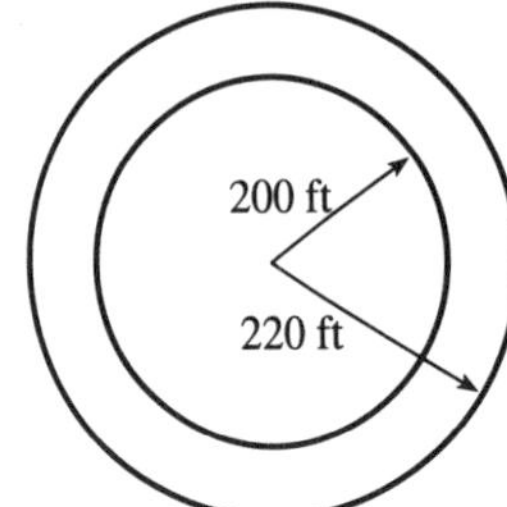

a. What is the distance around the inside of the track?

b. What is the distance around the outside of the track?

c. How much farther will a runner have to run on the outside of the track versus the inside?

15. Calculate the circumference of each wheel with the given radius.

a. 2 in

b. 4 in

c. 8 in

d. 16 in

e. What happens to the circumference when you double the radius?

16. Calculate the circumference of each wheel with the given diameter.

a. 1 ft

b. 2 ft

c. 4 ft

d. 8 ft

e. What happens to the circumference when you double the diameter?

17. a. Find the distance around the track shown in the figure.

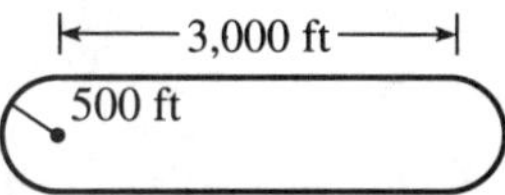

b. Express this distance in miles (round to tenths).

18. a. Find the distance around the track shown in the figure.

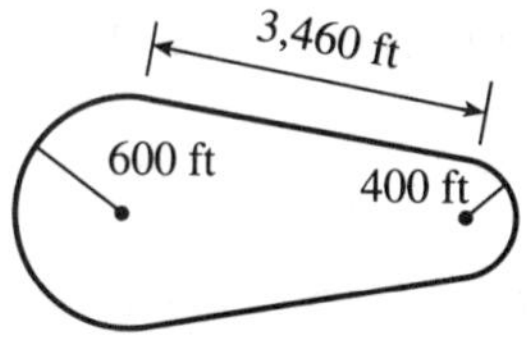

b. Express this distance in miles (round to tenths).

3 *Find the area. If not provided, draw and label a diagram.*

19.

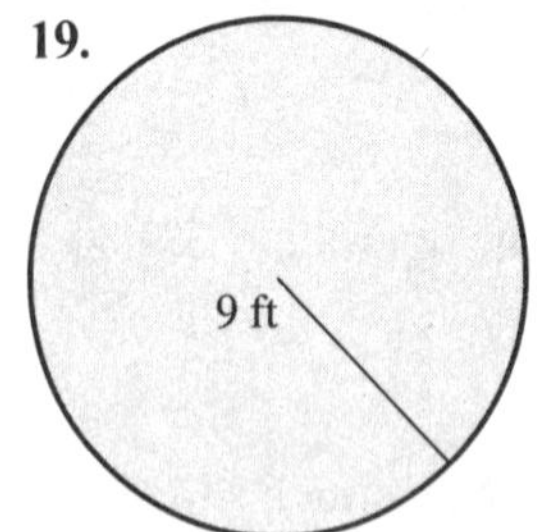

20.

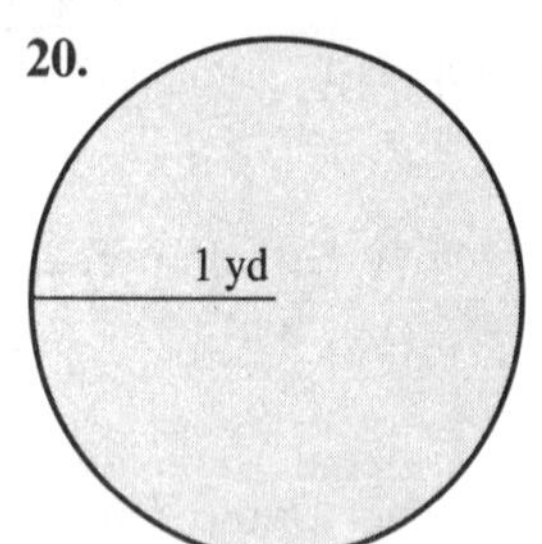

21.

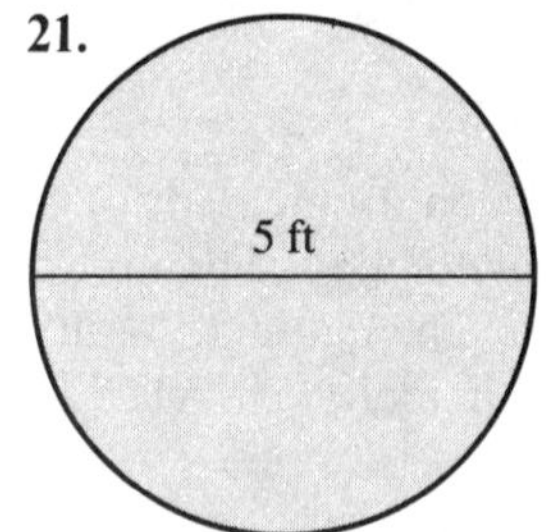

22.

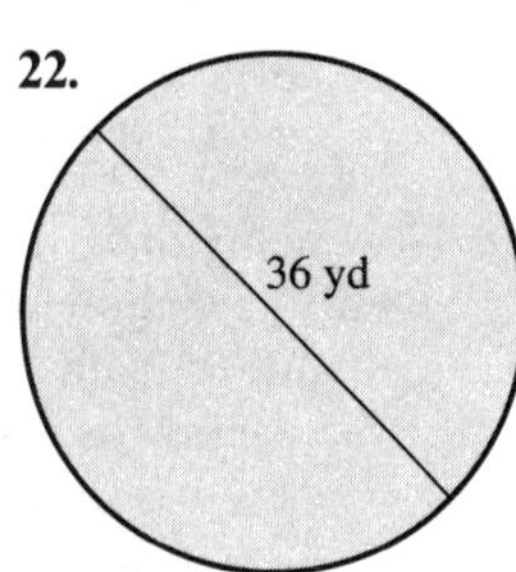

23. Find the area of a circle with a radius of 2 in.

24. What is the area of a circle with a 7-ft radius?

25. What is the area of a 30-yard circle?

26. Find the area of a 3-mile circle?

27. Find the area of a circle with a 2 ft 6 in radius
   a. in square feet.
   b. in square inches.

28. Find the area of a circle with a 10 yd 2 ft radius
   a. in square feet.
   b. in square inches.

4

29. How many sq ft of glass are needed in the construction of the window shown in the figure? The top of the window is a semicircle (half a circle).

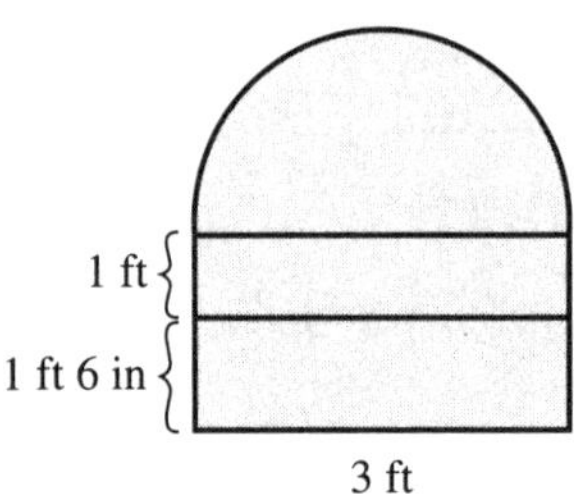

30. Find the area of the shaded region. Each circle has a radius of 1 in.

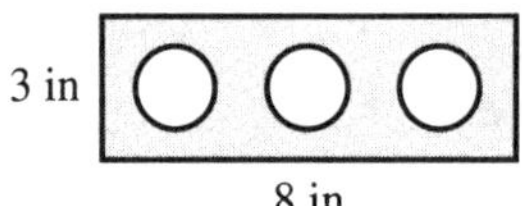

31. a. Find the area of the circular racetrack.

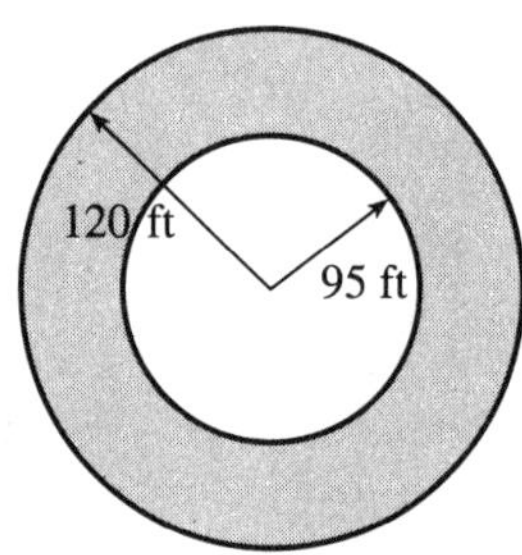

   b. Find the cost of asphalting the track at $70 per 100 sq ft.

32. The driving circle at a busy intersection will be landscaped at a cost of $1.75 per sq ft. Find the total cost.

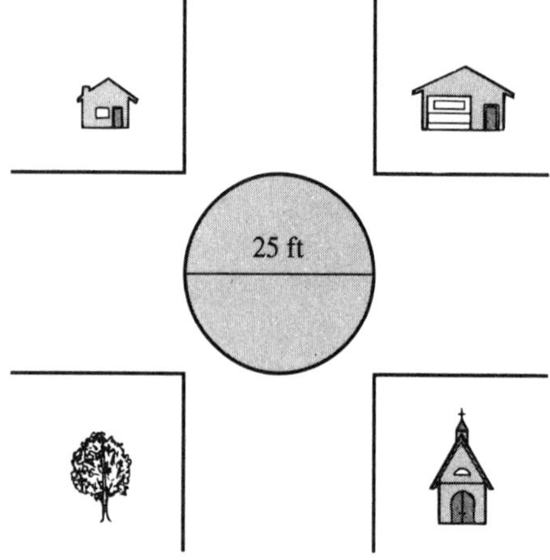

33. A radio station can be heard 600 miles away on a clear broadcasting day. How many square miles are in the station's listening area on such a day?

34. A giant sequoia has a base diameter of 18 ft.
   a. Find the approximate circumference at the base of the tree.
   b. Find the approximate area of the tree's footprint (ground area covered by the base of the tree).

35. a. Find the area of the ice-skating rink in the figure.

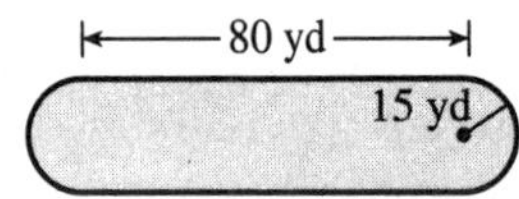

b. In one week it costs $0.075 per sq yd to maintain the rink. What is the weekly maintenance cost?

36. Find the area of the shaded portion of the figure. The small circle has a 1 mm radius.

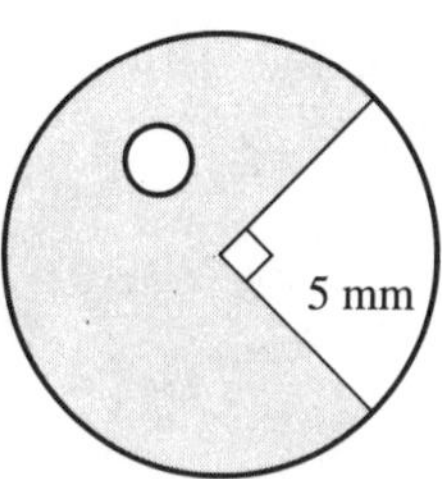

SKILLSFOCUS (Section 1.7) *Round to hundredths.*

37. 5.609 **38.** 17.9958 **39.** 0.0729 **40.** 9.0909

EXTEND YOUR THINKING

SOMETHING MORE

41. The Earth The diameter of the Earth at the equator is 7,926 miles.

a. What is the circumference of the Earth at the equator? Use $\pi = 3.14159$. Round to units.

b. The diameter of the Earth through the poles is 7,900 miles. How much longer is the equatorial circumference than the polar circumference?

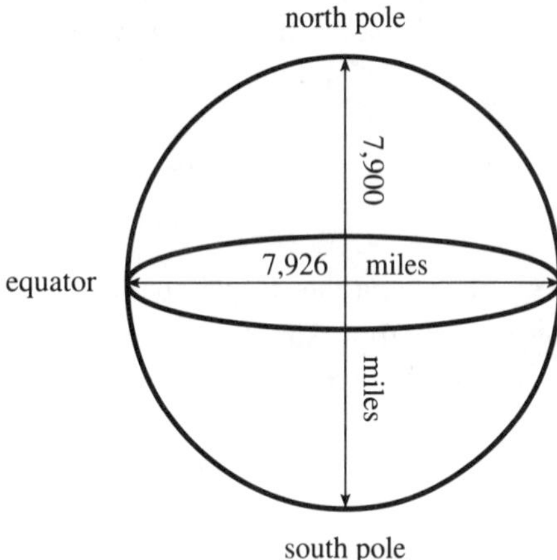

42. Circles and the 10-Speed Bike Most 10-speed bikes have 2 circular sprockets connected to the front pedals (sprockets have teeth, and hold the chain). They have 5 circular sprockets connected to the rear wheel. This gives a total of $2 \times 5 = 10$ possible gears. A chain connects a front sprocket to a rear one.

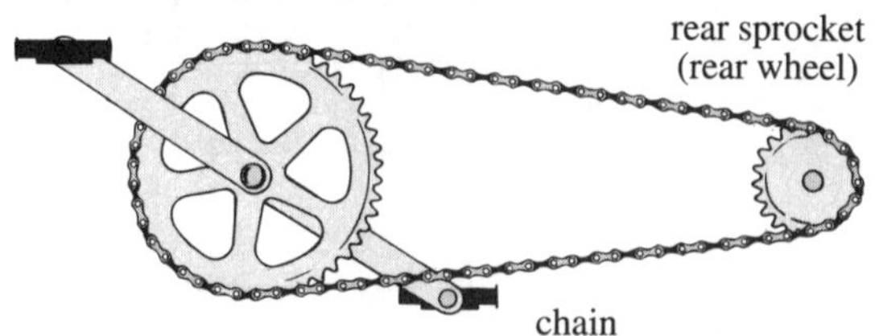

To determine how far a bike travels when the pedals (and so the front sprocket) make 1 revolution, evaluate the following formula.

$$\text{distance traveled on 1 revolution of the pedals} = \frac{\text{(number of teeth in front sprocket)}}{\text{(number of teeth in rear sprocket)}} \cdot \frac{\pi D}{12} \text{ ft}$$

This is the *gear ratio.* It tells you how many times the rear wheel rotates with one revolution of the pedals.

πD is the circumference of the rear wheel. D is the diameter in inches. Divide by 12 to change inches to feet.

With a 27″ bike (27″ means 27 inches), the diameter of the rear wheel is approximately 27 inches. How far will the bike go in 10th gear with one revolution of the pedals? In 10th gear, the chain is around the larger sprocket in the front, and the smallest in the rear. If the larger front sprocket has 52 teeth, and the smallest rear sprocket has 14 teeth, the bike will go

$$\frac{52}{14} \cdot \frac{3.14 \cdot 27}{12} = \frac{52 \cdot 3.14 \cdot 27}{14 \cdot 12}$$

$\doteq 26$ ft with one revolution of the pedals in 10th gear.

Note, in 10th gear, the gear ratio is $\frac{52}{14} \doteq 3.7$.

This means the rear wheel rotates about 3.7 times with one revolution of the pedals.

a. In first gear, the chain is around the smaller sprocket in the front (39 teeth) and the largest in the rear (28 teeth). How far will the 27″ bike go in 1st gear with one revolution of the pedals?

b. About how many times farther will this bike go in 10th gear versus 1st gear on one revolution of the pedals?

c. Suppose you ride 1 mile on your 27″ bike. How many revolutions of the pedals do you make in 10th gear? 1st gear? Round to tens.

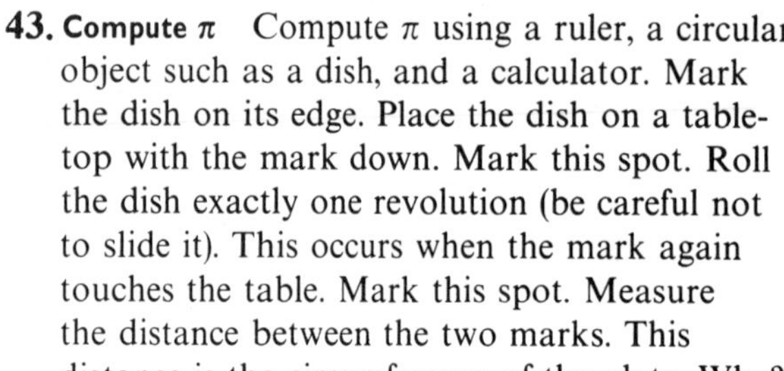

43. Compute π Compute π using a ruler, a circular object such as a dish, and a calculator. Mark the dish on its edge. Place the dish on a tabletop with the mark down. Mark this spot. Roll the dish exactly one revolution (be careful not to slide it). This occurs when the mark again touches the table. Mark this spot. Measure the distance between the two marks. This distance is the circumference of the plate. Why?

Next, measure the diameter of the plate. Fix one end of the ruler on the mark at the edge of the plate. Move the ruler back and forth on the opposite side of the plate. The longest measurement you can get is the diameter of the plate. Why?

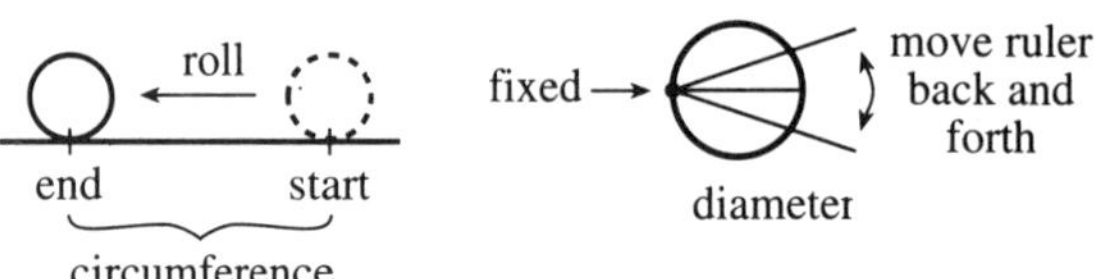

Estimate π by dividing the circumference by the diameter.

$$\pi = \frac{\text{circumference}}{\text{diameter}}$$

Use your calculator. Round to two places. How far is your value from 3.14?

44. Pizza Complete the chart for three different sizes of pepperoni pizza. The measure for a round pizza is the length of the diameter. Which pizza is the best buy per square inch?

Size of Pizza	*Cost*	*Area*	*Cost per sq. in.*
9" round	\$5.60		
12" round	\$8.50		
2 ft by 1 ft	\$10.99		

WRITING TO LEARN ♦♦♦

45. Explain clearly what π means. Use the word **ratio** in your explanation.

46. Explain the difference between area and circumference.

47. Explain how circumference and perimeter are alike. Explain how they are different.

6.7 INTRODUCTION TO PROPORTIONS

OBJECTIVE

Define proportions and cross product.

NEW VOCABULARY

proportion | false proportion | cross product | true proportion | cross product property

Defining Proportion and Cross Product

A **proportion** is a statement that two ratios are equal. For example,

$$\frac{1}{3} = \frac{2}{6}$$

is a proportion. This is read "1 is to 3 as 2 is to 6." It is a statement that the ratio $\frac{1}{3}$ equals the ratio $\frac{2}{6}$. You can write this proportion in *colon form* as 1:3::2:6. 1 and 6 are called the *extremes.* 3 and 2 are called the *means.*

A proportion is either true or false. A **true proportion** has equal cross products. **Cross products** are found by multiplying in a crosswise fashion.

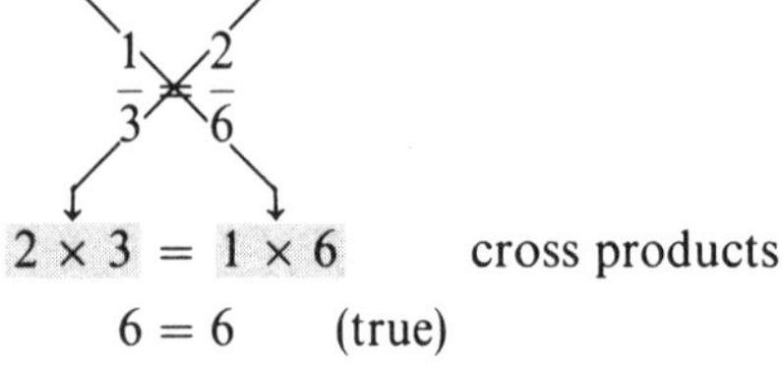

$$2 \times 3 = 1 \times 6 \qquad \text{cross products}$$

$$6 = 6 \qquad \text{(true)}$$

The cross products are 2×3 and 1×6. $\frac{1}{3} = \frac{2}{6}$ is a true proportion because both cross products equal 6. You also say it is true because the product of the means (2×3) equals the product of the extremes (1×6). A **false proportion** has unequal cross products.

> **The Cross Product Property**
>
> A true proportion has equal cross products. This is called the **cross product property**, and is pictured as follows.
>
> $$\frac{a}{b} = \frac{c}{d}$$
>
> $$b \cdot c = a \cdot d$$

EXAMPLE 1 Determine if the proportion is true or false.

a. $\frac{2}{5} \stackrel{?}{=} \frac{4}{10}$

$$4 \times 5 \stackrel{?}{=} 2 \times 10$$

$$20 = 20 \qquad \text{(true)}$$

b. $\frac{6}{16} \stackrel{?}{=} \frac{9}{24}$

$$9 \times 16 \stackrel{?}{=} 6 \times 24$$

$$144 = 144 \qquad \text{(true)}$$

c. $\frac{1}{2} \stackrel{?}{=} \frac{6}{8}$

$$6 \times 2 \stackrel{?}{=} 1 \times 8$$

$$12 \neq 8 \qquad \text{(false)}$$

Since the cross products are unequal, write $\frac{1}{2} \neq \frac{6}{8}$. ■

▶ **You Try It** Determine if each proportion is true or false.

1. $\frac{7}{10} \stackrel{?}{=} \frac{3}{5}$ 2. $\frac{8}{20} \stackrel{?}{=} \frac{6}{15}$ 3. $\frac{21}{56} \stackrel{?}{=} \frac{39}{104}$ 4. $\frac{64}{80} \stackrel{?}{=} \frac{16}{24}$

OBSERVE A proportion is true if both of its ratios reduce to the same lowest terms answer. The proportion in Example 1a is true because $\frac{4}{10}$ can be reduced to $\frac{2}{5}$. The proportion in Example 1b is true because $\frac{6}{16}$ and $\frac{9}{24}$ both reduce to $\frac{3}{8}$. In Example 1c, the proportion is false because $\frac{6}{8}$ reduces to $\frac{3}{4}$, and $\frac{1}{2} \neq \frac{3}{4}$.

CAUTION Cross multiplication is only used with proportions. It is never used when multiplying fractions.

EXAMPLE 2 Is the proportion $\frac{4.5}{6} \stackrel{?}{=} \frac{0.9}{1.2}$ true or false?

$$\frac{4.5}{6} \stackrel{?}{=} \frac{0.9}{1.2}$$

$$6 \times 0.9 \stackrel{?}{=} 4.5 \times 1.2$$

$$5.4 = 5.4$$

The proportion is true. ■

EXAMPLE 3 Is the proportion $\frac{2\frac{1}{2}}{\frac{3}{5}} \stackrel{?}{=} \frac{5}{\frac{1}{3}}$ true or false?

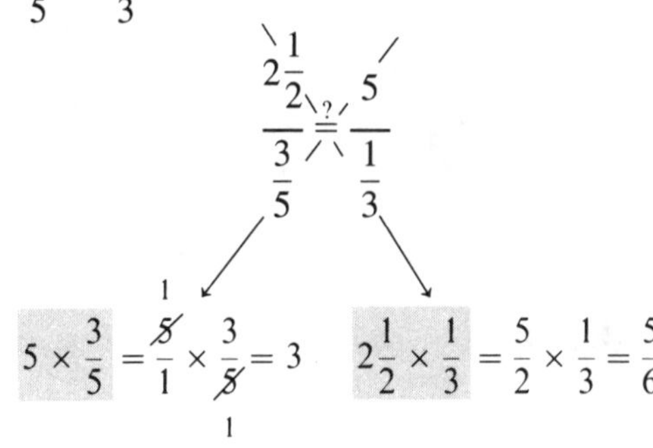

$$5 \times \frac{3}{5} = \frac{5}{1} \times \frac{3}{5} = 3 \qquad 2\frac{1}{2} \times \frac{1}{3} = \frac{5}{2} \times \frac{1}{3} = \frac{5}{6}$$

$$3 \neq \frac{5}{6}$$

The proportion is false. ■

▶ **You Try It** Determine if each proportion is true or false.

5. $\frac{0.8}{3.5} \stackrel{?}{=} \frac{4.2}{9.25}$ 6. $\frac{4\frac{4}{5}}{6} \stackrel{?}{=} \frac{1\frac{1}{2}}{1\frac{7}{8}}$

▶ **Answers to You Try It** 1. $35 \neq 30$, false 2. $120 = 120$, true 3. $2{,}184 = 2{,}184$, true 4. $1{,}536 \neq 1{,}280$, false 5. $7.4 \neq 14.7$, false 6. $9 = 9$, true

SECTION 6.7 EXERCISES

Cross multiply to determine if each proportion is true or false.

1. $\frac{1}{2} \stackrel{?}{=} \frac{3}{6}$ **2.** $\frac{3}{4} \stackrel{?}{=} \frac{9}{12}$ **3.** $\frac{2}{5} \stackrel{?}{=} \frac{8}{20}$ **4.** $\frac{4}{8} \stackrel{?}{=} \frac{2}{6}$

5. $\frac{5}{7} \stackrel{?}{=} \frac{2}{3}$ **6.** $\frac{1}{5} \stackrel{?}{=} \frac{1}{6}$ **7.** $\frac{7}{8} \stackrel{?}{=} \frac{21}{24}$ **8.** $\frac{12}{27} \stackrel{?}{=} \frac{20}{45}$

9. $\frac{7}{10} \stackrel{?}{=} \frac{11}{15}$ **10.** $\frac{3}{5} \stackrel{?}{=} \frac{5}{8}$ **11.** $\frac{14}{35} \stackrel{?}{=} \frac{6}{15}$ **12.** $\frac{10}{18} \stackrel{?}{=} \frac{30}{54}$

13. $\frac{60}{72} \stackrel{?}{=} \frac{45}{54}$ **14.** $\frac{200}{650} \stackrel{?}{=} \frac{80}{260}$ **15.** $\frac{24}{77} \stackrel{?}{=} \frac{16}{44}$ **16.** $\frac{32}{45} \stackrel{?}{=} \frac{132}{145}$

17. $\frac{8}{13} \stackrel{?}{=} \frac{72}{117}$ **18.** $\frac{28}{84} \stackrel{?}{=} \frac{33}{99}$ **19.** $\frac{7}{5} \stackrel{?}{=} \frac{10}{7}$ **20.** $\frac{12}{17} \stackrel{?}{=} \frac{15}{19}$

21. $\frac{3.6}{6} \stackrel{?}{=} \frac{6}{10}$ **22.** $\frac{2}{0.4} \stackrel{?}{=} \frac{0.5}{0.1}$ **23.** $\frac{100}{2.4} \stackrel{?}{=} \frac{25}{0.8}$ **24.** $\frac{9.6}{8} \stackrel{?}{=} \frac{3.6}{2.8}$

25. $\frac{\frac{1}{2}}{0.2} \stackrel{?}{=} \frac{8}{5}$ **26.** $\frac{\frac{1}{4}}{0.6} \stackrel{?}{=} \frac{1}{2}$ **27.** $\frac{\frac{5}{8}}{\frac{2}{5}} \stackrel{?}{=} \frac{\frac{4}{8}}{9}$ **28.** $\frac{\frac{9}{4}}{\frac{3}{4}} \stackrel{?}{=} \frac{6}{\frac{2}{3}}$

29. $\frac{\frac{3}{5}}{2} \stackrel{?}{=} \frac{6}{20}$ **30.** $\frac{16}{\frac{4}{9}} \stackrel{?}{=} \frac{9}{\frac{1}{4}}$ **31.** $\frac{100}{\frac{1}{2}} \stackrel{?}{=} \frac{50}{\frac{1}{4}}$ **32.** $\frac{\frac{1}{8}}{\frac{1}{4}} \stackrel{?}{=} \frac{\frac{1}{6}}{\frac{1}{3}}$

SKILLSFOCUS (Section 2.1) *Multiply.*

33. 2.4×0.68 **34.** 4.5×8.2 **35.** 0.7×1.9 **36.** 20×4.3

EXTEND YOUR THINKING

SOMETHING MORE

37. A blueprint is drawn according to the scale

$\frac{1}{2}$ inch: 6 feet.

A wall $3\frac{5}{8}$ inches long on the blueprint is actually 38 feet long. Was the wall constructed according to blueprint specifications? (Hint: Make two ratios, and see if they form a true proportion.)

38. A blueprint is drawn according to the scale

1 inch: 8 feet.

A floor $7\frac{1}{4}$ inches long on the blueprint is actually 58 feet long. Was the floor constructed according to blueprint specifications?

▶ TROUBLESHOOT IT

Find and correct the error.

39. $\frac{5}{8} \times \frac{3}{10} = \frac{5}{8} \times \frac{3}{10} = \frac{24}{50} = \frac{12}{25}$

40. $\frac{12}{21} \stackrel{?}{=} \frac{7}{4} \rightarrow 12 \times 7 \stackrel{?}{=} 21 \times 4 \rightarrow 84 = 84$. Therefore, $\frac{12}{21} = \frac{7}{4}$ is true.

WRITING TO LEARN ▶▶▶▶

41. What is a proportion? Create an example for your explanation.

42. Make up your own examples of a true proportion and a false proportion. Then explain why each proportion is true or false.

6.8 SOLVING PROPORTIONS

OBJECTIVE

Solve proportions using cross multiplication.

Solving Proportions A proportion has four numbers. If you are only given three of the numbers, you can solve for the fourth. For example, solve for x.

$$\frac{2}{3} = \frac{x}{6}$$

The letter x represents the number that will make this a true proportion. Even though x is unknown, if this is to be a true proportion the cross products must be equal.

$$\frac{2}{3} = \frac{x}{6}$$

$$3 \cdot x = 2 \cdot 6$$

$$3 \cdot x = 12$$

Divide both sides of the equation by 3, the number multiplied by x.

$$\frac{\overset{1}{\cancel{3}} \cdot x}{\underset{1}{\cancel{3}}} = \frac{12}{3}$$

$$x = 4$$

The answer is $x = 4$. To check, replace x with 4 in the original proportion. The proportion is true if the cross products are equal.

Check: $$\frac{2}{3} \stackrel{?}{=} \frac{4}{6}$$

$$3 \cdot 4 \stackrel{?}{=} 2 \cdot 6$$

$$12 = 12 \quad \text{(true)}$$

> To solve a proportion for the missing number x
>
> **1.** Cross multiply. Set the two cross products equal to each other.
> **2.** Solve this equation for x. Do this by dividing both sides of the equation by the number multiplied by x.
> **3.** Check your answer. In the original proportion, replace x with the number found in Step 2, and cross multiply. If the cross products are equal, your answer is correct.

EXAMPLE 1 Solve for x. $\frac{3}{8} = \frac{x}{32}$

$$\frac{3}{8} = \frac{x}{32}$$

Cross multiply. Set the two cross products equal to each other.

$$8 \cdot x = 3 \cdot 32$$

$$\frac{\overset{1}{\cancel{8}} \cdot x}{\underset{1}{\cancel{8}}} = \frac{3 \cdot \overset{4}{\cancel{32}}}{\underset{1}{\cancel{8}}}$$ Divide both cross products by 8, the number multiplied by x. Reduce.

$$x = 12$$ The solution.

Check: Replace x with 12 in the original proportion.

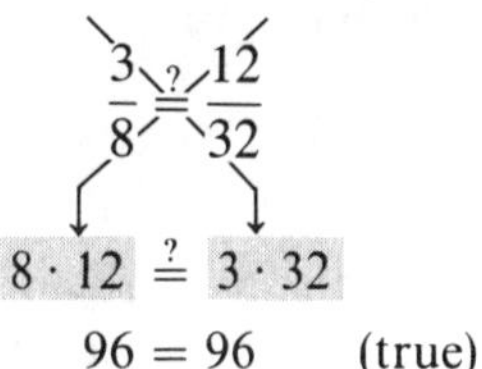

$$8 \cdot 12 \stackrel{?}{=} 3 \cdot 32$$

$$96 = 96 \qquad \text{(true)}$$ ■

▶ **You Try It** **1.** Solve for y. $\dfrac{y}{12} = \dfrac{14}{21}$

EXAMPLE 2 Solve for x. $\dfrac{15}{x} = \dfrac{24}{40}$

$$\frac{15}{x} = \frac{24}{40}$$ Cross multiply. Set the two cross products equal to each other.

$$24 \cdot x = 15 \cdot 40$$

$$\frac{\overset{1}{\cancel{24}} \cdot x}{\underset{1}{\cancel{24}}} = \frac{\overset{5}{\cancel{15}} \cdot \overset{5}{\cancel{40}}}{\underset{\underset{1}{\cancel{3}}}{\cancel{24}}}$$ Divide both sides of the equation by 24, the number multiplied by x. Reduce.

$$x = 25$$ The solution.

Check: Replace x with 25 in the original proportion

$$\frac{15}{25} \stackrel{?}{=} \frac{24}{40}$$

$$25 \cdot 24 \stackrel{?}{=} 15 \cdot 40$$

$$600 = 600 \qquad \text{(true)}$$ ■

▶ **You Try It** Solve.

2. $\dfrac{18}{x} = \dfrac{63}{49}$

3. $\dfrac{27}{2} = \dfrac{45}{y}$

OBSERVE You may reduce one of the ratios in a proportion before solving for x. In Example 2, reduce $\dfrac{24}{40}$. Then cross multiply.

$$\frac{15}{x} = \frac{\overset{3}{\cancel{24}}}{\underset{5}{\cancel{40}}} \longrightarrow \frac{15}{x} = \frac{3}{5}$$ Cross multiply. Set the cross products equal to each other.

$$3 \cdot x = 15 \cdot 5$$

$$\frac{\overset{1}{\cancel{3}} \cdot x}{\underset{1}{\cancel{3}}} = \frac{\overset{5}{\cancel{15}} \cdot 5}{\underset{1}{\cancel{3}}}$$

$x = 25$ Same answer as in Example 2.

The check for each of the following examples is left as an exercise for the student.

EXAMPLE 3 Solve for n. $\frac{n}{10} = \frac{16}{28}$

$$\frac{n}{10} = \frac{\overset{4}{\cancel{16}}}{\underset{7}{\cancel{28}}} \longrightarrow \frac{n}{10} = \frac{4}{7}$$ Cross multiply.

$$10 \cdot 4 = 7 \cdot n$$

$$\frac{10 \cdot 4}{7} = \frac{\overset{1}{\cancel{7}} \cdot n}{\underset{1}{\cancel{7}}}$$ Divide each side by 7, the number multiplied by n.

$$5\frac{5}{7} = n$$ ■

▶ **You Try It** 4. Solve. $\frac{x}{7} = \frac{4}{9}$

EXAMPLE 4 Solve for y and round to hundredths. $\frac{17}{11} = \frac{9}{y}$

$$\frac{17}{11} = \frac{9}{y}$$ Cross multiply. Set the cross products equal to each other.

$$17 \cdot y = 11 \cdot 9$$

$$\frac{\overset{1}{\cancel{17}} \cdot y}{\underset{1}{\cancel{17}}} = \frac{99}{17}$$ Divide each side by 17. $17\overline{)99.000}$ quotient $5.823 \doteq 5.82$

$y \doteq 5.82$ rounded to the nearest hundredth ■

CAUTION When you round your answer, your check will be approximately correct. Checking Example 4 gives the following.

$$\frac{17}{11} \overset{?}{=} \frac{9}{5.82} \longrightarrow 11 \cdot 9 \overset{?}{\doteq} 17 \cdot 5.82 \longrightarrow 99 \doteq 98.94$$

▶ **You Try It** 5. Solve and round to tenths. $\frac{15}{s} = \frac{22}{17}$

EXAMPLE 5 Solve for x. $\frac{x}{6} = \frac{3\frac{1}{2}}{10}$

$$\frac{x}{6} = \frac{3\frac{1}{2}}{10}$$ Cross multiply.

$$10 \cdot x = 6 \cdot 3\frac{1}{2}$$ Multiply: $6 \cdot 3\frac{1}{2} = \frac{6}{1} \cdot \frac{7}{2} = 21$.

$$10 \cdot x = 21$$

$$\frac{\overset{1}{\cancel{10}} \cdot x}{\underset{1}{\cancel{10}}} = \frac{21}{10}$$ Divide each side by 10.

$$x = 2\frac{1}{10} \quad \text{or} \quad 2.1$$ ■

EXAMPLE 6 Solve for t. $\frac{4}{t} = \frac{5.2}{3.51}$

$$\frac{4}{t} = \frac{5.2}{3.51}$$ Cross multiply.

$$5.2 \cdot t = 4 \cdot 3.51$$

$$\frac{\overset{1}{\cancel{5.2}} \cdot t}{\underset{1}{\cancel{5.2}}} = \frac{14.04}{5.2}$$ Divide both sides by 5.2. $5.2\overline{)14.04} \longrightarrow 52\overline{)140.4}$ = 2.7

$$t = 2.7$$ ■

▶ You Try It Solve.

6. $\frac{m}{5\frac{1}{4}} = \frac{8}{30}$

7. $\frac{7.5}{0.18} = \frac{w}{0.045}$

▶ Answers to You Try It 1. $y = 8$ 2. $x = 14$ 3. $y = 3\frac{1}{3}$ 4. $x = 3\frac{1}{9}$ 5. $s = 11.6$ 6. $m = 1\frac{2}{5}$ or 1.4 7. $w = 1.875$

SECTION 6.8 EXERCISES

Solve each proportion, and check your answer.

1. $\frac{4}{9} = \frac{x}{36}$

2. $\frac{y}{24} = \frac{7}{8}$

3. $\frac{20}{t} = \frac{4}{3}$

4. $\frac{7}{10} = \frac{42}{n}$

5. $\frac{8}{h} = \frac{12}{9}$

6. $\frac{x}{25} = \frac{40}{50}$

7. $\frac{27}{40} = \frac{y}{120}$

8. $\frac{55}{72} = \frac{165}{z}$

9. $\frac{4}{21} = \frac{s}{168}$

10. $\frac{x}{6} = \frac{63}{42}$

11. $\frac{14}{18} = \frac{49}{x}$

12. $\frac{y}{56} = \frac{10}{35}$

13. $\frac{2}{3} = \frac{x}{27}$

14. $\frac{30}{x} = \frac{12}{20}$

15. $\frac{75}{120} = \frac{5}{x}$

16. $\frac{x}{385} = \frac{200}{350}$

17. $\frac{x}{15} = \frac{16}{12}$

18. $\frac{35}{21} = \frac{x}{36}$

19. $\frac{88}{33} = \frac{56}{y}$

20. $\frac{60}{x} = \frac{50}{20}$

21. $\frac{x}{12} = \frac{10}{14}$

22. $\frac{55}{44} = \frac{z}{22}$

23. $\frac{17}{51} = \frac{100}{v}$

24. $\frac{120}{t} = \frac{74}{37}$

25. $\frac{60}{92} = \frac{x}{138}$

26. $\frac{64}{40} = \frac{96}{x}$

27. $\frac{45}{35} = \frac{x}{14}$

28. $\frac{3}{25} = \frac{60}{x}$

29. $\frac{x}{8} = \frac{132}{24}$

30. $\frac{4}{x} = \frac{28}{63}$

31. $\frac{9}{h} = \frac{60}{200}$

32. $\frac{500}{800} = \frac{x}{240}$

33. $\frac{40}{9} = \frac{x}{5}$

34. $\frac{18}{y} = \frac{4}{7}$

35. $\frac{\frac{1}{2}}{4} = \frac{x}{6}$

36. $\frac{x}{\frac{3}{4}} = \frac{10}{4}$

37. $\frac{x}{8} = \frac{3\frac{1}{2}}{7}$

38. $\frac{2\frac{1}{4}}{4} = \frac{y}{10}$

39. $\frac{5}{3\frac{1}{3}} = \frac{y}{4}$

40. $\frac{12\frac{1}{2}}{100} = \frac{5}{y}$

41. $\frac{2.4}{6} = \frac{x}{10}$

42. $\frac{y}{2.5} = \frac{4}{10}$

43. $\frac{t}{3.6} = \frac{4.5}{6}$

44. $\frac{x}{12} = \frac{4.5}{2.25}$

45. $\frac{1.4}{0.4} = \frac{0.7}{x}$

46. $\frac{0.4}{t} = \frac{0.29}{2.03}$

47. $\frac{0.2}{0.8} = \frac{s}{0.5}$

48. $\frac{p}{0.75} = \frac{0.7}{1}$

49. $\frac{2.07}{k} = \frac{15.9}{4.93}$ (round to thousandths)

50. $\frac{7.84}{5} = \frac{4.2}{h}$ (round to tenths)

SKILLSFOCUS (Section 4.7) Order the fractions from smallest to largest.

51. $\frac{1}{4}, \frac{5}{16}, \frac{3}{8}$

52. $\frac{7}{10}, \frac{19}{30}, \frac{13}{20}$

53. $\frac{5}{6}, \frac{11}{12}, \frac{7}{9}, \frac{2}{3}$

54. $\frac{4}{7}, \frac{5}{14}, \frac{1}{2}, \frac{15}{28}$

EXTEND YOUR THINKING ▶▶▶▶

▶ TROUBLESHOOT IT

Find and correct the error.

55. $\frac{x}{16} = \frac{12}{5}$

$$\frac{x}{\cancel{16}_{4}} = \frac{\cancel{12}^{3}}{5}$$

$5 \cdot x = 3 \cdot 4$

$x = 2.4$

56. $\frac{10}{y} = \frac{50}{80}$

$$\frac{10}{y} = \frac{\cancel{50}^{5}}{\cancel{80}_{8}}$$

$8 \cdot y = 5 \cdot 10$

$y = 6.25$

WRITING TO LEARN ▶▶▶▶

57. Explain in writing each step used to solve $\frac{6}{10} = \frac{x}{45}$.

58. Explain your answer to each problem. Use your own examples to demonstrate your point.

a. (True/False) Every ratio can be expressed as a fraction.

b. (True/False) A false proportion has unequal ratios.

c. (True/False) A mixed number cannot be written as a ratio.

6.9 APPLICATIONS

OBJECTIVE

Solve applications using proportions.

NEW VOCABULARY

proportional

Solving Applications

Proportions are used to solve word problems whose quantities are **proportional**. Two quantities are proportional if doubling one means the other will double, tripling one means the other will triple, halving one means the other will halve, and so on.

For example, gallons of gas and miles are proportional. If you get 25 miles on 1 gallon, you can expect to get 50 miles on 2 gallons driving at the same speed under the same conditions. Ingredients in a recipe are proportional. If you halve one ingredient, the rest are halved to maintain the same taste.

Not all units are proportional. A person's age and height are not proportional. If a 20-year-old woman is 5 feet tall, at double the age (40 years) she will not be double the height (10 feet).

When deciding to use proportions to solve a problem, ask the following question. If one quantity is increased or decreased, will the other increase or decrease by the same factor? If the answer is yes, use proportions to solve the problem.

EXAMPLE 1 Ron drove 120 miles on 6 gallons of gas. How many gallons will he use on a 300-mile trip?

Let x = number of gallons used on a 300-mile trip.

Ron drove 120 miles on 6 gallons of gas. How many gallons will he use on a 300-mile trip?

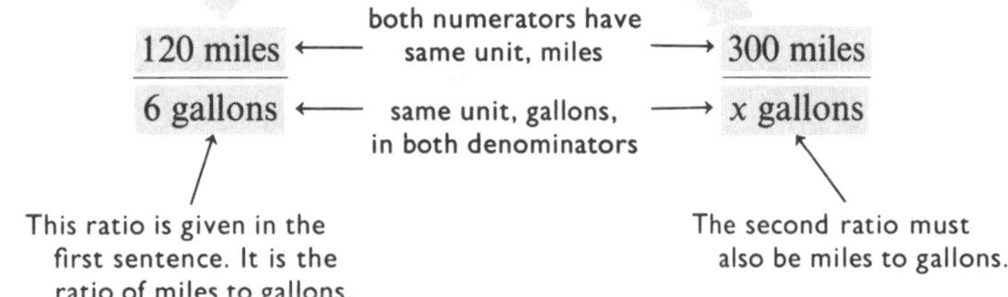

Set the two ratios equal to make a proportion.

$$\frac{120 \text{ miles}}{6 \text{ gallons}} = \frac{300 \text{ miles}}{x \text{ gallons}}$$

You may drop the units to solve the proportion.

$$\frac{120}{6} = \frac{300}{x}$$

$$120 \cdot x = 6 \cdot 300 \qquad \text{Cross multiply.}$$

$$\frac{\overset{1}{\cancel{120}} \cdot x}{\underset{1}{\cancel{120}}} = \frac{\overset{1}{\cancel{6}} \cdot \overset{15}{\cancel{300}}}{\underset{\underset{1}{\cancel{20}}}{\cancel{120}}} \qquad \text{Divide both sides by 120. Reduce.}$$

$$x = 15 \text{ gallons}$$

Ron will use 15 gallons of gas on the 300-mile trip. ■

OBSERVE Either of the following proportions could have been used to solve Example 1. Both give the same cross products as in Example 1.

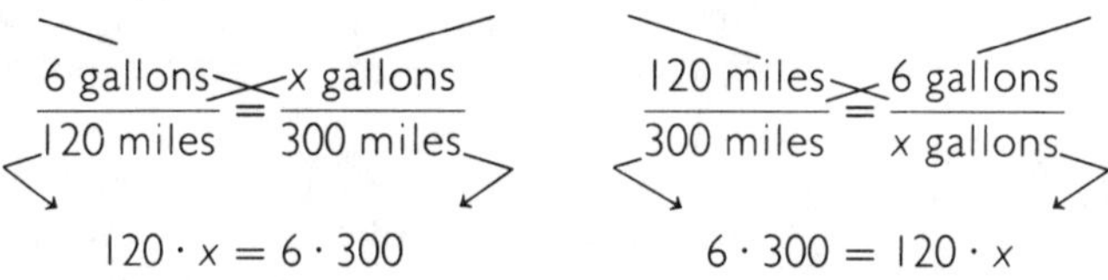

CAUTION The next proportion could not be used to solve Example 1, because the ratios are unlike. A ratio of gallons to miles is unequal to a ratio of miles to gallons. In addition, the cross products are not the same as in Example 1.

$$\frac{6 \text{ gallons}}{120 \text{ miles}} \neq \frac{300 \text{ miles}}{x \text{ gallons}}$$

$$120 \cdot 300 \neq 6 \cdot x$$

▶ **You Try It** **1.** Donald paid \$98 for 7 shirts. What will he pay for 18 shirts?

To solve a word problem using proportions

1. Let x represent the unknown amount.
2. Use the information in the problem to make two ratios.
3. The first ratio is often given in the statement of the word problem. Write it in fraction form. Write the unit name next to each number in the ratio.
4. Write the second ratio so both numerators have the same name, and both denominators have the same name.

$$\frac{\text{miles}}{\text{gallons}} \quad \frac{\text{miles}}{\text{gallons}}$$

numerators have same name

denominators have same name

5. Make a proportion by setting the ratios equal. Solve the proportion for x.

EXAMPLE 2 5 pens cost \$8. How much will 12 pens cost?

Let x = the cost of 12 pens.

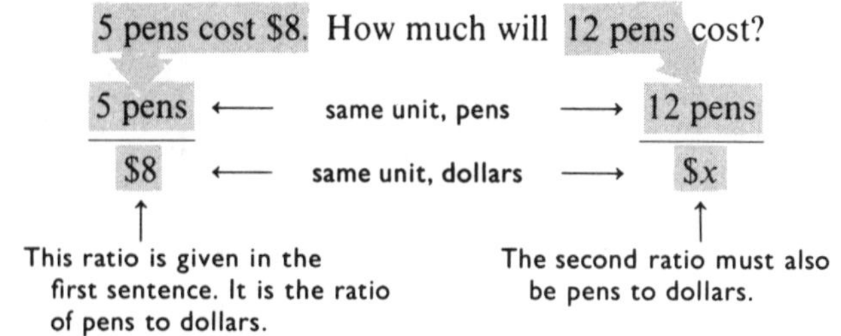

Write the proportion by setting the two ratios equal.

$$\frac{5 \text{ pens}}{\$8} = \frac{12 \text{ pens}}{\$x}$$

$$5 \cdot x = 8 \cdot 12 \qquad \text{Cross multiply.}$$

$$\frac{\overset{1}{\cancel{5}} \cdot x}{\underset{1}{\cancel{5}}} = \frac{96}{5} \qquad \text{Divide each side by 5.}$$

$$x = \$19.20$$

The cost of 12 pens is \$19.20. ■

▶ **You Try It** 2. Melissa made \$49.26 for 3 hours work. How many hours must she work to make \$410.50?

EXAMPLE 3 The scale for a blueprint is 1 inch represents 6 feet. What is the actual length of a wall that measures $4\frac{3}{8}$ inches on the blueprint?

Let x = actual length of the wall.

$$\frac{1 \text{ in (blueprint)}}{6 \text{ ft (actual)}} = \frac{4\frac{3}{8} \text{ in (blueprint)}}{x \text{ ft (actual)}}$$

Write this ratio from the first sentence: "The scale for a blueprint is 1 inch represents 6 feet."

Also write this ratio as blueprint inches to actual feet.

$$\frac{1}{6} = \frac{4\frac{3}{8}}{x}$$

$$1 \cdot x = 6 \cdot 4\frac{3}{8} \qquad \text{Cross multiply.}$$

$$x = \frac{\overset{3}{\cancel{6}}}{1} \cdot \frac{35}{\underset{4}{\cancel{8}}} = \frac{105}{4} = 26\frac{1}{4}$$

$$x = 26\frac{1}{4} \text{ ft}$$

The actual wall length is $26\frac{1}{4}$ feet. ■

▶ **You Try It** 3. The scale for a blueprint is $\frac{1}{8}$ inch represents 4 feet. What is the actual width of a room that measures $1\frac{3}{4}$ inches on the blueprint?

EXAMPLE 4 A rancher owns 6,000 cows. He randomly tests 160 for a disease and finds that 36 are infected. How many of his 6,000 cows can the rancher expect are infected?

36 infected out of 160 cows is this ratio.

Let x = the expected number of infected cows in the herd of 6,000.

$$\frac{36 \text{ infected}}{160 \text{ cows}} = \frac{x \text{ infected}}{6{,}000 \text{ cows}}$$

$$160 \cdot x = 36 \cdot 6{,}000 \qquad \text{Cross multiply.}$$

$$\frac{\overset{1}{\cancel{160}} \cdot x}{\underset{1}{\cancel{160}}} = \frac{\overset{9}{\cancel{36}} \cdot \overset{150}{\cancel{6{,}000}}}{\underset{\underset{1}{\cancel{40}}}{\cancel{160}}} \qquad \text{Divide each side by 160. Reduce.}$$

$$x = 1{,}350$$

The rancher expects about 1,350 of his 6,000 cows are infected. ■

▶ **You Try It** **4.** The population of a city is 250,000 people. 400 people are randomly tested and 24 are found to have a rare virus. How many city residents do you expect have the virus?

EXAMPLE 5 Diane reads 25 pages of a novel in 40 minutes. How long will it take her to read a 325-page novel if she reads at the same rate?

Let x = number of minutes to read 325 pages.

Reading 25 pages in 40 minutes is this ratio.

Reading 325 pages in x minutes is this ratio.

$$\frac{25 \text{ pages}}{40 \text{ minutes}} = \frac{325 \text{ pages}}{x \text{ minutes}}$$

$$40 \cdot 325 = 25 \cdot x \qquad \text{Cross multiply.}$$

$$\frac{40 \cdot \overset{13}{\cancel{325}}}{\underset{1}{\cancel{25}}} = \frac{\overset{1}{\cancel{25}} \cdot x}{\underset{1}{\cancel{25}}} \qquad \text{Divide each side by 25. Reduce.}$$

$$520 \text{ minutes} = x$$

Diane will read the novel in about 520 minutes, or 8 hours 40 minutes. ■

▶ **You Try It** **5.** A child reads 9 pages of a book in 6 minutes. How long will it take to finish the book if it has 51 pages?

EXAMPLE 6 **Unit Price** A 15-ounce box of raisins costs \$1.39. Find the unit price per ounce.

Let x = the price per ounce.

15-ounce box costs \$1.39 (= 139¢) → 139¢; 1 ounce costs x¢ → x¢

$$\frac{15 \text{ oz}}{139¢} = \frac{1 \text{ oz}}{x¢}$$

$$15 \cdot x = 1 \cdot 139$$

$$x = \frac{139}{15} \doteq 9.3¢ \text{ per oz}$$ ■

▶ **You Try It** 6. A 36-ounce jar of peanut butter costs \$2.79. Find the unit price per ounce. Round to tenths.

Proportions can be used to find parts per hundred, parts per thousand, and parts per million (or ppm).

EXAMPLE 7 Of 263,000 people infected with a virus, 18,410 died. How many people died per 100 infected?

x = number dead per 100 infected

$$\frac{18{,}410 \text{ dead}}{263{,}000 \text{ infected}} = \frac{x \text{ dead}}{100 \text{ infected}}$$

$$263{,}000x = 100 \cdot 18{,}410$$

$$x = \frac{1{,}841{,}\cancel{000}}{263{,}\cancel{000}}$$

$$x = 7$$

There were 7 deaths per 100 people infected. ■

OBSERVE Parts per 100 is called percent. 7 deaths per 100 means 7% died. Percents are studied in the next chapter.

▶ **You Try It** 7. Of 42,600 eligible voters, 30,672 voted. How many voters per 100 voted?

SOMETHING MORE

Shadows On a sunny day, the length of an object's shadow is proportional to its height. If a pole is twice as tall as you, its shadow will be twice as long as yours. If your pet is one-third your height, its shadow will be one-third as long as yours. You can use this fact to find the heights of buildings, towers, trees, and so on.

To find the height of a building, use a yardstick to measure the length of its shadow. Immediately afterward (why?), hold the yardstick vertically on the ground. Mark the end of its shadow.

Then measure the shadow's length. Use the proportion below to find the height of the building.

$$\frac{\text{yardstick height}}{\text{yardstick shadow length}} = \frac{\text{building height}}{\text{building shadow length}}$$

a. A building casts a shadow 136 feet long. A yardstick casts a shadow 2 feet long. How tall is the building? (Let x = the height of the building.)

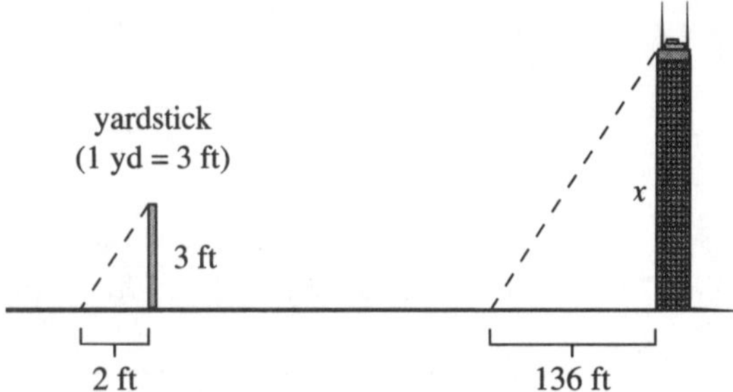

b. A flagpole casts a shadow 60 feet long. At the same time, a yardstick casts a 4-foot shadow. How high is the pole?

c. A tree casts a shadow 55 feet long. At the same time, a yardstick has a 2 foot 6 inch shadow. How tall is the tree?

d. You Be the Judge Susan claims the procedure used in this problem only gives accurate answers when the ground over which the shadows lie is perfectly level. Do you agree? Explain your decision.

▶ **Answers to You Try It** 1. \$252 2. 25 hr 3. 56 ft 4. 15,000 people 5. 34 min 6. 7.8¢/oz 7. 72 per 100 voted

▶ **Answers to Something More** a. 204 ft b. 45 ft c. 66 ft

$\left(\text{Note: since 6 in} = \frac{6}{12}\text{ ft} = \frac{1}{2}\text{ ft, write 2 ft 6 in as } 2\frac{1}{2}\text{ ft}\right)$

SECTION 6.9 EXERCISES

Use proportions to solve each word problem.

1. Suzanne drove 140 miles on 5 gallons of gas. How far can she drive on 12 gallons?

2. Tom drove 57 miles on 3 gallons of gas. How far will 8 gallons take him?

3. If 8 favors cost $28.48, how many can you buy for $74.76?

4. If 6 burgers cost $7.74, how many can you buy for $19.35?

5. The scale for a blueprint is 1 inch represents 20 feet. What is the actual length of a kitchen wall measuring $\frac{5}{8}$ inch on the blueprint?

6. The scale for a blueprint is $\frac{1}{2}$ inch represents 4 feet. What is the actual length of a living room floor measuring $2\frac{3}{8}$ inches?

7. Of 480 people who voted today, 250 voted for Candidate A. If 30,240 people will vote today, how many will vote for Candidate A if voting patterns do not change?

8. 270 geese were randomly examined. 95 were contaminated with crude oil. How many geese in a population of 8,100 can you expect to be contaminated by oil?

9. If 2 pounds of beef are needed to feed 6 adults, how many pounds will be needed for a cookout with 27 adults?

10. If 9 tablespoons of coffee are needed to serve 6 people, how many people can be served with 24 tablespoons of coffee?

11. If a car travels 140 miles in 3 hours, at the same speed how long will it take to travel 560 miles?

12. An express train travels 195 miles in two hours. At that rate, how long will it take to go 1,755 miles?

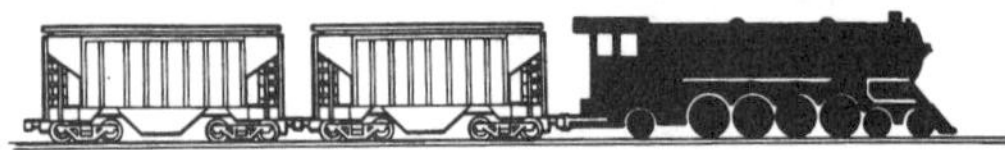

13. The scale for a map is 1 inch equals 13.5 miles. What is the actual distance between two cities that are $8\frac{3}{4}$ inches apart on the map?

14. The scale for a globe is 660 miles equals 1 inch. Find the actual distance between New York and Hawaii if they are $7\frac{3}{4}$ inches apart on the globe.

15. Ed drove 264 miles on 14.1 gallons of gas. If he plans to drive 9,300 miles on a cross-country vacation, how many gallons of gas will he need, to the nearest gallon?

16. Tammy gets 16.3 miles on each gallon of gas. How many gallons will she need to drive to her mother's house 450 miles away? Round to the nearest gallon.

17. A baseball team won 16 of their first 27 games. If they play 162 games in one season, how many games will the team win if they keep winning at the same rate?

18. A mining company extracts 30 pounds of iron from 100 pounds of iron ore. How many pounds of iron can be extracted from 3,600 pounds of iron ore?

19. If an average family of 4 produces 58 pounds of garbage a week, about how much garbage is produced weekly in a city of 1,200,000 people?

20. A doctor found that 3 out of every 200 people have contracted a virus. About how many of the 256,000,000 people in the United States are infected with the virus?

21. A 20-ounce jar of honey costs \$1.69. What is the unit price per pound rounded to the nearest tenth of a cent?

22. A 54-ounce can of meat costs \$8.49. What is the unit price per pound rounded to the nearest tenth of a cent?

23. 3 ounces of sirloin steak contain 75 milligrams of cholesterol. How much cholesterol is in an 8-ounce cut of sirloin?

24. 55 milligrams of cholesterol are contained in $3\frac{1}{2}$ ounces of brook trout. How much cholesterol is in an 8-ounce fillet of trout?

25. The recipe for fixing two servings of oatmeal calls for $1\frac{1}{2}$ cups of water, $\frac{1}{2}$ teaspoon salt, and $\frac{2}{3}$ cup of oats.

a. If you use $1\frac{2}{3}$ cups of oats, how much salt and water must be used to keep the recipe in proportion?

b. How many servings will you now be making?

26. To make 1 quart of instant milk you need $1\frac{1}{3}$ cups of instant milk mix and $3\frac{3}{4}$ cups of water.

a. If you use 4 cups of instant milk mix, how much water will you need?

b. How many quarts will you now be making?

27. Candidate A received 71,592 votes out of 125,600 votes cast. How many votes per 100 did Candidate A receive?

28. In a population of 255 million people, 765 were infected with a rare disease. How many people per million were infected?

SKILLSFOCUS (Section 2.8) *Change each fraction to a decimal.*

29. $\frac{3}{5}$

30. $\frac{7}{10}$

31. $\frac{5}{8}$

32. $\frac{8}{11}$ (round to thousandths)

EXTEND YOUR THINKING ▶▶▶

▶ SOMETHING MORE

33. The Capitol Building in Washington, DC, casts a 500-foot shadow. A yardstick has a 5-foot shadow at the same time of day. Calculate the height of the Capitol.

34. At the moment a plane is directly overhead, its shadow is 1,200 feet away. Calculate the altitude of the plane if a yardstick casts a 1 foot 9 inch shadow at the same time of day.

35. **Constructing a Proportion Slide** A proportion slide is a ruler, yardstick, meterstick, or other straight measuring device with a thick slide that moves back and forth. The slide does not have to be attached to the ruler, and can be something as simple as an eraser or building block.

If you know the distance to an object, you can use the proportion slide to calculate its height. (Or, if you know the height you can calculate the distance.) Place the ruler under your eye. Point it toward the object whose height you are measuring. Move the slide back and forth until the height of the slide matches the height of the object being measured. Use the following proportion to calculate the object's height.

$$\frac{\text{distance from eye to slide}}{\text{height of slide}} = \frac{\text{distance from eye to object}}{\text{height of object}}$$

a. You point your proportion slide at a lamp 18 feet away. Move the slide until its height matches the lamp's height. At that point, the distance from your eye to the slide is 6 inches. The slide is $1\frac{1}{4}$ inches high. How tall is the lamp? (Let x = the height of the lamp.)

slide is $1\frac{1}{4}$ in high

x

6 in

18 ft

b. Amy points her proportion slide at the Washington Monument in Washington, DC. She moves the slide until its height matches the monument's height, at the $8\frac{5}{16}$ inch mark on her ruler. The slide is $\frac{7}{8}$ inch high. Amy is exactly 1 mile (= 5,280 feet) away from the monument. Calculate the height of the Washington Monument to the nearest ten feet.

c. Nicole knows a tower is 200 feet high. She sights the tower with her proportion slide. She finds the $\frac{1}{2}$-inch slide sitting on the $8\frac{1}{4}$ inch mark on her ruler. Calculate how far Nicole is from the tower.

36. A dining room measures $5\frac{1}{2}$ inches by $3\frac{7}{8}$ inches on a blueprint. If the scale is $\frac{1}{4}$ inch equals 1 foot, what are the actual dimensions of the room?

37. a. Immediately after jogging, Sam's heart beats 26 times in 10 seconds. What is his heart rate per minute?

b. After 1 minute, he counts 19 heartbeats in 10 seconds. What is his heart rate per minute?

c. What is his recovery rate in the first minute?

38. ***Voyager II*** The *Voyager II* spacecraft is expected to be returning data to Earth until the year 2020, when its nuclear power source is expected to be exhausted. It was launched in 1977. By 1989 it had traveled a total of 4,400,000,000 miles. How far from Earth can we expect *Voyager II* to be by 2020? Round to hundred millions.

WRITING TO LEARN ▶▶▶▶

39. Write a proportion word problem whose answer is \$3.60, rounded to the nearest cent. The problem is about candy and a child named Ted. Solve the problem, explaining each step.

▶ YOU BE THE JUDGE

40. Is this proportion valid? Explain your decision.

$$\frac{7 \text{ pens}}{\$4.95} = \frac{\$x}{12 \text{ pens}}$$

CHAPTER 6 REVIEW

VOCABULARY AND MATCHING

New words and phrases introduced in this chapter are shown in the left-hand column. Match each term on the left with the phrase or sentence on the right that best describes it.

A. ratio	______	also written 4:6 and 4/6
B. order in a ratio	______	a statement that two ratios are equal
C. 4 to 6	______	price for 1 unit of an item
D. terms of the ratio	______	ratio or rate whose second term is 1
E. reducing ratios	______	$\frac{24}{40} = \frac{35}{60}$, for example
F. rate	______	also means $6 \div 4$
G. per	______	increase one quantity, and you increase the other by the same factor
H. 6 to 4	______	cross products are equal for a true proportion
I. unit ratio or unit rate	______	numbers are written in the same order as the names
J. unit price	______	for $\frac{6}{4} = \frac{3}{2}$, they are $6 \cdot 2$ and $3 \cdot 4$
K. comparing ratios	______	word often used when writing rates, as in mpg or mph
L. proportion	______	a comparison of two quantities
M. true proportion	______	when 6 to 4 becomes 3 to 2, or 100:30 becomes 10:3
N. false proportion	______	requires rewriting both ratios with the same second term
O. cross products	______	$\frac{15}{24} = \frac{35}{56}$, for example
P. cross product property	______	what 4 and 6 are in the ratio 4 to 6
Q. proportional	______	a ratio whose names cannot be eliminated

REVIEW EXERCISES

6.1 Solving Equations

Solve each equation. Check each answer.

1. $a + 6 = 15$ **2.** $k - 12 = 2$ **3.** $v - 7 = 0$

4. $8x = 48$ **5.** $-2 = 4y$ **6.** $\frac{1}{3}n = 6$

6.2 The Meaning of Ratio

7. A school has 163 students, 17 teachers, and 4 administrators. Write the following ratios.

a. students to teachers

b. teachers to administrators

c. teachers and administrators to students

8. There are $30,615 of investment to $12,059 of profit. Write the ratio of investment to profit in three ways.

Reduce each ratio to lowest terms.

9. 50 to 35 **10.** 2,000:6,000 **11.** 75/125

12. Joan reads a 640-page novel in the same time it takes her to read 160 pages of a science text. What is the lowest ratio of novel pages to science pages?

13. Mike is 6 feet tall. His left-to-right arm extension measures 72 inches. What is Mike's height to arm extension ratio in lowest terms?

14. Cancel the units in each ratio and reduce to lowest terms.

a. 10 miles to 13 miles
b. 4 feet to 10 inches
c. 2 minutes to 40 seconds
d. 3 feet 9 inches to 2 feet

6.3 Ratio as Rate

Reduce each rate to lowest terms.

15. 450 miles to 100 hours

16. $36 per 12 toys

17. 240 miles per 15 gallons

18. $87 per 9 hours

19. In 21 working days a salesman sold 30 cars. Write the rate of cars sold per day in lowest terms.

20. In 56 minutes Greg can run 8 miles. What is the lowest terms rate of minutes per mile?

Change the unit in each rate as indicated. Reduce.

21. 45 miles per 2 hours: change hours to minutes

22. $6 per 8 gallons: change dollars to cents

23. $336 per 2 weeks: change weeks to days

6.4 Rewriting Ratios Involving Fractions and Decimals as Ratios of Whole Numbers

Rewrite each ratio as a ratio of whole numbers.

24. $\frac{5}{8}$ to 4 **25.** 3 to $2\frac{1}{4}$ **26.** $5\frac{1}{3}$ to $2\frac{3}{4}$

27. 0.8 to 0.05 **28.** 6.04 to 3.2 **29.** 0.024 to 0.0003

30. The ratio of cups of water to cups of flour is $2\frac{1}{2}$ to $4\frac{1}{3}$. Rewrite this as a ratio of whole numbers.

31. A table takes 4.5 hours to make and 7.75 hours to finish. Write the ratio of make time to finish time in terms of whole numbers.

6.5 Applications

32. Find the unit price if 6 pounds of apples cost $2.34.

33. A 7.41-pound turkey breast costs $10.27. Find the unit price to the nearest tenth of a cent.

34. David drove 278 miles on 16.5 gallons of gas
    A. He drove 312 miles on 17.2 gallons of gas
    B. Which gave the better gas mileage?

35. 2 liters (67.6 fluid ounces) of soda costs $1.09. A six-pack of 12-ounce cans is selling for $1.69. Which is the better buy?

36. Fran bought a 40-ounce jar of honey for $2.89. Bob bought the 20-ounce jar for $1.89. Who made the better buy?

37. The jumbo 96-ounce box of detergent sells for $7.69. The regular 52-ounce size sells for $3.89 Which box of detergent is the better buy?

6.6 Circles

Find the circumference of each circle.

38.

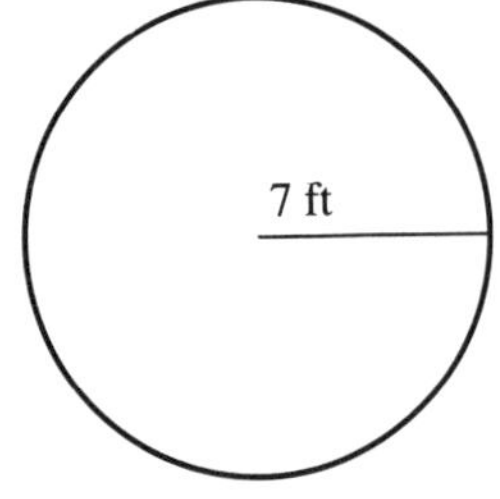

39.

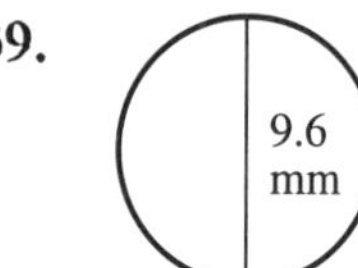

Find the area of each circle.

40.

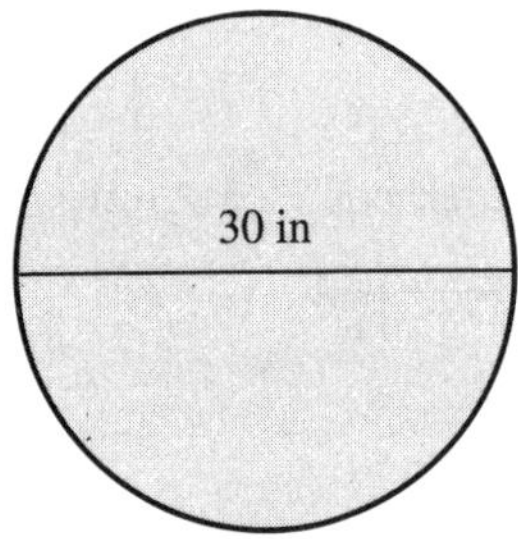

41.

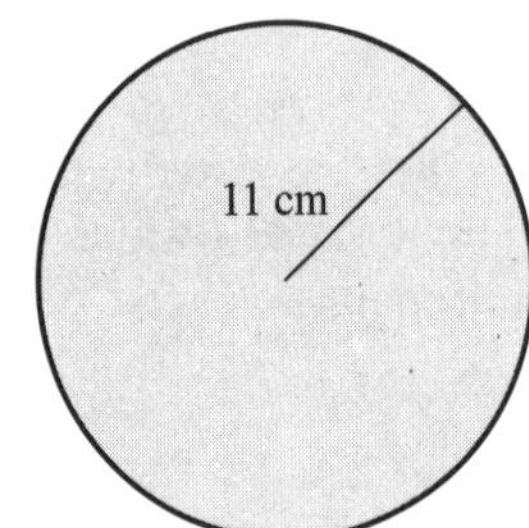

42. Find the circumference of a circle with a 50-ft radius.

43. Find the area of a circle with a 6.8-yd radius. Round to tenths.

44. Find the area of the circle if the square has a perimeter of 20 in.

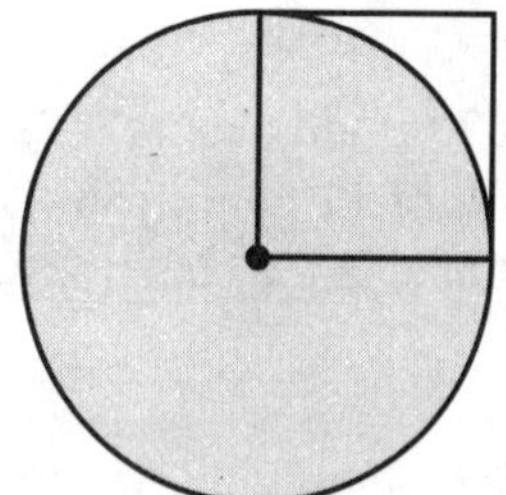

6.7 Introduction to Proportions

Determine if each proportion is true or false.

45. $\frac{5}{9} \stackrel{?}{=} \frac{4}{7}$

46. $\frac{15}{40} \stackrel{?}{=} \frac{21}{56}$

47. $\frac{0.9}{1.5} \stackrel{?}{=} \frac{0.21}{0.35}$

48. $\frac{\frac{2}{5}}{4} \stackrel{?}{=} \frac{1}{15}$

6.8 Solving Proportions

Solve the proportion. Check each answer.

49. $\frac{x}{6} = \frac{6}{9}$

50. $\frac{10}{16} = \frac{x}{24}$

51. $\frac{8}{10} = \frac{6}{x}$

52. $\frac{5}{x} = \frac{0.7}{1.4}$

53. $\frac{x}{5} = \frac{8}{7}$ (round to tenths)

54. $\frac{4.62}{0.9} = \frac{3.8}{x}$ (round to hundredths)

55. $\frac{5\frac{2}{3}}{4} = \frac{x}{1\frac{7}{8}}$

56. $\frac{48}{x} = \frac{7\frac{1}{2}}{5\frac{3}{4}}$

6.9 Applications

57. The daily dosage for a certain medicine is 3 ounces for each 60 pounds of body weight. What is the daily dosage for a man weighing 200 pounds?

58. Four tablespoons of weed killer are mixed with 3 quarts of water. How many quarts of water must be mixed with 10 tablespoons of weed killer?

59. The property tax on a $50,000 home is $720. At the same rate, what is the property tax on a home worth $130,000?

60. The fuel mixture for a gas saw is 1 ounce of oil for every $2\frac{1}{2}$ pints of gas. How much oil must be mixed with 4 pints of gas?

61. The scale for a blueprint is $\frac{1}{4}$ inch equals 2 feet. What is the actual length of a wall that measures $2\frac{5}{8}$ inches on the blueprint?

62. The scale for a map is 1 inch equals 7.2 miles. What is the distance between two cities $4\frac{1}{2}$ inches apart on the map?

63. If 1 gallon of paint covers 450 square feet of wall space, how many gallons are needed for 2,950 square feet?

64. 52 out of 800 computers tested were found to be defective. At this rate, how many of the 22,000 computers assembled at the same time are expected to be defective?

7

Percents

OIL

The following table presents world crude oil production and consumption by major producing region.

Region	*World Production (57.6 million barrels per day)*	*World Consumption (55.2 million barrels per day)*	*World Reserves (541.4 million barrels per day)*
Middle East	37.0%	2.0%	58.4%
The former Soviet Union and Eastern Europe	16.8%	15.8%	11.9%
Africa	9.9%	1.9%	9.7%
United States	18.9%	31.2%	6.5%
Western Europe	0.7%	26.2%	3.09%
Other Western Hemisphere	6.4%	8.0%	4.1%
Caribbean	6.4%	0.9%	2.9%
Japan	0.03%	9.1%	0.01%
Other Eastern Hemisphere	3.87%	4.9%	3.4%

SOURCE: Federal Energy Administration, Project Independence Blueprint.

From the table, the United States consumes 31.2% of the 55.2 million barrels used each day in the world. How many barrels is this? The United States has about 6% of the world's population. Estimate how many times more oil the United States consumes when compared to its size. These questions will be answered in this chapter.

Percents are widely used in business, science, and everyday living to make comparisons. It is easier to make comparisons using percents because percents are based on 100.

For example, suppose 3,700 votes were cast in an election. 1,887 votes were for candidate Paul. This does not immediately tell you if Paul won. But if you know that 1,887 out of 3,700 is 51 out of every 100 votes cast, then Paul was victorious. 51 votes out of every 100 is 51% of the votes.

In this chapter you will learn to make percent conversions and solve percent problems. You will then study some of the many practical applications that use percents, including commission, discount, advertising, interest, and finding a monthly payment.

7.1 THE MEANING OF A PERCENT

OBJECTIVES	NEW VOCABULARY
1 Define percent.	percent
2 Define complement of a percent.	base
3 Define percent greater than 100%.	complement

1 Defining Percent

Percent means parts per 100. A **percent** is a number of parts out of 100 equal parts.

A percent is written as a number followed by the percent symbol, %. 36% means 36 parts out of 100 equal parts.

FIGURE 7.1
36% shaded.

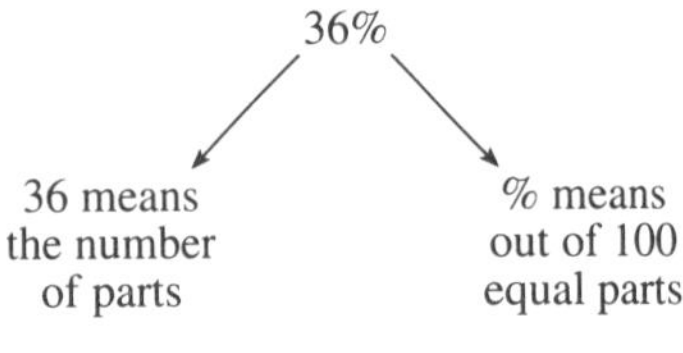

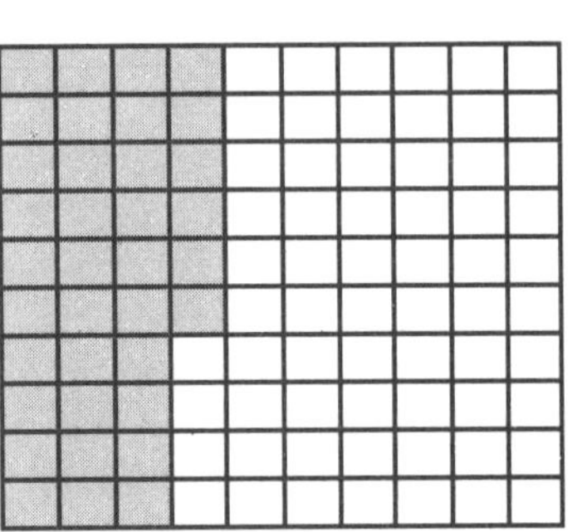

The **base** is the entire population being considered. The base represents 100%, or 100 parts out of a total of 100 equal parts. In Figure 7.1, the base is represented by a large rectangle. It is divided into 100 equal parts, or boxes, of which 36 are shaded. 36% of the figure is shaded.

OBSERVE Since percent means parts per 100, a percent is a ratio of parts to 100 equal parts. 36% is the ratio of 36 parts to 100 parts.

EXAMPLE 1 Explain what each percent means. Represent each percent as a shaded region in Figure 7.2.

a. 50% means 50 parts out of 100 equal parts. This is half of the boxes in Figure 7.2.

b. 1% means 1 part out of 100 equal parts. This is 1 of the 100 boxes in Figure 7.2.

c. $6\frac{1}{4}\%$ means $6\frac{1}{4}$ parts out of 100 equal parts. This is 6 boxes plus $\frac{1}{4}$ of a seventh box in Figure 7.2.

d. $\frac{1}{2}\%$ means $\frac{1}{2}$ of 1 part out of 100 equal parts. This is $\frac{1}{2}$ of 1 box in Figure 7.2.

e. 7.3% means 7.3 parts out of 100 equal parts. This is 7 boxes plus $0.3 = \frac{3}{10}$ of an eighth box in Figure 7.2. That is, divide the eighth box into 10 equal pieces, and take 3. ■

FIGURE 7.2

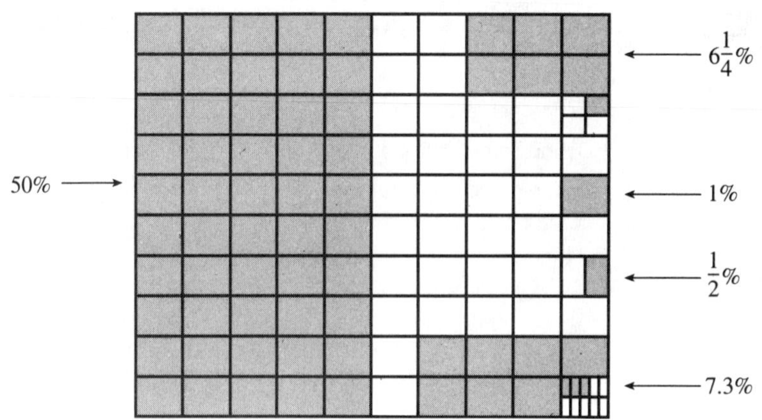

OBSERVE To understand percents, it may be helpful to use one dollar (100¢) to represent 100%. Then 36% is 36¢ out of one dollar. 1% is one cent. What is $\frac{1}{2}\%$? It is $\frac{1}{2}$ of one penny. $\frac{1}{2}\%$ is not 50¢, a common wrong answer.

▶ You Try It Explain what each percent means. How many boxes in Figure 7.2 does each percent represent?

1. 70% **2.** 3% **3.** $16\frac{2}{3}\%$

4. $\frac{1}{5}\%$ **5.** 2.8%

---

2 Complement of a Percent

36% of the boxes in Figure 7.1 are shaded. Subtract 36% from 100%.

100%	−	36%	=	64%
whole		shaded		unshaded

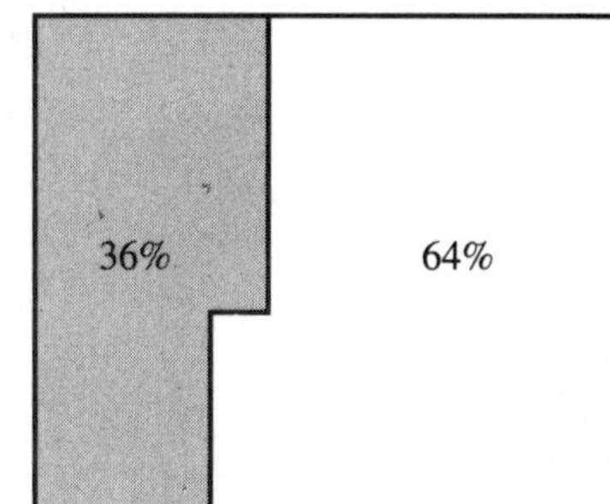

The remaining 64% of the boxes are unshaded. 64% is called the complement of 36%. It is the rest of the whole. A percent plus its complement equals 100%: 36% + 64% = 100%.

The **complement** of a percent is found by subtracting that percent from 100%.

EXAMPLE 2 Find the complement of each percent.

a. The complement of 80% is 100% − 80% = 20%.

b. The complement of $37\frac{1}{2}\%$ is $100\% - 37\frac{1}{2}\% = 62\frac{1}{2}\%$.

c. The complement of 4.7% is 100% − 4.7% = 95.3%. ■

You Try It Find the complement of each percent.

6. 3% 7. 60% 8. $12\frac{3}{4}\%$

9. $\frac{1}{2}\%$ 10. 72.4%

3 Percents Greater than 100%

A percent greater than 100% means one or more wholes of 100 equal parts each, plus a portion of a final 100 equal parts.

EXAMPLE 3 Explain what 150% means.

150% means 100% plus 50%. This is one whole of 100 equal parts, plus 50 parts of a second whole.

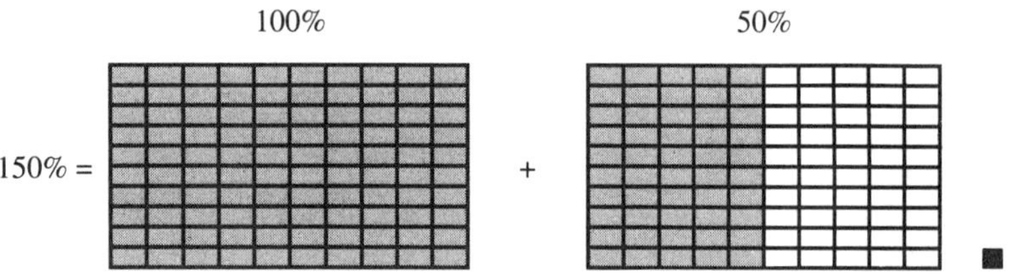

CAUTION 150% makes sense in some contexts, but not in others. To say *150% of the students in your class passed the exam* does not make sense. If 100% pass, then everyone passes. No one is left.

However, suppose it costs you $10 to build a chair. *You sell it for a 150% profit.* What do you sell it for? You sell it for $25. This includes the $10 it cost you, plus 100% of $10 ($10), plus another 50% of $10 ($5), or $25.

EXAMPLE 4 Explain what 325% means.

325% means 100% + 100% + 100% + 25%. This is 3 wholes of 100 equal parts each, plus 25 parts out of a fourth whole.

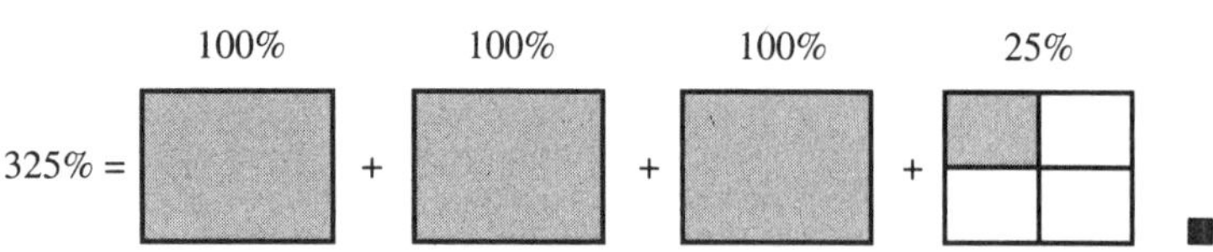

You Try It Explain what each percent means.

11. 175% 12. 200% 13. 430% 14. 101%

▶ **Answers to You Try It** 1. 70 parts out of 100 equal parts (70 boxes in Figure 7.2) 2. 3 parts out of 100 equal parts (3 boxes) 3. $16\frac{2}{3}$ parts out of 100 equal parts $\left(16 \text{ boxes plus } \frac{2}{3} \text{ of a 17th box}\right)$ 4. $\frac{1}{5}$ of 1 part out of 100 equal parts $\left(\frac{1}{5} \text{ of 1 box}\right)$ 5. 2.8 parts out of 100 equal parts $\left(2 \text{ boxes plus } \frac{8}{10} \text{ of a third box}\right)$ 6. 97% 7. 40% 8. $87\frac{1}{4}\%$ 9. $99\frac{1}{2}\%$ 10. 27.6% 11. 1 whole of 100 equal parts plus 75 parts of a second whole. 12. 2 wholes of 100 equal parts each 13. 4 wholes of 100 equal parts each, plus 30 parts of a fifth whole 14. 1 whole of 100 equal parts, plus 1 part out of a second whole

SECTION 7.1 EXERCISES

1 *Fill in the blanks. Draw and shade a diagram to represent each percent.*

1. 40% represents ______ parts out of ______ equal parts.

2. 5% means ______ parts out of ______ equal parts.

3. ______% represents 85 parts out of 100 equal parts.

4. ______% means 72 parts out of 100 equal parts.

5. ______% represents 12.5 parts out of 100 equal parts.

6. 6.25% means ______ parts out of ______ equal parts.

7. $\frac{1}{4}$% represents ______ of one part out of ______ equal parts.

8. ______% means $33\frac{1}{3}$ parts out of ______ equal parts.

2 *Find the complement of the percent.*

9. 70%

10. 23%

11. 61%

12. 20%

13. 80%

14. 0%

15. $83\frac{1}{3}$%

16. $37\frac{1}{2}$%

17. $\frac{1}{2}$%

18. 10.6%

19. 0.08%

20. 65.7%

21. If 10% of the politicians smoke cigars, then ______% do not smoke cigars.

22. If 63% of the cattle are branded, then ______% are not branded.

23. If 96.1% of the weight of a cucumber is water, then ______% of the cucumber is not water.

24. If 3% of the bacteria die in one hour, then ______% will still be alive.

25. If ______% of a paycheck goes toward purchasing food, then 82% goes toward nonfood items.

26. If ______% of a solution is acid, then $66\frac{2}{3}\%$ is not acid.

Draw and shade a diagram to represent each percent.

27. 125%

28. 250%

29. 400%

30. 320%

31. 190%

32. 675%

SKILLSFOCUS (Section 1.6) *Change each decimal to a fraction and reduce.*

33. 0.64

34. 0.9

35. 0.06

36. 1.25

EXTEND YOUR THINKING ▶▶▶▶

▶ SOMETHING MORE

37. A soap commercial claims its product is 99 and 44 one hundredths percent pure. What percent is not pure?

▶ TROUBLESHOOT IT

Find and correct the error.

38. $\frac{3}{4}\%$ is 75 parts out of 100 equal parts.

39. The complement of 0.6% is 100% − 0.6% = 94%.

WRITING TO LEARN ▶▶▶▶

40. Explain how a percent is a ratio. Use your own example.

41. A coach said his team gave 150% in a championship game.
   a. Explain why this does not make sense.
   b. What is the coach trying to say?

42. It costs a pharmacy 10¢ to make a pill. They sell the pill for a 400% profit. What is the cost of one pill? Explain.

▶ YOU BE THE JUDGE

43. Tomas claims 120% has a complement? Is he correct? Explain your decision.

7.2 DECIMALS AND PERCENTS

OBJECTIVES

1. Change a decimal to a percent.
2. Change a percent to a decimal.

Any number can be written in three ways: as a fraction, as a decimal, and as a percent. In this section and the next, you will learn how to change a percent to a fraction or decimal. You do this before you use a percent to solve a word problem. You also learn how to change a fraction or decimal to a percent. You do this in problems that ask you to find percent.

1 Changing a Decimal to a Percent

Percent means hundredths, or parts per 100. To change a decimal to a percent, express the decimal as parts per 100. For example, change 0.42 to a percent.

$$0.42 = 42 \text{ hundredths} = 42 \text{ parts out of } 100 \text{ equal parts} = 42\%$$

decimal point moved two places to the right (note, 42% = 42.%)

42 parts out of 100 equal parts means 42%, so 0.42 = 42%. Notice that the decimal point moved two places to the right.

EXAMPLE 1 Change each decimal to a percent.

a. $0.24 = \frac{24}{100} = 24 \text{ hundredths} = 24\%$

decimal point moved two places to the right

b. $0.5 = \frac{5}{10} = \frac{5 \times 10}{10 \times 10} = \frac{50}{100} = 50 \text{ hundredths} = 50\%$

decimal point moved two places to the right

c. $0.742 = \frac{742}{1{,}000} = \frac{742 \div 10}{1{,}000 \div 10} = \frac{74.2}{100} = 74.2 \text{ hundredths} = 74.2\%$

decimal point moved two places to the right

d. $3.68 = 3\frac{68}{100} = \frac{368}{100} = 368 \text{ hundredths} = 368\%$

decimal point moved two places to the right ■

> To change a decimal to a percent
>
> 1. Move the decimal point exactly two places to the right.
> 2. Add the percent symbol, %.

EXAMPLE 2

a. 0.28 = 0 28.% or 28%

b. 0.0638 = 0 06.38% or 6.38%

c. 0.7 = 0.70 = 0 70.% or 70%

d. $0.3\frac{1}{2} = 0.35 = 0\ 35.\%$ or 35%

e. 1.27 = 1 27.% = 127%

f. 1 = 1.00 = 1 00.% or 100% ■

From Example 2f, 1 equals 100%. Therefore,

$$2 = 2 \times 1 = 2 \times 100\% = 200\%;$$
$$5 = 5 \times 1 = 5 \times 100\% = 500\%;$$
$$24 = 24 \times 1 = 24 \times 100\% = 2{,}400\%; \text{ etc.}$$

▶ You Try It Change each decimal to a percent.

1. 0.19	**2.** 0.2	**3.** 0.84	**4.** 0.06	**5.** 2.58
6. 0.2153	**7.** 4.6	**8.** 0.007	**9.** 3	**10.** $0.9\frac{1}{2}$

2 Changing a Percent to a Decimal

Percent means hundredths, or parts per 100. 63% means 63 hundredths.

$$63\% = 63 \text{ hundredths} = \frac{63}{100} = 0.63$$

decimal point moved two places left

Division by 100 moves a decimal point two places to the left (Section 2.5). Since 63 divided by 100 is 0.63, 63% = 0.63.

EXAMPLE 3 Change each percent to a decimal.

a. $25\% = 25 \text{ hundredths} = \frac{25}{100} = 0.25$ (decimal point moved two places left)

b. $4\% = 4 \text{ hundredths} = \frac{4}{100} = 0.04$ (decimal point moved two places left)

c. $106\% = 106 \text{ hundredths} = \frac{106}{100} = 1.06$ (decimal point moved two places left)

d. $8.8\% = 8.8 \text{ hundredths} = \frac{8.8}{100} = 0.088$ (decimal point moved two places left) ■

To change a percent to a decimal

1. Move the decimal point exactly two places to the left.
2. Drop the % symbol.

EXAMPLE 4 Change each percent to a decimal.

a. $28\% = 28.\% = .28\not\% = 0.28$

b. $0.05\% = 00.05\%$
$= .00\,05\not\% = 0.0005$

c. $260\% = 260.\% = 2.60\not\% = 2.60$

d. $3\frac{1}{4}\% = 3.25\%$
$= 03.25\% = .03\,25\not\% = 0.0325$ ■

▶ You Try It Change each percent to a decimal.

11. 42%	**12.** 75%	**13.** 9%	**14.** 265%	**15.** 600%
16. 2.5%	**17.** 0.003%	**18.** 19.37%	**19.** $12\frac{1}{2}\%$	**20.** 0.4%

▶ Answers to You Try It 1. 19% 2. 20% 3. 84% 4. 6% 5. 258% 6. 21.53% 7. 460% 8. 0.7% 9. 300% 10. 95% 11. 0.42 12. 0.75 13. 0.09 14. 2.65 15. 6 16. 0.025 17. 0.00003 18. 0.1937 19. 0.125 20. 0.004

SECTION 7.2 EXERCISES

1 *Change the following decimals to percents.*

1. 0.62	**2.** 0.33	**3.** 0.05	**4.** 0.01
5. 0.212	**6.** 0.683	**7.** 0.018	**8.** 0.007
9. 0.0608	**10.** 0.0009	**11.** 0.1004	**12.** 0.0908
13. 0.3	**14.** 0.8	**15.** 0.50001	**16.** 0.00702
17. 4	**18.** 7	**19.** 2.5	**20.** 9.8
21. 12	**22.** 50	**23.** 750	**24.** 4400
25. 3.0007	**26.** 5.732	**27.** 12.55	**28.** 60.6
29. $0.62\frac{1}{2}$	**30.** $0.6\frac{1}{4}$	**31.** $4\frac{1}{2}$	**32.** $3.72\frac{1}{4}$

2 *Change the following percents to decimals.*

33. 48%	**34.** 62%	**35.** 5%	**36.** 2%
37. 72.2%	**38.** 60.9%	**39.** 200%	**40.** 350%
41. 234.7%	**42.** 600.4%	**43.** 4,000%	**44.** 1,245%
45. 0.04%	**46.** 0.25%	**47.** 0.0006%	**48.** 0.0012%
49. 80.91%	**50.** 69.331%	**51.** 2.3%	**52.** 5.007%
53. 290%	**54.** 385%	**55.** 2,263.7%	**56.** 100.01%
57. $6\frac{1}{2}\%$	**58.** $22\frac{1}{2}\%$	**59.** $35\frac{1}{4}\%$	**60.** $4\frac{1}{4}\%$
61. $12\frac{3}{8}\%$	**62.** $7\frac{1}{10}\%$		

SKILLSFOCUS (Section 2.8) *Change each fraction to a decimal.*

63. $\frac{4}{5}$

64. $\frac{3}{8}$

65. $\frac{7}{10}$

66. $\frac{1}{6}$ (round to thousandths)

EXTEND YOUR THINKING ▶▶▶▶

▶ TROUBLESHOOT IT

Find and correct the error.

67. 0.004% = 0 00.4% = 0.4

68. 250 = 250. = 2.50 % = 2.5%

WRITING TO LEARN ▶▶▶▶

69. Change 0.84 to a percent. Write an explanation for each step.

70. Change 240% to a decimal. Write an explanation for each step.

7.3 FRACTIONS AND PERCENTS

OBJECTIVES

1. Change a fraction to a percent.
2. Change a percent to a fraction.
3. Change a decimal percent to a fraction.

1 Changing a Fraction to a Percent

To change a fraction to a percent, first change the fraction to a decimal (Section 2.8). Then change the decimal to a percent (Section 7.2). For example, change $\frac{3}{4}$ to a percent.

$$\frac{3}{4} \xrightarrow{\text{change to a decimal}} \begin{array}{r} 0.75 \\ 4\overline{)3.00} \\ \underline{2\,8} \\ 20 \\ \underline{20} \\ 0 \end{array} \xrightarrow{\text{change to a percent}} 0.75 = 0\,75.\% = 75\%$$

> **To change a fraction to a percent**
>
> 1. Change the fraction to a decimal by dividing the numerator by the denominator.
> 2. Change the decimal to a percent.

EXAMPLE 1 Change each fraction to a percent.

a. $$\frac{3}{5} \xrightarrow{\text{change to a decimal}} \begin{array}{r} 0.6 \\ 5\overline{)3.0} \\ \underline{3\,0} \\ 0 \end{array} \xrightarrow{\text{change to a percent}} 0.6 = 0\,60.\% = 60\%$$

Therefore, $\frac{3}{5} = 60\%$.

b. $$\frac{1}{8} \xrightarrow{\text{change to a decimal}} \begin{array}{r} 0.125 \\ 8\overline{)1.000} \\ \underline{8} \\ 20 \\ \underline{16} \\ 40 \\ \underline{40} \\ 0 \end{array} \xrightarrow{\text{change to a percent}} 0.125 = 0\,12.5\% = 12.5\%$$

Therefore, $\frac{1}{8} = 12.5\%$. $\left(\text{NOTE, } 12.5\% = 12\frac{1}{2}\%\right)$ ■

▶ **You Try It** Change each fraction to a percent.

1. $\frac{2}{5}$ **2.** $\frac{3}{8}$ **3.** $\frac{9}{20}$

When changing a fraction to a decimal, the decimal may repeat. In this case, you can carry the division to two decimal places. Write any remainder over the divisor. Then change to a percent.

EXAMPLE 2 Change $\frac{2}{3}$ to a percent.

$$\begin{array}{r} 0.66\frac{2}{3} \\ 3\overline{)2.00} \\ \underline{1\,8} \\ 20 \\ \underline{18} \\ 2 \end{array} \qquad 0.66\frac{2}{3} = 66\frac{2}{3}\%$$

Therefore, $\frac{2}{3} = 66\frac{2}{3}\%$ ■

▶ **You Try It** Change each fraction to a percent.

4. $\frac{1}{3}$

5. $\frac{7}{11}$

When changing a fraction to a decimal, round the decimal to any place value you choose. Then change to a percent.

EXAMPLE 3 Change $\frac{3}{7}$ to a percent. Round to thousandths.

$$\begin{array}{r} 0.4285 \\ 7\overline{)3.0000} \\ \underline{2\,8} \\ 20 \\ \underline{14} \\ 60 \\ \underline{56} \\ 40 \\ \underline{35} \\ 5 \end{array} \qquad 0.4285 \doteq 0.429 = 42.9\%$$

To round to thousandths, add zeros to the ten-thousandths place.

Therefore, $\frac{3}{7} = 42.9\%$. ■

▶ **You Try It** Change each fraction to a percent. Round each decimal to the indicated place before changing to a percent.

6. $\frac{8}{9}$ (thousandths)

7. $\frac{5}{24}$ (ten-thousandths)

To change a mixed number to a percent, first change the mixed number to an improper fraction.

EXAMPLE 4 Change $3\frac{17}{20}$ to a percent.

$$3\frac{17}{20} = \frac{77}{20} \longrightarrow 20\overline{)77.00}$$

$$\begin{array}{r} 3.85 \\ 20\overline{)77.00} \\ \underline{60} \\ 17\,0 \\ \underline{16\,0} \\ 1\,00 \\ \underline{1\,00} \\ 0 \end{array}$$

$3.85 = 3\,85.\% \quad \text{or} \quad 385\%$

Therefore, $3\frac{17}{20} = 385\%$. ■

▶ **You Try It** **8.** Change $1\frac{4}{5}$ to a percent.

---

2 Changing a Percent to a Fraction

Percent means hundredths, or parts per 100. To change a percent to a fraction, write the percent as hundredths. Then simplify.

$$72\% = 72 \text{ hundredths} = \frac{72}{100} = \frac{\overset{18}{\cancel{72}}}{\underset{25}{\cancel{100}}} = \frac{18}{25}$$

Drop % symbol, write number that remains over 100.

EXAMPLE 5 Change each percent to a fraction.

a. $25\% = 25 \text{ hundredths} = \frac{25}{100} = \frac{\overset{1}{\cancel{25}}}{\underset{4}{\cancel{100}}} = \frac{1}{4}$ **b.** $9\% = 9 \text{ hundredths} = \frac{9}{100}$

c. $300\% = 300 \text{ hundredths} = \frac{300}{100} = \frac{\overset{3}{\cancel{300}}}{\underset{1}{\cancel{100}}} = \frac{3}{1} = 3$ ■

> To change a percent to a fraction
>
> **1.** Drop the % symbol.
> **2.** Write the number that remains over 100.
> **3.** Simplify.

EXAMPLE 6 Change each percent to a fraction.

a. $79\% = \frac{79}{100}$

b. $60\% = \frac{60}{100} = \frac{\overset{3}{\cancel{60}}}{\underset{5}{\cancel{100}}} = \frac{3}{5}$

c. $2\% = \frac{2}{100} = \frac{\overset{1}{\cancel{2}}}{\underset{50}{\cancel{100}}} = \frac{1}{50}$

d. $100\% = \frac{100}{100} = 1$

e. $160\% = \frac{160}{100} = \frac{\overset{8}{\cancel{160}}}{\underset{5}{\cancel{100}}} = \frac{8}{5} \text{ or } 1\frac{3}{5}$ ■

▶ **You Try It** Change each percent to a fraction.

9. 30%	**10.** 75%	**11.** 80%	**12.** 6%
13. 200%	**14.** 650%	**15.** 21%	**16.** 1%

EXAMPLE 7 Change $16\frac{2}{3}\%$ to a fraction.

$$16\frac{2}{3}\% = \frac{16\frac{2}{3}}{100} = 16\frac{2}{3} \div 100 = \frac{50}{3} \div \frac{100}{1} = \frac{\overset{1}{\cancel{50}}}{3} \times \frac{1}{\underset{2}{\cancel{100}}} = \frac{1}{6}$$

Fraction bar means divide $16\frac{2}{3}$ by 100.

To divide fractions, invert the divisor and multiply.

Suppose $16\frac{2}{3}\%$ of a paycheck is tax. Since $16\frac{2}{3}\% = \frac{1}{6}$, this means $\frac{1}{6}$ of the paycheck, or $1 out of every $6 earned is tax. In this example the fraction form of the percent is easier to understand. ■

EXAMPLE 8 Change $\frac{1}{4}\%$ to a fraction. Interpret the answer.

$$\frac{1}{4}\% = \frac{\frac{1}{4}}{100} = \frac{1}{4} \div \frac{100}{1} = \frac{1}{4} \times \frac{1}{100} = \frac{1}{400}$$

$\frac{1}{4}\% = \frac{1}{400}$ and can be interpreted to mean 1 part out of 400 equal parts. If $\frac{1}{4}\%$ of a population lives to be 100 years old, then 1 person in 400 lives to that age. ■

▶ **You Try It** Change each percent to a fraction.

17. $4\frac{1}{6}\%$ **18.** $\frac{5}{8}\%$

---

3 Changing a Decimal Percent to a Fraction

Change 12.7% to a fraction.

$$12.7\% = .12\ 7\cancel{\%} = .127 = 127 \text{ thousandths} = \frac{127}{1{,}000}$$

First change the percent to a decimal.

Then change the decimal to a fraction.

> To change a decimal percent to a fraction
>
> 1. Change the percent to a decimal.
> 2. Change the decimal to a fraction.
> 3. Reduce.

EXAMPLE 9 Change each percent to a fraction.

a. $3.5\% = .03\,5\% = .035 = 35$ thousandths $= \dfrac{35}{1{,}000} = \dfrac{\overset{7}{\cancel{35}}}{\underset{200}{\cancel{1{,}000}}} = \dfrac{7}{200}$

b. $72.25\% = .72\,25\% = .7225 = 7{,}225$ ten-thousandths

$= \dfrac{7{,}225}{10{,}000} = \dfrac{\overset{289}{\cancel{7{,}225}}}{\underset{400}{\cancel{10{,}000}}} = \dfrac{289}{400}$ Reduce using 25.

c. $0.0008\% = .00\,0008\% = .000008 = 8$ millionths

$= \dfrac{8}{1{,}000{,}000} = \dfrac{\overset{1}{\cancel{8}}}{\underset{125{,}000}{\cancel{1{,}000{,}000}}} = \dfrac{1}{125{,}000}$

d. $1.2\frac{1}{6}\% = .01\,2\frac{1}{6}\% = .012\frac{1}{6} = 12\frac{1}{6}$ thousandths

$= \dfrac{12\frac{1}{6}}{1{,}000} = \dfrac{73}{6} \div \dfrac{1{,}000}{1} = \dfrac{73}{6} \times \dfrac{1}{1{,}000} = \dfrac{73}{6{,}000}$ ■

EXAMPLE 10 In some states a blood alcohol level of 0.1% or higher is considered proof of intoxication. Express this level as a fraction.

$$0.1\% = 0.00\,1\% = 0.001 = \frac{1}{1{,}000}$$

Intoxicated means $\frac{1}{1{,}000}$ (or more) of your blood volume is alcohol. ■

▶ **You Try It** Change each percent to a fraction.

19. 7.5% **20.** 6.25% **21.** 0.02% **22.** $2.4\frac{1}{3}\%$

23. A machine can produce parts with measurements accurate to 2.5% of a centimeter. Express this accuracy as a fraction.

▶ **Answers to You Try It** 1. 40% 2. 37.5% 3. 45% 4. $33\frac{1}{3}\%$ 5. $63\frac{7}{11}\%$ 6. 88.9% 7. 20.83% 8. 180% 9. $\frac{3}{10}$ 10. $\frac{3}{4}$ 11. $\frac{4}{5}$ 12. $\frac{3}{50}$ 13. 2 14. $6\frac{1}{2}$ 15. $\frac{21}{100}$ 16. $\frac{1}{100}$ 17. $\frac{1}{24}$ 18. $\frac{1}{160}$ 19. $\frac{3}{40}$ 20. $\frac{1}{16}$ 21. $\frac{1}{5{,}000}$ 22. $\frac{73}{3{,}000}$ 23. accurate to $\frac{1}{40}$ of a centimeter

SOMETHING MORE

Deception with Percents? Advertisements have been known to lead the unwary into thinking a product is good based on research the consumer seldom sees. A TV ad used some time ago to sell shaving equipment claimed:

57% of the people who use Smoothe say they like it as well as or better than their current blade.

This boast makes one feel that more people prefer Smoothe. Further thought, however, could lead to something quite different.

Determining what the claim means depends on how you interpret "as well as or better than." When someone uses a Smoothe blade, there are three distinct possibilities.

1. like Smoothe better than their current blade
2. like Smoothe as well as their current blade
3. like Smoothe less than their current blade

The following table presents three possible ways to interpret 57%.

liked Smoothe "as well as or better than" their current blade	liked Smoothe better	as well as	liked Smoothe less	liked current blade "as well as or better than" Smoothe
57% (53% + 4%)	53%	4%	43%	47% (4% + 43%)
57% (7% + 50%)	7%	50%	43%	93% (50% + 43%)
57% (0% + 57%)	0%	57%	43%	100% (57% + 43%)

For instance, 57% liked Smoothe "as well as or better than" could mean 50% liked it "as well as" and 7% liked it "better than." Then 100% − 57% = 43% liked it "less." In other words, 43% liked the competing blade better. This means 50% + 43% = 93% liked their current blade "as well as or better than" Smoothe, an unexpected turn of the tables on the Smoothe ad.

The extreme case comes from the bottom row of the table. Assume 0% liked Smoothe "better" and 57% liked it "as well as." Then the remaining 43% liked it "less." From this you conclude 57% + 43% = 100% liked the other brand "as well as or better than" Smoothe!

You have no way of knowing how 57% was actually split between the "as well as" and the "better than" people when Smoothe consumer tested its blade. However, a company may very well take its test results and word them in an ad so that their product is viewed in the best possible light. 57% may give a false impression or mislead consumers, even though it may be a perfectly accurate number derived from consumer data.

a. 18% of the people tested liked CRX better than and 34% liked it as well as their current brand. What percent liked CRX "as well as or better than" their current brand?

b. What percent liked their current brand "as well as or better than" CRX?

▶ **Answers to Something More** a. 18% + 34% = 52% liked CRX as well as or better than their current brand. b. 100% − 52% = 48% liked their current brand better. Therefore, 34% + 48% = 82% liked their current brand as well as or better than CRX.

SECTION 7.3 EXERCISES

1 *Change each fraction to a percent.*

1. $\frac{1}{4}$ **2.** $\frac{1}{3}$ **3.** $\frac{4}{5}$ **4.** $\frac{1}{7}$

5. $\frac{5}{8}$ **6.** $\frac{1}{6}$ **7.** $\frac{4}{11}$ **8.** $\frac{2}{9}$

9. $\frac{5}{12}$ **10.** $\frac{5}{6}$ **11.** $\frac{9}{10}$ **12.** $\frac{7}{9}$

13. $\frac{7}{5}$ **14.** $\frac{21}{10}$ **15.** $\frac{15}{8}$ **16.** $\frac{7}{2}$

17. $2\frac{1}{4}$ **18.** $5\frac{3}{8}$ **19.** $1\frac{1}{2}$ **20.** $3\frac{23}{25}$

21. $4\frac{1}{40}$ **22.** $9\frac{11}{50}$ **23.** $6\frac{1}{3}$ **24.** $2\frac{4}{15}$

25. $0.2\frac{1}{6}$ **26.** $0.4\frac{5}{9}$ **27.** $0.04\frac{2}{3}$ **28.** $0.36\frac{4}{11}$

29. Change $\frac{5}{16}$ into a percent.

30. Change $4\frac{2}{7}$ into a percent.

31. One-twentieth of the people in a village are hunters. What percent are hunters?

32. Twenty-four twenty-fifths of the weight of a head of lettuce is water. What percent of the lettuce is water?

33. Sam lost $\frac{2}{27}$ of his body weight playing football. What percent weight loss is this?

34. About $\frac{7}{10}$ of the Earth's surface is covered with water. What percent is this?

2 *Change each percent to a fraction. Reduce to lowest terms.*

35. 55%

36. 38%

37. 90%

38. 20%

39. 69%

40. 12%

41. 8%

42. 94%

43. 180%

44. 240%

45. 700%

46. 450%

47. $66\frac{2}{3}\%$

48. $12\frac{1}{2}\%$

49. $83\frac{1}{3}\%$

50. $28\frac{4}{7}\%$

51. $237\frac{1}{2}\%$

52. $444\frac{4}{9}\%$

53. $\frac{1}{5}\%$

54. $\frac{7}{10}\%$

55. $\frac{1}{7}\%$

56. $\frac{8}{11}\%$

57. $5\frac{1}{3}\%$ of the taxes collected in our city pays for sanitation. Change this percent to a fraction and interpret the answer.

58. $32\frac{3}{4}\%$ of a paycheck is tax. Change this percent to a fraction and interpret the answer.

59. $82\frac{1}{2}\%$ of your students study a foreign language. Change this percent to a fraction, and interpret the answer.

60. $2\frac{4}{5}\%$ of the bulbs are defective. Change this percent to a fraction and interpret the answer.

61. If 98% of a product is pure,
a. what percent is not pure?
b. what fraction of the product is not pure?

62. 42% of a family's income is spent on rent, and $20\frac{1}{2}\%$ is spent on food.
a. What percent of the income is spent on rent and food?
b. What percent is spent on items other than rent and food?
c. What fraction is spent on items other than rent and food?

3 *Change the decimal percent to a fraction.*

63. 2.2%

64. 0.3%

65. 16.25%

66. 8.4%

67. 42.5%

68. 90.05%

69. 62.5%

70. 1.025%

71. 0.162%

72. 0.05%

73. 0.00002%

74. 0.0006%

75. 140.5%

76. 330.75%

77. $5.2\frac{1}{3}\%$

78. $0.4\frac{1}{7}\%$

---

79. 51.3% of the babies born in the United States are female. What fraction of the babies are female?

80. 5.26% of each dollar made is profit. What fraction of each dollar is this?

81. 0.9% of a substance is niacin. What fraction is niacin?

82. 71.4% of the Earth is covered with water. What fraction of the Earth is this?

83. 8.125% of the budget is spent on defense. What fraction of the budget is spent on defense?

84. Inflation rose 0.75% this month. By what fraction did inflation rise this month?

Any number can be written in three ways: as a fraction, as a decimal, and as a percent. Write the given number in the other two forms. Round decimals to the ten-thousandths place.

	Fraction	Decimal	Percent
85.		0.36	
86.			72%
87.	$\frac{5}{4}$		
88.			45%
89.		1.5	
90.	$\frac{9}{11}$		
91.	$\frac{6}{10}$		

	Fraction	Decimal	Percent
92.			$2\frac{1}{6}\%$
93.		0.008	
94.	$1\frac{6}{7}$		
95.			$166\frac{2}{3}\%$
96.			375%
97.		0.6225	
98.	$\frac{17}{20}$		

SKILLSFOCUS (Section 6.1) *Solve each equation. Round to hundredths.*

99. $8b = 28$ **100.** $120r = 45$ **101.** $0.72y = 400$ **102.** $57.3m = 9.6$

EXTEND YOUR THINKING ▶▶▶▶

▶ SOMETHING MORE

103. 12% of the women tested liked MBQ better than and 51% liked it as well as their current brand. What percent liked MBQ "as well as or better than" their current brand? What percent liked their current brand "as well as or better than" MBQ?

104. Using a Calculator to Change a Fraction to a Percent You can change a fraction to a percent by using the [%] key on your calculator. For example, to change $\frac{3}{4}$ to a percent, enter the following sequence of keypresses.

3 [÷] 4 [%]

The calculator displays 75, meaning 75%. Use your calculator to change the following fractions to percents.

a. $\frac{1}{2}$ **b.** $\frac{5}{8}$

c. $\frac{3}{5}$ **d.** $\frac{2}{3}$

e. $5\frac{1}{4}$

▶ TROUBLESHOOT IT

Find and correct the error.

105. $\frac{2}{25} \longrightarrow 25\overline{)2.00}$ with quotient 0.8; $2\,00$, remainder 0. $= 0.80.\%$ or 80%. So, $\frac{2}{25} = 80\%$.

106. $0.6\% = \frac{0.6}{100} = \frac{6}{10} = \frac{3}{5}$

WRITING TO LEARN ▶▶▶▶

107. Explain how to change $\frac{7}{20}$ to a percent, as you would explain it to a classmate. Write a reason for each step.

108. Explain how to change 0.8 to a fraction, as you would explain it to a classmate. Write a reason for each step.

7.4 IDENTIFYING THE THREE NUMBERS IN A PERCENT PROBLEM

OBJECTIVE

Identify the three numbers in any percent problem.

THE PERCENT SENTENCE

Let's examine three ordinary problems involving percents and try to extract the essential question in each of them.

First, suppose a store is having a 20% off sale on everything and you are interested in purchasing an item that costs $78. To determine the amount you would save, you need to answer the question:

What is 20% of $78?

Next, suppose you are earning $6.50 an hour and your boss says you will get a raise of $.40 per hour. To determine the percent of your raise in salary, you need to answer the question:

$.40 is what percent of $6.50?

Finally, suppose your friend, who is a real estate salesperson, tells you she made a commission of $500 on a recent sale and you know she gets 2% of the selling price. To determine the amount of the sale, you need to answer the question:

$500 is 2% of what number?

Take another look at these three questions:

1. What is 20% of 78?
2. 0.40 is what percent of 6.50?
3. 500 is 2% of what number?

If you examine these three sentences very carefully, you will see that the structure of each sentence is the same. In each case, the sentence is: a number (either known or unknown) is a percent (either known or unknown) of another number (either known or unknown). In fact, **every** percent question can be written as a sentence of this type. If we use A to represent the first number, B% to represent the percent, and C to represent the second number, then the sentence is

A is B% of C

Because English is a flexible language, you may also see the above three sentences written in a slightly different order but with the same meaning, as follows:

1. 20% of 78 is what number?
2. What percent of 6.50 is 0.40?
3. 2% of what number is 500?

So using A, B%, and C, our percent sentence could also be written:

B% of C is A

In this section, we will focus on identifying the three numbers in this percent sentence.

Identifying the Three Numbers

Given a simple percent sentence like the ones above, we will give names to each of the three numbers and discuss how to identify each one.

- The easiest number is the **PERCENT**, which is the number with the percent sign. If the percent is the unknown number, then the sentence will read "what percent".
- Next, the number you are finding a percent *of* is the **WHOLE**.
- Finally, the number you get from taking a percent of the whole is the **PART.**

Thus, our two versions of the basic percent sentence can now be written

THE PART IS A PERCENT OF THE WHOLE.

or

A PERCENT OF THE WHOLE IS THE PART.

NOTE: If you read other mathematics books you may notice that the use of the words PART, PERCENT, and WHOLE is not universal. Some books call the PERCENT the RATE, some call the PART the AMOUNT, some call the WHOLE the BASE. Each word has its advantages and disadvantages. The important thing is to be able to distinguish between the PART and the WHOLE since these two numbers play very different roles in each problem.

Also, within the context of a particular problem, the PART and the WHOLE have names that are specific to that context. One common example is buying something on sale, as in problem 1 above. In this situation, the PART is called the DISCOUNT and the WHOLE is called the ORIGINAL PRICE. Thus the percent sentence, in the context of buying something on sale, becomes

THE DISCOUNT IS A PERCENT OF THE ORIGINAL PRICE.

We will study applications like these in later sections. For now, we will focus on basic percent sentences and identify the PART, PERCENT, and WHOLE in each one.

EXAMPLE 1 In the percent sentence: 8 is 10% of 80.
identify the PERCENT, WHOLE, and PART, and write the sentence another way with the same meaning.

The number with the percent sign is 10%, so the PERCENT is 10%.
The number we are taking 10% *of* is 80, so the WHOLE is 80.
The number you get from taking 10% of 80 is 8, so the PART is 8.
Writing the sentence another way: 10% of 80 is 8.

EXAMPLE 2 In the percent sentence: 80% of 2000 is 1600.
identify the PERCENT, WHOLE, and PART, and write the sentence another way with the same meaning.

The number with the percent sign is 80%, so the PERCENT is 80%.
The number we are taking 80% *of* is 2000, so the WHOLE is 2000.
The number you get from taking 80% of 2000 is 1600, so the PART is 1600.
Writing the sentence another way: 1600 is 80% of 2000.

EXAMPLE 3 In the percent sentence: 60 is 150% of 40.
identify the PERCENT, WHOLE, and PART, and write the sentence another way with the same meaning.

The number with the percent sign is 150%, so the PERCENT is 150%.
The number we are taking 150% *of* is 40, so the WHOLE is 40.
The number you get from taking 150% of 40 is 60, so the PART is 60.
Writing the sentence another way: 150% of 40 is 60.

OBSERVE In this last example, notice that the "PART" is larger than the "WHOLE". The reason for this is that the percent is larger than 100%, so we are actually taking *more* than all of 40. It is important to remember that the "WHOLE" is the number you are finding a percent *of*, and it is not necessarily the larger number. Remember that when you studied fractions, proper fractions represented *part* of a whole whereas improper fractions represented a whole or *more* than a whole. The answer to taking a percent of a number can be *smaller* than the whole (if the percent is smaller than 100%) or it can be *larger* than the whole (if the percent is larger than 100%).

♦You Try It In each percent sentence, identify the PERCENT, WHOLE, and PART, and write the sentence another way with the same meaning.

1. 17.25 is 30% of 57.5. **2.** 9.5% of 250 is 23.75. **3.** 51 is 170% of 30.

EXAMPLE 4 In the percent sentence: What is 34% of 160?
identify the PERCENT, WHOLE, and PART, and write the sentence another way with the same meaning.

The number with the percent sign is 34%, so the PERCENT is 34%.
The number we are taking 34% *of* is 160, so the WHOLE is 160.
The number we get from taking 34% of 160 is unknown, so the PART must be given a variable name. Let's call the PART p.
Writing the sentence another way: 34% of 160 is p.

EXAMPLE 5 In the percent sentence: 92% of what number is 54.6?
identify the PERCENT, WHOLE, and PART, and write the sentence another way with the same meaning.

The number with the percent sign is 92%, so the PERCENT is 92%.
The number we are taking 92% *of* is unknown, so the WHOLE must be given a variable name. Let's call the WHOLE w.
The number we get from taking 92% of w is 54.6, so the PART is 54.6.
Writing the sentence another way: 54.6 is 92% of w.

EXAMPLE 6 In the percent sentence: 80 is what percent of 20?
identify the PERCENT, WHOLE, and PART, and write the sentence another way with the same meaning.

The number with the percent sign is missing, so the PERCENT must be given a variable name. Let's call the PERCENT $x\%$.
The number we are taking $x\%$ *of* is 20, so the WHOLE is 20.
The number we get from taking $x\%$ of 20 is 80, so the PART is 80.
Writing the sentence another way: What percent of 20 is 80?

OBSERVE Notice that in this last example, the PART, 80, is *larger* than the WHOLE, 20. Although the percent is unknown, you should expect that it is *larger* than 100%.

♦ You Try It In each percent sentence, identify the PERCENT, WHOLE, and PART. Use a letter to represent any unknowns. Write the sentence another way with the same meaning.

4. What is 83% of 245?
5. $8\frac{1}{2}\%$ of what number is 50?
6. 0.42 is what percent of 6?
7. What percent of 5 is 12?

♦ Answers to You Try It **1.** Percent 30%, Whole 57.5, Part 17.25. 30% of 57.5 is 17.25. **2.** Percent 9.5%, Whole 250, Part 23.75. 23.75 is 9.5% of 250. **3.** Percent 170%, Whole 30, Part 51. 170% of 30 is 51. **4.** Percent 83%, Whole 245, Part p. 83% of 245 is p. **5.** Percent $8\frac{1}{2}\%$, Whole w, Part 50. 50 is $8\frac{1}{2}\%$ of w. **6.** Percent $x\%$, Whole 6, Part 0.42. $x\%$ of 6 is 0.42. **7.** Percent $x\%$, Whole 5, Part 12. 12 is $x\%$ of 5.

SECTION 7.4 EXERCISES

Identify the Percent, Whole, and Part. Write the sentence another way with the same meaning.

1. 27 is 60% of 45.

2. 4 is 5% of 80.

3. 20 is 40% of 50.

4. 15 is 75% of 20.

5. 50% of 1,800 is 900.

6. 16% of 400 is 64.

7. 10% of 6,000 is 600.

8. 4.2% of 2,010 is 84.42.

9. 54 is 120% of 45.

10. 250% of 30 is 75.

Identify the Percent, Whole, and Part. Use a variable name of your choice to name any unknown numbers. Use the variable to write the sentence another way with the same meaning. Do not solve.

11. What is 14% of 60?

12. What is 61% of 12?

13. 6 is 40% of what number?

14. 500 is 130% of what number?

15. What percent of 84 is 11?

16. What percent of 6 is 10?

17. 62% of 95 is what number?

18. 200% of 9 is what number?

19. 2% of what number is 0.5?

20. $5\frac{1}{2}\%$ of what number is 60?

21. 13 is what percent of 15?

22. 8,000 is what percent of 80?

23. What percent of \$90 is \$15?

24. 25% of what number is 9,600?

25. 0.036 is 12% of what number?

26. 60 is what percent of 90?

27. 12% of 0.036 is what number?

28. 90 is what percent of 60?

SKILLSFOCUS (Sections 4.4 and 4.5) *Add or subtract.*

29. $\frac{5}{8}+\frac{4}{5}$ **30.** $\frac{11}{12}-\frac{2}{3}$ **31.** $2\frac{1}{4}-1\frac{1}{3}$ **32.** $4\frac{4}{5}+6\frac{3}{4}$

EXTEND YOUR THINKING ♦ ♦ ♦

♦ TROUBLESHOOT IT

Find and correct the error.

33. In the percent sentence: 42 is what percent of 6?, the Percent is x%, the Whole is 42, and the Part is 6.

WRITING TO LEARN ♦ ♦ ♦

34. Explain in your own words how to identify the Percent, Whole, and Part in a percent sentence.

35. If the Percent is larger than 100%, explain what relationship must exist between the Part and the Whole. Give two examples.

7.5 MENTAL COMPUTATIONS INVOLVING COMMON PERCENTS

OBJECTIVES

1 Solve mentally problems using these common percents: 50%, 100%, 200%, 10%, and 25%,

2 Find the missing percent in a problem that involves one of these common percents.

3 Solve applications involving common percents.

1 COMMON PERCENTS

Consider the following situations:

1. You are shopping in a department store that is having a one-day sale in which everything in the store is 25% off.
2. You have just finished a meal in a restaurant and decide to leave a 15% tip for the waitress.
3. Your boss has just told you that you will be receiving a 10% raise.

Each of these situations involves a computation with a percent and most likely you would not have a pencil, paper, or calculator handy. It would be very useful to be able to do these computations mentally, both in order to deal effectively with situations like these and also to be able to check more complicated percent problems. First, we need to identify the most common percents. They are: 10%, 25%, 50%, 100%, and 200%. We will begin with the simplest ones: 50%, 100%, and 200%.

Fifty Percent

The fraction form of 50% is $\frac{1}{2}$, so verbally 50% means "half". Finding 50% of a number means finding "half" of the number. Remember that to find half of a number is the same as dividing by 2.

EXAMPLE 1 What is 50% of 48?

50% of 48 means **half** of 48, which is 24.

Answer: 50% of 48 is **24**.

EXAMPLE 2 What is 50% of 500?

50% of 500 means **half** of 500, which is 250.

Answer: 50% of 500 is **250**.

EXAMPLE 3 What is 50% of 45?

50% of 45 means **half** of 45, which is 22.5.

Answer: 50% of 45 is **22.5**.

EXAMPLE 4 16 is 50% of what number?

This problem is slightly different because now we know the result, not the original number. We can rephrase this sentence: 16 is **half** of ______ . If 16 is half of this unknown number, then the unknown number **must** be **double** 16, so the answer is 32.

Answer: 16 is 50% of **32**.

EXAMPLE 5 50% of what number is 80?

This problem is similar to the last one but the sentence is written differently. We know that **half** of ______ is 80. If half of the unknown number is 80, then the unknown number must be **double** 80, so the answer is 160.

Answer: 50% of **160** is 80.

♦You Try It

1. What is 50% of 400?

2. What is 50% of 75?

3. 50% of what number is 30?

4. 500 is 50% of ______ ?

5. 50% of 4.2 is what number?

One Hundred Percent 100% is equivalent to the whole number 1. 100% means the whole or "all", so to take 100% of a number means to take "all" of the number. When the percent is 100%, the part and the whole have the same value.

EXAMPLE 6 What is 100% of 82?

100% of 82 means all of 82, so the answer is 82.

Answer: **82** is 100% of 82.

EXAMPLE 7 What is 100% of 3000?

100% of 3000 means all of 3000, so the answer is 3000.

Answer: **3000** is 100% of 3000.

EXAMPLE 8 100% of what number is 3.6?

All of the unknown number is 3.6, so the unknown number must be 3.6.

Answer: 100% of **3.6** is 3.6.

♦You Try It

6. What is 100% of 99?

7. 100% of 4.1 is _____?

8. 51 is 100% of what number?

9. 100% of _____ is 19?

Two Hundred Percent

200% is equivalent to the whole number 2. 200% means "double", so to find 200% of a number means to "double" the number. Remember that to double a number is to multiply the number by 2.

EXAMPLE 9 What is 200% of 30?

To find 200% of 30 means to double 30, so the answer is 60.

Answer: **60** is 200% of 30.

EXAMPLE 10 200% of 4.1 is what number?

To find 200% of 4.1 means to double 4.1, so the answer is 8.2.

Answer: 200% of 4.1 is **8.2**.

EXAMPLE 11 30 is 200% of what number?

This problem is different from the previous two. Here, we know that 30 is double the unknown number, so the unknown number must be half of 30. The answer is 15.

Answer: 30 is 200% of **15**.

EXAMPLE 12 200% of what number is 14?

Again, we know that if you double an unknown number you get 14, so the unknown number must be half of 14. The answer is 7.

Answer: 200% of **7** is 14.

♦You Try It

10. What is 200% of 16? **11.** 200% of 60 is what number?

12. 24 is 200% of what number? **13.** 200% of ______ is 800?

Ten Percent

10% is equivalent to the fraction $\frac{1}{10}$ and the decimal 0.1. To find 10% of a number means to find $\frac{1}{10}$ of the number or multiply the number by $\frac{1}{10}$. Remember, though, that multiplying by $\frac{1}{10}$ is the same as dividing by 10 and that to divide a number by 10, move its decimal point one place to the left. Therefore, to find 10% of a number means to divide the number by 10 by moving its decimal point one place to the left.

EXAMPLE 13 What is 10% of 80?

To find 10% of 80 means to divide 80 by 10, so the answer is 8.

Answer: **8** is 10% of 80.

EXAMPLE 14 10% of $43.50 is what number?

To find 10% of $43.50, divide $43.50 by 10 (move the decimal point one place to the left). The answer is $4.35.

Answer: 10% of $43.50 is **$4.35**.

EXAMPLE 15 What is 10% of 65?

To find 10% of 65, divide 65 by 10. The answer is 6.5.

Answer: **6.5** is 10% of 65.

EXAMPLE 16 7 is 10% of what number?

This example is different from the previous three. Here we know that 7 is the number you get when you divide the unknown number by 10. Then the unknown number must be 10 times 7 or 70.

Answer: 7 is 10% of **70**.

EXAMPLE 17 92 is 10% of what number?

Again, 92 is the number you get when you divide the unknown number by 10, so the unknown number must be 10 times 92 or 920.

Answer: 92 is 10% of **920**.

EXAMPLE 18 10% of what number is 1.07?

If 10% of the unknown number is 1.07, the unknown number must be 10 times 1.07 or 10.7.

Answer: 10% of **10.7** is 1.07.

♦You Try It

14. What is 10% of 90?

15. What is 10% of 86?

16. 10% of 560 is what number?

17. 6 is 10% of what number?

18. 7.4 is 10% of what number?

19. 10% of _____ is 13?

20. 10% of 13 is _____ ?

Twenty-five Percent

25% is equivalent to the fraction $\frac{1}{4}$ or the decimal 0.25. To find 25% of a number means to find $\frac{1}{4}$ of the number or to multiply the number by $\frac{1}{4}$. Remember that multiplying a number by $\frac{1}{4}$ is the same as dividing the number by 4. Therefore, to find 25% of a number is to divide the number by 4.

EXAMPLE 19 What is 25% of 36?

To find 25% of 36 means to divide 36 by 4, so the answer is 9.

Answer: **9** is 25% of 36.

EXAMPLE 20 What is 25% of 2000?

To find 25% of 2000 means to divide 2000 by 4, so the answer is 500.

Answer: **500** is 25% of 2000.

OBSERVE You may have noticed that 25% is half of 50%. Another way of finding 25% of a number is to find 50% (half) and then take half of that answer. It is sometimes quicker to do these two calculations mentally than to try to divide by 4. In this example, 50% of 2000 is 1000, and half of 1000 is 500.

EXAMPLE 21 25% of 60 is what number?

To find 25% of 60 means to divide 60 by 4, so the answer is 15.

Alternatively, 50% of 60 is 30 and half of 30 is 15, so 25% of 60 is 15.

Answer: 25% of 60 is **15**.

EXAMPLE 22 6 is 25% of what number?

This example is different from the previous three. Here, 6 is the number you get when you take 25% of the unknown number. Another way to say this is that 6 is $\frac{1}{4}$ of the unknown number. If you get 6 when you divide the unknown number by 4, then the unknown number must be 4 times 6 or 24.

Answer: 6 is 25% of **24**

EXAMPLE 23 20 is 25% of what number?

20 is $\frac{1}{4}$ of the unknown number, so the unknown number must be 4 times 20 or 80.

Answer: 20 is 25% of **80.**

EXAMPLE 24 25% of what number is 1?

$\frac{1}{4}$ of the unknown number is 1, so the unknown number must be 4 times 1 or 4.

Answer: 25% of **4** is 1.

♦ **You Try It**

21. What is 25% of 88?

22. What is 25% of 96?

23. 25% of 500 is ______ ?

24. 12 is 25% of what number?

25. 80 is 25% of ______ ?

26. 25% of what number is 50?

2 FINDING THE MISSING PERCENT

Now that you are familiar with finding common percents of numbers, we will investigate examples in which the part and the whole are both known and you need to determine the percent. In these examples, you need to understand clearly which number is the part and which number is the whole so that you can compare them. Recall that when you **compare** two numbers, you are writing a **ratio**. If you compare the Part to the Whole, you can think of the ratio $\frac{\text{PART}}{\text{WHOLE}}$. This can help you to see the percent. Here are some observations based on the examples so far.

- If the part is equal to the whole, then the ratio $\frac{\text{PART}}{\text{WHOLE}}$ equals 1 and the percent is 100%.
- If the part is $\frac{1}{10}$ of the whole, then the ratio $\frac{\text{PART}}{\text{WHOLE}}$ reduces to $\frac{1}{10}$ and the percent is 10%.
- If the part is half of the whole, then the ratio $\frac{\text{PART}}{\text{WHOLE}}$ reduces to $\frac{1}{2}$ and the percent is 50%.
- If the part is double the whole, then the ratio $\frac{\text{PART}}{\text{WHOLE}}$ equals 2 and the percent is 200%.
- If the part is $\frac{1}{4}$ of the whole, then the ratio $\frac{\text{PART}}{\text{WHOLE}}$ reduces to $\frac{1}{4}$ and the percent is 25%.

EXAMPLE 25 40 is what percent of 80?

The part is 40, the whole is 80. 40 is **half** of 80 and the ratio $\frac{\text{PART}}{\text{WHOLE}} = \frac{40}{80} = \frac{1}{2}$ so the percent is 50%.

Answer: 40 is **50%** of 80.

EXAMPLE 26 80 is what percent of 40?

Although the numbers are the same, their roles in the sentence have changed. Now, the part is 80 and the whole is 40. Here 80 is **double** 40 and the ratio

$\frac{\text{PART}}{\text{WHOLE}} = \frac{80}{40} = \frac{2}{1} = 2$, so the percent is 200%.

Answer: 80 is **200%** of 40.

EXAMPLE 27 5 is what percent of 20?

The part is 5, the whole is 20. Notice that you can multiply 5 by 4 to get 20, so 5 is $\frac{1}{4}$ of 20. Also the ratio $\frac{\text{PART}}{\text{WHOLE}} = \frac{5}{20} = \frac{1}{4}$. The percent is 25%.

Answer: 5 is **25%** of 20.

EXAMPLE 28 7 is what percent of 70?

The part is 7, the whole is 70. If you multiply 7 by 10, you get 70, so 7 is $\frac{1}{10}$ of 70. Also the ratio $\frac{\text{PART}}{\text{WHOLE}} = \frac{7}{70} = \frac{1}{10}$. The percent is 10%.

Answer: 7 is **10%** of 70.

♦You Try It

27. 9 is what percent of 18?

28. 2 is what percent of 20?

29. 4 is what percent of 16?

30. 20 is _____% of 10.

31. 5 is _____ % of 10.

32. What percent of 200 is 100?

33. What percent of 45 is 45?

3 APPLICATIONS

Let's return to the situations described at the beginning of the section.

Sale Price In the first situation, you are in a store in which everything is on sale for 25% off. Suppose you want to estimate the sale price of several items. The first item is a shirt with a price tag of \$32.95. How can you easily estimate the sale price of this shirt? First, you will need to find the amount of the discount, which is 25% of \$32.95. To do this mentally, you need to round \$32.95 to a convenient whole dollar amount. Since 25% is $\frac{1}{4}$, you will need to divide by 4, so it is convenient to round \$32.95 to \$32. The discount, then, is approximately $\frac{1}{4}$ of \$32 or \$8. Remember that this is the **discount** only. To find the sale price, you need to subtract the discount from the original price. The sale price will be approximately \$32 – \$8 or \$24.

♦ **You Try It** *Estimate the sale price of each of the following items in the same store (25% off).*

34. A television with a price tag of $355.

35. A pair of shoes that cost $59.

36. An alarm clock that costs $35.99.

Tipping The second situation involves tipping in a restaurant. A tip of 15% of the total bill (excluding tax) is usual. Notice that 15% was not one of the common percents that we covered. What is an easy way to compute a 15% tip on a dinner bill of $63.50? We know that 10% is easy to find, and 5% is half of 10%. Also, it is easier to calculate these percents mentally if we round $63.50 to $64 first. So we will find 10% of $64 first. Then 5% of $64 will be half of that. Finally, we add the two numbers together.

10% of $64 = $6.40

5% of $64 = $3.20

15% of $64 = $6.40 + $3.20 = $9.60

So the tip should be approximately $9.60.

♦ **You Try It** *Find the approximate tip on each dinner bill.*

37. A pizza dinner for four for $28.75.

38. A French dinner for two for $66.80.

39. A seafood dinner for a family of three for $48.56.

Pay Raise The final situation involves a wage increase of 10%. Suppose you earn $6.40 per hour and have been told that you will be receiving a 10% raise in pay. What is your new hourly rate? First, you need to (mentally) compute the raise, which is 10% of $6.40, or $.64. Next, the raise is added to the current rate of pay, so the your hourly rate is $6.40 + $.64 = $7.04.

♦ **You Try It** *Determine the new rate of pay.*

40. You earn $14,600 per year and you receive a 10% raise.

41. You earn $8.90 per hour and you receive a 10% raise.

42. You earn $22,310 per year and you receive a 10% raise.

♦**Answers to You Try It** **1.** 200 **2.** 37.5 **3.** 60 **4.** 1,000 **5.** 2.1 **6.** 99 **7.** 4.1 **8.** 51 **9.** 19 **10.** 32 **11.** 120 **12.** 12 **13.** 400 **14.** 9 **15.** 8.6 **16.** 56 **17.** 60 **18.** 74 **19.** 130 **20.** 1.3 **21.** 22 **22.** 24 **23.** 125 **24.** 48 **25.** 320 **26.** 200 **27.** 50% **28.** 10% **29.** 25% **30.** 200% **31.** 50% **32.** 50% **33.** 100% **34.** about $270 **35.** about $45 **36.** about $27 **37.** about $4.50 **38.** about $10.50 **39.** about $7.50 **40.** $16,060 **41.** $9.79 **42.** $24,541

SECTION 7.5 EXERCISES

1 *Do each of the following problems mentally.*

1. What is 50% of 40?

2. What is 50% of 2,300?

3. ____ is 50% of 6.4?

4. 50% of 96 is _____?

5. 50% of 88 is what number?

6. 18 is 50% of what number?

7. 42 is 50% of what number?

8. 3,000 is 50% of ________ ?

9. 50% of what number is 11?

10. 50% of _________ is 90?

11. What is 100% of 74?

12. ____ is 100% of 2,300?

13. 100% of 17.2 is what number?

14. 100% of 0.054 is ________ ?

15. 100% of 32 is what number?

16. 8 is 100% of what number?

17. 87 is 100% of what number?

18. 23 is 100% of __________ .

19. 100% of _____ is 0.56.

20. 100% of what number is 3.214?

21. What is 200% of 51?

22. What is 200% of 93?

23. ____ is 200% of 3,000?

24. 200% of 34 is what number?

25. 200% of 1.2 is _____ ?

26. 20 is 200% of what number?

27. 86 is 200% of what number?

28. 200% of what number is 2,400?

29. 200% of _____ is 4.8?

30. 200% of _____ is 800?

31. What is 10% of 70?

32. What is 10% of 3,200?

33. _____ is 10% of 78?

34. 10% of 56 is what number?

35. 10% of 244 is ______ ?

36. 12 is 10% of what number?

37. 82 is 10% of what number?

38. 10% of what number is 5.3?

39. 10% of _____ is 9?

40. 10% of ______ is 4.4?

41. What is 25% of 100?

42. What is 25% of 8,800?

43. ______ is 25% of 4.04?

44. 25% of 8 is what number?

45. 25% of 36 is _____ ?

46. 3 is 25% of what number?

47. 15 is 25% of _______ ?

48. 100 is 25% of what number?

49. 25% of what number is 40?

50. 25% of _______ is 0.5?

2 *Compare the part and the whole to find the missing percent.*

51. 16 is what percent of 32?

52. 16 is what percent of 16?

53. 16 is what percent of 8?

54. 16 is what percent of 160?

55. 16 is what percent of 64?

56. 32 is what percent of 16?

57. 8 is what percent of 16?

58. 3.2 is what percent of 32?

59. 3.4 is what percent of 3.4?

60. 7 is what percent of 28?

61. What percent of 84 is 84?

62. What percent of 650 is 65?

63. What percent of 600 is 300?

64. What percent of 300 is 600?

65. What percent of 80 is 20?

3 *Use your knowledge of common percents to solve each application.*

66. A CD-player with a regular price of $240 is on sale at one store for 10% off. What is the sale price?

67. A different store has the same $240 CD-player on sale for 25% off. What is the sale price at this store?

68. Two couples have had dinner together at a Chinese restaurant and the bill is $49.80. How much tip should they leave?

69. Harry paid $13,400 for his car when it was new. After three years, it is worth only 50% of its original value. What is its dollar value after three years?

70. Sarah purchased some stock at $42 per share. After one month, the price had increased 10%. What is the new value of the stock per share?

71. When a new diet cookie first came into the market, the price was $1.20 per box. The cookie proved so popular that after six months its price rose by 100%. What is the new price of a box of these cookies?

72. A pair of shoes that normally sells for $60 is on sale for 25% off. What is the sale price?

73. A seasonal furniture store has an end-of-summer sale on patio furniture. A table and four chairs has a regular price of $640 and it is on sale for 25% off. What is the sale price?

SKILLSFOCUS (Section 5.5) *Use the rule for order of operations to simplify each expression.*

74. $-4 \cdot 3 - 2 \cdot 5$ **75.** $\frac{-72}{8} + \frac{-80}{10}$ **76.** $-3 + 2(1-4)$ **77.** $\frac{5-3(8)}{(5-3)8}$

EXTEND YOUR THINKING ♦ ♦ ♦

♦ TROUBLESHOOT IT

Find and correct the error.

78. 24 is what percent of 12? Answer: 50%

♦ SOMETHING MORE

79. Penny took a test with 200 questions. She got 75% of them right. How many questions did she get right? How many questions did she get wrong?

80. In some countries, the inflation rate is 200%. That means, for example, that if a loaf of bread costs $1.20, its price will go up 200%. What will the new price of the bread be?

WRITING TO LEARN ♦ ♦ ♦

81. Explain how to use common percents to compute 75% of 200.

82. Explain how to use common percents to compute 1% of 500.

7.6 SOLVING THE PERCENT EQUATION

OBJECTIVES

1. Translate the percent sentence into an equation.
2. Find the Part.
3. Find the Whole.
4. Find the Percent.

Where are we now and where are we going? All of the problems and topics covered so far in this chapter have focused on **skills** needed to become comfortable with percents. We continue to build these skills with a final section on solving percent problems. Once these skills have been mastered you will be ready to examine many of the interesting examples and uses of percents in everyday life.

1 TRANSLATING THE PERCENT EQUATION

Now that you are familiar with the basic percent sentence and can solve some percent problems mentally, it is time to consider percent problems that require a precise answer and that cannot be computed mentally with ease. Fortunately, the percent sentence is a simple one and translates easily into an equation. Once the equation is written, you can use your equation-solving skills (Section 6.1) to complete the solution. Remember that the basic percent sentence is

PART is a PERCENT of WHOLE.

or

PERCENT of WHOLE is PART.

depending on the way the sentence is written.

Remember that to find a *fraction of* a number, you multiply the fraction times the number.

For example, to find $\frac{3}{4}$ of 80, you multiply $\frac{3}{4} \times 80$. The same is true for finding a percent of a number. If you need to find a *percent of* a number, you multiply the percent times the number. Now we have

PART is PERCENT × WHOLE

or

PERCENT × WHOLE is PART

The word "is" translates into mathematics as " = ", so we now have an equation.

> PART = PERCENT × WHOLE
>
> or
>
> PERCENT × WHOLE = PART

Since we cannot multiply a percent times a number, the Percent must be changed to a fraction or a decimal before solving the equation.

EXAMPLE 1 Translate the percent sentence: What number is 23% of 86? into an equation. Do not solve.

Identify the three numbers:
Percent $= 23\% = 0.23$
Whole $= 86$
Part $= x$

$$\text{PART} = \text{PERCENT} \times \text{WHOLE}$$
$$x = 0.23 \times 86$$

Answer: Sentence: What number is 23% of 86?
Equation: $x = 0.23 \times 86$

EXAMPLE 2 Translate the percent sentence: 2% of what number is 57.2? into an equation. Do not solve.

Identify the three numbers:
Percent $= 2\% = 0.02$
Whole $= x$
Part $= 57.2$

$$\text{PERCENT} \times \text{WHOLE} = \text{PART}$$
$$(0.02)\,(x) = 57.2$$

or more simply

$$0.02x = 57.2$$

Answer: Sentence: 2% of what number is 57.2?
Equation: $0.02x = 57.2$

EXAMPLE 3 Translate the percent sentence: 16 is what percent of 90? into an equation. Do not solve.

Identify the three numbers:
Percent $= x$ (Here x represents the percent written as a decimal)
Whole $= 90$
Part $= 16$

$$\text{PART} = \text{PERCENT} \times \text{WHOLE}$$
$$16 = (x)\,(90)$$

In Section 1.2 we noted that the order in which we multiply numbers does not change the answer and this property of multiplication is called the *commutative* property. That means that $(x)(90)$ is the same as $(90)(x)$, which is the same as $90x$. The equation is now written

$$16 = 90x$$

Answer: Sentence: 16 is what percent of 90?
Equation: $16 = 90x$

♦You Try It *Translate each sentence into an equation. Do not solve.*

1. 67% of 14 is what number?

2. 43 is 40% of what number?

3. What percent of 78 is 90?

4. What is 120% of 25?

2 FINDING THE PART Now we will solve these percent problems by translating to an equation and then solving the equation. We begin by solving problems in which the Part is missing.

EXAMPLE 4 What is 47% of 62?

Identify the three numbers:
Percent = 47% = 0.47
Whole = 62
Part = x

PART = PERCENT × WHOLE

$$x = 0.47 \times 62$$

To solve this equation means to calculate the value that represents x. In this equation we only need to do the indicated multiplication. We get

$$x = 29.14$$

Answer: **29.14** is 47% of 62.

Check: We can use our knowledge of common percents to check answers. 47% is close to 50% and 62 is close to 60. 50% of 60 is 30, so the answer should be close to 30.

EXAMPLE 5 130% of 14 is what number?

Identify the three numbers:
Percent = 130% = 1.30
Whole = 14
Part = x

PERCENT × WHOLE = PART

$$1.30 \times 14 = x$$
$$18.2 = x$$

Answer: 130% of 14 is **18.2** .

Check: First, since 130% is larger than 100%, the Part should be larger than the Whole. To get a more precise estimate, 100% of 14 is 14 and 25% of 14 is about 4, so 130% of 14 is about 18.

EXAMPLE 6 What is $33\frac{1}{3}\%$ of 72?

Identify the three numbers:

Percent = $33\frac{1}{3}\% = \frac{1}{3}$ Note that in this case, it is easier to use the fraction form of the percent.

Whole = 72
Part = x

PART = PERCENT × WHOLE

$$x = \frac{1}{3} \times 72$$

$$x = \frac{72}{3}$$

$$x = 24$$

Answer: **24** is $33\frac{1}{3}$% of 72.

Check: $33\frac{1}{3}$% is about 30% and 72 is about 70. 10% of 70 is 7, so 30% of 70 is 21. The answer should be close to 21.

♦ **You Try It** *Solve each percent problem by writing and solving an equation.*

5. What is 19% of 145?

6. 150% of 0.39 is what number?

7. Find 0.2% of 14.

8. $66\frac{2}{3}$% of 960 is what number?

---

3 FINDING THE WHOLE

Now let's look at some examples where the Whole is missing.

EXAMPLE 7 15 is 30% of what number?

Identify the three numbers:
Percent = 30% = 0.30
Whole = x
Part = 15

$$\text{PART} = \text{PERCENT} \times \text{WHOLE}$$
$$15 = (0.30)(x)$$
$$15 = 0.30x$$

To solve this equation, we need to *divide* both sides of the equation by 0.30:

$$\frac{15}{0.30} = \frac{0.30x}{0.30}$$

$$50 = x$$

Answer: 15 is 30% of **50**.

Check: To check this, check to see that 30% of 50 does give 15. 10% of 50 is 5, so 30% of 50 is 3 × 5 = 15.

EXAMPLE 8 110% of what number is 979?

Identify the three numbers:
Percent = 110% = 1.10
Whole = x
Part = 979

$$\text{PERCENT} \times \text{WHOLE} = \text{PART}$$
$$(1.10)(x) = 979$$
$$1.10x = 979$$

Divide both sides by 1.10: $\frac{1.10x}{1.10} = \frac{979}{1.10}$

$x = 890$

Answer: 110% of **890** is 979.

Check: Since 110% is larger than 100%, the Whole should be smaller than the Part. Now check to see that 110% of 890 does give 979. 890 is close to 900. 100% of 900 is 900 and 10% of 900 is 90, so 110% of 900 is 900 + 90 = 990. The part should be close to 990.

EXAMPLE 9 $12 is $2\frac{1}{2}$% of how much?

Identify the three numbers:

Percent $= 2\frac{1}{2}\% = 2.5\% = 0.025$

Whole $= x$

Part $= 12$

$$\text{PART} = \text{PERCENT} \times \text{WHOLE}$$
$$12 = (0.025)(x)$$
$$12 = 0.025x$$

Divide both sides by 0.025 $\frac{12}{0.025} = \frac{0.025x}{0.025}$

$480 = x$

Answer: $12 is $2\frac{1}{2}$% of **$480**.

Check: Check to see that $2\frac{1}{2}$% of 480 does give 12. 10% of 480 is 48, so 1% of 480 is 4.8 and 2% of 480 is 4.8 + 4.8 = 9.6. Another $\frac{1}{2}$% of 480 would be 2.4, so $2\frac{1}{2}$% of 480 is 9.6 + 2.4 or 12.

♦ **You Try It** Solve each percent problem by writing and solving an equation.

9. 60 is 40% of what number?

10. 8.3% of what number is 7.47?

11. 160% of what number is 90?

12. 50 is $33\frac{1}{3}$% of what number?

4 FINDING THE PERCENT

Finally, we look at examples where the Percent is missing.

EXAMPLE 10 16 is what percent of 80?

Identify the three numbers:

Percent $= x$ (**Remember** that x represents the percent written as a **decimal**.)
Whole $= 80$
Part $= 16$

$$\text{PART} = \text{PERCENT} \times \text{WHOLE}$$
$$16 = (x)(80)$$

Remember that $(x)(80)$ is the same as $(80)(x)$, or $80x$.

$$16 = 80x$$

To solve this equation, divide both sides of the equation by 80:

$$\frac{16}{80} = \frac{80x}{80}$$

Dividing 16 by 80 gives 0.2, so

$$0.2 = x$$

Here is where it is very important to remember that we used x to represent the **decimal** version of the percent. To get the final answer, we must change this decimal to a percent (multiply by 100):

$$20\% = x$$

Answer: 16 is **20%** of 80.

Check: To check this, compare the Part, 16, to the Whole, 80. The ratio $\frac{\text{PART}}{\text{WHOLE}}$ is $\frac{16}{80}$, which reduces to $\frac{1}{5}$, which is 20%. Alternatively, check to see that 20% of 80 does give 16. 10% of 80 is 8 so 20% of 80 is $2 \times 8 = 16$.

OBSERVE When looking for the Percent, it is often easier to think of *comparing* the Part to the Whole or writing the ratio, $\frac{\text{PART}}{\text{WHOLE}}$. When this fraction is treated as a division, the answer is the decimal representation of the Percent. Multiplying by 100 gives the Percent.

EXAMPLE 11 What percent of 9,320 is 6,240? Round your answer to the nearest whole percent.

Identify the three numbers:
Percent $= x$ (as a decimal)
Whole $= 9{,}320$
Part $= 6{,}240$

$$\text{PERCENT} \times \text{WHOLE} = \text{PART}$$
$$(x)(9{,}320) = 6{,}240$$
$$9{,}320x = 6{,}240$$

To solve this equation, divide both sides by 9,320: $\frac{9{,}320x}{9{,}320} = \frac{6{,}240}{9{,}320}$

The division to four places gives: $x = 0.6695$

Before rounding, change this decimal representation of x to a percent: $x = 66.95\%$

Now round to the nearest whole percent: $x \doteq 67\%$
(Recall that $\doteq$ means "is approximately equal to")

Answer: Approximately **67%** of 9,320 is 6,240.

Check: To check, compare the Part, 6,240, to the Whole, 9,320. The closest "easy" numbers are 6,000 compared to 10,000 or 60 compared to 100, which is 60%. So the Percent should be "near" 60%. Alternatively, check that 67% of 9,320 is 6,240. Since 67% is close to $66\frac{2}{3}\%$, which is $\frac{2}{3}$, we can find $\frac{2}{3}$ of 9,320:

$\frac{2}{3} \times \frac{9320}{1} = \frac{18640}{3} \doteq 6213$. This tells us our answer is reasonable.

EXAMPLE 12 450 is what percent of 200?

Identify the three numbers:
Percent $= x$ (as a decimal)
Whole $= 200$
Part $= 450$

Notice here that we are taking a percent *of* 200, so 200 is the Whole even though it is the smaller number. Since the Part is larger than the Whole, we can expect the Percent to be larger than 100%.

$$\text{PART} = \text{PERCENT} \times \text{WHOLE}$$
$$450 = (x)(200)$$
$$450 = 200x$$

To solve this equation, divide both sides by 200:

$$\frac{450}{200} = \frac{200x}{200}$$
$$2.25 = x$$

Remember that this is the **decimal** representation of the percent. Multiplying by 100 to get the percent: $225\% - x$

Answer: 450 is **225%** of 200.

Check: Check to see that 225% is reasonable. 200% of 200 is $2 \times 200 = 400$ and 25% of 200 is 50, so 225% of 200 is 450.

♦ You Try It

13. 88 is what percent of 160? **14.** What percent of 720 is 240?

15. What percent of 413 is 25? (Give your answer to the nearest whole percent.)

16. 8,800 is what percent of 8,000? **17.** \$0.36 is what percent of \$0.75?

◆**Answers to You Try It** **1.** $0.67 \times 14 = x$ **2.** $43 = 0.40x$ **3.** $78x = 90$ **4.** $x = 1.20 \times 25$ **5.** 27.55 **6.** 0.585 **7.** 0.028 **8.** 640 **9.** 150 **10.** 90 **11.** 56.25 **12.** 150 **13.** 55% **14.** $33\frac{1}{3}\%$ **15.** 6% **16.** 110% **17.** 48%

SECTION 7.6 EXERCISES

1 2 3 4 *Translate each percent sentence into an equation and solve.*

1. 42% of 600 is what number?

2. 6% of 40 is what number?

3. 87% of 380 is what amount?

4. 2.2% of 750 is what amount?

5. What number is 7.2% of \$65?

6. What number is 100% of 25.3 pounds?

7. $12\frac{1}{2}\%$ of 640 men is how many men?

8. $83\frac{1}{3}\%$ of \$420 is how much?

9. What is 0.06% of 22,000 pounds?

10. Find 0.3% of 0.85 grams?

11. 240% of \$550 is how much money?

12. 338% of 62.5 is what number?

13. 55% of what number is 44?

14. 27% of what number is 162?

15. 83% of what number is 41.5?

16. $37\frac{1}{2}\%$ of what number is 15?

17. 20 is $22\frac{2}{9}\%$ of what number?

18. 50 is $83\frac{1}{3}\%$ of what quantity?

19. 60 people is 0.3% of how many people?

20. \$126 is 140% of how much money?

21. 275% of what number is 330?

22. 103% of what number is 49.44?

23. 0.1% of how many liters is 0.01 liter?

24. 38% of how many cars is 171 cars?

25. What percent of 300 is 90?

26. What percent of 20 is 12?

27. What percent of \$65 is \$19.50?

28. What percent of 680 children is 51 children?

29. 667 men is what percent of 2,900 men?

30. 14 is what percent of 30?

31. 45 senators is what percent of 75 senators?

32. 13.2 is what percent of 13.2?

33. 0.8 is what percent of 0.5?

34. 230 is what percent of 57.5?

35. What percent of \$5.10 is \$1.70?

36. What percent of 0.4 is 0.016?

37. 2.9% of what number is 10.15?

38. 38 is 20% of what number?

39. What is $18\frac{2}{11}\%$ of 33?

40. 185% of \$340 is how much?

41. 54 is what percent of 72?

42. What percent of 8 is 3?

43. $\frac{1}{3}$ is 20% of what number?

44. 140% of what number is $\frac{7}{8}$?

45. 7.2 is what percent of 9?

46. $\frac{3}{7}$ is what percent of $\frac{4}{7}$?

47. What is 80% of \$80?

48. 150% of \$25 is how much?

SKILLSFOCUS (Section 1.8) *Add.*

49. 2.7 + 30 + 5.06 + 0.91

50. 600.3 + 25.08 + 4 + 0.6

51. 0.07 + 0.008 + 1 + 0.95

52. 200.003 + 20.03 + 2.3 + 2,000

▶ YOU BE THE JUDGE

53. Paula paid \$200 for a painting. She sold it for 150% of what she paid for it. What did she sell the painting for? Explain how you arrived at your answer.

54. Suppose 200 people are in a train station. Phil says 150% of them are holding train tickets to New York. Is Phil making sense? Explain your decision.

7.7 APPLICATIONS OF PERCENT

OBJECTIVES

1 To solve application problems involving percents.

2 To solve problems involving circle graphs.

1 APPLICATIONS

References to percent appear every day in newspaper articles and advertisements, percents are used to calculate interest on loans, discounts at sales, increases in profits, sales tax on purchases, and raises in salary, to name just a few applications. As an educated adult and informed consumer, you need to be skilled at percent calculations in order to understand what you read and make intelligent choices. In this section we will examine a few of these applications of percent.

EXAMPLE 1 The Microsoft Corporation dominates the computer software applications market, holding about 75% of the \$1.3 billion business. What is the dollar amount of Microsoft's share of the market?

In situations like this, the information you need is often imbedded in several sentences of explanatory text. While the text is essential to your understanding of the situation, it can get in the way of your ability to solve a particular percent problem. The first thing to do is to **write a percent sentence** similar to the ones we studied in sections 7.4, 7.5, and 7.6. The sentence should **reflect the situation under discussion.** Recall that the sentence is

PART is a PERCENT of WHOLE.

In applications, each of the three numbers may have its own name that makes the sentence more meaningful. In this situation, after reading the problem carefully, the sentence is:

Microsoft's share is 75% of the total software market

Examining the information more carefully, we see that \$1.3 billion is the *total* market, or the Whole, in this situation, and Microsoft's share is the (unknown) Part, x. The sentence's final form is

x is 75% of \$1.3 billion

To summarize

Percent $= 75\% = 0.75$

Whole $= \$1.3$ billion

Part $= x$

$$\text{PART} = \text{PERCENT} \times \text{WHOLE}$$

$$x = 0.75 \times \$1.3 \text{ billion}$$

Multiplying, we get $x = \$0.975$ billion

Since 0.975 billion is the same as 975 million, an appropriate answer would be:

Microsoft's share of the software applications market is \$975 million.

Check: Estimate the answer by finding 75% of 1 billion. 50% of 1 billion is 500 million and 25% of 1 billion is 250 million, so 75% of 1 billion is 750 million. Thus the answer should be more than 750 million but less than 1.3 billion.

♦ You Try It

1. At the beginning of a semester, a biology professor had 32 students in an introductory course. At the end of the semester, 75% of the students passed the course and 25% of those who passed the course earned a CRHH grade.
   a. How many students passed the course?
   b. How many students earned a CRHH grade?

EXAMPLE 2 A television set costs \$349.99. Find the New Jersey sales tax and the total cost of this set.

It is interesting to notice that this problem doesn't mention percent at all! To solve this problem, you need to know a few things about sales tax in general, and about sales tax in New Jersey, in particular. First, you probably know, from experience, that sales tax is a percent of the cost. In fact that is one of the special percent sentences:

SALES TAX is a PERCENT of COST

In New Jersey, the sales tax rate is 6%, so in New Jersey, the sentence is

SALES TAX is 6% of COST

In this problem, we know
Percent = 6% = 0.06
Whole = cost of TV = \$349.99
Part = sales tax = x

PART = PERCENT × WHOLE

$x = 0.06 \times 349.99$

$x = 20.9994$

Since sales tax is money, we should round to the nearest penny: $x = 21.00$

Thus, the sales tax on the television is \$21.00. The total cost is the price plus the tax, or \$349.99 + 21.00 = \$370.99.

Check: To estimate the tax, find 5% of \$350. 10% of 350 is 35, so 5% of 350 is about 18, so the tax is *about* \$18 and the final cost is *about* \$368.

♦ You Try It

2. A bread machine costs \$140. Find the New Jersey sales tax and the final cost.

EXAMPLE 3 In Craig's family there are five birthdays in December, so he is shopping (in Monmouth County) for everyone at once. He buys his brother a shirt for \$24.99, his daughter two CD's at \$13.99 each and a sweatsuit for \$54, his father a drill for \$84.99, his wife a necklace for \$32.99, and his boss a book for \$22.99. Find the total cost of Craig's purchases, including sales tax.

This problem is similar to example 1 except that clothing is not taxable in New Jersey. To compute the sales tax on Craig's purchases, we need to distinguish the taxable and the non-taxable items.

Taxable Items	
2 CD's at \$13.99 each	= \$27.98
drill at \$84.99	= 84.99
necklace at \$32.99	= 32.99
book at \$22.99	= 22.99
Total taxable	= \$168.95

Non-Taxable Items	
shirt at \$24.99	= \$24.99
sweatsuit at \$54	= 54.00
Total non-taxable	= \$78.99

So far, Craig's purchases total \$168.95 + \$78.99 = \$247.94. However, sales tax is computed only on \$168.95. Our next step is to compute the sales tax:

SALES TAX is 6% of \$168.95

Percent = 6% = 0.06
Whole = \$168.95
Part = Tax = x

PART = PERCENT WHOLE

$$x = 0.06 \times \$168.95$$
$$x = 10.137$$

Again, round to the nearest penny. Tax = \$10.14

The sales tax on Craig's taxable purchases is \$10.14 and we can now compute his total cost:

Total taxable	\$168.95
Total non-taxable	78.99
Sales Tax	10.14
Total cost	\$258.08

Thus, the total cost of Craig's purchases, including sales tax, is \$258.08.

Check: We can check this by using rounding to estimate the totals. For the taxable items, the CD's are *about* \$15 each or \$30, the drill is *about* \$85, the necklace is *about* \$30, and the book is *about* \$20, for a total of *about* \$165. The tax on \$165 is *about* half of 10% or *about* \$9. For the non-taxable items, the shirt is *about* \$25 and the sweatsuit is *about* \$55, for a total of *about* \$80. The total, then, would be *about* \$165 + 9 + 80 or \$254.

♦ **You Try It**

3. A local sporting goods store is having a sale on skis and Jodi has decided to purchase a set of skis for \$149.99 and some ski clothing as well. She purchases a ski suit for \$84.99 and a pair of gloves for \$24.99. Find the total cost of her purchases, including the tax.

EXAMPLE 4 A Brookdale student was discussing her scores on the New Jersey College Basic Skills Placement Test (NJCBSPT) with her counselor. She was told that she had gotten 21 questions correct on the Computation section of the test and that this amounted to a score of 70%. How many questions are there on this part of the test?

This situation involves taking a test with a total number of questions and getting a percent of those questions correct. From experience, the percent sentence for this situation is

NUMBER CORRECT is a PERCENT of TOTAL QUESTIONS

In this case, we know:

Percent = 70% = 0.70
Number Correct = 21
Total Questions = Unknown = x

PART = PERCENT × WHOLE

$$21 = (0.70)(x)$$
$$21 = 0.70x$$

Divide both sides by 0.70: $\frac{21}{0.70} = \frac{0.70x}{0.70}$

$30 = x$

Thus, there are 30 questions on the Computation Section of the NJCBSPT.

Check: Check to see that 21 is 70% of 30. 10% of 30 is 3, so 70% of 30 is $7 \times 3 = 21$.

♦ **You Try It** 4. The GED (General Educational Development) Test is a test used as an alternative to a high school diploma for students who dropped out of high school. On average, about 72 percent of those who take the test pass it. If about 576,000 people passed the test last year, how many took it?

EXAMPLE 5 One of the top ranking players on a women's college soccer team scored seven goals in a 19 - 8 victory over another team. What percent of her team's goals did this player score?

In this situation, the percent sentence is

PLAYER'S GOALS are a PERCENT of TOTAL TEAM GOALS

Reading the information carefully, we know that the player scored 7 goals. To determine the total team goals, notice that the game was a *victory* for the player's team and the score was 19 to 8, so the winning team scored 19 goals. Therefore, the total number of goals is 19. Now the sentence becomes

7 is what percent of 19

We know
Percent = unknown = x (as a decimal)
Whole = total goals = 19
Part = player's goals = 7

PART = PERCENT × WHOLE

$7 = (x)(19)$

$7 = 19x$

Divide both sides by 19: $\frac{7}{19} = \frac{19x}{19}$

The result of the division to five places is $0.36842 = x$

Change the decimal to a percent $36.842\% = x$

In this situation it makes sense to round to the nearest whole percent, so the player scored approximately 37% of her team's goals.

Check: To compare 7 to 19, estimate by comparing 7 to 20, which is the same as 35 to 100 or 35%.

EXAMPLE 6 The American Heart Association has recommended that Americans should limit the amount of fat in their diet. Specifically, no more than 30% of their total calories should be from fat. In order to follow this recommendation, it is necessary to know what percent of a particular food is fat. To do that, you can use the food labeling information on the product but you need to know one more thing:

One gram of fat = 9 calories

We will determine the percent of fat in condensed tomato soup. According to the label, one serving has a total of 90 calories and contains 2 grams of fat. If we want to know what percent of the soup is fat, we need to know the number of fat calories so that we can compare them to the total calories. If one gram of fat is 9 calories then 2 grams of fat is 2×9 or 18 calories. So the tomato soup has 18 fat calories out of a total of 90 calories. In this situation the percent sentence is

FAT CALORIES is a PERCENT of TOTAL CALORIES

In this example, 18 calories is a percent of 90 calories.
We know:
Percent = unknown = x (as a decimal)
Whole = 90 calories
Part = 18 calories

$$\text{PART} = \text{PERCENT} \times \text{WHOLE}$$
$$18 = (x)(90)$$
$$18 = 90x$$

Divide both sides by 90:
$$\frac{18}{90} = \frac{90x}{90}$$
$$0.2 = x$$

Change the decimal to a percent:
$$20\% = x$$

Thus, 20% of the calories in tomato soup are fat calories.

Check: To compare 18 to 90, estimate by comparing 20 to 100, which is 20%.

♦ **You Try It** **5.** Milk that is labeled 2% Milk contains the following nutritional information: in a one cup serving there are 100 calories and 2 grams of fat. Determine fat calories as a percent of total calories in 2% milk.

EXAMPLE 7 Sharon is a chemistry student who is working with a solution that is labeled "5% Hydrochloric Acid". Suppose she pours 600 ml of this solution into a beaker.
a. What percent of the solution is acid?
b. What percent is water?
c. How much acid is in the beaker?
d. How much water is in the beaker?
e. If Sharon adds 200 ml of water to the beaker, is it still a 5% solution? Explain.

a. Since the solution is labeled "5% Hydrochloric Acid", then **5%** of the solution is acid.

b. Using complements, if 5% of the solution is acid, the 100% – 5% = 95% of the solution must be water.

c. If 5% of the solution is acid, we can use the percent sentence

5% of TOTAL SOLUTION is ACID

We know:
Percent = 5% = 0.05
Total Liquid = 600 ml
Amount of Acid = x

$$5\% \times \text{TOTAL SOLUTION} = \text{ACID}$$
$$0.05 \times 600 \text{ ml} = x$$
$$30 = x$$

There are **30 ml** of acid in the beaker.

d. If the beaker has a total of 600 ml of solution and 30 ml are acid, then we can find the amount of water by subtraction:

$$\text{Water} = \text{Total} - \text{Acid}$$
$$\text{Water} = 600 \text{ ml} - 30 \text{ ml}$$
$$\text{Water} = 570 \text{ ml}$$

There are **570 ml** of water in the beaker.

e. If Sharon adds 200 ml of water to the beaker, the amount of **water** in the beaker changes, and the total amount of solution changes, but the amount of **acid** in the beaker does not change.
We have:
Amount of acid in beaker = 30 ml
Amount of water in beaker = 570 ml + 200 ml = 770 ml
Total amount of solution in beaker = 600 ml + 200 ml = 800 ml

The solution is no longer a 5% solution because the *same* amount of acid, 30 ml, is now a percent of a *larger* total, 800 ml. To see what percent solution Sharon has now, use the percent sentence

PERCENT of TOTAL SOLUTION is ACID

Percent = x (as a decimal)

$$\text{PERCENT} \times \text{TOTAL SOLUTION} = \text{ACID}$$
$$(x)(800) = 30$$
$$800x = 30$$

Divide both sides by 800 $$\frac{800x}{800} = \frac{30}{800}$$

$$x = 0.0375$$

Change the decimal to a percent: $x = 3.75\%$

So by adding 200 ml of water, Sharon has made a **3.75%** solution.

♦ **You Try It** 6. Saline solution is a mixture of sodium chloride (salt) and water. Saline is commonly used to mix medications and this solution is 0.9% sodium chloride. In a bottle of 500 ml of this saline solution,
a. What percent is sodium chloride?
b. What percent is water?
c. How much sodium chloride is in the bottle?
d. How much water is in the bottle?

EXAMPLE 8 **Chapter Problem** Read the problem at the beginning of this chapter.

a. The United States consumes 31.2% of the 55.2 million barrels of oil used worldwide each day. How many barrels is this?

Percent = 31.2% = 0.312
Whole = total barrels used worldwide = 55.2 million barrels
Part = number of barrels consumed by the United States = x

$$\text{PART} = \text{PERCENT} \times \text{WHOLE}$$

$$x = 0.312 \times 55.2 \text{ million}$$

$$x \doteq 17.2 \text{ million}$$

The United States consumes approximately 17.2 million barrels of oil each day.

b. The United States has about 6% of the world's population. It consumes 31.2% of the oil used in the world each day. About how many times more oil does the United States consume daily when compared to its population size?

Find how many times 6% is contained in 31.2%. This is a division question.

$$\frac{31.2\%}{6\%} = \frac{0.312}{0.06} \doteq 5$$

The United States consumes about 5 times its share of oil.

♦ **You Try It**

7. Refer to the table at the beginning of this chapter.
   a. The Middle East produces 37% of the 57.6 million barrels of oil produced daily in the world. How many barrels is this?
   b. About how many times more oil does the Middle East produce daily compared to the United States?

---

2 CIRCLE GRAPHS

Circle graphs, sometimes called *pie charts*, are often used to show how a particular amount subdivides into categories. The circle represents the whole amount and each category is represented by a wedge of the "pie". The size of the wedge corresponds to the ratio or percent each category is of the total. In the circle graph in Figure 7.3, the circle represents an entire budget of approximately $29,300,000. Each wedge is a percent of this total budget. Since the wedges use up the entire circle, the sum of all the percents in a circle graph should be 100%.

FIGURE 7.3 Domestic Market Share of Fast-Food Hamburger Chains in 1992

TOTAL SALES = $87 BILLION

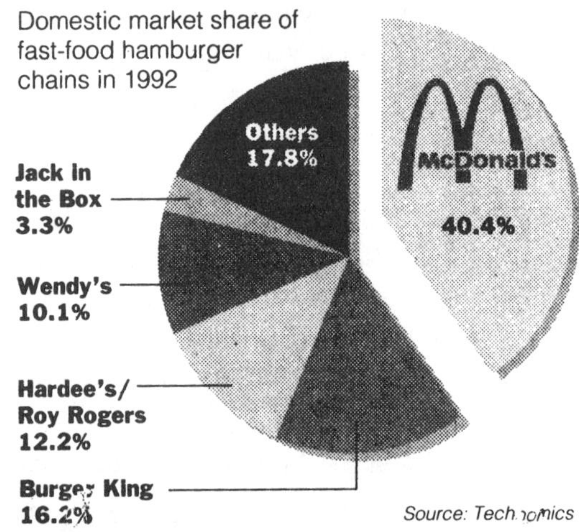

EXAMPLE 9 Use the circle graph in Figure 7.3 to answer each question.

a. What was McDonald's share of the total sales in 1992?

McDonald's earned 40.4% of the \$87 billion total, so we know:

Percent = 40.4% = 0.404
Whole = Total Sales = \$87 billion
Part = McDonald's share = x

$$\text{PART} = \text{PERCENT} \times \text{WHOLE}$$
$$x = 0.404 \times 87 \text{ billion}$$
$$x \doteq 35.1 \text{ billion}$$

McDonald's share of the \$87 billion total was about \$35.1 billion.

b. About how many times larger was McDonald's share than Wendy's?

One way to answer this is to find Wendy's share first:

Wendy's Share = 10.1% of Total

Wendy's Share = x
Total = \$87 billion

$$x = 0.101 \times \$87 \text{ billion}$$
$$x \doteq 8.8 \text{ billion}$$

Wendy's share was \$8.8 billion.
Now to see how many times larger McDonald's share was than Wendy's, we need to divide:

$$\frac{35.1 \text{ billion}}{8.8 \text{ billion}} \doteq 4$$

McDonald's earned about four times as much as Wendy's.

Notice, however, that we could have gotten this answer directly from the percent share for each:

$$\frac{40.4\%}{10.1\%} = 4$$

♦ **You Try It** **8.** Use the circle graph in Figure 7.3 to answer each question.
a. What was Burger King's share of the total sales?
b. About how many times larger was Burger King's share than Jack in the Box?

♦**Answers to You Try It** **1. a.** 24 students passed the course. **b.** 6 students earned CRHH. **2.** The sales tax is \$8.40 and the total cost is \$148.40. **3.** The tax on the skis is \$9.00 and the total cost is \$268.97. **4.** About 800,000 people took the test. **5.** 2% milk is 18% fat. **6. a.** 0.9% is sodium chloride **b.** 99.1% is water **c.** 4.5 ml is sodium chloride **d.** 495.5 ml is water **7. a.** The Middle East produces 21.3 million barrels of oil. **b.** $\frac{37\%}{18.9\%} \doteq 2$. The Middle East produces about twice the oil the United States produces daily **8. a.** Burger King's share was \$14.1 billion. **b.** Burger King earned about five times as much as Jack in the Box.

SECTION 7.7 EXERCISES

For each problem, write the percent sentence, translate into an equation, and then solve. Be sure to give a complete answer.

1. Thirty-eight percent of the employees for the Mutual Insurance Company are salespeople. Mutual employs 5,500 people. How many are in sales?

2. There are 40 pizza shops in town. Thirty percent of them advertise in the local newspaper. How many advertise in the local paper?

3. Maria claims that 10% of the women who belong to her country club are doctors. If 23 women are doctors, how many women belong to the club?

4. In the school, 760 students are science majors. This is 95% of all the students attending the school. How many students attend the school?

5. In a survey of 300 children, 213 preferred playing with toy widgets. What percent of the children preferred the widget?

6. Sam received a check for $615.25. He used $516.81 to purchase an oak desk. What percent of his check was spent on the desk?

7. The tax on the purchase of a stereo is $44.00. If this is 5% of the purchase price, what was the price of the stereo?

8. A football team won 10 games, which was 62.5% of the games the team played. How many games did the team play?

9. A congresswoman polled 1,500 of her constituents on the question of a tax increase. 645 favored a tax increase.

a. What percent of those polled favored the increase?

b. What percent did not favor the increase?

10. In a class of 30 students, 6 received a grade of A.

a. What percent of the class received an A grade?

b. What percent received a different grade?

11. $\frac{2}{3}\%$ of an ore is gold. If 600 pounds of the ore is mined, how many pounds of gold can you expect?

12. 85% of a donation of $4,200 will be used to clothe the needy. How much of the donation will be used to clothe the needy?

13. A builder constructed a home and sold it for \$145,000. If 16% of the sale price of the home is the builder's profit, how much profit did the builder make?

14. A rat weighing 36 ounces ate 16 ounces of food today. What percent of the rat's body weight does this food represent?

15. In a class of 35 students, 14 students received a grade of B. What percent of the class received a grade of B?

16. John lost 10 pounds on a fat free diet. His goal is to lose 25 pounds. What percent of his goal has John reached?

17. Sal uses 38% of his net salary to pay his home mortgage. If Sal's net salary is \$1,450, how large is his mortgage payment?

18. Alberta sold her home for \$66,500. If 40% of this is profit, how much profit has she made on the sale of her home?

19. George weighs 92% of what his doctor says is his ideal weight. If George weighs 138 pounds, what is his ideal weight?

20. If 0.17% of the weight of the human body is chlorine, what would a body weigh that contained 0.306 pound of chlorine?

21. A 3-ounce piece of lean beef contains 77 milligrams of cholesterol. A 3-ounce piece of beef liver contains 483% of the cholesterol in lean beef. How much cholesterol is contained in 3 ounces of beef liver?

22. One frankfurter contains 27 milligrams of cholesterol. If one egg contains 1,000% of the cholesterol in a frankfurter, how many milligrams of cholesterol are in one egg?

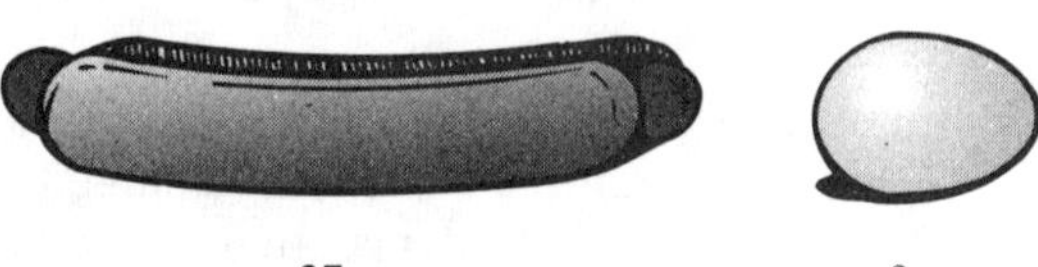

Reword each problem to contain the phrase "percent of," then solve.

23. Ed won \$180 on a game show. He spent \$65 on a fancy dinner afterward. What percent did he spend on the dinner?

24. There are 4,500 cats in this city. 750 are strays. What percent are strays?

25. Last week 1,500 men were unemployed. 70% found work. How many found a job?

26. Gary purchased 175 pounds of bananas to sell to his customers. 8% spoiled before they were sold. How many pounds spoiled?

27. Cathy sold her cabin cruiser for an unknown sum of money. She used 20%, or $35,000, to buy a catamaran. She banked the rest. How much did Cathy get for the cabin cruiser?

28. A solution is 12% acid. If there are 15 liters of acid in the solution, how many liters of solution are there?

29. Lucy purchased 80 flowering plants and 5% died. How many died?

30. 4,500 people work for Munks Stores. If 3.2% will retire this year, how many Munk employees will retire this year?

31. Paul and Kim dined out this evening. Their dinner bill was $37.60. They left a 15% tip. How large was the tip?

32. Mike and Debbie want to leave a 20% tip for their waiter. If the bill was $62.40, how much of a tip do they leave?

2 *Use Figure 7.4to answer questions 33-37.*

FIGURE 7.4
Family Budget for a Monthly Income of $2,800

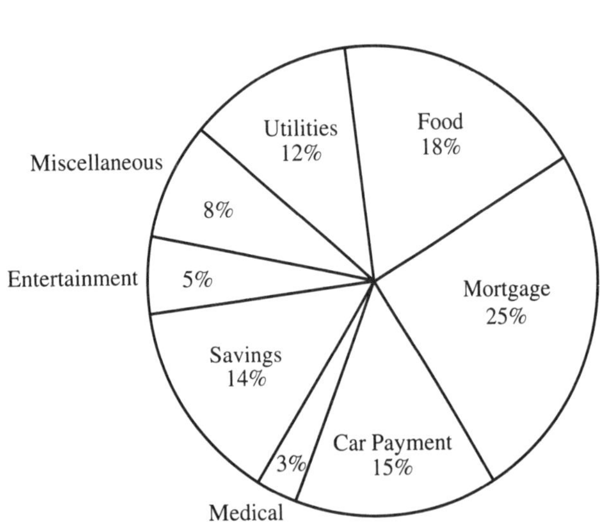

33. How much is budgeted for utilities each month?

34. What is the ratio of money budgeted for entertainment to money budgeted for the mortgage?

35. How much more money is budget for food than for utilities each month?

36. What fraction of the budget is the mortgage?

37. Suppose the monthly income increases to \$3,200. If the percents in Figure 7.4 do not change, how much additional money will go into savings each month?

SKILLSFOCUS (Section 2.5) *Divide. Round to hundredths.*

38. $7.6 \div 0.19$

39. $48.9 \div 3$

40. $0.8 \div 1.7$

41. $4 \div 6.3$

EXTEND YOUR THINKING ▶▶▶▶

▶SOMETHING MORE

42. Your medical insurance covers 80% of the cost of your surgery. You must pay the rest, which amounts to \$450. What is the total charge for the surgery?

43. In a group, 20% are adults, and the rest are children. Seventy percent of the adults and forty percent of the children will be attending a seminar. What percent of the group will attend the seminar?

44. A farmer sold 50 pounds of broccoli for \$20 to a neighbor. The neighbor sold it to a wholesaler for \$30. The wholesaler increased the price by 25% and sold it to a retailer. The retailer marked the price up by 40% to sell to his customers. What price per pound will the retailer charge?

7.8 COMMISSION, DISCOUNT, AND PERCENT INCREASE AND DECREASE

OBJECTIVES

1. Solve commission problems.
2. Solve discount problems.
3. Solve percent increase and decrease problems.

NEW VOCABULARY

commission	discount
straight commission	percent increase
salary plus commission	percent decrease

1 COMMISSION Although many employees earn an hourly wage, some, primarily salespersons, work on a commission basis. This means that all or part of their salary is a percent of their total sales. The amount earned is called the **commission** and the percent is called the **commission rate**. The percent sentence is

COMMISSION is a PERCENT of TOTAL SALES.

If the salesperson works on a **straight commission** then the person's total salary is from commission. However, some employees work on a **salary plus commission** basis. That means that they get a fixed salary (no matter what their total sales) plus a commission.

EXAMPLE 1 Ted works at a local car dealership and earns a 20% commission on new cars that he sells. What commission does he earn if he sells a new car for $18,450?

Reading the problem carefully, we see that Ted's commission rate is 20%, the selling price of the car is $18,450, and we need to calculate Ted's commission.

We know
Percent = 20% = 0.20
Total Sales = 18,450
Commission = x

The sentence — COMMISSION is a PERCENT of TOTAL SALES.
leads to the equation — COMMISSION = PERCENT × TOTAL SALES
so we have

$$x = 0.20 \times 18{,}450$$
$$x = 3{,}690$$

Thus, Ted's commission is $3,690.

Check: To estimate, round 18,450 to 18,500. 10% of 18,500 is 1850, so 20% of 18,500 is 2 × 1850 or 3700.

♦ **You Try It** 1. Alma works for a large insurance company and earns a commission based on her clients' yearly premiums. When she sells an ordinary life insurance policy, she earns, for the first year, a commission of 55% of the premium. She has just sold an ordinary life insurance policy worth $250,000 and the yearly premium is $5,025. What will Alma earn for the first year of this policy?

EXAMPLE 2 Jennifer is a salesperson at an electronics store and she specializes in selling large-screen televisions. She earns a commission of $6\frac{1}{2}\%$ on every large-screen television she sells. If she wants to earn no less than \$700 per week, what would her total sales need to be? If the average price of a large-screen TV is \$2,000, how many would she have to sell each week to meet her goals?

There are two questions here and we will focus first on the question of finding the total sales if she wants her weekly commission to be \$700.

We know

Percent $= 6\frac{1}{2}\% = 6.5\% = 0.065$
Total Sales $=$ unknown $= x$
Commission $= \$700$

The sentence — COMMISSION is a PERCENT of TOTAL SALES.
leads to the equation — COMMISSION = PERCENT × TOTAL SALES

$$700 = (0.065)(x)$$
$$700 = 0.065x$$

Divide both sides by 0.065:
$$\frac{700}{0.065} = \frac{0.065x}{0.065}$$
$$10{,}769.23077 = x$$

In this situation, it is reasonable to round to the nearest hundred dollars, so Jennifer should have about **\$10,800** in total sales of large-screen televisions in order to earn \$700 per week.

We now need to find out how many large-screen televisions Jennifer must sell in order to have \$10,800 in sales each week. Since each large screen television costs about \$2,000, we want to know how many \$2,000's are in \$10,800, so we need to divide 10,800 by 2,000:

$$10{,}800 \div 2{,}000 = 5.4$$

Of course, it is impossible to sell 5.4 televisions but from this answer we can see that Jennifer must sell at least **six** large-screen televisions each week to meet her income goals.

♦ **You Try It** **2.** Alfonso earns an 8% straight commission on his sales at an appliance store. Last month he earned \$2,700. What were his total sales for the month?

EXAMPLE 3 A real estate agency has just sold a house for \$255,000, earning a commission of \$17,850. What is the agency's commission rate?

In this case, we know the commission, \$17,850, and the total sales, \$255,000, and we want to find the commission rate.

We know
Percent $= x$ (as a decimal)
Total Sales $= \$255{,}000$
Commission $= \$17{,}850$

The sentence leads to the equation

COMMISSION is a PERCENT of TOTAL SALES.
COMMISSION = PERCENT × TOTAL SALES

$$17{,}850 = (x)(255{,}000)$$
$$17{,}850 = 255{,}000x$$

Divide both sides by 255,000: $$\frac{17{,}850}{255{,}000} = \frac{255{,}000x}{255{,}000}$$

$$0.07 = x$$

Change the decimal to a percent: $7\% = x$

Thus, the agency earned a **7%** commission on the sale of the house.

♦ **You Try It** **3.** A salesperson earns a commission of $648 during a month when her total sales were $14,400. What is her commission rate?

EXAMPLE 4 A shoe salesman earns a flat salary of $1,600 per month plus a 4% commission on his total sales. One month his total sales were $7,655. What did he earn that month?

In this situation the salesman's earnings come from two sources: a flat salary and a commission of total sales. We already know the monthly salary, so before we can find the total monthly income, we need to find the salesman's commission for the month.

Outline:
1. Find the commission
2. Find the total monthly income

Estimate:
To estimate the commission, find 5% of $7,600. 10% of 7,600 is 760, so 5% of 7,600 is 380. The salesman's commission is *about* $380 and his flat salary is $1,600, so his total income for the month is *about* $380 + 1,600 = $1,980.

Calculate:
1. Find the commission. We know:

   Percent = 4% = 0.04
   Total Sales = $7,655
   Commission = x

   COMMISSION is a PERCENT of TOTAL SALES.
   COMMISSION = PERCENT × TOTAL SALES

   $$x = 0.04 \times 7{,}655$$
   $$x = 306.2$$

   The commission is $306.20.

2. Find the total monthly income.

   Total monthly income = Salary + Commission
   Total monthly income = $1,600 + $306.20
   Total monthly income = $1,906.20

This shoe salesman earned a total monthly income of $1,906.20.

♦ **You Try It**

4. Nancy works for a computer software company, selling software to businesses in New Jersey. She earns a monthly salary of \$2,300 plus a commission of 6% of her total sales. One month her total sales were \$32,000. Find her total income for that month.

2 DISCOUNT

This is probably one of the more familiar applications of percent since you probably buy things on sale all the time. When you buy something on sale, you get a **discount**, which is a **percent** of the **original price**. The percent sentence in this situation is

DISCOUNT is a PERCENT of ORIGINAL PRICE.

The percent is usually called the **discount rate.**

EXAMPLE 5 A television set that normally sells for \$699.99 is on sale for 15% off. What is the sale price?

It is important to recognize that in this problem we have **two** things to do, find the discount and then find the sale price. We should outline this problem and estimate the answer before doing the calculations.

Outline:

1. Find the discount.
2. Find the sale price.

Estimate:

1. The discount is 15% of *about* \$700. 10% of 700 is 70, so 5% of 700 is 35. Then 15% of 700 is 70 + 35 or 105.
2. The sale price is *about* 700 – 100 = 600.

Calculate:

1. To find the discount, we know:

Percent = 15% = 0.15
Original Price = \$699.99
Discount = x

DISCOUNT is a PERCENT of ORIGINAL PRICE.
DISCOUNT = PERCENT × ORIGINAL PRICE

$$x = 0.15 \times 699.99$$
$$x = 104.9985$$

Since this is money, round to the nearest penny: $x = \$105.00$

Thus, the discount is \$105.00.

2. To find the sale price, subtract the discount from the original price:

Sale price = original price – discount
Sale price = \$699.99 – 105.00
Sale price = \$594.99

Thus, the sale price of the television is \$594.99.

OBSERVE When you are given the rate of discount, it is natural to find the discount first and then subtract from the original price to find the sale price. However, we can get the sale price more directly by using **percent complements**. If you are getting 15% **off** of the price of the television, then you will **pay** 85% of the original price of the television. In other words, in this problem

SALE PRICE is 85% of ORIGINAL PRICE

To see that it works, Sale price = 0.85×699.99

Sale price = 594.9915

To the nearest penny, the sale price is $594.99, which is the same answer we got before.

♦ **You Try It** **5.** A refrigerator that normally sells for $850 is on sale for 20% off. Find the sale price. If you bought the refrigerator on sale, what would your final cost be, including sales tax?

EXAMPLE 6 A leather coat that usually sells for $420 is on sale for $315. What is the discount rate?

Here we need to be rather careful. If we want to find the discount rate, we should use the percent sentence:

DISCOUNT is a PERCENT of ORIGINAL PRICE.

but in order to use this sentence to find the discount rate (percent), we need to know the **discount** and the **original price**. In the statement of the problem, we are given the **original price** ($420) and the **sale price** ($315). We do not have the **discount**. However, we can easily find the discount by subtracting the sale price from the original price. Once we have the discount we can use the percent sentence above to find the discount rate.

Outline:

1. Find the discount.

   Method: Discount = Original Price − Sale Price

2. Find the discount rate.

   Method: Discount = Percent × Original Price

Estimate:

1. The discount is *about* 420 − 320 or $100.
2. The discount rate is the discount, $100, compared to the original price, *about* $400, or 25%.

Calculate:

1. Find the discount: Discount = Original Price − Sale Price

   Discount = $420 − $315

   Discount = $105

2. Find the discount rate:

   Discount = Percent × Original Price

   Discount = 105
   Percent = x (as a decimal)
   Original Price = 420

$$105 = (x)(420)$$
$$105 = 420x$$

Divide both sides by 420: $\dfrac{105}{420} = \dfrac{420x}{420}$

$$0.25 = x$$

Change the decimal to a percent: $25\% = x$

Thus, the leather coat was on sale for **25%** off.

♦ **You Try It**

6. A queen size bedspread that regularly sells for \$70 is on sale for \$39. What is the discount rate?

EXAMPLE 7 Doug bought a Star Trek model for his cousin at a 15% discount and saved \$3.75. What was the original price of the model? What was the sale price?

We can use the percent sentence

DISCOUNT is a PERCENT of ORIGINAL PRICE

We know:

Percent = 15% = 0.15
Original Price = x
Discount = \$3.75

So we have

$$\text{DISCOUNT} = \text{PERCENT} \times \text{ORIGINAL PRICE}$$
$$3.75 = (0.15)(x)$$
$$3.75 = 0.15x$$

Divide both sides by 0.15:

$$\frac{3.75}{0.15} = \frac{0.15x}{0.15}$$
$$25 = x$$

Thus, the original price of the model was \$25.00.

To answer the second question, the sale price is the discount subtracted from the original price:

Sale Price = Original Price − Discount
Sale Price = 25.00 − 3.75
Sale Price = \$21.25

So Doug paid \$21.25 for the model on sale.

♦ **You Try It**

7. Marla bought a silver necklace at a 30% discount and saved \$12. What was the original price of the necklace? What did Marla pay for the necklace on sale?

3 PERCENT INCREASE AND DECREASE

The next application of percents is one of the most commonly encountered in the news -- percent increase or decrease. Unlike the other applications we have worked with in this section, percent increase or decrease is very general. It applies to any situation in which there has been a change, either an increase or a decrease. For example, if you get a raise in pay, the percent of your raise is the **percent increase** in salary. If the enrollment at a school goes down, the percent is the **percent decrease** in enrollment. If a company's sales go down, the percent is the **percent decrease** in sales, or the **percent loss**. The percent sentence for these situations is

The INCREASE is a PERCENT of the ORIGINAL AMOUNT.

or

The DECREASE is a PERCENT of the ORIGINAL AMOUNT.

EXAMPLE 8 In 1985, the Federal Government counted 22,250 new cases of tuberculosis. In 1992, the count was 26,700 new cases. What was the percent increase in the number of new cases of tuberculosis?

Since we are looking for the percent increase, we will use the sentence

The INCREASE is a PERCENT of the ORIGINAL AMOUNT.

Reading the information carefully, we know

Number in 1985 = 22,250 = original amount
Number in 1992 = 26,700 = new number

The percent sentence requires the **increase** and the **original amount**. We have the original amount and the new amount, so we need to first find the actual **increase** in the number of cases of tuberculosis.

Outline:

1. Find the actual increase in new tuberculosis cases.
   Method: Increase = New number − Old number
2. Find the percent increase.
   Method: Increase = Percent × Original Amount

Calculation:

1. Find the increase

   Increase = New number − Old number
   Increase = 26,700 − 22,250
   Increase = 4,450

2. Find the percent increase

   INCREASE is a PERCENT of ORIGINAL AMOUNT
   INCREASE = PERCENT × ORIGINAL AMOUNT

   We know
   Percent = x (as a decimal)
   Original Amount = 22,250
   Increase = 4,450

$$4{,}450 = (x)(22{,}250)$$
$$4{,}450 = 22{,}250x$$

Divide both sides by 22,250: $$\frac{4{,}450}{22{,}250} = \frac{22{,}250x}{22{,}250}$$

$$0.2 = x$$

Change the decimal to a percent: $20\% = x$

Thus, there was a **20%** increase in the number of new tuberculosis cases from 1985 to 1992.

♦ **You Try It**

8. As part of a crackdown on violent crime, government officials are considering a sharp increase in the Federal licensing fee for gun dealers. In 1968 the fee for a three-year license was set at \$30. In February 1994, the fee for a three-year license rose to \$200. A proposal is under consideration to raise the fee even further to \$600.

a. Find the percent increase in the fee from 1968 to 1994.
b. Find the percent increase in the fee from 1994 to the \$600 amount.
c. Find the percent increase in the fee from 1968 to the \$600 amount.

EXAMPLE 9 Last year, Harold was laid off from his job at an automobile factory where he had earned \$36,400 a year. He has found a job as a clerical worker, but the salary is only \$24,500 a year. What is the percent decrease in Harold's salary?

Since we are looking for the percent decrease, we can use the percent sentence

The DECREASE is a PERCENT of the ORIGINAL AMOUNT.

In this problem, we know Harold's old salary and his new salary, so we need to find the decrease in salary first.

Outline:

1. Find the actual decrease in Harold's salary.
   Method: Decrease = Old salary – New salary
2. Find the percent decrease.
   Method: Decrease = Percent × Original Amount

Calculation:

1. Find the decrease

$$\text{Decrease} = \text{Old salary} - \text{New salary}$$
$$\text{Decrease} = 36{,}400 - 24{,}500$$
$$\text{Decrease} = 11{,}900$$

2. Find the percent decrease

DECREASE is a PERCENT of ORIGINAL AMOUNT
DECREASE = PERCENT × ORIGINAL AMOUNT

We know
Percent = x (as a decimal)
Original Amount = 36,400 (Harold's old salary)
Decrease = 11,900

$$11{,}900 = (x)(36{,}4000)$$
$$11{,}900 = 36{,}400x$$

Divide both sides by 36,400: $$\frac{11{,}900}{36{,}400} = \frac{36{,}400x}{36{,}400}$$

$$0.326923 = x$$

Change the decimal to a percent: $$32.6923\% = x$$

An answer to the nearest whole percent is appropriate in this situation. There was a **33%** decrease in Harold's salary.

♦ **You Try It** **9.** As part of a fare war, one airline company lowered the fare on one route from \$350 to \$260. Find the percent decrease in the price of this ticket.

EXAMPLE 10 A retail store purchases a shipment of shirts, with each shirt priced at \$12, and marks them up 60% for retail sale. At the end of the season, the shirts that remain in the store are marked down 60%. Is this final price \$12? Why or why not?

Remember that when you calculate a percent, you find a **percent of a whole or base number**. Even if the percent is the same, if the whole or base number changes, the result will be different. In this example, the shirts are marked **up** by 60% of \$12, but they are marked **down** by taking 60% of the marked-up amount. Therefore the final price will not be the same. We can verify this by doing the calculations.

For the Mark-Up:

Increase in Price = 60% of \$12
Increase in Price = 0.60 × \$12
Increase in Price = \$7.20

New Price = Old Price + Increase
New Price = \$12.00 + 7.20
New Price = \$19.20

For the Mark-Down:

Decrease in Price = 60% of \$19.20
Decrease in Price = 0.60 × \$19.20
Decrease in Price = \$11.52

New Price = Old Price – Decrease
New Price = \$19.20 – 11.52
New Price = \$7.68

Since the 60% mark-**up** was based on only \$12, while the 60% mark-**down** was based on a higher amount, \$19.20, the final answer is much lower than the original \$12.

♦ **You Try It**

10. a. During a prolonged hospital stay, Janice's weight rose from 115 pounds to 140 pounds. What was the percent increase in her weight?

b. Now that Janice has recovered from her illness, she wants to get her weight back down to 115 pounds. What percent decrease will this be?

EXAMPLE 11 The teachers at a local school have negotiated a 6% raise in pay next year. One of the teachers earns \$29,400 this year. What will he earn next year?

Since a raise is an increase, we can use the percent increase sentence

INCREASE is a PERCENT of ORIGINAL AMOUNT.

or we can write the sentence specifically for this situation:

RAISE is a PERCENT of OLD SALARY.

We know
Percent = 6% = 0.06
Old Salary = \$29,400
Raise = x

RAISE is a PERCENT of OLD SALARY
RAISE = PERCENT × OLD SALARY
$x = 0.06 \times 29{,}400$
$x = \$1{,}764$

The raise in pay is \$1,764.

New Salary = Old Salary + Raise
New Salary = \$29,400 + 1,764
New Salary = \$31,164

The teacher's salary next year will be \$31,164.

OBSERVE We could compute the teacher's new salary in one step instead of two. Next year, the teacher will earn his old salary plus an increase of 6%. If we think of his old salary as **100%**, then next year he will earn 100% of his old salary plus 6% of his old salary, or a total of **106%** of his old salary. This gives us a new equation

New Salary = 106% of Old Salary
New Salary = 1.06 × 29,400
New Salary = \$31,164

This method gives the same answer with fewer steps.

♦ **You Try It** **11.** Suppose you earn \$6.10 per hour and your boss gives you a 7% raise in pay. Calculate your new hourly rate in two ways; using the method of example 11, and using the alternate method described above.

SOMETHING MORE

Real Estate Commission Sam works for the Sage Real Estate Agency. He has agreed to sell his next door neighbor's house. Sam then becomes the *listing agent* for the home, and Sage is the *listing agency*. If Sam sells this home, the agency makes a 7% commission on the selling price of the property. Sam and the agency split the commission, each getting $3\frac{1}{2}$%.

a. Sam sells the home for \$88,500. What commission is made?
b. How much of this commission does Sam get?
c. How much does Sage get?

William works for the Winter Real Estate Agency, a strong Sage competitor. Because Sage and Winter both belong to a *Multiple List Service*, the two agents can sell homes listed with either agency. Suppose William sells Sam's neighbor's home for \$95,000. A 7% commission is made on this sale. But because Winter sold a home listed for sale by the Sage agency, the two agencies split the 7% commission, each getting $3\frac{1}{2}$%.

Because Sam is the listing agent, Sam and Sage evenly split this $3\frac{1}{2}$%, each getting $1\frac{3}{4}$%. Because William sold the home, he and Winter split their $3\frac{1}{2}$%, each getting $1\frac{3}{4}$%.

d. How much commission is made on the sale of this home?
e. How much do Sam and Sage each receive?
f. How much do William and Winter each receive?

♦**Answers to You Try It** **1.** Alma earns a commission of \$2,763.75. **2.** Alfonso's total sales were \$33,750. **3.** The commission rate is 4.5%. **4.** The commission is \$1,920. The total income is \$4,220. **5.** The sale price is \$680. The tax is \$40.80. The final cost is \$720.80. **6.** The discount rate is approximately 44%. **7.** The original price of the necklace was \$40. The sale price was \$28. **8. a.** 567% increase **b.** 200% increase **c.** 1900% increase **9.** Approximately 26% **10. a.** Janice's weight increased approximately 22%. **b.** Janice's weight will decrease approximately 18% **11.** Your new hourly rate is \$6.53.

♦**Answers to Something More** **a.** \$6,195 **b.** \$3,097.50 **c.** \$3,097.50 **d.** \$6,650 **e.** \$1,662.50 **f.** \$1,662.50

SECTION 7.8 EXERCISES

1

1. Jack sells boats. He makes a 4.7% commission on each sale. What is his commission on the sale of a $92,000 powerboat?

2. A townhouse sold for $56,600. If the agency who sold the home made a 6% commission on the sale, what commission did they earn?

3. Arnold is paid a 15% commission on everything he sells. If he needs $1,200 per month to meet his financial obligations, how much merchandise must he sell monthly?

4. What sales must Janet make in one week to earn $1,170 in commission? Her commission rate is 7.8%.

5. Barb works for a marina. She works on an 8% commission. This month she sold a $28,000 catamaran and a $46,000 speedboat. What was Barb's total commission for the month?

6. Sally made a 3.5% commission on the sale of a $349,000 mansion. What commission did Sally make?

7. A real estate agency sold a 3-acre piece of land. The agency made a $5,450 commission on the sale. If the rate of commission was 10%, what did the land sell for?

8. A 20% commission was made on the sale of a diamond ring. If the commission was $255, what did the ring sell for?

9. A $2.78 commission was made on a $55.60 sale. What was the commission rate?

10. What is the commission rate if a $6,507 commission is made on the sale of a $144,600 home?

11. Ed is paid a fixed salary of $200 per month plus a 12% commission on sales. What is Ed's July salary if his sales for July totaled $5,620?

12. Irma sold $2,250 in jewelry this week. If her salary is $125 per week plus a 5.4% commission on sales, what did Irma earn this week?

13. Jerry made $920 this week selling cars. If he is paid a fixed salary of $200 per week plus a 4% commission on sales, what were his total car sales for the week?

14. Howard earns a fixed salary of $40 per week plus a 16% commission on sales. What were his sales for the week if his salary was $200?

2

15. A $900 sofa is being sold at a 35% discount.
    a. What is the amount of discount?
    b. What is the sale price of the sofa?

16. A $450 pair of skis is on sale at a 40% discount.
    a. What is the amount of discount?
    b. What is the sale price of the skis?

17. The Charles Men's Shop is having a storewide clearance sale. Every item is being discounted 40%. For what sale price can you purchase the following items?
    a. A pair of slacks regularly selling for $32.
    b. A tie that regularly sells for $11.50.
    c. A suit priced at $249.99 (round to the nearest cent).

18. To celebrate customer appreciation day, a department store is offering 20% off on every item it stocks. For what price can you purchase each item?

    a. A $40 pair of slacks.

    b. A watch that regularly sells for $29.95.

    c. A basketball net for $3.69 (round to the nearest cent).

19. Frances works for a local department store. She wants to buy a TV regularly priced at $438. The TV is on sale for 20% off. Frances gets another 20% off of the sale price because she is an employee of the store. What will she pay for the TV set?

20. Jason works for a department store. A coat regularly selling for $180 is on sale for 25% off. Jason gets an additional 15% off of the sale price because he is a store employee. What does Jason pay for the coat?

21. Fred puts a microwave oven on sale for $238. If the regular price is $340, what percent discount is Fred offering?

22. Alice sells a boom box for $180. Her competition sells the same one for $135. What discount rate must Alice offer to meet her competitor's price?

23. A stereo is discounted 20%, giving a savings of $173.

    a. What is the original selling price of the stereo?

    b. What is the sale price?

24. A set of china is discounted 60%, resulting in a savings of $330.

    a. What was the original price of the china?

    b. What is the sale price?

3

25. If George pays his electric bill before November 2, he must pay $55.60. After that date he must pay $58.38. What percent increase is this?

26. A wool sweater regularly selling for $48 was reduced to $16 because of a flaw. What percent reduction is this?

27. A home selling for $130,000 was reduced to $117,000 for a quick sale. By what percent was the home reduced in price?

28. A book originally selling for $14 was sold in a clearance sale for $10.50. What percent decrease in price is this?

29. George made $800 last month. He estimates he will make $1,000 this month.

    a. What is the expected percent increase in earnings?

    b. George only made $940 this month. What percent increase did he actually make?

30. Bonnie made $50 last summer selling lemonade. This summer she expects to make $80.

    a. What percent increase in sales does Bonnie expect to make?

    b. This summer Bonnie only made $70. What percent increase in sales did she actually make over last summer?

31. **Depreciation** Depreciation is the loss in value of an item due to old age or obsolescence. Suppose a new car depreciates 30% in the first year.

    a. How much will a new $23,400 car depreciate in its first year?

    b. What will it then be worth?

32. **Appreciation** Appreciation is the increase in value of an item over time. A \$180,000 home appreciates 16% in one year.

 a. How much has the home appreciated?

 b. What is the home then worth?

33. Wolfe's Bakery purchased a \$3,600 oven. It is expected to depreciate 20% in the first year.

 a. How much will the oven depreciate in the first year?

 b. What will it then be worth?

34. A car dealer pays the manufacturer \$14,000 for a new car. The dealer increases this price by 22% to get the base sticker price.

 a. How much of an increase is this?

 b. What is the base sticker price?

35. A house sold for \$18,000 nine years ago. Today it sells for \$44,000. What percent increase in value is this?

36. According to the U.S. Bureau of Statistics, a certain basket of groceries cost \$10 in 1967. In 1987, the same groceries cost \$29.60. What is the percent increase in cost?

37. a. Willy weighed 144 pounds in the spring. Due to a hectic summer, his weight dropped to 120 pounds. What percent decrease in weight is this?

 b. Willy weighed 120 pounds in the fall. During the winter, his weight rose to 144 pounds. What was Willy's percent increase in weight?

38. a. A \$250 investment grew to \$400 in one year. What was the percent gain?

 b. A \$400 investment was worth \$250 in one year. What was the percent loss?

39. If George makes \$20,500 a year, and receives a 5.5% increase in salary, what is his new yearly salary?

40. A company pays its employees \$9.40 per hour and decides to increase the hourly wage by 5%.

 a. What is the new hourly wage?

 b. If an employee works a 40-hour week, how much more money will the employee be making per week?

SKILLSFOCUS (Section 2.7) *Evaluate each formula.*

41. Find M if $M = 6S$ and $S = 7$.

42. Find Y if $Y = 2L + 2W$ and $L = 3$ and $W = 8$.

43. Find I if $I = PRT$ and $P = 100$, $R = 0.06$, and $T = 2$.

44. Find D if $D = 16T^2$ and $T = 5$.

EXTEND YOUR THINKING ▶▶▶

▶SOMETHING MORE

45. Alida sold a home for \$175,000. She was the listing agent for the home, and her company was the listing agency. She earned a 7% commission on the sale.

 a. How much commission was made on the sale?

 b. How much did Alida get? How much did her company get?

46. Connie, who works for CDE Real Estate, sold a townhome for \$72,450. An 8% commission was made on the sale. The listing agent for the home was Tara, who works for TUV, the listing agency.

 a. What commission was made on the sale?

 b. How much did Connie, CDE, Tara, and TUV each receive?

47. **Lots for Sale** Anita owned two building lots. She sold them for $24,000 each. She sold one for 25% less than she originally paid for it. She sold the other for 25% more than she originally paid for it.

a. What did Anita originally pay for each lot?

b. Did she make a profit or loss on this transaction, or did she break even?

48. **Baseball Averages** This year Ben has a baseball batting average of .320. Since .320 = 32%, this means he got a hit 32% of the time he came to bat.

a. Ben came to bat 650 times. How many hits did he get?

b. Next season, Ben wants to increase his batting average to 0.380. If he comes to bat 650 times next season, how many more hits will he need?

c. A baseball player started his season with 5 hits in his first 20 at bats. His current batting average is $\frac{5}{20} = .250$. How many hits must he make in the next 30 times at bat to raise his average to .360?

49. **Flogging** In 1989, Hong Kong decided to abolish the use of flogging as a form of corporal punishment. The punishment, in which a male offender was beaten at most 18 times with a cane in the presence of a doctor, declined from 476 floggings in 1952 to only 2 in 1988. What was the percent decrease in floggings in this period of time?

50. **Inflation** Inflation refers to a time of rising prices for goods and services. Suppose an economic study concludes that a family of four needs $36,000 a year to maintain a moderate standard of living. Over the next year, the inflation rate is 4.2%. This means the overall price of goods and services in the economy increases 4.2%.

a. How much more money will the family of four need per year to maintain their standard of living?

b. What yearly income would they then have?

c. The inflation rate for the following year is 5.5%. What increase in annual income is needed for the family to continue to maintain their standard of living?

WRITING TO LEARN ▶▶▶▶

51. Write a commission word problem about Mary who makes an 8.6% commission on each yacht she sells. Then write the solution.

▶ YOU BE THE JUDGE

52. Which is more, $\frac{1}{2}$ of a million dollars, or $\frac{1}{2}$% of a million dollars? Explain your decision.

53. Ron earns $100 per week. His employer increased his salary by 10%. A month later, Ron's employer decreased his salary by 10%. Ron claims this is unfair. Is he right? Explain your decision.

54. **Farmer Prints His Own Money** Recently an evening news program reported that a farmer, unable to get a loan from a bank, decided to print his own money. Actually, to raise money, the farmer printed a note valued at $9, and sold it for $9. Later, when the farmer harvested his crops, the owner of the note could use it to purchase $10 worth of produce from the farmer. The farmer raised money to stay in business, and at the same time paid out a profit in food to his investors.

a. The reporter for this story said that the farmer paid a 10% profit in food on each one of his notes. Is this correct? If not, what do you think is the actual profit?

b. Give an example of a sale price and a redemption value for this note that results in a 10% profit being made.

7.9 SIMPLE AND COMPOUND INTEREST

OBJECTIVES

1. Calculate simple interest.
2. Application: calculate a monthly payment.
3. Calculate compound interest.

NEW VOCABULARY

interest	maturity value
principal	installment buying
time	finance charge
simple interest	compound interest

1 Calculating Simple Interest

Interest is what you pay to use someone else's money. You pay interest on money you borrow to buy a car, and on purchases made with a credit card.

You earn interest for letting other people use your money. You earn interest on bank deposits, certificates of deposit, and bonds.

Principal is the amount on which interest is calculated. It is the amount borrowed or invested. The rate is the percent at which interest is calculated. The **time**, or term, is the total time over which interest is calculated. Interest calculated only on the principal is called **simple interest**.

> **simple interest = principal × rate × time**
>
> $$\mathbf{I = P \times R \times T}$$
>
> I = interest (amount earned or charged)
> P = principal (amount borrowed or deposited)
> R = rate of interest (per year)
> T = time (in years)
>
> The *rate R* and *time T* must be expressed in the *same time unit.*

EXAMPLE 1 How much will \$2,000 earn at 7% simple interest for 3 years?

I = interest earned
P = \$2,000
R = 7%
T = 3 years

$I = P \times R \times T$

$I = \$2,000 \times 0.07 \times 3$

$I = \$420$ earned in interest ■

▶ You Try It 1. How much will \$800 earn at 6% simple interest for 2 years?

EXAMPLE 2 Lucy borrowed \$846 at 18% for 7 months. How much interest will she pay?

I = ?
P = \$846
R = 18%
$T = \frac{7}{12}$ years

{The rate is 18% per year. The time, 7 months, must also be expressed in years. Since 12 months = 1 year, 7 months = $\frac{7}{12}$ years.

$I = P \times R \times T$

$I = \$846 \times 0.18 \times \frac{7}{12} = \frac{\$846}{1} \times \frac{0.18}{1} \times \frac{7}{12} = \frac{\$1065.96}{12}$

$I = \$88.83$

Lucy will pay \$88.83 in interest to borrow \$846 for 7 months. ■

▶ **You Try It** 2. Tomas borrowed $1,248 at 14% simple interest for 5 months. How much interest will he pay?

2 Application: Calculating a Monthly Payment

Maturity value, S, is total principal plus interest.

maturity value = principal + interest

$$S = P + I$$

In Example 1, the principal is $2,000. The interest earned is $420. The maturity value is S = P + I = $2,000 + $420 = $2,420.

With many purchases, such as a car or stereo, you can go home with it now and pay for it later. This is called **installment buying**. You often pay for the item in equal monthly payments. This payment includes an interest charge, called a **finance charge**.

To figure a monthly payment

1. Subtract the down payment or trade-in from the total cost. This gives the principal to be borrowed.
2. Compute the interest on this principal. For the time T, use the *total length of time* over which payments will be made.
3. Add principal to interest to get maturity value.
4. To find the monthly payment, divide maturity value by the number of months over which payments will be made.

$$\text{Monthly Payment} = \frac{\text{Maturity Value}}{\text{Number of Months}}$$

EXAMPLE 3 Alberto borrows $600 to finance the purchase of a microwave oven. The simple interest charge is 12% over 24 months. Find the monthly payment.

Step #1: Calculate the interest charge on $600 at 12% for 24 months. The time for the loan is 24 months = 24 ÷ 12 = 2 years.

I = ?
P = $600
R = 12%
T = 2 years

$$I = P \times R \times T$$
$$I = \$600 \times 0.12 \times 2$$
$$I = \$144$$

Step #2: maturity value = P + I = $600 + $144 = $744

Step #3: $\text{monthly payment} = \frac{\text{maturity value}}{\text{number of payments}} = \frac{\$744}{24} = \$31$

Alberto must pay $31 per month for 24 months. ■

▶ **You Try It** 3. Barbara borrowed $1,800 to finance the purchase of a large screen TV. The simple interest charge is 10% for 36 months. Find the monthly payment

EXAMPLE 4 Maria purchased a stereo system priced at \$1,620. She made a \$180 down payment and financed the rest at 14.5% for 18 months.

a. What is her monthly payment?

Step #1: Find the amount to be borrowed.

\$1,620	cost
− \$180	down payment
\$1,440	principal to be borrowed

Step #2: Calculate the interest on \$1,440 at 14.5% for 18 months.

$$I = ?$$
$$P = \$1{,}440$$
$$R = 14.5\% = 0.145$$
$$T = 1.5 \text{ yrs } \left(18 \text{ months} = \frac{18}{12} = 1.5 \text{ years}\right)$$

$$I = P \times R \times T$$
$$I = \$1{,}440 \times 0.145 \times 1.5$$
$$I = \$313.20$$

Step #3: Find the maturity value.

$$\text{maturity value} = P + I = \$1{,}440 + \$313.20 = \$1{,}753.20$$

Step #4: Find the monthly payment.

$$\text{monthly payment} = \frac{\text{maturity value}}{\text{number of months}} = \frac{\$1{,}753.20}{18}$$
$$= \$97.40$$

Maria must pay \$97.40 per month for 18 months.

b. What is the total cost of the stereo on the installment plan?

The total cost of the \$1,620 stereo on the installment plan is

\$1,753.20	P + I
+ 180.00	down
\$1,933.20	

■

You Try It **4.** Carl bought a riding mower for \$3,200. He put \$200 down, and financed the rest over 15 months at 9%.

a. What is his monthly payment?

b. What is the total cost of the mower on the installment plan?

---

3 Calculating Compound Interest

Simple interest is calculated one time on the original principal for the entire term of the loan.

EXAMPLE 5 Calculate simple interest on \$1,000 at 10% for 2 years.

I = ?
P = \$1,000
R = 10%
T = 2 years

$$I = P \times R \times T$$
$$I = \$1{,}000 \times 0.10 \times 2$$
$$I = \$200$$

The total interest earned is \$200. ■

Compound interest allows you to earn interest on previously earned interest. This is possible because interest is calculated more than once. Savings accounts often earn compound interest. Solve Example 5 again using compound interest.

EXAMPLE 6 Calculate the compound interest earned on \$1,000 at 10% compounded annually for 2 years.

Compounded annually means calculate the interest once a year. Then add this interest to the principal before calculating interest again.

First year: I = ?
P = \$1,000
R = 10%
T = 1 year

$$I = P \times R \times T$$
$$I = \$1{,}000 \times 0.10 \times 1$$
$$I = \$100 \text{ earned interest}$$

The interest earned in the first year is \$100. It is added to the old principal \$1,000 to get the new principal \$1,100.

Second year: I = ?
P = \$1,100 (new principal)
R = 0.10
T = 1 year

$$I = P \times R \times T$$
$$I = \$1{,}100 \times 0.10 \times 1$$
$$I = \$110 \text{ earned interest}$$

The interest earned in the second year is \$110. After 2 years the total interest earned is \$100 + \$110 = \$210. ■

In Example 5, \$1,000 earned \$200 in 2 years at 10% simple interest. In Example 6, \$1,000 earned \$210 in 2 years at 10% compounded annually. \$10 more was earned with compound interest. The reason is that interest was earned on interest. The extra \$10 in interest was earned on the \$100 in interest made in the first year.

You Try It

5. Calculate simple interest on \$100 at 20% for 2 years.
6. Calculate compound interest on \$100 at 20% compounded annually for 2 years. How much more interest is earned here when compared to Problem 5?

compounded annually—means interest is calculated 1 time a year (T = 1 year)

compounded semiannually—interest is calculated every 6 months $\left(T = \frac{6}{12} = 0.5 \text{ year}\right)$

compounded quarterly—interest is calculated every 3 months $\left(T = \frac{3}{12} = 0.25 \text{ year}\right)$

compounded monthly—interest is calculated each month $\left(T = \frac{1}{12} \text{ year}\right)$

compounded daily—interest is calculated every day $\left(T = \frac{1}{365} \text{ year}\right)$

EXAMPLE 7 How much interest will \$15,000 earn at 8% compounded quarterly for 1 year.

Quarterly means interest is calculated every 3 months. In 1 year, interest will be calculated 4 times. 3 months $= \frac{3}{12}$ year = 0.25 year.

first 3 months:

$$I = P \times R \times T = \$15{,}000 \times 0.08 \times 0.25 = \$300$$
$$\text{new } P = \$15{,}000 + \$300 = \$15{,}300$$

second 3 months:

$$I = P \times R \times T = \$15{,}300 \times 0.08 \times 0.25 = \$306$$
$$\text{new } P = \$15{,}300 + \$306 = \$15{,}606$$

third 3 months:

$$I = P \times R \times T = \$15{,}606 \times 0.08 \times 0.25 = \$312.12$$
$$\text{new } P = \$15{,}606 + \$312.12 = \$15{,}918.12$$

fourth 3 months:

$$I = P \times R \times T = \$15{,}918.12 \times 0.08 \times 0.25 \doteq \$318.36$$
$$\text{maturity value} = \$15{,}918.12 + \$318.36 = \$16{,}236.48$$

Find the total interest earned as follows.

\$16,236.48	maturity value
−\$15,000.00	original principal
\$1,236.48	compound interest earned ■

total compound interest earned = maturity value − original principal

♦ **You Try It** 7. How much interest will $12,000 earn at 15% compounded semiannually for 2 years?

SOMETHING MORE

Annual Yield You invest $1,000 at 8% compounded monthly. A bank tells you the *annual yield* at this rate is 8.2995%. This means to find the annual interest, just take 8.2995% of $1,000.

$$I = P \times \text{Annual Yield} = \$1{,}000 \times 0.082995 \doteq \$83.00$$

8.2995% compounded annually earns the same amount of interest in one year as 8% compounded monthly. They are equivalent rates. The annual yield is given to consumers so they can easily and quickly figure the yearly interest earned on an investment.

Solve each problem.

a. A $5,000 CD earns 9% compounded daily, with an annual yield of 9.4162%. What interest is earned in the first year?

b. Suppose $2,500 is invested at 12% compounded monthly with an annual yield of 12.6825%. What interest is earned in the first year?

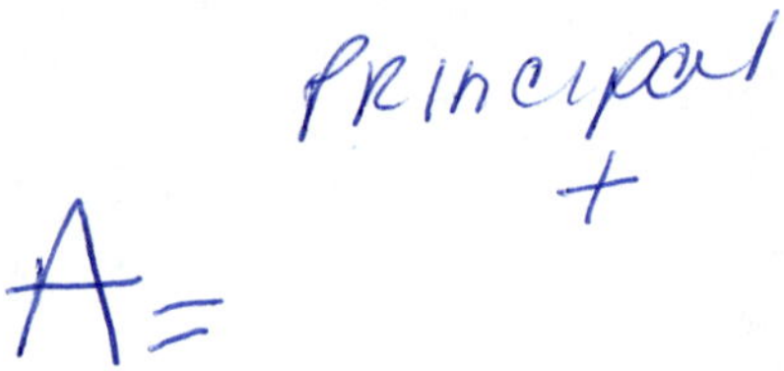

♦ **Answers to You Try It** **1.** I = $96 **2.** $T = \frac{5}{12}$; I = $72.80 **3.** $T = \frac{36}{12} = 3$; I = $540; S = $2,340; monthly payment = $65 **4. a.** $T = \frac{15}{12} = 1.25$; I = $337.50; maturity value = $3,337.50; monthly payment = $222.50 **b.** total cost = $3,537.50 **5.** I = $40 **6.** first year: I = $20; new P = $120; second year: I = $24; total interest earned = $44; $4 more in interest is earned **7.** T = 0.5 yr; first 6 mo: I = $900; new P = $12,900; second 6 mo: I = $967.50; new P = $13,867.50; third 6 mo: I = $1,040.06; new P = $14,907.56; fourth 6 mo: I = $1,118.07; maturity value = $16,025.63; total interest earned = $4,025.63

♦ **Answers to Something More** **a.** $470.81 **b.** $317.06

SECTION 7.9 EXERCISES

1 *Find the simple interest. Round answers to the nearest cent.*

1. P = \$800, R = 10%, T = 1 year
2. P = \$450, R = 11%, T = 3 years
3. P = \$4,500, R = 8%, T = 4 years
4. P = \$260, R = 14%, T = 2 years
5. P = \$569.95, R = 7.5%, T = 2 years
6. P = \$934.56, R = 10%, T = 2 years
7. P = \$4,000, R = 16%, T = 12 months
8. P = \$350, R = $9\frac{1}{2}$%, T = 24 months
9. P = \$700, R = 6%, T = 18 months
10. P = \$2,420, R = 18%, T = 42 months
11. What interest do you earn if you invest \$5,000 at 15% for 2 years?
12. What interest is earned if you deposit \$2,150 at 6.75% for 4 years?
13. What interest do you pay on a \$600 loan at 13.5% for 1 year?
14. What interest do you pay on a \$3,500 loan at 12% for 3 years?
15. \$1,500 is borrowed at 6.5% for 30 months. Find the interest.
16. \$780 is invested at 8.8% for 48 months. Find the interest.

---

2 *Find the monthly payment.*

17. Daniele borrowed \$4,000 at 14% simple interest for 12 months to build a deck. Calculate her monthly payment.
18. Susan purchased a computer for \$1,540. She financed it at 16% simple interest for 24 months. Find her monthly payment.
19. A couple borrowed \$3,000 for a vacation to Disneyworld. Find their monthly payment if they are paying 20% simple interest for 6 months.
20. Ed borrowed \$1,620 to buy office equipment. If he pays the money back at 8% simple interest over 9 months, find his monthly payment.
21. Jana purchased a car for \$14,600. She put \$2,600 down and financed the rest at 10% simple interest over 24 months. What is her monthly payment?
22. A \$748 TV can be purchased for \$148 down with the rest financed at 11% simple interest for 15 months. Find the monthly payment.

23. A motorboat costs $12,000. It may be purchased for 15% down, with the rest paid in 36 monthly payments at 16% simple interest.
   a. What is the finance charge?

   b. What is the monthly payment (round to the nearest cent)?

24. A jeweled watch priced at $2,400 is sold on the installment plan for 20% down with the rest paid in 18 monthly payments at 16% simple interest.
   a. What is the finance charge?

   b. What is the monthly payment (round to the nearest cent)?

25. A lot is on sale for $46,000. The owner will finance the sale for $6,000 down, with the remainder paid in 60 monthly payments at 10% simple interest.
   a. What is the finance charge?

   b. What is the monthly payment?

   c. What is the total cost of the lot on the installment plan?

26. A necklace is on sale for $640. The store owner will finance the purchase for $\frac{1}{4}$ down with the balance paid in 16 equal monthly installments at 12% simple interest.
   a. What is the finance charge?

   b. What is the monthly payment?

   c. What is the total cost on the installment plan?

3 *Find the maturity value and the total interest earned.*

27. $800 at 6% compounded annually for 2 years

28. $500 at 9%, compounded annually for 3 years

29. $2,400 at 12%, compounded quarterly for 1 year

30. $250 at 8% compounded quarterly for 1 year

31. $3,750 at 12% compounded monthly for 5 months

32. $600 at 24% compounded monthly for 3 months

33. $40,000 at 9% compounded semiannually for 30 months

34. $25,000 at 7% compounded semiannually for 2 years

35. Ed deposits $5,000 in an account earning 8% compounded quarterly.

a. Find his maturity value after one and one-half years.

b. What total interest did he earn?

36. Simone invested $3,000 in a CD earning 12% compounded semiannually for 4 years.

a. How much will she have when her CD matures?

b. What total interest will she have earned?

37. John invested $2,000 in a bond earning 7.2% compounded annually.

a. What will the bond be worth when it matures in 7 years?

b. How much interest will have been earned?

38. Regina deposited $45,000 at 10.91% compounded annually.

a. How much will she have in 4 years?

b. How much interest will she have earned?

SKILLSFOCUS (Section 3.4) *Divide.*

39. $\frac{8}{9} \div \frac{2}{9}$

40. $\frac{6}{15} \div \frac{7}{10}$

41. $4 \div \frac{5}{8}$

42. $3\frac{4}{7} \div 2$

EXTEND YOUR THINKING

SOMETHING MORE

43. $4,500 is invested at 8% compounded monthly. The annual yield is 8.2995%. Find the first year's interest.

44. A certificate of deposit for $17,200 earns 12% compounded monthly. The annual yield is 12.6825%. What interest is earned in the first year?

45. Find the maturity value and the amount of interest earned in each case if $1,000 is deposited for 1 year at 12% compounded

a. annually.

b. semiannually.

c. quarterly.

d. monthly.

e. What conclusion do you draw about the amount of interest earned when you compound more often?

46. Find the maturity value and amount of interest earned if each sum of money is deposited at 10% compounded annually for 4 years.

a. \$1,000

b. \$2,000

c. \$4,000

d. \$8,000

e. If the principal is doubled, does the interest earned also double?

47. Find the maturity value and total interest earned if you deposit \$1,000 at 12% compounded quarterly for each period of time.

a. 3 months

b. 6 months

c. 9 months

d. 12 months

e. Does adding 3 months to the term each time increase the interest earned by a fixed amount each time?

48. Compute the maturity value and interest earned if you invest \$1,000 for 5 years at each compounded annual rate.

a. 2%

b. 4%

c. 8%

d. 16%

e. If the rate doubles, does the interest earned also double?

WRITING TO LEARN ▶▶▶▶

49. In your own words, explain the meaning of the terms *simple interest*, *principal*, *rate*, *time*, and *maturity value*.

50. Write the formula for interest. Clearly define all variables used. Give the proper units (dollars, years, %) for each of the variables.

51. Explain the difference between simple interest and compound interest.

CHAPTER 7 REVIEW

VOCABULARY AND MATCHING

New words and phrases introduced in this chapter are shown in the left column. Match each term on the left with the word or sentence on the right that best describes it.

A. percent	_____	remaining number after percent and whole are identified
B. whole	_____	salary that is made up entirely of commissions
C. complement	_____	paid to borrow money; earned when you invest it
D. rate	_____	interest earns interest
E. part	_____	income earned, expressed as a percent of what you sell
F. percent sentence	_____	number of parts out of 100 parts
G. commission	_____	quantity on which interest is calculated
H. straight commission	_____	interest calculated only on the principal for the entire term of the loan
I. salary plus commission	_____	found by subtracting a percent from 100%
J. discount	_____	original principal plus all interest earned
K. sale price	_____	difference as a percent of the original number
L. percent increase or decrease	_____	original selling price minus discount
M. interest	_____	base amount, follows the words "percent of"
N. principal	_____	earn a fixed salary plus commission on sales made
O. time, or term	_____	number with the percent symbol
P. simple interest	_____	Part is Percent of Whole, or Percent of Whole is Part.
Q. maturity value	_____	original selling price minus sale price
R. compound interest	_____	entire time over which simple interest is calculated

REVIEW EXERCISES

7.1 The Meaning of a Percent

1. 30% means _______ parts out of _______ equal parts.

2. _______ parts out of _______ equal parts is 7.5%.

3. 99 parts out of 100 equal parts is _______ %.

4. _______ % means 70 parts out of 100 equal parts.

5. The complement of 28% is _______ %.

6. The complement of 3.7 % is _____ %.

7. _______ % is the complement of $22\frac{2}{9}$%.

8. _______ % is the complement of $\frac{1}{4}$%.

7.2 Decimals and Percents

Change each decimal to a percent.

9. 0.37 **10.** 0.009 **11.** 1.4 **12.** 2.0137 **13.** 4 **14.** 0.875

Change each percent to a decimal.

15. 91% **16.** 5% **17.** 215% **18.** 0.0635% **19.** 7.2% **20.** $6\frac{1}{2}\%$

7.3 Fractions and Percents

Change each fraction to a percent.

21. $\frac{3}{5}$ **22.** $\frac{5}{8}$ **23.** $\frac{7}{3}$ **24.** $\frac{9}{11}$ **25.** $\frac{9}{10}$ **26.** $\frac{11}{12}$

Change each percent to a fraction and reduce.

27. 45% **28.** 17% **29.** $88\frac{8}{9}\%$ **30.** $27\frac{3}{11}\%$ **31.** 137.5%

32. 14.4% **33.** 8.25% **34.** 0.6% **35.** $\frac{1}{3}\%$ **36.** $1\frac{5}{8}\%$

7.4 Identifying the Three Numbers in a Percent Word Problem

Identify the Percent, Whole, and Part. Do not solve.

37. 35% of 400 cars is 140 cars.

38. 115% of $240 is $276.

39. 38 is 10% of 380.

40. 173.6 grams is 155% of 112 grams.

41. What percent of 60 is 50?

42. 30 is what percent of 24?

43. 20 is 40% of what number?

44. 74% of what number is 333?

45. 105% of 45 is what number?

46. What is 0.75% of $40?

7.5 Mental Computations Involving Common Percents

Solve each problem mentally.

47. Find 50% of 940.

48. 100% of 2,300 is what number?

49. 24 is what percent of 240?

50. What is 10% of 65?

51. 200% of what number is 80?

52. 70 is what percent of 140?

53. What percent of 23 is 23?

54. What is 25% of 440?

55. What is a 15% tip on a dinner bill of $82?

56. What is the discount on a $45 shirt, at 20% off?

7.6 Solving the Percent Equation

Solve each problem by writing and solving an equation.

57. 42% of what number is $147?

58. 114% of what number is 627?

59. What percent of 70 is 50?

60. 36 is what percent of 24?

61. 120% of 45 is what number?

62. What is 1.25% of $40?

7.7 Percent Applications

63. Suppose 64 people belong to the country club. If 24 are senior citizens, what percent are seniors?

64. A new car is advertised to get 40 miles to a gallon of gas. George bought the car and found he got 28 miles to the gallon. What percent of the advertised miles did George actually get?

65. There are 750 yachts in the marina. If 38% of them exceed 36 feet in length, how many exceed 36 feet?

66. Dale had $48.50 in his wallet. he spent 16% of his money on lunch. How much did he spend on lunch?

67. Sol's mortgage payment is $650. If this is 40% of his monthly net pay, what is Sol's monthly net pay?

68. Suppose 17% of a small town's yearly budget is spent on education. If $391,000 is spent on education, how large is the town's annual budget?

69. Suppose 1,650 plants were grown in an experimental soil. If 56% died before maturity, how many died before maturity?

70. The museum had 4,000 visitors today. Of those, 1,600 reside out of state. What percent reside out of state?

Use Figure 7.5 to answer questions 71 - 75.

FIGURE 7.5

Grade Distribution for Credit Courses in the Fall Semester at Norwood College

Circle = 14,291 grades assigned

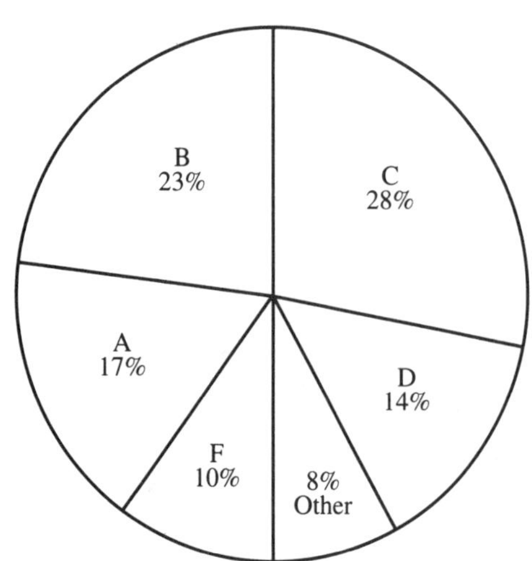

71. About how many C grades were assigned? Round to tens.

72. What is the ratio of C grades to D grades?

73. How many more F grades were assigned than "Other" grades? Round to tens.

74. What percent of the grades were C or higher?

75. How many of the grades were passing grades? Round to tens. (Assume that "Other" grades are not passing.)

7.8 Commission, Discount, and Percent Increase and Decrease

76. Harry works for an 11% commission on sales. If he sells $6,000 in merchandise, what will his commission be?

77. Jack made $870 in commissions this month. If he is paid a 15% straight commission, what were Jack's total sales for the month?

78. Al is paid a monthly salary of $440 plus a 7.5% commission on sales. If he sold $5,400 in merchandise this month, what was his salary?

79. Edna works on salary plus commission. She made $1,160 this week. If her weekly salary is $260 and she earns a 9% commission, what were her total sales for the week?

80. A ten-speed bike usually sells for $180. It is being sold this week at a 30% discount. Find the sale price.

81. A motorcycle is discounted 30%, giving a savings of $1,440.
a. What was the original price of the motorcycle?
b. What is the sale price?

82. Dave sold $1,340 in books yesterday, and $1,400.30 in books today. Find Dave's percent increase in sales.

83. Cal's Car Wash served 450 cars last week, but only 414 cars this week. What percent decrease is this?

84. Isadore makes $34,560 per year. He is getting a 4.2% increase in salary.
a. How much of an increase will Isadore get?
b. What will his new annual salary be?

85. A man originally weighing 465 pounds had a 60% decrease in weight over a two-year period.
a. How much weight did he lose?
b. What was his new weight?

7.9 Simple and Compound Interest

86. Find the simple interest if $4,000 earns 12% over 3 years.

87. What interest does $360 earn at 8.5% simple interest over 6 years?

88. How much interest will Gail pay if she borrows $1,450 at 5% simple interest for 18 months?

89. What will $66,000 earn at 4.8% simple interest for 9 months?

90. What monthly payment will be made if $1,200 is borrowed at 16% simple interest for 12 months?

91. Lucy is planning to borrow $2,800 for a trip to Hawaii. She plans to pay the money back at 9% simple interest over 5 months. Find her monthly payment.

92. What will $250 earn at 8% compounded annually for 3 years?

93. What will $40,000 earn at 6% compounded semiannually for 2 years?

94. What will $9,400 earn at 6% compounded monthly for half a year?

95. Ida deposits $15,000 in an account earning 12% compounded quarterly.

a. Find her maturity value after one and one-half years.

b. What total interest did she earn?

APPENDIX A
DIVISION OF WHOLE NUMBERS; ESTIMATION

OBJECTIVES	NEW VOCABULARY
1 Divide using a one digit divisor.	remainder
2 Divide using a divisor with two or more digits.	trial divisor
3 Cross out zeros and divide by powers of 10.	
4 Estimate solutions to problems	

1 Divide Using a One Digit Divisor

Multiplication is a shortcut for repeated addition. Division is a shortcut for repeated subtraction. For example, show $15 \div 5 = 3$ using repeated subtraction.

```
 15
- 5
---
 10
- 5
---
  5
- 5
---
  0
```

There are 3 fives in 15. So, $15 \div 5 = 3$

```
     3
5 ) 15
   -15  ← 3 × 5 = 15
   ---
     0  ← Remainder = 0
```

In the next two examples, place-value names are used to stress the importance of keeping like named digits in the same vertical column.

EXAMPLE 1 Find $68 \div 4$.

```
     tens ones
      1        ← Step 1: Divide 4 into 6 tens. 6 tens ÷ 4 gives 1 ten. Write 1 ten
4 )   6   8              directly above 6 tens in the quotient.
     -4
      2        ← Step 2: 1 ten × 4 = 4 tens. Write 4 tens directly below 6 tens.
               ← Step 3: Subtract the tens. 6 − 4 = 2. Write 2 directly below
                         the 4 in the tens column.

      1   7
4 )   6   8
     -4        ← Step 4: Bring down the next digit, 8 ones.
      2   8    ← Step 5: Divide 4 into 28 ones. 28 ones ÷ 4 = 7 ones. Write
     -2   8              7 ones in the quotient directly above 8 ones.
          0    ← Step 6: 7 ones × 4 = 28 ones. Write 28 below 28.
               ← Step 7: Subtract 28 − 28 = 0. There are no more digits left to
                         bring down in the dividend. The division is done.
```

Therefore, $68 \div 4 = 17$. The check is: $17 \times 4 = 68$. ■

> In long division, every digit you bring down from the dividend must have a digit written above it in the quotient.

EXAMPLE 2 Find 2,695 ÷ 7.

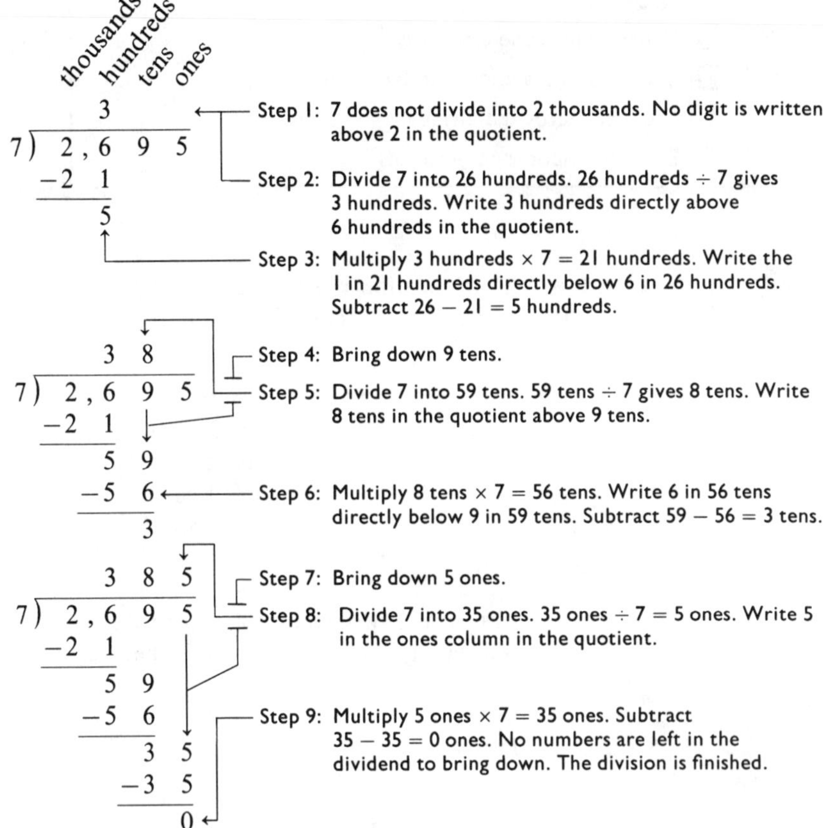

Therefore, 2,695 ÷ 7 = 385. Check: 385 × 7 = 2,695. ■

▶ **You Try It** Divide and check.

1. 91 ÷ 7 **2.** 135 ÷ 5 **3.** 836 ÷ 4 **4.** 6,201 ÷ 9

Use repeated subtraction to find 29 ÷ 6.

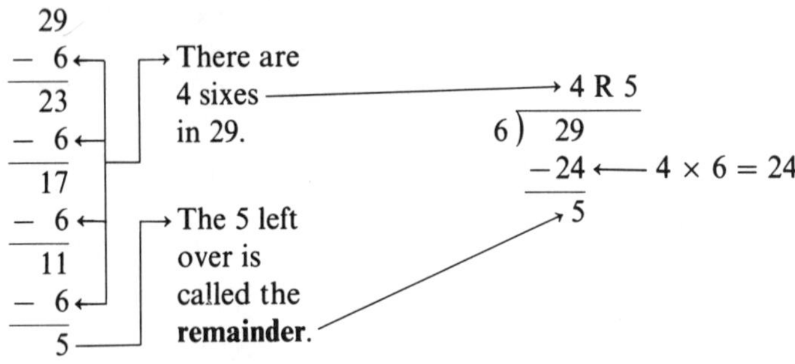

The answer to 29 ÷ 6 is 4 with a remainder of 5, written 4 R 5. To check the answer, use the following rule.

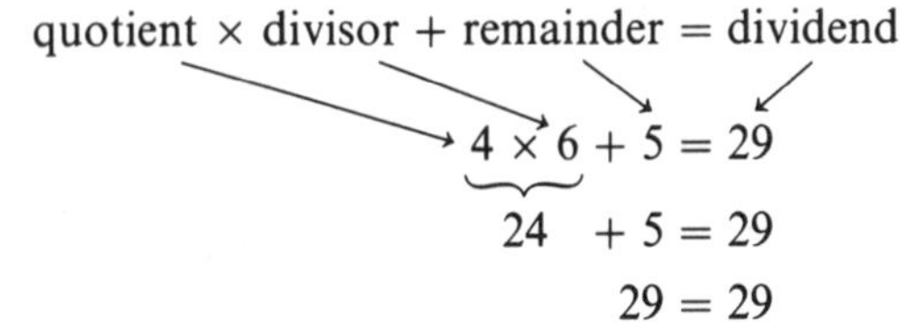

> Check long division using the rule
>
> quotient × divisor + remainder = dividend

EXAMPLE 3 Divide $523 \div 8$.

$$\begin{array}{r} 65\text{ R }3 \\ 8\overline{)\,523} \\ -48 \\ \hline 43 \\ -40 \\ \hline 3 \end{array}$$

$-48 \longleftarrow 6 \times 8 = 48$

$-40 \longleftarrow 5 \times 8 = 40$

Check: $\underbrace{65 \times 8} + 3$
$= 520 + 3$
$= 523$

Therefore, $523 \div 8 = 65$ R 3. ■

OBSERVE In Example 3, you can write the answer 65 R 3 as $65\frac{3}{8}$ by writing the remainder 3 over the divisor 8. 3/8 is called the fraction form for a remainder.

▶ **You Try It** Divide and check.

5. $488 \div 7$

6. $941 \div 6$

Each time you bring down a digit when dividing, you must write a digit above it in the quotient. As shown in the next two examples, the digit you write in the quotient may be 0.

EXAMPLE 4 What is $43 \div 4$?

$$\begin{array}{r} 10\text{ R }3 \\ 4\overline{)\,43} \\ -4 \\ \hline 03 \\ -\ 0 \\ \hline 3 \end{array}$$

$03 \longleftarrow$ Bring down the 3 ones. Divide 4 into 3. Since 3 ones ÷ 4 gives 0 ones, *you must write 0 ones in the quotient above 3 ones*. 0 is a placeholder. It says there are no ones in the quotient. The remainder is 3.

Therefore, $43 \div 4 = 10$ R 3.

Check: $10 \times 4 + 3$
$= 40 + 3$
$= 43$ ■

EXAMPLE 5 Find $\frac{23{,}027}{5}$.

$$\begin{array}{r} 4{,}605\text{ R }2 \\ 5\overline{)\,23{,}027} \\ -20 \\ \hline 30 \\ -30 \\ \hline 02 \\ -\ 0 \\ \hline 27 \\ -25 \\ \hline 2 \end{array}$$

$02 \longleftarrow$ Bring down the 2 tens. 2 tens ÷ 5 gives a quotient of 0. *You must write 0 in the quotient above the 2.* 0 is a placeholder, meaning there are no tens in the quotient.

$-0 \longleftarrow$ Multiply 0 tens × 5 = 0 tens. Write 0 tens below 2 tens and subtract.

Therefore, $\frac{23{,}027}{5} = 4{,}605$ R 2. Check: $4{,}605 \times 5 + 2$

$= 23{,}025 + 2$

$= 23{,}027$ ■

▶ **You Try It** Divide and check.

7. $65 \div 6$

8. $21{,}549 \div 7$

---

2 Divide Using a Divisor with Two or More Digits

When the divisor has two or more digits, use a **trial divisor** to help you find the quotient. Use the trial divisor found by rounding the divisor to the digit farthest left.

EXAMPLE 6

Divisor		Trial Divisor
54	⟶	50
783	⟶	800
17	⟶	20 ■

EXAMPLE 7 Divide $379 \div 54$.

$54\overline{)379}$ 54 does not divide into 3. Do not write a digit above 3. 54 does not divide into 37. Do not write a digit above 7. 54 does divide into 379.

$54\overline{)379}$ ⟶ Think: $50\overline{)379}$ with 7 above. The trial divisor is 50. 5 into 37 goes 7 times. Try 7 in the quotient.

$$\begin{array}{r} 7 \text{ R } 1 \\ 54\overline{)\ 379} \\ -378 \\ \hline 1 \end{array}$$

Therefore, $379 \div 54 = 7$ R 1. Check: $7 \cdot 54 + 1$

$= 378 + 1$

$= 379$ ■

EXAMPLE 8 Find $37\overline{)1{,}526}$.

$37\overline{)1{,}526}$ ⟶ Think: $40\overline{)152}$ with 3 above. The trial divisor is 40. 4 into 15 goes 3 times. Try 3 in the quotient.

$$\begin{array}{r} 3 \\ 37\overline{)\ 1{,}526} \\ -1\ 11 \\ \hline 41 \end{array}$$

$-1\ 11$ ⟵ $3 \times 37 = 111$

41 ⟶ Since $41 > 37$, the trial divisor 3 is too small. Try 4.

$$\begin{array}{r} 41 \text{ R } 9 \\ 37\overline{)\ 1{,}526} \\ -1\ 48 \\ \hline 46 \\ -37 \\ \hline 9 \end{array}$$

$-1\ 48$ ⟵ $4 \times 37 = 148$

46 ⟵ 37 divides into 46, 1 time

Therefore, $1{,}526 \div 37 = 41$ R 9. Check: $37 \times 41 + 9$

$= 1{,}517 + 9$

$= 1{,}526$ ■

▶ **You Try It** Divide and check.

9. $379 \div 76$ **10.** $3{,}279 \div 43$

3 Crossing Out Zeros and Dividing By Powers of 10

Suppose both divisor and dividend are whole numbers with zeros on their right sides. Simplify long division by first crossing out an equal number of zeros from the right sides of both divisor and dividend.

$$\begin{array}{r} 6 \\ 50\overline{)\;300} \\ -300 \\ \hline 0 \end{array} \longrightarrow 5\not{0}\overline{)\;30\not{0}} \text{ or } 5\overline{)\;30} \quad (\text{quotient } 6)$$

Crossing out one zero on the right side of both divisor and dividend gives the same answer.

Zeros may be crossed out regardless of how the division is written.

$$300 \div 50 = 30\not{0} \div 5\not{0} = 6$$

$$300/50 = 30\not{0}/5\not{0} = 6$$

$$\frac{300}{50} = \frac{30\not{0}}{5\not{0}} = 6$$

EXAMPLE 9

a. $8{,}000 \div 200 = 8{,}0\not{0}\not{0} \div 2\not{0}\not{0} = 80 \div 2 = 40$

b. $\frac{72{,}000}{9{,}000} = \frac{72{,}\not{0}\not{0}\not{0}}{9{,}\not{0}\not{0}\not{0}} = \frac{72}{9} = 8$

c. $60{,}000\overline{)\;4{,}800{,}000} = 6\not{0}{,}\not{0}\not{0}\not{0}\overline{)\;4{,}80\not{0}{,}\not{0}\not{0}\not{0}} = 6\overline{)\;480}$ (quotient 80) ■

▶ **You Try It** Divide by first crossing out zeros.

11. $9{,}000 \div 30$ **12.** $\frac{56{,}000}{7{,}000}$ **13.** $3{,}000\overline{)\;120{,}000}$

What pattern do you see when you divide by powers of 10 (10, 100, 1,000, and so on)?

EXAMPLE 10

a. $\frac{56{,}000}{10} = \frac{56{,}00\not{0}}{1\not{0}} = \frac{5{,}600}{1} = 5{,}600$

b. $\frac{56{,}000}{100} = \frac{56{,}0\not{0}\not{0}}{1\not{0}\not{0}} = \frac{560}{1} = 560$

c. $\frac{56{,}000}{1{,}000} = \frac{56{,}\not{0}\not{0}\not{0}}{1{,}\not{0}\not{0}\not{0}} = \frac{56}{1} = 56$ ■

> To divide by a power of 10, cross out the same number of zeros on the right side of the dividend as are in the power of 10.

▶ **You Try It** Divide by first crossing out zeros.

14. $\frac{82{,}500}{100}$ **15.** $\frac{400{,}000}{10{,}000}$ **16.** $\frac{680}{10}$

4 Estimation

The rules for estimating a multiplication are used to estimate division.

> To estimate an answer to a multiplication or division problem
>
> 1. Round each number to the place value of its digit farthest left.
> 2. Multiply or divide the rounded numbers.

EXAMPLE 11 Estimate $836 \div 38$.

$836 \div 38$ — round to hundreds (836), round to tens (38) → $800 \div 40$

Estimate $836 \div 38$ by $800 \div 40$.

$$40\overline{)800} = 4\not{0}\overline{)80\not{0}} = 4\overline{)80}\ \ 20$$

Estimate $836 \div 38$ to be about 20. (Actual answer = 22.) ■

EXAMPLE 12 The cost to take 47 people on a weekend bus trip to Niagara Falls is \$2,838. Estimate the cost per person.

The cost per person is \$2,838 ÷ 47.

round to thousands (\$2,838), round to tens (47) → \$3,000 ÷ 50

Estimate \$2,838 ÷ 47 by \$3,000 ÷ 50.

$$\$3{,}000 \div 50 = \frac{\$3{,}00\not{0}}{5\not{0}} = \frac{\$300}{5} = \$60 \text{ estimated cost per person}$$ ■

EXAMPLE 13 The defense department plans to spend \$729,837,406 to build 184 high-tech tanks. Estimate the cost per tank.

The cost per tank is \$729,837,406 ÷ 184.

round to hundred millions (\$729,837,406), round to hundreds (184) → \$700,000,000 ÷ 200

$$= \frac{\$700{,}000{,}0\not{0}\not{0}}{2\not{0}\not{0}} = \frac{\$7{,}000{,}000}{2} = \$3{,}500{,}000$$

The estimated cost is about \$3,500,000 per tank. ■

▶ **You Try It**

17. Estimate $783 \div 18$.

18. A plane was chartered by 27 people for a total of \$3,217. Estimate the cost per person.

19. The county budgeted \$5,875,418 to hire 191 new teachers. Estimate the amount budgeted per teacher.

▶ **Answers to You Try It** 1. 13 2. 27 3. 209 4. 689 5. 69 R 5 6. 156 R 5 7. 10 R 5 8. 3,078 R 3 9. 4 R 75 10. 76 R 11 11. 300 12. 8 13. 40 14. 825 15. 40 16. 68 17. 40 18. \$100 19. \$30,000

APPENDIX A EXERCISES

1 *Find the quotient and remainder for each division, and check.*

1. $14 \div 3$ **2.** $17 \div 2$ **3.** $21 \div 4$ **4.** $11 \div 5$

5. $\frac{26}{8}$ **6.** $\frac{19}{6}$ **7.** $\frac{32}{8}$ **8.** $\frac{27}{3}$

9. $38 \div 9$ **10.** $45 \div 6$ **11.** $50 \div 7$ **12.** $61 \div 8$

13. $\frac{3{,}052}{8}$ **14.** $\frac{8{,}407}{2}$ **15.** $513 \div 6$ **16.** $944 \div 9$

17. $6{,}304 \div 7$ **18.** 7,003/8 **19.** $4\overline{)20{,}107}$ **20.** $9\overline{)60{,}003}$

2 *Find the quotient and remainder. Check each answer.*

21. $96 \div 12$ **22.** $90 \div 15$ **23.** $127 \div 31$ **24.** $217 \div 43$

25. $678 \div 52$ **26.** $690 \div 31$ **27.** $1{,}598 \div 47$ **28.** $4{,}066 \div 19$

29. $5{,}007 \div 76$ **30.** $7{,}009 \div 84$ **31.** $17{,}385 \div 65$ **32.** $56{,}803 \div 67$

33. $67{,}021 \div 52$ **34.** $28{,}209 \div 28$ **35.** $89{,}264 \div 603$

36. $48{,}167 \div 380$ **37.** $340 \div 10$ **38.** $430 \div 10$

3

39. $600 \div 40$ **40.** $800 \div 50$ **41.** $700 \div 100$

42. $400 \div 100$ **43.** $4{,}500 \div 150$ **44.** $7{,}940 \div 20$

45. $34{,}600 \div 200$ **46.** $77{,}000 \div 1{,}000$ **47.** $30{,}000 \div 1{,}000$

48. $390{,}000 \div 100$ **49.** $1{,}600{,}000 \div 80{,}000$ **50.** $3{,}600{,}000 \div 150{,}000$

4 *Estimate the answer.*

51. $942 \div 87$ **52.** $384 \div 21$ **53.** $870 \div 33$ **54.** $610 \div 28$

55. $3{,}280 \div 615$ **56.** $8{,}617 \div 276$ **57.** $7{,}724 \div 43$ **58.** $5{,}290 \div 98$

59. A lottery jackpot of $2,238,162 was shared equally by 37 people. About how much will each person receive?

60. A 472-page book contains 184,692 words. Estimate the number of words per page.

61. A company paid a total of $6,728,544 for 72 tractors. Someone claimed this amounts to about $68,000 per tractor.

a. Estimate the cost of each tractor.

b. Is the claim approximately correct?

62. A total of $12,569 is spent to train 215 students. A paper claims this amounts to about $500 per student.

a. Estimate the answer.

b. Does the paper's claim seem correct?

SKILLSFOCUS (Section 1.8) *Multiply.*

63. $68 \cdot 406$

64. $314 \cdot 200$

65. $5{,}319 \cdot 1{,}000$

66. Estimate $894 \cdot 37$

67. Estimate $139 \cdot 422$

EXTEND YOUR THINKING ▶▶▶▶

▶ SOMETHING MORE

Figure the answer mentally.

68. $\dfrac{(61 + 61 + 61)}{3}$

69. $\dfrac{(18 + 18 + 18 + 18 + 18)}{18}$

▶ TROUBLESHOOT IT

Find and correct the error.

70.
$$\begin{array}{r} 3\,2\,4 \\ 3\overline{)\;9{,}0\,7\,2} \\ -9\phantom{{,}0\,7\,2} \\ \hline 0\,0\,7 \\ -6 \\ \hline 1\,2 \\ -1\,2 \\ \hline 0 \end{array}$$

71. $\dfrac{60{,}200}{100} = \dfrac{6\not{0}{,}20\not{0}}{1\not{0}\not{0}} = 620$

72. Estimate $261 \div 46 \doteq 200 \div 50$
$= 4$

WRITING TO LEARN ▶▶▶▶

73. Explain the procedure for crossing out zeros in a division. Then explain how you would apply it to the following problem.

$$\frac{450{,}000}{30{,}000} =$$

ANSWER SECTION

CHAPTER 1

Section 1.1 (p. 11)

1. The unemployment rate in 1970 was about 5%. **3.** In 1990 the unemployment rate was closest to the 1950 rate. **5.** The lowest unemployment rate occurred in 1945. It was about 2%. **7. a.** The unemployment rate in 1929 was about 3.5%; in 1932 it was about 23.5%. **b.** The 1932 rate was about 7 times greater than the 1929 rate. **9.** About 2,900 calories are recommended daily for a 21-year-old man. **11. a.** Men are expected to have the most calories per day in the age range 19 to 22. **b.** 2,900 calories **c.** Calories needed by men increases from age 11 to 19, stays the same until age 22, then decreases from age 22 to 76+. **13. a.** In the age ranges from 19 to 22 and 23 to 50, the difference between calories needed by men and women is the greatest. **b.** The difference is about 800 calories per day. **15.** Two eighty-year-old grandparents should consume a total of about 3,650 calories per day. **17.** About 51% of college graduates in 1990 were women. **19.** Between 1940 and 1950 the percent of women college graduates dropped 17%. **21.** In 1970 41% of college graduates were women. **23.** The upward trends occurred from 1900 to 1940 and again from 1950 to 1990. **25.** The total sales in March 1990 were $20,000. **27. a.** In 1990 sales were greatest in August. **b.** Sales for August 1990 were $60,000. **29.** $30,000 more in sales were made in July 1991 than July 1990. **31. a.** Jan to June and July to Aug **b.** June to July and Aug to Nov **c.** Nov to Dec
33. The population of Europe in 1850 was 265 million. **35.** Europe had a 1989 population about 7 times its 1650 population. **37.** To borrow $75,000 at 10.5% for 30 years, an annual income of $29,403 is needed. **39.** An additional $10,123 is needed. **41.** Their interest rate is 10%. **43.** 58% of parents feel competent to help their children in math. **45.** 53% of black parents, 50% of Hispanic parents, and 34% of white parents expect their kids to earn advanced degrees. **47.** The proposed bonus award is $5,000.

Section 1.2 (p. 25)

1. $0 < 8$ **3.** $101 > 100$ **5.** $6 < 6\frac{1}{3}$ **7.** $99\frac{1}{4} < 100$

9. $77.78 < 78$ **11.** $11 + 12\frac{1}{2}$ **13.** $\frac{18}{6}$ or $18 \div 6$

15. 3(18) **17.** $2h$, where h = height **19.** 19 + 7 **21.** $n + 8$, where n = a number **23.** 18 −5 **25.** $\frac{11}{x}$ or $11 \div x$, where x = a number **27.** $8 > n$, where n = a number **29.** $3 < 3.3$ **31.** $4 < 5 < 6$ **33.** $a = b \cdot c$ **35.** $56 = 8 \times 7$ **37.** $14 = x - 23$, where x is some number **39.** 9 **41.** 8 **43.** 6 **45.** 60 **47.** 48 **49.** 3 **51.** 4 **53.** True **55.** False **57.** True **59.** True **61. a.** $19{,}500 < 25{,}000$ **b.** $27{,}000 > 25{,}000$

c. $11{,}500 > 6{,}000$ **63.** Answers may vary: $100\frac{1}{2}$,

100.1, 100.7 **65.** $\frac{7}{8}$ is between 0 and 1.

67. Answers may vary: $8 + x$, $22x$

Section 1.3 (p. 35)

1. 60 **3.** 200 **5.** 1,780 **7.** 500 **9.** 4,400 **11.** 3,500 **13.** 0 **15.** 4,000 **17.** 5,000 **19.** 1,000 **21.** 60,000 **23.** 4,000,000 **25.** 99,990 **27.** 2,500,000 **29.** 70; 100; 0 **31.** 1,660; 1,700; 2,000 **33.** 7,290; 7,300; 7,000 **35.** 34,610; 34,600; 35,000 **37.** 262,910; 262,900; 263,000 **39.** 112,000 **41.** $12,500,000 **43.** 90 **45.** 700 **47.** 14,600 **49.** 20 **51.** 5,000 **53.** 700 **55.** 1,500 **57.** 290 **59.** 3,100 **61.** $2,100 **63.** 12,000 miles **65.** $1,780 **67. a.** 270 **b.** 300 **69. a.** 2,900 **b.** 3,000 **71. a.** 7,000 **b.** 0 **73.** 6 **75.** 8 **77.** 650 **81. a.** 0 **b.** 8,000

Section 1.4 (p. 45)

1. The truck can carry 1,320 more pounds. **3. a.** The total area is 7,449,803 square miles. **b.** Canada has more area. **c.** The difference in area is 212,263 square miles. **5. a.** 6,845 pounds were loaded onto the truck. **b.** The truck can carry 1,655 more pounds. **c.** No; 950 lb + 1,340 lb = 2,290 lb, which is more than 1,655 pounds. **7. a.** His checks totaled $444. **b.** David's new balance is $116. **9.** Audrey's new balance is $5. **11.** Lauren will make $3.000 in eight weeks. **13.** Ed makes $453 each week. **15.** Irma will get 504 miles on a full tank. **17.** 67 gallons of gas is needed. **19.** Yes; it would take her 44 minutes to seed the whole lawn. **21.** The novel contains about 500 × 500 = 250,000 words. **23.** Sandy will make $351. **25.** Joyce will make $876. **27.** The total cost is $420. **29.** Each woman pays $8,162. **31.** The monthly payment was $47. **33.** The monthly payment was $250. **35.** The average is 8. **37.** Ron sold an average of 10 vacuum cleaners per month. **39.** The average selling price was $126,700.

41. A = 150 sq. ft. **43.** A = 224 sq ft **45. a.** \$800 **b.** \$850 **47. a.** In 1991, Middlesex County had an average salary closest to the state average. **b.** In 1991, Monmout and Middlesex Counties had average annual salaries closest to each other. **49.** The farmer can expect to make \$7,280.

Section 1.5 (p. 61)

1.  **3.**  **5.** $2\frac{1}{4}$ **7.** $7\frac{1}{7}$

9. $\frac{43}{10}$ **11.** $\frac{65}{8}$ **13.** $\frac{4}{5}=\frac{8}{10}=\frac{16}{20}$ (There are many other choices.) **15.** $\frac{2}{9}=\frac{6}{27}=\frac{12}{54}$ (There are many other choices.) **17.** $\frac{3}{5}$ **19.** $\frac{3}{4}$ **21.** $\frac{6}{8}=\frac{3}{4}$ **23.** $\frac{5}{4}$

25. $\frac{12}{35}$ **27.** $\frac{3}{4}$ **29.** $\frac{7}{7}=1$ **31.** $9\frac{2}{5}$ **33.** $\frac{10}{21}$

35. $\frac{3}{6}=\frac{1}{2}$ **37.** $11\frac{1}{2}$ **39.** $\frac{4}{10}=\frac{2}{5}$

Section 1.6 (p. 69)

1. hundredths **3.** tens **5.** tenths **7.** hundreds

9. $\frac{21}{100}$ **11.** $2\frac{1}{100}$ **13.** $4\frac{61}{1000}$ **15.** $100\frac{1}{10}$

17. $300\frac{1}{100}$ **19.** $21\frac{1}{1000}$ **21.** sixty-seven and six tenths **23.** Six hundred seventy-six thousandths **25.** Twenty and two hundredths **27.** Ninety and nine tenths **29.** 0.042 **31.** 100.003 **33.** 0.086 **35.** 0.011 **37.** 0.203 **39.** 203,000 **41.** 13,000 **43.** 5,000

Section 1.7 (p. 77)

1. 4.7 **3.** 260 **5.** 4.48 **7.** 76 **9.** 90.027 **11.** 500 **13.** 6.5 **15.** 27 **17.** 500 **19.** 3.009 **21.** 0.3 **23.** 684 **25.** 0 **27.** 0.066 **29.** 427.0 **31.** 2.0048 **33.** 0.000571 **35.** 5.0 **37.** 0.01 **39.** 600.000 **41.** \$4,572.05 **43.** \$0.50 **45.** 80 **47.** 3,100 **49.** 20 **51.** \$300 **53.** $0.8 > 0.089$ **55.** $0.0971 < 0.103$ **57.** $0.00091 < 0.002$ **59.** 0.6, 0.603, 0.6104 **61.** 3.0063, 3.026, 3.03 **63.** 0.2042, 0.205, 0.254, 0.26 **65.** 4, 4.0069, 4.03, 4.1 **67.** 0.0061, 0.019, 0.02, 0.1 **69.** 7.018, 7.07, 7.089, 7.09, 7.2 **71.** Connie gets \$3 change. **73.** Your quiz average is 81.

Section 1.8 (p. 86)

1. 14.32 **3.** 696.44 **5.** 1.094 **7.** 14.068 **9.** 60.942 **11.** 613.2358 **13.** 605.3694 **15.** 1.4 **17.** 9.65 **19.** 9.03 **21.** 4.327 **23.** 16.748 **25.** 0.001 **27.** 3.2 **29.** 6.528 **31.** 1.97 **33.** 62.125 feet **35.** 13.375 inches **37.** 334.7 **39.** 60.05 **41.** 0.94 **43.** Bill spent \$60.38. **45.** The change is \$20.33. **47.** Casey's batting average is 0.032 higher this year. **49.** The balance at the end of the day was \$174.19. **51. a.** Anna ran 108.7 miles this week. **b.** She must run 31.3 more miles to make 140 miles. **53.** \$66.63 **55.** \$2,793.83 **57. a.** He needs to borrow about \$9,000. **b.** He needs to borrow exactly \$9,872.47. **59.** Alan's net pay is \$1,270.97. **61. a.** Her new balance is about \$370. **b.** Her new balance is exactly \$363.63. **63. a.** Alicia bought 4.48 pounds of lunch meat. **b.** The total cost was \$22.45. **65. a.** The total is about \$400. **b.** The total is about \$430. **d.** Erica spent \$425.45. **67.** The penny is the least expensive to make. **69.** It costs 0.8 cents more to make a quarter than a nickel. **71.** P = 20 yd **73.** P = 130 m **75.** P = 40 ft

77. P = 120 ft **79.** P = 33.48 m **81.** $\frac{109}{1000}$ **83.** The diameter is 2 inches. **85. a.** P = 48 ft **b.** P = 96 ft

Review (p. 91)

1. a. About 50 people per 100,000 died from heart disease in 1900. **b.** About 300 people per 100,000 died from heart disease in 1990. **c.** About 6 times more people died from heart disease in 1990 than in 1900. **3.** Cancer and heart disease resulted in more deaths in 1990 than than 1900. **5.** The average cost of one share of XYZ stock in May was about \$39. **7. a.** The average value of one share dropped the most between May and June. **b.** The value dropped about \$9. **9.** Teresa made about \$1,440 profit. **11.** 40,666 people were in the survey. **13.** The highest amount of smokers was in the age group 25 - 44. **15.** 25.5% of USA adults smoked in 1990.

17. $72.31 > 72$ **19.** $14\frac{1}{4} < 15$ **21.** $6 < 11$ **23.** $\frac{7}{x}$ or $7 \div x$, where x represents a number **25.** 2 **27.** 9 **29.** False **31.** True **33.** 12,000 **35.** 12,350 **37.** Nina's average grade is 85. **39.**

41. $\frac{1}{4}=\frac{2}{8}=\frac{10}{40}$ (There are many other choices.) **43.** $5\frac{1}{2}$ **45.** $\frac{29}{7}$ **47.** $\frac{12}{13}$ **49.** $\frac{7}{11}$ **51.** $\frac{14}{33}$ **53.** $\frac{6}{7}$

55. Thirty-four and six hundredths **57.** $34\frac{6}{100}=34\frac{3}{50}$ **59.** 0.003 **61.** 123.5 **63.** 0.02, 0.2, 0.2002, 2.002 **65.** 47.695 **67.** Dion will get \$.14 change.

CHAPTER 2

Section 2.1 (p. 109)

1. Factors: 7.2 and 3.1; Product: 22.32 **3.** Factors: 11.1 and 3; Product: 33.3 **5.** Factors: 0.001 and 0.02; Product: 0.00002 **7.** Factors: 8.5 and 450; Product: 3,825 **9.** Factors: 99.2 and 81.3; Product: 8,064.96 **11.** between **13.** between **15.** between **17.** smaller than either **19.** $3.358 \times 8.601 \doteq 3 \times 9 = 27$ **21.** $0.580(0.911) \doteq 0.6(0.9) = 0.54$ **23.** $320(0.125) \doteq 320(0.1) = 32$ **25.** $0.752(0.073) \doteq 0.8(0.1) = 0.08$ **27.** 0.18 **29.** 1.62 **31.** 11.218 **33.** 1655.9348 **35.** 0.0072 **37.** 0.00001 **39.** 19.22 **41.** 3.888 **43.** 10 **45.** 0.00393 **47.** 1 **49.** 403.6 **51.** The trucker will pay $161.88. **53. a.** 28.5 hours × \$9.48 $\doteq$ 30 × 9 = \$270 **b.** Jayne earned \$270.18. **55.** Alice will make \$4,206.80. **57.** The seamstress will make \$56.98. **59.** Ed paid \$4.78. **61.** A half-hour call cost \$11.47. **63. a.** Suzanne earns \$235.52 for a 16-hour week. **b.** She earns \$552.00 for a 37.5-hour week. **c.** She earns \$883.20 for a 52.5-hour week. **65. a.** Lucy paid a total of \$8,548.20. **b.** The interest charge was \$2,348.20. **67.** \$6.30 is saved by paying cash. **69.** You owe \$199.30 for renting the car. **71.** 207 × \$317.62 $\doteq$ 200 × \$300 = \$60,000. **73.** It would cost \$42.12 to tile the floor. **75.** The carpet will cost \$1,151.52. **77.** False **79.** True **81.** The difference in area is 306 sq ft.

Section 2.2 (p. 119)

1. Factors: 8 and 7 **3.** Factors: 5, x, and y. **5.** Terms: 7, x, and t. **7.** Base is 4, exponent is 3 **9.** Base is 11, exponent is 2 **11.** Base is 0.5, exponent is 3 **13.** Base is 5.5, exponent is 3 **15.** Base is 1, exponent is 17 **17.** 6^4 **19.** 7^3 **21.** 10^6 **23.** 12^2 **25.** 9 **27.** 125 **29.** 343 **31.** 400 **33.** 0.04 **35.** 0.000004 **37.** 64 **39.** 243 **41.** 0.027

43. $\frac{9}{100}$ **45.** $\frac{1}{8}$ **47.** $\frac{4}{25}$ **49.** 1,849 **51.** 1,048,576 **53.** 2,825,761 **55.** 14,348,907 **57.** 95,443.993 **59.** 3.01 is between 3 and 4.

Section 2.3 (p. 127)

1. \$3,240,000,000,000 **3.** \$6.05 million = \$6,050,000 **5.** 10^{100} **7.** $10^1 = 10$ (ten), $10^2 = 100$ (one hundred), $10^3 = 1{,}000$ (one thousand), $10^4 = 10{,}000$ (ten thousand), $10^5 = 100{,}000$ (one hundred thousand), $10^6 = 1{,}000{,}000$ (one million) **9.** 1,354.6 **11.** 3,560 **13.** 3.598 **15.** 4,200 **17.** 420,000 **19.** \$1,436,000 **21.** The total cost is \$150.00 **23.** The area is 827,000 sq ft. **25.** 3.5¢ **27.** 23.5¢ **29.** 45 ÷ 7 **33.** 2,700,000,000 **35.** Each person would pay \$16,920.

Section 2.4 (p. 135)

1. The square root of 36 is 6. **3.** 4 **5.** 2 **7.** 10 **9.** 7 **11.** 12 **13.** 0.1 **15.** 0.08 **17.** 30 **19.** 40 **21.** 9 **23.** 15 **25.** 18 **27.** 66 **29.** 23 **31.** 0.14 **33.** 3.2 **35.** 102 **37.** 53 **39.** 0.09 **41.** 1.96 **43.** The difference is \$448 million.

Section 2.5 (p. 149)

1. Divisor: 3, dividend: 180, quotient: 60 **3.** divisor: 0.5, dividend: 4.5, quotient: 9 **5.** larger than 1 **7.** equal to 1 **9.** smaller than 1 **11.** larger than 1 **13.** 0.8 **15.** 1.4 **17.** 2.16 **19.** 78.5 **21.** 4 **23.** 12 **25.** 0.7 **27.** 40 **29.** 0.014 **31.** 6.4 **33.** 0.875 **35.** 0.72 **37.** 0.048 **39.** 0.1 **41.** 0.186 **43.** 6.86 **45.** 4.9 **47.** 31.0 **49.** 0.22 **51.** 5.06 **53.** 78.3 **55.** 4.02 **57.** 0.288 **59.** Tom makes \$5.26 per hour. **61.** You make about \$547.36 per week. **63.** Each brick costs \$0.40. **65.** Jeff gets about 23.0 miles per gallon. **67.** Each payment will be \$136.84. **69. a.** Each bar of the three-pack costs \$0.33 or 33¢. **b.** Each bar of the twelve-pack costs \$0.30 or 30¢. **c.** The twelve pack will save you 3¢ per bar. **71. a.** It will cost \$5.55 to mail the 32-ounce package. **b.** You could mail a 40-ounce package for \$6.91. **73.** The cost of the brick for the house will be \$7,620. **75. a.** 14 toys can be purchased. **b.** \$5.51 will be left over. **77.** The test average is about 85.9. **79.** 1.26 **81.** 0.4175 **83.** 0.69301 **85.** 0.593016 **87.** 0.00608 **89.** 0.0000028 **91.** \$2.56 **93.** One toy costs \$17.60. **95.** Each ticket holder will receive \$12,503.41. **97.** \$0.49 **99.** \$0.025 **101.** The cost of four cans is \$3.16. **103.** 216 **105.** 9 **107.** \$691 million = \$691,000,000 **111.** You would save \$16.00. **113.** Riverfront Stadium is the least expensive. **115.** False **117.** False

Section 2.6 (p. 165)

1. 38 **3.** 50 **5.** 2 **7.** 1 **9.** 12 **11.** 10 **13.** 44 **15.** 8 **17.** 1.75 **19. a.** 17 **b.** 25 **c.** 17 **21. a.** 25 **b.** 25 **c.** 1 **23. a.** 100 **b.** 50 **c.** 50 **25.** The car rental will cost \$165.85. **27.** The yearly subscription will save you \$72.75. **29.** Samira will have \$55 left. **31.** The product of 5 and 7 increased by 2; 2 more than the product of 5 and 7. **33.** 16 divided by the sum of 4 and 9. **35.** 5 less than the product of 2 and 12; The product of 2 and 12 decreased by 5. **37.** Three more than twice a number. **39.** Twice the sum of some number and 3; Double the sum of a number and 3. **41.** The sum of a number and 2, divided by 9. **43.** The product of 5 and a number is less than 12. **45.** 6 is less than the quotient of a number and 13.

47. $4(5 + 7)$ **49.** $\frac{80}{5}+9$ **51.** $10(25 - 7)$

53. $7 < (10)(25)$ **55.** $9 > \frac{42}{6}$ **57.** $8(n + 15)$, where n is the number **59.** $4 \cdot 5-3 \cdot 2$ **61.** $\frac{x}{y}+5$, where x and y are any numbers **63.** $9 \cdot 4+9 \cdot 5$ **65.** $6 \cdot 8+6 \cdot 16$ **67.** $4x+4 \cdot 2$ **69.** $25n+25 \cdot 20$ **71.** $8(4 + 2)$ **73.** $40(15 + 12)$ **75.** $3(x + 4)$ **77.** $16(t + a)$ **79.** 7,310 **81.** 3.487

Section 2.7 (p. 177)

1. P = 34 ft **3.** P = 28.8 m **5.** P = 20 cm **7.** P = 72 ft **9.** P = 39.2 ft **11.** P = 32 in **13.** P = 46 in **15.** V = 24,000 cu. cm. **17.** V = 175.5 cu. ft. **19.** V = 1 cu. ft. **21.** P = 864 m; Cost = $2,548.80 **23.** A = 1,728 sq. ft.; She will need 2 bags. **25. a.** V = 75 cu. ft. **b.** She will have to pay $750.00. **27.** c = 5 in. **29.** c = 25 cm. **31. a.** c = 82 mm **b.** P = 180 mm. **33.** 343 **35.** 0.000009 **37.** 165 Pulitzer Prizes were won in all. **39.** 36 < 37 **41.** Motor vehicles are the leading cause of accidental deaths. **43.** The 1992 rate is less than half that of 1922.

Section 2.8 (p. 194)

1. 0.8 **3.** 0.45 **5.** 0.08 **7.** 3.48 **9.** 4.75 **11.** 0.875 **13.** 0.15 **15.** 2.0625 **17.** 0.11 **19.** 0.1 **21.** 0.32 **23.** 1.67 **25.** 2.286 **27.** 4.17 **29.** 0.267 **31.** 0.182 **33.** The carpet will cost $634.95. **35.** $12\frac{3}{4}$ gallons will cost $14.79. **37.** The cost of 1,000 folders is $35.00. **39.** The total cost is $9.48. **41.** 4.359 **43.** 6.6 **45.** 26.2107 **47.** 36.22 **49.** 7.35 **51.** 11.2 **53.** 0.707 **55.** 0.707 **57.** $4<\sqrt{18}<5$ **59.** $6<\sqrt{43}<7$ **61.** $4<\sqrt{20}<5$ **63.** $3<\sqrt{12}<4$ **65.** $c \doteq 27.8$ in.

67. a. $c \doteq 5.57$ ft **b.** $P \doteq 12.92$ ft. **69.** $\frac{6}{11}$ or $.\overline{54}$

Review (p. 199)

1. 1.296 **3.** 179.52 **5.** $6.78 \times 9.87 \doteq 7 \times 10 = 70$ **7.** $0.042 \times 8.01 \doteq 0.04 \times 8 = 0.32$ **9.** A = 63,372.22 sq ft **11.** $5xy$ has three factors but there are many examples. **13.** 9 **15.** 49 **17.** 100 **19.** 3,375 **21.** 999,000,000 **23.** 95.3 **25.** 89,000 **27.** 5 **29.** 0.2 **31.** 23 **33.** 0.8 **35.** 80 **37.** 0.47 **39.** less than 1 **41.** equal to 1 **43.** 121¢ **45.** 20 **47.** 121 **49.** 55 **51.** Twice the sum of 32 and 5 **53.** Four less than the product of 5 and 2.

55. $2(25 + x)$ **57.** $\frac{32}{2}-4$ **59.** $3 \cdot 6+3 \cdot 11$ **61.** $3(5 + 7)$ **63.** P = 12.4 in **65.** c = 35 m

67. $0.\overline{3}$ **69.** $\frac{4}{9}>0.4$ **71.** 11.09

CHAPTER 3

Section 3.1 (p. 211)

1. 1, 2, 3, 6 composite **3.** 1, 2, 4, 8, 16 composite **5.** 1, 2, 7, 14 composite **7.** 1, 13 prime **9.** 1, 2, 13, 26 composite **11.** 1, 3, 11, 33 composite **13.** 1, 2, 4, 5, 8, 10, 16, 20, 40, 80 composite **15.** 1, 2, 3, 4, 6, 8, 12, 24 composite **17.** 1, 79 prime **19.** 1, 2, 71, 142 composite **21.** 1, 3, 17, 51 composite **23.** 1, 59 prime **25.** 3 **27.** 11 **29.** 97 **31.** 4 **33.** 10 **35.** 9 **37.** False **39.** 2^2 **41.** prime **43.** $2 \cdot 7$ **45.** 2^4 **47.** $3 \cdot 11$ **49.** $2^4 \cdot 5$ **51.** 3^4 **53.** $2^4 \cdot 3^2$ **55.** $3 \cdot 17$ **57.** $2 \cdot 3 \cdot 11$ **59.** $7 \cdot 11$ **61.** $2 \cdot 3 \cdot 5 \cdot 7$ **63.** 17.68 **65.** 19.54 **67. a.** No **b.** Yes **c.** No **d.** No **e.** Yes, Yes

Section 3.2 (p. 217)

1. $\frac{3}{4}$ **3.** $\frac{2}{3}$ **5.** $\frac{3}{7}$ **7.** $\frac{3}{4}$ **9.** $\frac{3}{5}$ **11.** $\frac{3}{4}$ **13.** $\frac{1}{4}$ **15.** $\frac{3}{5}$ **17.** $\frac{4}{5}$ **19.** $\frac{3}{7}$ **21.** $\frac{3}{4}$ **23.** $\frac{3}{5}$ **25.** $\frac{4}{5}$ **27.** $\frac{2}{5}$ **29.** $\frac{3}{8}$ **31.** $\frac{17}{22}$ **33.** $\frac{19}{26}$ **35.** $\frac{3}{7}$ **37.** $\frac{3}{5}$ **39.** $\frac{1}{3}$ **41.** $\frac{1}{3}$ **43.** $\frac{1}{4}$ **45.** $\frac{1}{6}$ **47.** $\frac{7}{10}$ **49. a.** $\frac{6}{7}$ **b.** $\frac{1}{7}$ **51.** $\frac{18}{48}=\frac{3}{8}$ **53.** $4\frac{4}{5}$ **55.** $5\frac{5}{7}$ **57.** $1\frac{2}{9}$ **59.** $1\frac{8}{11}$ **61.** $1\frac{7}{8}$ **63.** $7\frac{1}{2}$ **65.** $11\frac{2}{9}$ **67.** 50.05 **69.** 0.046 **71.** $\frac{36}{54}=\frac{36 \div 9}{54 \div 9}=\frac{4}{6}=\frac{2}{3}$

Section 3.3 (p. 223)

1. $\frac{2}{5}$ **3.** $\frac{32}{63}$ **5.** $\frac{3}{7}$ **7.** $\frac{3}{7}$ **9.** $\frac{14}{25}$ **11.** 1 **13.** $\frac{5}{12}$ **15.** $\frac{15}{56}$ **17.** $\frac{6}{25}$ **19.** $\frac{4}{9}$ **21.** $\frac{5}{54}$ **23.** $\frac{1}{3}$ **25.** $3\frac{3}{4}$ **27.** $4\frac{2}{5}$ **29.** $7\frac{7}{8}$ **31.** $10\frac{1}{2}$ **33.** $13\frac{17}{27}$ **35.** $12\frac{4}{5}$ **37.** $4\frac{1}{2}$ **39.** 16 **41.** 28 **43.** $91\frac{2}{3}$ **45.** 33 **47.** 1 **49.** $\frac{9}{25}$ **51.** $8\frac{1}{36}$ **53.** $\frac{216}{343}$ **55.** $\frac{81}{10{,}000}$ **57.** $7\frac{58}{81}$ **59.** $11\frac{1}{9}$ **61.** $\frac{49}{64}$ **63.** 6 **65.** 1 **67.** $\frac{3}{8} \times \frac{1}{4}=\frac{3}{32}$ **69.** $3\frac{1}{4} \times \frac{2}{5}=1\frac{3}{10}$

Section 3.4 (p. 229)

1. $\frac{7}{2}$ **3.** $\frac{1}{8}$ **5.** 10 **7.** $\frac{15}{11}$ **9.** $\frac{8}{41}$

11. $\frac{4}{7}$ **13.** $1\frac{1}{20}$ **15.** $1\frac{5}{7}$ **17.** $\frac{14}{15}$

19. $\frac{14}{15}$ **21.** $1\frac{1}{6}$ **23.** $\frac{10}{11}$ **25.** $1\frac{1}{3}$

27. $1\frac{1}{4}$ **29.** $\frac{4}{15}$ **31.** 9 **33.** 5 **35.** 1

37. $1\frac{17}{28}$ **39.** $1\frac{4}{5}$ **41.** 18 **43.** 3

45. $2\frac{3}{16}$ **47.** 5 **49.** $\frac{1}{4}$ **51.** 4

53. $\frac{1}{3}$ **55.** $\frac{306}{473}$ **57.** $1\frac{1}{8}$ **59.** $9\frac{1}{5}$

61. $5\frac{3}{4}$ **63.** $7\frac{1}{2}$ **65.** $\frac{31}{5}$ **67.** $\frac{16}{9}$ **69.** $\frac{7}{1}, \frac{14}{2}$

71. $\frac{1}{1}, \frac{2}{2}$ **73.** $\frac{7}{9} \div \frac{1}{6} = \frac{7}{\cancel{9}_3} \times \frac{\cancel{6}^2}{1} = \frac{14}{3} = 4\frac{2}{3}$

75. $\frac{1}{2} \div 2 = \frac{1}{2} \div \frac{2}{1} = \frac{1}{2} \times \frac{1}{2} = \frac{1}{4}$

Section 3.5 (p. 239)

1. 200 **3.** 200 men

5. $\frac{2}{9}$ of $12 = \frac{2}{\cancel{9}_3} \times \frac{\cancel{12}^4}{1} = \frac{8}{3} = 2\frac{2}{3}$ gal

7. 8 **9.** 4 **11.** 20 **13.** 26 lb

15. 75 hamburgers **17.** $15\frac{3}{4}$ **19.** $\frac{5}{7}$

21. $1\frac{3}{16}$ **23.** 48 lb **25.** 32 tiles **27.** $\$8\frac{3}{4}$

29. 5 **31.** $\frac{1}{4}$ of a pie each **33.** $6\frac{2}{5}$ hr

35. $98\frac{1}{3} \div 7\frac{1}{2} = 13\frac{1}{9}$; you must make 14 trips.

37. 124 cups **39.** **a.** 7 months **b.** 17 months

41. Add $1\frac{1}{4}$ in to itself 24 times $= 24 \times 1\frac{1}{4} =$

$\frac{\cancel{24}^6}{1} \times \frac{5}{\cancel{4}_1} = 30$ in

43. 100 oz **45.** $715

47. How many $\frac{3}{16}$ are in 24 =

$24 \div \frac{3}{16} = \frac{\cancel{24}^8}{1} \times \frac{16}{\cancel{3}_1} = 128$ ties

49. 3 out of 80 is $\frac{3}{80}$.

$\frac{3}{80}$ of $72{,}000 = \frac{3}{80} \times \frac{72{,}000}{1} = 2{,}700$ will have the disease. **51.** **a.** $2,136 **b.** $534

53. 60 sq. ft. **55.** 13 sq. yd. **57.** 162 sq. m

59. 1.5 sq. mi. **61.** 121.5 sq. in.

63. Area = 56 sq. in. Perimeter = 51.6 in

65. Area = 40 sq. ft. Area = 5760 sq. in.

67. 26.3 **69.** 35 **71.** **a.** 60,000 oz.

b. $320,000 **73.** 84 mi. **75.** Utilities cost $600; rent costs $560; entertainment costs $160; food costs $320 and other costs $280. **77.** 4800 people

Review (p. 247)

1. 6 **3.** $\frac{2}{4}$ **5.** $\frac{0}{4}$ **7.**

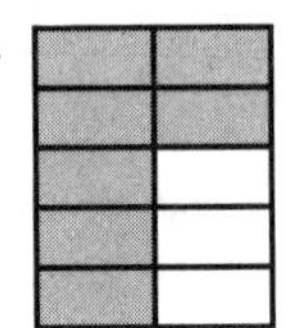

9. 2 **11.** 0 **13.** $\frac{21}{5}$ **15.** $\frac{38}{7}$

17. $4\frac{3}{4}$ **19.** $5\frac{5}{8}$ **21.** 1, 7, 49 composite

23. 1, 2, 4, 5, 7, 8, 10, 14, 20, 28, 35, 40, 56, 70, 140, 280 composite **25.** 1, 167 prime **27.** 1, 2, 47, 94 composite **29.** $3^2 \cdot 13$ **31.** $2^3 \cdot 7$ **33.** $5 \cdot 73$

35. $3 \cdot 7 \cdot 13$ **37.** $\frac{2}{3}$ **39.** $\frac{7}{8}$ **41.** $1\frac{23}{45}$ **43.** $\frac{17}{19}$

45. $\frac{12}{35}$ **47.** $\frac{125}{324}$ **49.** $8\frac{1}{4}$ **51.** $30\frac{3}{4}$

53. $\frac{343}{729}$ **55.** $\frac{169}{400}$ **57.** $2\frac{1}{2}$ **59.** $\frac{5}{6}$

61. $4\frac{4}{5}$ **63.** 15 **65.** $\frac{4}{17}$ **67.** 112 quarts

69. $31\frac{1}{4}$ **71.** 34 plums **73.** 28 wires **75.** $31\frac{1}{2}$ sq.ft.

CHAPTER 4

Section 4.1 (p. 257)

1. $\frac{7}{9}$ **3.** $\frac{1}{2}$ **5.** $1\frac{3}{10}$ **7.** $\frac{1}{3}$ **9.** $\frac{5}{6}$ **11.** $\frac{17}{25}$

13. $1\frac{3}{4}$ **15.** 2 **17.** $2\frac{1}{5}$ **19.** 2 **21.** $7\frac{3}{5}$

23. 5 **25.** $12\frac{2}{3}$ **27.** $7\frac{1}{5}$ **29.** $18\frac{1}{2}$ **31.** 10

33. 49 **35.** $\frac{3}{7}$ **37.** $\frac{1}{4}$ **39.** $\frac{1}{2}$ **41.** $\frac{1}{10}$

43. $\frac{1}{3}$ **45.** $\frac{1}{3}$ **47.** $1\frac{2}{7}$ **49.** $\frac{1}{3}$

51. $2\frac{2}{3}$ **53.** $1\frac{1}{2}$ **55.** $8\frac{1}{2}$ **57.** $\frac{1}{5}$

59. $9\frac{5}{12}$ **61.** 49 **63.** 16

65. 7 **67.** 9 **69.** 12 **71.** $\frac{5}{18}$

73. $\frac{5}{8} + \frac{7}{8} = \frac{12}{8} = 1\frac{1}{2}$

Section 4.2 (p. 263)

1. equivalent **3.** not equivalent **5.** not equivalent
7. equivalent **9.** not equivalent **11.** equivalent
13. not equivalent **15.** not equivalent
17. not equivalent **19.** equivalent

21. $\frac{18}{24}$ **23.** $\frac{20}{24}$ **25.** $\frac{56}{64}$ **27.** $\frac{18}{48}$

29. $\frac{16}{30}$ **31.** $\frac{18}{21}$ **33.** $\frac{2}{6}, \frac{3}{9}, \frac{4}{12}, \frac{5}{15}, \frac{6}{18}$

35. $\frac{14}{20}, \frac{21}{30}, \frac{28}{40}, \frac{35}{50}, \frac{42}{60}$ **37.** $\frac{6}{8}, \frac{9}{12}, \frac{12}{16}, \frac{15}{20}, \frac{18}{24}$ **39.** 4

41. 16 **43.** 16 **45.** 10 **47.** 6 **49.** 10
51. 36 **53.** 56 **55.** 175 **57.** 81 **59.** 66
61. 153 **63.** 21 **65.** 16 **67.** $2^3 \cdot 5$ **69.** $2 \cdot 3^2$
71. $3 \cdot 11$

Section 4.3 (p. 271)

1. 16 **3.** 60 **5.** 30 **7.** 24 **9.** 60 **11.** 80
13. 36 **15.** 210 **17.** 60 **19.** 18 **21.** 40
23. 28 **25.** 50 **27.** 90 **29.** 180 **31.** 126
33. 320 **35.** 1,200 **37.** 30 **39.** 140 **41.** 12
43. 60 **45.** 360 **47.** 1,020 **49.** 48

51. $\frac{25}{6}$ **53.** $\frac{27}{4}$ **55.** LCD for $\frac{3}{4}$ and $\frac{5}{8}$ is 8

Section 4.4 (p. 277)

1. $1\frac{1}{6}$ **3.** $\frac{17}{20}$ **5.** $\frac{5}{9}$ **7.** $1\frac{1}{12}$ **9.** $1\frac{1}{2}$

11. $\frac{11}{14}$ **13.** $1\frac{1}{4}$ **15.** $\frac{5}{6}$ **17.** $\frac{5}{36}$ **19.** $\frac{119}{160}$

21. $\frac{7}{144}$ **23.** $\frac{55}{108}$ **25.** $1\frac{2}{15}$ **27.** $\frac{57}{80}$

29. $\frac{13}{32}$ **31.** $\frac{11}{12}$ **33.** $1\frac{9}{16}$ **35.** $\frac{127}{150}$

37. $1\frac{65}{72}$ **39.** $1\frac{79}{120}$ **41.** $5\frac{5}{6}$ **43.** $9\frac{5}{21}$

45. $9\frac{5}{8}$ **47.** $6\frac{23}{24}$ **49.** $6\frac{1}{4}$ **51.** $18\frac{1}{8}$

53. $7\frac{5}{7}$ **55.** $11\frac{31}{40}$ **57.** $39\frac{1}{10}$ **59.** $12\frac{7}{9}$

61. $14\frac{47}{50}$ **63.** $13\frac{3}{14}$ **65.** $11\frac{11}{12}$ **67.** $8\frac{13}{24}$

69. 456.1 **71.** 56,000

73. Final addition should be $\frac{6}{15} + \frac{10}{15} = \frac{16}{15} = 1\frac{1}{15}$

Section 4.5 (p. 283)

1. $\frac{1}{4}$ **3.** $\frac{5}{8}$ **5.** $\frac{3}{8}$ **7.** $\frac{1}{6}$ **9.** $\frac{1}{24}$ **11.** $\frac{21}{40}$

13. $\frac{13}{24}$ **15.** $\frac{5}{36}$ **17.** $\frac{1}{12}$ **19.** $\frac{19}{36}$ **21.** $\frac{11}{35}$

23. $\frac{25}{54}$ **25.** $\frac{32}{105}$ **27.** $\frac{1}{4}$ **29.** $\frac{1}{5}$ **31.** $\frac{41}{180}$

33. $2\frac{1}{2}$ **35.** $3\frac{5}{21}$ **37.** $3\frac{3}{7}$ **39.** $2\frac{1}{8}$ **41.** $8\frac{1}{12}$

43. $1\frac{19}{24}$ **45.** $4\frac{3}{4}$ **47.** $6\frac{2}{9}$ **49.** $1\frac{1}{3}$ **51.** $14\frac{5}{8}$

53. $\frac{2}{7}$ **55.** $3\frac{1}{40}$ **57.** $11\frac{9}{10}$ **59.** $7\frac{7}{9}$ **61.** $\frac{17}{50}$

63. $7\frac{19}{42}$ **65.** $18\frac{5}{12}$ **67.** $2\frac{71}{84}$ **69.** 4

71. $35 + 26 = 61$ **73.** $52 - 37 = 15$

75.
$$\begin{array}{r} 3\frac{8}{40} = 2\frac{48}{40} \\ -1\frac{15}{40} = 1\frac{15}{40} \\ \hline 1\frac{33}{40} \end{array}$$

Section 4.6 (p. 289)

1. $30\frac{1}{10}$ gal **3.** $\frac{13}{24}$ in **5.** $39\frac{5}{24}$ ft

7. a. $16\frac{7}{8}$ gal **b.** $\$20\frac{1}{4}$

9. a. $\$46\frac{3}{4} + \$1\frac{3}{8} = \$48\frac{1}{8}$ **b.** $160 \times \$1\frac{3}{8} = \$220$ more

11. $1\frac{7}{12}$ cups of nuts **13.** $1\frac{23}{30}$ **15.** $\frac{1}{12}$

17. $8\frac{3}{5}$ gal **19.** $2\frac{1}{4}$ in $+ \frac{7}{8}$ in $+ \frac{7}{8}$ in $= 4$ in

21. $\frac{1}{3} + \frac{2}{5} = \frac{11}{15}$, Cy owns $1 - \frac{11}{15} = \frac{15}{15} - \frac{11}{15} = \frac{4}{15}$

23. $1\frac{11}{24}$ qt **25.** $\frac{1}{8} + \frac{1}{10} = \frac{9}{40}$, $1 \div \frac{9}{40} = 4\frac{4}{9}$ hr

27. $a = 6\frac{3}{8}$ ft, $b = 1\frac{1}{4}$ ft

29. a. $\frac{2}{5} + \frac{4}{15} + \frac{1}{6} = \frac{5}{6}$ used. Therefore, $\frac{1}{6}$ are hanging.

b. $\frac{1}{6}$ of $60 = \frac{1}{6} \times \frac{60}{1} = 10$ balloons **31.** 19 **33.** 28

35. a. Fill $\frac{1}{2}$ cup. Pour into $\frac{1}{3}$ cup until full. This leaves $\frac{1}{6}$ of a cup of oil in the $\frac{1}{2}$ cup measure because $\frac{1}{2} - \frac{1}{3} = \frac{1}{6}$. **b.** $\frac{1}{3} - \frac{1}{4} = \frac{1}{12}$

37. a. $\frac{15}{16}$ of an acre each **b.**

Section 4.7 (p. 299)

1. $1\frac{1}{10}$ 3. 1 5. $1\frac{34}{45}$ 7. $\frac{56}{75}$ 9. $9\frac{1}{7}$ 11. 5

13. $8\frac{1}{6}$ 15. 72 17. $1\frac{1}{4}$ 19. 7 21. $\frac{1}{8}$

23. 4 25. 11 27. $8\frac{1}{2}$ 29. 170 31. $1\frac{1}{3}$

33. 15 35. $\frac{1}{80}$ 37. $2\frac{3}{7}$ 39. $6\frac{3}{4}$ 41. $\frac{77}{96}$

43. $3\frac{1}{4} + 1\frac{5}{8} + 2 = 6\frac{7}{8}$ oz of drug,

$6\frac{7}{8} \div \frac{2}{3} = 10\frac{5}{16}$, 10 doses can be made

45. Each gets \$315 47. $\frac{3}{5} < \frac{5}{8}$ 49. $\frac{9}{10} < \frac{11}{12}$

51. $\frac{3}{4} < \frac{4}{5} < \frac{5}{6}$ 53. $\frac{1}{5} < \frac{7}{15} < \frac{2}{4}$ 55. $\frac{1}{8} < \frac{5}{16} < \frac{8}{20}$

57. $\frac{1}{4} < \frac{3}{5} < \frac{5}{8} < \frac{2}{3}$ 59. $\frac{7}{12} < \frac{5}{8} < \frac{3}{4} < \frac{5}{6}$ 61. $\frac{5}{8}$ in

63. Company A; $\frac{7}{60}$ is higher 65. $1\frac{1}{4}$ or 1.25 67. $1\frac{2}{5}$ or 1.4 69. 0.31 71. 3.01 73. 1.5 75. 3.09

77. The cheese will cost \$10.71. 79. a. The discount is \$164.75. b. The sale price is \$494.25. 81. 340

83. 64 85. Al made \$60; Bob made \$61. Bob made the greater commission.

Review (p. 303)

1. $1\frac{1}{4}$ 3. 1 5. $5\frac{3}{5}$ 7. $5\frac{7}{12}$ 9. $\frac{2}{9}$ 11. $\frac{1}{5}$

13. $3\frac{1}{2}$ 15. $3\frac{1}{3}$ 17. not equivalent

19. equivalent 21. $\frac{12}{30}$ 23. $\frac{21}{30}$

25. $\frac{2}{16}, \frac{3}{24}, \frac{4}{32}, \frac{5}{40}$ 27. $\frac{6}{10}, \frac{9}{15}, \frac{12}{20}, \frac{15}{25}$

29. $\frac{25}{30}$ 31. $\frac{63}{72}$ 33. 36 35. 160 37. 72

39. 168 41. $1\frac{13}{60}$ 43. $\frac{7}{10}$ 45. $11\frac{13}{15}$ 47. $9\frac{53}{72}$

49. $\frac{7}{36}$ 51. $\frac{29}{84}$ 53. $\frac{31}{45}$ 55. $2\frac{7}{10}$ 57. $3\frac{5}{12}$ ft

59. $23\frac{1}{20}$ mi 61. \$3,050 63. 8 65. 9

67. 21 69. 0.628 or $\frac{157}{250}$ 71. 26.37 or $26\frac{37}{100}$

73. $\frac{1}{6} < \frac{1}{4} < \frac{3}{8}$ 75. $\frac{3}{10} < \frac{8}{25} < \frac{7}{20}$

CHAPTER 5

Section 5.1 (p. 313)

1. +40° 3. −35 ft 5. +\$110 7. −23
9. −\$4,000 11. 13 13. 15 15. 8 17. 16
19. 10 21. > 23. < 25. < 27. >
29. = 31. > 33. > 35. < 37. −8
39. 45 41. −6 43. 6 45. −7 47. 0.75
49. 0.833... 51. $|-9| + |-6| = 9 + 6 = 15$

Section 5.2 (p. 321)

1. 8 plus 5; 13 3. negative 6 plus 4; −2 5. 3 added to negative 4; −1 7. negative 5 added to 16; 11
9. 6 plus negative 4; 2 11. 8 plus negative 8; 0
13. negative 7 added to negative 13; −20 15. negative 9 added to positive 5; −4 17. 0 19. −6 21. −12
23. −5 25. 0 27. 7 29. −1 31. 11
33. −22 35. 6 37. 3 39. −5 41. −18
43. 16 45. −37 47. −10 49. −13 51. 6
53. 0 55. −13 57. −15 59. −1 61. 0
63. −1 65. −58 67. 12° 69. 9 71. −\$15
73. > 75. < 77. $-5 + 10 = +5$

Section 5.3 (p. 327)

1. 3 3. −4 5. −10 7. −10 9. 11
11. −5 13. −20 15. 0 17. −5 19. 1
21. −8 23. 18 25. 0 27. −97 29. 100
31. 30 33. 8 35. 6 37. 50 39. 1
41. 0 43. 15 45. 0 47. 3 49. 10
51. −6 53. 8 55. \$197 57. \$1
59. rise = $17° - (-21°) = 17° + (+21°) = 38°$ 61. 65 ft
63. 2.45367 65. 0.00546 67. 0.671

Section 5.4 (p. 333)

1. 35 3. 6 5. −30 7. −16 9. −20
11. −9 13. 0 15. 56 17. 15 19. 36
21. −24 23. −27 25. 0 27. 27 29. −72
31. 24 33. −16 35. 48 37. −30 39. 0
41. −420 43. −48 45. 35 47. −6 49. −8
51. 5 53. −8 55. −12 57. Not Possible
59. −8 61. −5 63. 9 65. 0 67. 35
69. −8 71. −6 73. Not Possible 75. 8
77. 0 79. −7 81. 7 83. −20 85. 15
87. $\frac{2}{3}$ 89. $\frac{8}{3}$ 91. $(-2)(-3)(-7) = (+6)(-7) = -42$

Section 5.5 (p. 339)

1. 25. 3. 36 5. 100 7. −36 9. −27
11. 16 13. −27 15. −16 17. −36 19. 3
21. 54 23. −25 25. −8 27. 0 29. −21
31. 45 33. 6 35. 8 37. −4 39. 19
41. 32 43. −1 45. 20 47. 1 49. 1
51. 4 53. 10 55. 13 57. 5 59. 96
61. 49 63. 7 65. 24 67. −10 69. −14
71. −2 73. 1 75. −12 77. 15.2 feet
79. 5 square inches 81. 10

83. $(-4)(-4)(-4) = (+16)(-4) = -64$
85. $(-100) \div (10) \cdot (-10) = (-10) \cdot (-10) = 100$

Section 5.6 (p. 347)

1. 460 **3.** 592 **5.** 80,000 **7.** 993,200
9. 60,020,000 **11.** 7,300,000,000 **13.** 25,000
15. 5.6×10^1 **17.** 9.02×10^2 **19.** 5.8×10^3
21. 3.86×10^1 **23.** 6×10^4 **25.** 7.8×10^0
27. 4.65×10^5 **29.** 2.55×10^8 **31.** 1.86×10^5
33. 4.5×10^9 **35.** 9.26×10^7 **37.** 4.3×10^{-1}
39. 6.3×10^{-2} **41.** 9.53×10^{-3} **43.** 5.2×10^{-5}
45. 1×10^{-1} **47.** 7×10^{-6} **49.** 6.045×10^{-3}
51. 5.6×10^{-4} **53.** 6.1×10^{-6} **55.** 9.25×10^{-2}
57. 1.57×10^{-5} **59.** 7.5×10^{-2} **61.** 6.624×10^{-27}
63. 0.036 **65.** 0.00703 **67.** 0.1026
69. 0.000 090 4 **71.** 0.048 **73.** 0.000 850 7
75. 0.000 000 000 0054 **77.** 0.000 001 077
79. 0.000 000 000 000 000 000 000 026 372 **81.** 19
83. 9

Review (p. 349)

1. −37° **3.** 3 **5.** 16 **7.** 6 **9.** 22 **11.** 2
13. −10 **15.** −14 **17.** −45 **19.** −22
21. −11 **23.** 22 **25.** −16° **27.** 1 **29.** 0
31. 30 **33.** −17 **35.** −10 **37.** −51 **39.** −7
41. 43° **43.** 25 **45.** 21 **47.** −80 **49.** 36
51. −60 **53.** −1 **55.** −20 **57.** 24
59. −450 **61.** 8 **63.** 5 **65.** −2 **67.** 8
69. −8 **71.** 9 **73.** 12 **75.** 81 **77.** −1,000
79. −19 **81.** 3 **83.** −48 **85.** 70,500
87. 0.000 018 **89.** 60.25 **91.** 9,400
93. 1.5×10^4 **95.** 9×10^{-1} **97.** 5.07×10^5
99. 4.8×10^{-6} **101.** 6.4×10^3 **103.** 1.1×10^{-4}

CHAPTER 6

Section 6.1 (p. 361)

1. a. $x = 9$ is not a solution. **b.** $x = 5$ is a solution.
3. a. $w = 5$ is a solution. **b.** $w = 1$ is not a solution.
5. a. $x = 1$ is not a solution. **b.** $x = 0$ is a solution.
7. $x = 9$ **9.** $w = -6$ **11.** $b = 16$ **13.** $t = -10$
15. $s = 16$ **17.** $s = 0$ **19.** $x = -2$ **21.** $a = -10$
23. $a = 10$ **25.** $n = 7$ **27.** $y = -37$ **29.** $k = -24$
31. $x = 10$ **33.** $y = -4$ **35.** $t = -6$ **37.** $k = 2$
39. $x = 0$ **41.** $t = 0$ **43.** $y = 0$ **45.** $x = 12$

47. $n = -24$ **49.** $y = 16$ **51.** $t = 0$ **53.** $x = \frac{1}{4}$

55. $y = \frac{1}{2}$ **57.** $x = 2$ **59.** $y = 6$ **61.** $t = 2$

63. $k = 2$ **65.** $x = -4$ **67.** $y = \frac{1}{2}$ **69.** $n = \frac{8}{3}$

71. $p = \frac{13}{5}$ **73.** $\frac{2}{35}$ **75.** $\frac{16}{25}$

Section 6.2 (p. 367)

1. a. 21 to 23 **b.** 23 to 21 **c.** 23 to 51
d. 51 to 44 **3. a.** 12 to 17 **b.** 23 to 24
c. 29 to 47 **d.** 24 to 17 **5.** 40 to 27, 40:27, $\frac{40}{27}$
7. 49 to 15, 49:15, $\frac{49}{15}$ **9.** 80 to 21, 80:21, $\frac{80}{21}$
11. a. $\frac{24}{5}$ **b.** 5:24 **c.** 24 to 29, or twenty-four to twenty-nine **13.** 5:3, $\frac{5}{3}$ **15.** 6 to 5, 6:5
17. 40 to 1, $\frac{40}{1}$ **19.** 1 to 64, 1:64 **21.** 1:1, $\frac{1}{1}$
23. 8 to 9, $\frac{8}{9}$ **25.** 11 to 8 **27.** 19 to 10 **29.** 4:1
31. 1:3 **33.** $\frac{10}{3}$ **35.** $\frac{4}{5}$ **37.** 7 to 4
39. 25 to 7 **41.** 18 to 5 **43.** 1:5 **45.** 6 to 1
47. 5 to 12 **49.** 9 to 10 **51.** 16 to 3 **53.** 5:2
55. 1 to 25 **57.** 5/12 **59.** 7 to 2 **61.** 4:1
63. 20 to 3 **65.** 26 to 3 **67.** 4 to 1 **69.** 71 to 12
71. 5 to 12 **73.** 1:48 **75.** $\frac{4}{6}, \frac{6}{9}, \frac{8}{12}$
77. $\frac{2}{20}, \frac{3}{30}, \frac{4}{40}$ **79.** 18:1 **81.** $\frac{1}{3}$
83. $15 \text{ to } 40 = \frac{15 \div 5}{40 \div 5} = 3 \text{ to } 8$

Section 6.3 (p. 373)

1. 10 engines per 3 cars **3.** \$40 per 3 hr
5. 50 mi per 3 gal **7.** 50 mi per 1 hr, or 50 mph
9. 16 customers per 7 toys **11.** 30 books per 7 readers
13. 20 mi per 1 gal, or 20 mpg **15.** 235 books per 1 hr, or 235 books per hr **17.** 9 to 5 **19.** 1 mi per 2 min
21. 60¢ per 1 lb, or 60¢ per lb **23.** 20 min per 1 box, or 20 min per box **25.** 125¢ per 3 oz
27. a. \$3 per 5 dozen **b.** 60¢ per 1 dozen, or 60¢ per dozen **c.** 5¢ per 1 egg, or 5¢ per egg **29. a.** \$5 per 1 lb, or \$5 per lb **b.** 500¢ per 1 lb, or 500¢ per lb
c. 125¢ per 4 oz **31. a.** \$360 per 1 acre, or \$360 per acre **b.** \$1 per 121 square feet **c.** 100¢ to 121 square feet **33.** 15 **35.** 1 mi per 12 min

Section 6.4 (p. 379)

1. 2 to 1 **3.** 2 to 3 **5.** 5 to 66 **7.** 25 to 36
9. 20:21 **11.** 6 to 1 **13.** 5 to 1 **15.** 5 to 1
17. 10 to 9 **19.** 2 to 1 **21.** 9 to 2 **23.** 1 to 1
25. $1\frac{1}{4} \text{ to } \frac{5}{8} = \frac{5}{4} \div \frac{5}{8} = \frac{2}{1}$ or 2 to 1
27. 23 mi per 8 hr **29.** 24 to 1 **31.** 20 to 3
33. 1 to 3,000 **35.** 500 to 9 **37.** 3 to 1
39. 10 to 9 **41.** 7 to 1 **43.** 100 to 1 **45.** 1 to 10
47. 1 to 300 **49.** 18 to 25 **51.** 50 to 3 **53.** 1 to 21
55. 29 gallons per \$35 **57.** $x = 6$ **59.** $t = 10$
61. $\frac{7}{2} \div \frac{7}{1} = \frac{\overset{1}{\cancel{7}}}{2} \times \frac{1}{\underset{1}{\cancel{7}}} = \frac{1}{2}$ or, 1 to 2

Section 6.5 (p. 385)

1. 3.75 to 1 **3.** 15 to 1 **5.** 2.1 children per family

7. $6\frac{2}{3}$ dozen eggs per chicken **9.** $376\frac{2}{3}$ people to 1 millionaire **11.** 250 to 1 **13.** \$3 per pound **15.** \$1.40 per gal **17.** \$5 per model **19.** \$5.08 per lb **21.** \$28,500 per acre **23.** 11.0¢ per oz **25.** 7.4¢ per egg **27.** 12.0¢ per oz **29.** Meg plants 22 bulbs per hour; Susan plants 24 bulbs per hour. Susan plants 2 more bulbs per hour. **31.** Last year Gail scored 19 points per game. This year she scored 21 points per game. She scored 2 more points per game this year. **33. a.** \$1.15 per gallon for Jayne, \$1.25 per gallon for her friend. Jayne paid the better price. **b.** Jayne paid 10¢ less per gallon. **35. a.** 6.6¢ per oz for Mike, 6.0¢ per oz for Debbie. Debbie made the better buy. **b.** Debbie paid 0.6¢ less per oz. **37. a.** The six-pack sells for 2.5¢ per oz, the 2-liter bottle sells for 1.5¢ per oz. The 2-liter bottle is the better buy. **b.** The 2-liter bottle costs 1.0¢ less per oz. **39. a.** In the former Soviet Union there were 34 people per sq mi, in China there are 304 people per sq mi. **b.** China has about 9 times more people per sq mi. **41.** $\frac{13}{30}$

43. $3\frac{4}{9}$

Section 6.6 (p. 395)

1. 78.5 ft **3.** 56.52 cm **5.** 12.56 ft **7.** 200.96 ft
9.a. 4 ft 6 in = 4.5 ft, $C = \pi D \doteq 3.14 \times 54 \text{ in} = 169.56 \text{ in}$
b. 4 ft 6 in = 4.5 ft, $C = \pi D \doteq 3.14 \times 4.5 \text{ ft} = 14.13 \text{ ft}$
11. 125.6 ft **13. a.** 549.5 ft **b.** \$13,627.60
15. a. 12.56 in **b.** 25.12 in **c.** 50.24 in **d.** 100.48 in **e.** The circumference doubles.

17. a.
$$\begin{aligned} C &\doteq 3{,}000 \text{ ft} + 3{,}000 \text{ ft} + 2 \times 3.14 \times 500 \text{ ft} \\ &= 3{,}000 \text{ ft} + 3{,}000 \text{ ft} + 3{,}140 \text{ ft} \\ &= 9{,}140 \text{ ft.} \end{aligned}$$

b. $9{,}140 \text{ ft} \div 5{,}280 \text{ ft/mi} \doteq 1.7 \text{ mi}$ **19.** 254.34 ft^2

21. 19.625 ft^2 **23.** 12.56 in^2 **25.** 706.5 yd^2

27. a. 19.625 ft^2 **b.** 2,826 in^2 **29.** 11.0325 ft^2

31. a.
$$\begin{aligned} A &\doteq 3.14\,(120 \text{ ft})^2 - 3.14\,(95 \text{ ft})^2 \\ &= 45{,}216 \text{ ft}^2 - 28{,}338.5 \text{ ft}^2 = 16{,}877.5 \text{ ft}^2 \end{aligned}$$

b. $\frac{\$70}{100 \text{ ft}^2} = \frac{x}{16{,}877.5 \text{ ft}^2}$, $x = \$11{,}814.25$

33. 1,130,400 mi^2 **35. a.** 3,106.5 yd^2 **b.** \$232.99, rounded to the nearest cent **37.** 5.61 **39.** 0.07 **41. a.** 24,900 mi **b.** polar circumference = 24,819 miles; equatorial circumference is 81 miles longer.

Section 6.7 (p. 403)

1. true **3.** true **5.** false **7.** true **9.** false
11. true **13.** true **15.** false **17.** true
19. false **21.** true **23.** false **25.** false
27. false **29.** true **31.** true **33.** 1.632

35. 1.33 **37.** No, $\frac{\frac{1}{2} \text{ in}}{6 \text{ ft}} = \frac{3\frac{5}{8} \text{ in}}{38 \text{ ft}}$ is a false proportion

39. $\frac{\cancel{5}^{\,1}}{8} \times \frac{3}{\cancel{10}_{\,2}} = \frac{3}{16}$ Do not cross multiply when you multiply fractions.

Section 6.8 (p. 409)

1. $x = 16$ **3.** $t = 15$ **5.** $h = 6$ **7.** $y = 81$
9. $s = 32$ **11.** $x = 63$ **13.** $x = 18$ **15.** $x = 8$
17. $x = 20$ **19.** $y = 21$ **21.** $x = 8\frac{4}{7}$ **23.** $v = 300$
25. $x = 90$ **27.** $x = 18$ **29.** $x = 44$ **31.** $h = 30$
33. $x = 22\frac{2}{9}$ **35.** $x = \frac{3}{4}$ **37.** $x = 4$ **39.** $y = 6$
41. $x = 4$ **43.** $t = 2.7$ **45.** $x = 0.2$
47. $s = 0.125$ **49.** $k \doteq 0.642$ **51.** $\frac{1}{4} < \frac{5}{16} < \frac{3}{8}$

53. $\frac{2}{3} < \frac{7}{9} < \frac{5}{6} < \frac{11}{12}$ **55.**
$$\begin{aligned} \frac{x}{16} &= \frac{12}{5} \\ 5 \cdot x &= 12 \cdot 16 \\ x &= \frac{192}{5} \\ x &= 38.4 \end{aligned}$$

Section 6.9 (p. 417)

1. 336 mi **3.** 21 favors **5.** $12\frac{1}{2}$ft
7. 15,750 voters **9.** 9 lb **11.** 12 hr
13. 118.125 mi **15.**
$$\begin{aligned} \frac{264 \text{ mi}}{14.1 \text{ gal}} &= \frac{9{,}300 \text{ mi}}{x \text{ gal}} \\ 264 \cdot x &= 9{,}300 \cdot 14.1 \\ x &\doteq 497 \text{ gal} \end{aligned}$$

17. 96 games **19.** about 17,400,000 lb
21. 1 lb = 16 oz,
$$\begin{aligned} \frac{20 \text{ oz}}{\$1.69} &= \frac{16 \text{ oz}}{x} \\ x &\doteq \$1.352 \text{ per pound} \end{aligned}$$
23. 200 mg

25. a. $1\frac{1}{4}$ tsp salt, $3\frac{3}{4}$ cups of water **b.** 5 servings

Review (p. 421)

1. $a = 9$ **3.** $v = 7$ **5.** $y = -\frac{1}{2}$ **7. a.** $\frac{163}{17}$ **b.** $\frac{17}{4}$

c. $\frac{21}{163}$ **9.** 10 to 7 **11.** 3/5 **13.** 1 to 1

15. 4.5 mph **17.** 16 mpg **19.** $1\frac{3}{7}$ cars per day

21. $\frac{3}{8}$ or .375 miles per minute **23.** \$24 per day

25. 4 to 3 or $\frac{4}{3}$ 27. $\frac{0.8}{0.05} = \frac{80}{5} = \frac{16}{1}$

29. $\frac{0.024}{0.0003} = \frac{240}{3} = \frac{80}{1}$ 31. $\frac{4.5}{7.75} = \frac{450}{775} = \frac{18}{31}$

33. $1.386 per lb or 138.6¢ per lb 35. The 2-liter bottle costs 1.6¢ per oz. The six-pack costs 2.3¢ per oz. The 2-liter bottle is the better buy. 37. The jumbo box is 8.0¢ per oz. The regular size is 7.5¢ per oz. The regular size is the better buy. 39. C $\doteq$ 30.144 mm

41. A $\doteq 379.94\ cm^2$ 43. A $\doteq 145.2\ yd^2$ 45. False 47. True 49. $x = 4$ 51. $x = 7.5$ 53. $x \doteq 5.7$

55. $x = \frac{85}{32} = 2.65625$ 57. The daily dosage for a 200-pound man is 10 ounces. 59. The property tax on a $130,000-home is $1,872.00. 61. The actual length 21 ft. 63. About 6.6 gallons are needed, but you wou need to buy 7 gallons.

CHAPTER 7

Section 7.1 (p. 431)

1. 40, 100

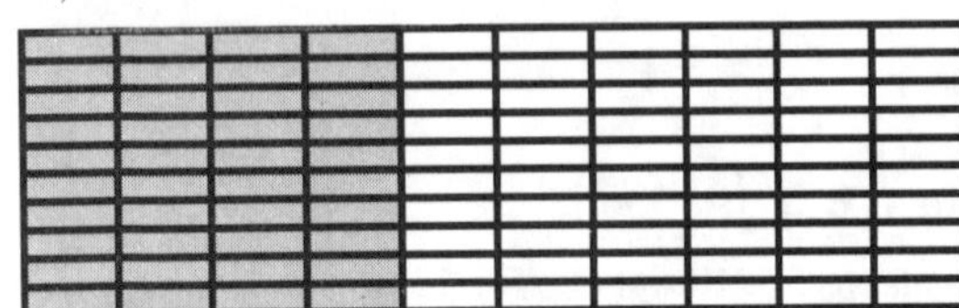

3. 85

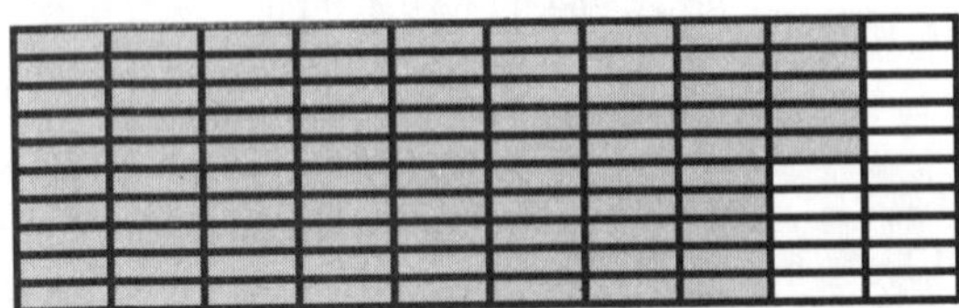

5. 12.5

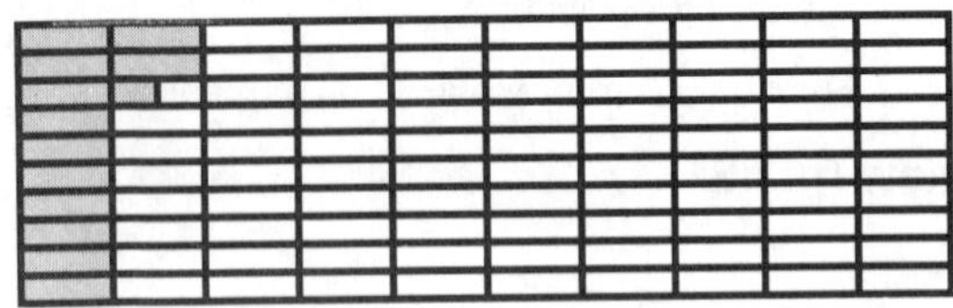

7. $\frac{1}{4}$, 100

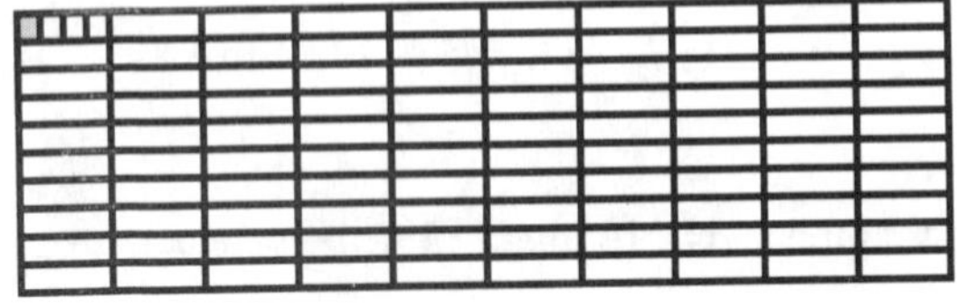

9. 30% 11. 39% 13. 20% 15. $16\frac{2}{3}\%$

17. $99\frac{1}{2}\%$ 19. 99.92% 21. 90 23. 3.9

25. 18 27.

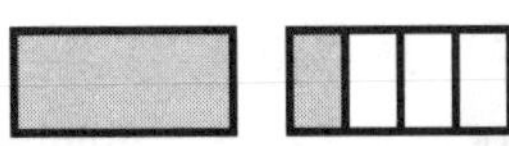

29.

31.

33. $\frac{16}{25}$ 35. $\frac{3}{50}$ 37. $\frac{56}{100}\%$ or 0.56%

39. 100% − 0.6% = 99.4%

Section 7.2 (p. 435)

1. 62% 3. 5% 5. 21.2% 7. 1.8%
9. 6.08% 11. 10.04% 13. 30% 15. 50.001%
17. 400% 19. 250% 21. 1,200%
23. 75,000% 25. 300.07% 27. 1,255%

29. $62\frac{1}{2}\%$, or 62.5% 31. 450% 33. 0.48

35. 0.05 37. 0.722 39. 2 41. 2.347 43. 40
45. 0.0004 47. 0.000006 49. 0.8091 51. 0.023

53. 2.9 55. 22.637 57. $0.06\frac{1}{2}$, or 0.065

59. $0.35\frac{1}{4}$, or 0.3525 61. 0.12375 63. 0.8

65. 0.7 67. 0.004% = .00 004% = 0.00004

Section 7.3 (p. 443)

1. 25% 3. 80% 5. $62\frac{1}{2}\%$ 7. $36\frac{4}{11}\%$

9. $41\frac{2}{3}\%$ 11. 90% 13. 140% 15. $187\frac{1}{2}\%$

17. 225% 19. 150% 21. 402.5% 23. $633\frac{1}{3}\%$

25. $21\frac{2}{3}\%$ 27. $4\frac{2}{3}\%$ 29. $31\frac{1}{4}\%$ 31. 5%

33. $7\frac{11}{27}\%$ 35. $\frac{11}{20}$ 37. $\frac{9}{10}$ 39. $\frac{69}{100}$

41. $\frac{2}{25}$ 43. $1\frac{4}{5}$ 45. 7 47. $\frac{2}{3}$ 49. $\frac{5}{6}$

51. $2\frac{3}{8}$ 53. $\frac{1}{500}$ 55. $\frac{1}{700}$

57. $5\frac{1}{3}\% = \frac{4}{75}$. $4 out of every $75 tax dollars pays for sanitation. 59. $82\frac{1}{2}\% = \frac{33}{40}$. 33 out of every 40 students studies a foreign language. 61. a. 2% b. $\frac{1}{50}$

63. $\frac{11}{500}$ 65. $\frac{13}{80}$ 67. $\frac{17}{40}$ 69. $\frac{5}{8}$ 71. $\frac{81}{50,000}$

73. $\frac{1}{5,000,000}$ 75. $1\frac{81}{200}$ 77. $\frac{157}{3,000}$ 79. $\frac{513}{1,000}$

81. $\frac{9}{1,000}$ 83. $\frac{13}{160}$ 85. $\frac{9}{25}$, 36% 87. 1.25, 125% 89. $1\frac{1}{2}$, 150% 91. 0.6, 60% 93. $\frac{1}{125}$, 0.8% 95. $1\frac{2}{3}$, 1.6667 97. $\frac{249}{400}$, 62.25%

99. $b = 3.5$ 101. $y \doteq 555.56$ 103. 63%; 88%

105. $\frac{2}{25} \rightarrow 25\overline{)2.00}$ = 8%, quotient .08; 2 00; 0

Section 7.4 (p. 451)

1. Percent = 60%, Whole = 45, Part = 27. 60% of 45 is 27. **3.** Percent = 40%, Whole = 50, Part = 20. 40 % of 50 is 20. **5.** Percent = 50%, Whole = 1,800, Part = 900. 900 is 50% of 1,800. **7.** Percent = 10%, Whole = 6,000, Part = 600. 600 is 10% of 6,000. **9.** Percent = 120%, Whole = 45, Part = 54. 120% of 45 is 54. **11.** Percent = 14%, Whole = 60, Part = p. 14% of 60 is p. **13.** Percent = 40%, Whole = w, Part = 6. 40% of w is 6. **15.** Percent = x%, Whole = 84, Part = 11. 11 is x% of 84. **17.** Percent = 62%, Whole = 95, Part = p. p is 62% of 95. **19.** Percent = 2%, Whole = n, Part = 0.5. 0.5 is 2% of n. **21.** Percent = x%, Whole = 15, Part = 13. x% of 15 is 13. **23.** Percent = n%, Whole = \$90, Part = \$15. \$15 is n% of 90. **25.** Percent = 12%, Whole = x, Part = 0.036. 12% of x is 0.036. **27.** Percent = 12%, Whole = 0.036, Part = x. x is 12% of 0.036. **29.** $\frac{57}{40}$ **31.** $\frac{11}{12}$ **33.** The Whole is 6 and the Part is 42.

Section 7.5 (p. 461)

1. 20 **3.** 3.2 **5.** 44 **7.** 84 **9.** 22 **11.** 74 **13.** 17.2 **15.** 32 **17.** 87 **19.** 0.56 **21.** 102 **23.** 6,000 **25.** 2.4 **27.** 43 **29.** 2.4 **31.** 7 **33.** 7.8 **35.** 24.4 **37.** 820 **39.** 90 **41.** 25 **43.** 1.01 **45.** 9 **47.** 60 **49.** 160 **51.** 50% **53.** 200% **55.** 25% **57.** 50% **59.** 100% **61.** 100% **63.** 50% **65.** 25% **67.** \$180 **69.** \$6,700 **71.** \$2.40 **73.** \$480 **75.** −17 **77.** $-\frac{19}{16}$ **79.** Penny got 150 questions right and 50 questions wrong.

Section 7.6 (p. 473)

1. $0.42\times600 = x$; $x = 252$ **3.** $0.87\times380 = x$; $x = 330.6$ **5.** $x = 0.072 \times 65$; $x = \$4.68$ **7.** $0.125 \times 640 = x$; $x = 80$ men **9.** $x = 0.0006 \times 22{,}000$; $x = 13.2$ lb **11.** $2.4 \times 550 = x$; $x = \$1{,}320$ **13.** $0.55x = 44$; $x = 80$ **15.** $0.83x = 41.5$; $x = 50$ **17.** $20 = \frac{2}{9}x$; $x = 90$ **19.** $60 = 0.003x$; $x = 20{,}000$ people **21.** $2.75x = 330$; $x = 120$ **23.** $0.001x = 0.01$; $x = 10$ liters **25.** $300x = 90$; $x = 30\%$ **27.** $65x = 19.50$; $x = 30\%$ **29.** $667 = 2900x$; $x = 23\%$ **31.** $45 = 75x$; $x = 60\%$ **33.** $0.8 = 0.5x$; $x = 160\%$ **35.** $5.10x = 1.70$; $x = 33\frac{1}{3}\%$ **37.** $0.029x = 10.15$; $x = 350$ **39.** $x = \frac{2}{11}\cdot 33$; $x = 6$ **41.** $54 = 72x$; $x = 75\%$ **43.** $\frac{1}{3} = 0.2x$; $x = 1\frac{2}{3}$ **45.** $7.2 = 9x$; $x = 80\%$ **47.** $x = 0.8 \times 80$; $x = \$64$ **49.** 38.67 **51.** 2.028

Section 7.7 (p. 483)

1. There are 2,090 people in sales. **3.** 230 women are in the club. **5.** 71% of the children preferred the widget. **7.** The stereo cost \$880. **9. a.** 43% favored the increase. **b.** 57% did not favor the increase. **11.** You can expect 4 lb of gold. **13.** The builder made \$23,200 profit. **15.** 40% of the class received a grade of B. **17.** Sal's mortgage payment is \$551. **19.** George's ideal weight is 150 lb. **21.** A 3-oz piece of beef liver has 371.91 mg of cholesterol. **23.** Ed spent $36\frac{1}{9}\%$ or approximately 36% on dinner. **25.** 1,050 men found jobs. **27.** Cathy sold the cabin cruiser for \$175,000. **29.** 4 plants died. **31.** They left \$5.64 for a tip. **33.** \$336 is budgeted for utilities each month. **35.** \$168 more is budgeted for food than for utilities. **37.** An additional \$56 would go to savings. **39.** 16.3 **41.** 0.63 **43.** 46% of the group will attend the seminar.

Section 7.8 (p. 497)

1. Jack's commission is \$4,324. **3.** Arnold must sell \$8,000 in merchandise monthly. **5.** Barb's commission was \$5,920. **7.** The land sold for \$54,500. **9.** The commission rate was 5%. **11.** Ed's commission is \$674.40. His July salary is \$874.40. **13.** Jerry's commission was \$720. His total sales were \$18,000. **15. a.** The discount is \$315. **b.** The sale price is \$585. **17. a.** \$19.20 **b.** \$6.90 **c.** \$149.99 **19.** The TV is on sale for \$350.40. Frances pays \$280.32. **21.** Fred is offering a discount of \$102, which is a 30% discount. **23. a.** The original price of the stereo is \$865. **b.** The sale price is \$692. **25.** 2.78 is x% of 55.60. The electric bill increased 5%. **27.** 13,000 is x% of 130,000. The price of the home was reduced 10%. **29. a.** 200 is x% of 800. He expects to earn 25% more. **b.** 140 is x% of 800. George earned 17.5% more. **31. a.** The car will depreciate \$7,020 in its first year. **b.** After one year the car will be worth \$16,380. **33. a.** The oven will depreciate \$720 the first year. **b.** After one year the oven will be worth \$2,880. **35.** 26,000 is x% of 18,000. The value has increased approximately 144%. **37. a.** 24 is x% of 144. Willy's weight decreased by about 17% (exactly $16\frac{2}{3}\%$). **b.** 24 is x% of 120. Willy's weight increased by 20%. **39.** George's raise is \$1,127.50. His new salary is \$21,627.50. **41.** M = 42 **43.** I = 12 **45. a.** \$12,250 **b.** Each received \$6,125. **47. a.** \$32,000 and \$19,200 **b.** She lost \$3,200 on the transaction. **49.** There was about a 99.6% decrease.

Section 7.9 (p. 507)

1. I = \$80 **3.** I = \$1,440 **5.** I = \$85.49 **7.** I = \$640 **9.** I = \$63 **11.** I = \$1,500 **13.** I = \$81 **15.** I = \$243.75 **17.** Daniele's monthly payment is \$380. **19.** The monthly payment is \$550. **21.** Jana borrowed \$12,000. The interest is \$2,400. The monthly payment is (\$12,000 + \$2,400) ÷ 24 = \$600. **23. a.** \$4,896 **b.** \$419.33 **25. a.** \$20,000 **b.** \$1,000 **c.** \$66,000 **27.** S = \$898.88, I = \$98.88 **29.** S ≐ \$2,701.22, I = \$301.22 **31.** S ≐ \$3,941.29, I = \$191.29 **33.** S ≐ \$49,847.28, I = \$9,847.28 **35. a.** S ≐ \$5,630.81 **b.** I = \$630.81 **37. a.** S ≐ \$3,253.82 **b.** I = \$1,253.82 **39.** 4

41. $6\frac{2}{5}$ **43.** I ≐ \$373.48 **45. a.** S = \$1,120 ; I = \$120 **b.** S = \$1,123.60 ; I = \$123.60 **c.** S ≐ \$1,125.51; I = \$125.51 **d.** S ≐ \$1,126.83 ; I = \$126.83 **e.** You earn more interest the more often you compound. **47. a.** S = \$1,030 ; I = \$30 **b.** S = \$1,060.90 ; I = \$60.90 **c.** S = \$1,092.73 ; I = \$92.73 **d.** S = \$1,125.51 ; I = \$125.51 **e.** No. Amount of interest earned in each additional quarter is more than previous quarter, but not by a fixed amount.

Review (p. 511)

1. 30, 100 **3.** 99% **5.** 72% **7.** $77\frac{7}{9}$% **9.** 37%

11. 140% **13.** 400% **15.** 0.91 **17.** 2.15

19. 0.072 **21.** 60% **23.** $233\frac{1}{3}$% **25.** 90%

27. $\frac{9}{20}$ **29.** $\frac{8}{9}$ **31.** $\frac{11}{8}$ or $1\frac{3}{8}$ **33.** $\frac{33}{400}$ **35.** $\frac{1}{300}$
37. Percent = 35%, Whole = 400, Part = 140
39. Percent = 10%, Whole = 380, Part = 38
41. Percent = x%, Whole = 60, Part = 50
43. Percent = 40%, Whole = x, Part = 20
45. Percent = 105%, Whole = 45, Part = x **47.** 470
49. 10% **51.** 40 **53.** 100% **55.** \$12.30
57. \$350 **59.** approximately 71.43% **61.** 54
63. 37.5% of country club members are seniors.
65. 285 yachts exceed 36 feet. **67.** Sol's monthly net pay is \$1,625. **69.** 924 plants died before maturity. **71.** About 4,000 C grades were assigned. **73.** About 290 more F grades were assigned than "Other" grades. **75.** About 11,720 of the grades were passing grades. **77.** Jack's sales were \$5,800. **79.** Commission = \$1,160 – \$260 = \$900. Edna's total sales were \$10,000. **81. a.** The original price was \$4,800. **b.** The sale price is \$3,360. **83.** There was an 8% decrease. **85. a.** He lost 279 pounds. **b.** He now weighs 186 pounds. **87.** I = \$183.60. **89.** I = \$2,376. **91.** Lucy's monthly payment is \$581. **93.** The total interest will be \$5,020.35. **95. a.** The maturity value is \$17,910.79. **b.** The total interest is \$2,910.79.

APPENDIX A (p. 523)

1. 4 R 2 **3.** 5 R 1 **5.** 3 R 2 **7.** 4 R 0 **9.** 4 R 2 **11.** 7 R 1 **13.** 381 R 4 **15.** 85 R 3 **17.** 900 R 4 **19.** 5,026 R 3 **21.** 8 R 0 **23.** 4 R 3 **25.** 13 R 2 **27.** 34 R 0 **29.** 65 R 67 **31.** 267 R 30 **33.** 1,288 R 45 **35.** 148 R 20 **37.** 34 R 0 **39.** 15 **41.** 7 **43.** 30 **45.** 173 **47.** 30 **49.** 20 **51.** 10 **53.** 30 **55.** 5 **57.** 200 **59.** estimate = \$2,000,000 ÷ 40 or about \$50,000 each **61. a.** estimate = \$7,000,000 ÷ 70 or about \$100,000 each **b.** No, it is too low. **63.** 27,608 **65.** 5,319,000 **67.** about 40,000 **69.** 5 **71.** $\frac{60{,}2\not{0}\not{0}}{1\not{0}\not{0}} = 602$

INDEX

N

O

P

Q

R

S